World Energy Council
CONSEIL MONDIAL DE L'ENERGIE

2004

SURVEY OF ENERGY RESOURCES

Elsevier Internet Homepage - http://www.elsevier.com
Consult the Elsevier homepage for full catalogue information on all books, major reference works, journals, electronic products and services.

Elsevier Titles of Related Interest

CLEVELAND
Encyclopaedia of Energy
(six volume set)
ISBN: 012176480X

ARENTSEN and KUNNEKE
National Reforms in European Gas
ISBN: 008-043687-0

EL-SAYED
The Thermoeconomics of Energy Conversions
ISBN: 0080442706

GALE
Greenhouse Gas Control Technologies
ISBN: 0080442765

Related Journals:
Elsevier publishes a wide-ranging portfolio of high quality research journals, encompassing the energy field. A sample journal issue is available online by visiting the Elsevier web site (details at the top of this page). Leading titles include:
Combustion & Flame
Energy Policy
Fuel
Nuclear Engineering & Design
Progress in Energy & Combustion Science
Renewable & Sustainable Energy Reviews
All journals are available online via ScienceDirect: www.sciencedirect.com

To contact the Publisher
Elsevier welcomes enquiries concerning publishing proposals: books, journal special issues, conference proceedings, etc. All formats and media can be considered. Should you have a publishing proposal you wish to discuss, please contact, without obligation, the publisher responsible for Elsevier's Composites and Ceramics programme:

Tony Roche
Senior Publishing Editor
Elsevier Ltd
The Boulevard, Langford Lane
Kidlington, Oxford
OX5 1GB, UK

Phone: +44 1865 843887
Fax: +44 1865 843931
E.mail: t.roche@elsevier.com

General enquiries, including placing orders, should be directed to Elsevier's Regional Sales Offices - please access the Elsevier homepage for full contact details (homepage details at the top of this page).

World Energy Council
CONSEIL MONDIAL DE L'ENERGIE

2004

SURVEY OF ENERGY RESOURCES

20th Edition

ELSEVIER

Amsterdam – Boston – Heidelberg – London – New York – Oxford – Paris – San Diego
San Francisco – Singapore – Sydney – Tokyo

WORLD ENERGY COUNCIL
Regency House
1–4 Warwick St.
London W1B 5LT
UK

ELSEVIER B.V.
Sara Burgerhartstraat 25
P.O. Box 211, 1000 AE Amsterdam
The Netherlands

ELSEVIER Inc.
525 B Street, Suite 1900
San Diego, CA 92101-4495
USA

ELSEVIER Ltd
The Boulevard, Langford Lane
Kidlington, Oxford OX5 1GB
UK

ELSEVIER Ltd
84 Theobalds Road
London WC1X 8RR
UK

20th edition 2004

ISBN: 0-08-044410-5

♾ The paper used in this publication meets the requirements of ANSI/NISO Z39.48-1992 (Permanence of Paper).

Printed in The Netherlands.

WEC MEMBER COMMITTEES 2004

Algeria
Angola
Argentina
Australia
Austria
Bangladesh
Belarus
Belgium
Bolivia
Botswana
Brazil
Bulgaria
Cameroon
Canada
China
Congo (Democratic Republic)
Côte d'Ivoire
Croatia
Czech Republic
Denmark
Ecuador
Egypt (Arab Republic)
El Salvador
Estonia
Ethiopia
Finland
France
Gabon
Georgia
Germany
Ghana
Greece
Guinea
Hong Kong, China
Hungary
Iceland
India
Indonesia
Iran (Islamic Republic)
Ireland
Israel
Italy
Japan
Jordan
Kenya
Korea (Republic)
Latvia
Lebanon
Libya/GSPLAJ
Lithuania
Luxembourg
Macedonia (Republic)
Mali
Mexico
Monaco
Mongolia
Morocco
Namibia
Nepal
Netherlands
New Zealand
Niger
Nigeria
Pakistan
Paraguay
Peru
Philippines
Poland
Portugal
Romania
Russian Federation
Saudi Arabia
Senegal
Serbia & Montenegro
Singapore
Slovakia
Slovenia
South Africa
Spain
Sri Lanka
Swaziland
Sweden
Switzerland
Syria (Arab Republic)
Taiwan, China
Tanzania
Thailand
Trinidad & Tobago
Tunisia
Turkey
Ukraine
United Kingdom
United States of America
Uruguay
Venezuela
Yemen

SURVEY OF ENERGY RESOURCES COMMITTEE MEMBERSHIP 2004

Executive Board

Chairman:	A. Clerici (Italy)
Joint Editors:	A.W. Clarke (United Kingdom)
	J.A. Trinnaman (United Kingdom)
Secretary:	E.V. Nekhaev (WEC Programme Manager)

Resource Consultants

Coal (including Lignite):	C. Copley *World Coal Institute*
Crude Oil and Natural Gas Liquids:	P.W. Riemer *World Petroleum Congress*
Oil Shale:	J.R. Dyni *U.S. Geological Survey*
Natural Bitumen and Extra-Heavy Oil:	R.F. Meyer & E. Attanasi *U.S. Geological Survey*
Natural Gas:	R. Aptroot *International Gas Union*
Uranium and Nuclear Energy:	H.-H. Rogner *International Atomic Energy Agency*
Hydropower:	R.M. Taylor *International Hydropower Association*
Peat:	R. Sopo *International Peat Society*
Wood Fuels:	M.A. Trossero & R. Drigo *Food and Agriculture Organization of the UN*
Bioenergy (other than Wood):	F. Rosillo-Calle *Imperial College, London*
Solar Energy:	C. Silvi *ISES Italia*
Geothermal Energy:	L.Y. Bronicki & M. Lax *WEC Israeli Member Committee*
Wind Energy:	S. Varming *ECON Analysis, Denmark*
Tidal Energy:	I.G. Bryden *The Robert Gordon University, Aberdeen*
Wave Energy:	T.W. Thorpe *Energetech Australia Pty. Ltd.*
Ocean Thermal Energy Conversion:	D.E. Lennard *Ocean Thermal Energy Conversion Systems Ltd.*

Acknowledgements

The WEC greatly appreciates the efforts of its Member Committees and those institutions and experts who have contributed to this Survey with data and their knowledge. Special thanks go to Alessandro Clerici, Chairman of the SER Executive Board, for guiding the production of the Survey and providing the Overview.

Finally, the WEC thanks the Joint Editors Judy Trinnaman and Alan Clarke for compiling, validating and formatting the data. Once again they have successfully and professionally completed this enormous task, both achieving an excellent quality and keeping to the planned schedule. The WEC is grateful to them for their knowledge, dedication, tenacity and inspiration.

Foreword

This 20th WEC *Survey of Energy Resources* contains a chapter for each energy resource, ranging from the conventional fossil fuels to the renewables, both new and traditional. Generally, the coverage of each resource comprises a commentary by a leading expert in the field, followed by definitions, data tables and country notes. The tables summarise the worldwide resources, reserves, production and consumption of fossil fuels and comparable data for non-fossil energy sources, as applicable. The country notes aim to highlight the main features of the resource and its utilisation.

- **Reserves/Resources** – where relevant, tables of fossil fuels provide reserve statistics (covered globally from WEC and non-WEC sources) and amounts in place (as reported by WEC Member Committees);
- **Tabulations** – data tables are arranged on a standard regional basis throughout;
- **Units** – where relevant, data have been provided in alternative units (cubic feet as well as cubic metres, barrels as well as tonnes) in order to facilitate use of survey data in an industry context;
- **References and Sources** – as far as possible, these have been consolidated in introductory notes to the data tables and country notes, or appended to the commentaries on each resource.

Any review of energy resources is critically dependent upon the availability of data, but reliable, comprehensive information does not always exist. While the basis of the compilation was input provided by WEC Member Committees, completion necessitated recourse to a multitude of national and international sources and in some cases estimation. Difficulties in obtaining information continue to be compounded by current trends in the energy sector. The availability of data has been reduced with the process of deregulation and privatisation, as data-reporting channels have been lost or specific items have become confidential. Moreover, problems in the quantification of energy resources persist, in particular for those universally-found resources: solar energy, wind power and bioenergy, owing to their evolutionary status and generally decentralised nature.

Although there will always be a problem with complete documentation of both solar and wind energy, the coverage of both has been much improved in this edition of the *Survey*. The utilisation of these two globally-available resources is expanding at both macro and micro levels but while the information regarding high-profile schemes (e.g. large offshore windfarms or solar roof programmes) is widely available, that for, say, isolated stand-alone wind turbines or PV installations for remote medical refrigeration is not and probably never will be.

Another problematical area is that of the definitions relating to resources and reserves: it is well recognised that each country tends to have its own notion of what constitutes resources and reserves. In this connection, we welcome a contribution from Dietmar Kelter, describing the work of the UN/ECE Ad Hoc Group of Experts which has been developing the UN Framework Classification for Reserves/Resources. In due course it is to be hoped that this work will provide the basis for a worldwide harmonisation of the relevant terminology. In the meantime, the resources and reserves specified in the present *Survey* conform as far as possible with the established definitions specified coal-fired power generation plants, their costs are high. This will inevitably inhibit their wider deployment in the regions and countries where the use of coal is expected to grow most. Moreover, the issue of the high CO_2 emissions of coal-fired plants compared to gas-fired combined cycle technologies remains unresolved.

Oil

The past couple of years have clearly demonstrated the volatile nature of oil and the world's continuing dependence on this leading energy resource. The doubling of oil prices during the last few years has not, however, been caused by dwindling reserves. The *Survey* demonstrates that global reserves of oil are still large enough to meet the demand for the next few decades, and the continuous improvement in exploration, processing, conversion and end-use technologies may extend this period even further. Concentration of oil resources in a few regions and long supply routes to the main markets are at the heart of the issue.

Natural Gas

Global natural gas reserves are large and currently yield a reserve/production ratio of 50 to 60 years. It is widely expected that in the coming 2-3 decades natural gas will overtake oil as the most important energy resource in the world. Few, however, realise that this would be a huge challenge, not least due to the enormous investment requirement. Where will the necessary investment come from? The most prominent project in the gas industry so far, the development of the Troll gas field in the Norwegian part of the North Sea, has cost billions of euros to implement. Such investment would have hardly been possible on the basis of spot market prices.

It is also expected that LNG will be playing an increasingly important role, particularly in supplying remote markets. Its market penetration will continue to grow, spurred by technological developments in the liquefaction/re-gasification processes and a reduction in transport costs.

Uranium and Nuclear

As a result of growing international efforts in nuclear disarmament at the end of the last century, nuclear fuel from surplus military plutonium entered the commercial market and began to curtail demand for freshly mined uranium. Currently, freshly mined uranium accounts for just over half of the global annual reactor fuel demand, with the balance provided by secondary sources. Known uranium reserves are more than adequate to cover the requirements of existing reactors during their lifetimes and beyond.

In mid-2004 the nuclear power industry celebrates its 50th anniversary. The first nuclear power plant in the world was commissioned in Obninsk, Russia in 1954. Nuclear power's share of worldwide electricity supplies has been steady at 16–17% for many years, but it is expected to decline as old plants are de-commissioned and only a few new ones built. Reactor safety, waste disposal and plant decommissioning are still matters of concern.

Demand for new nuclear power is coming primarily from Asia, while in Western Europe, the only new nuclear reactor to be constructed is a 1 600 MW_e European Pressurised Water Reactor ordered by the Finnish utility TVO.

Renewable Resources

Although the worldwide production of renewable energy is expected to grow quickly, its share of the global energy mix will hardly increase.

Hydropower is the largest and most important renewable resource and generates about 17% of the world's electricity. It is estimated that only 33% of the technically and economically feasible global potential of hydropower has been developed to date, although there are significant regional variations. In Europe and North America, the majority of sites have been developed, while considerable potential for new development remains in Africa, Asia and South America. Large hydropower schemes, however, often face challenges due to their environmental impacts and long-term returns on investment.

Non-hydro renewables are expected to make a growing contribution to global power generation, even if their total share is likely to reach only about 5% in 2030.

Biomass has the potential to become the world's largest and most sustainable renewable energy source. To progress from this "potential" stage, both production and end-use technologies must be modernised.

Wind is often considered to be the most advanced of the renewables, after hydropower. Offshore projects spur the development of larger machines and wind turbines of up to 5 MW are about to enter the market. However, the electricity systems with an increasing share of wind power in their fuel mix will have to face new challenges. Experience in those countries with a high share of wind in their electricity production (i.e. ~20% and above), demonstrates the problems of integrating an intermittent energy source into the grid and the implications this can have for the global power system performance, including the need for new concepts for power plant operation scheduling and system control.

Geothermal is an important renewable resource and it can be utilised for base-load electricity production. The best geothermal fields are located within well-defined belts of geologic activity. Geothermal energy converting systems are able to provide electricity with an annual capacity factor of over 90%.

Solar radiation, the earth's prime source of energy, is being increasingly utilised. While photovoltaic (PV) power generation is still the most expensive solar technology, costs are falling and its versatility enables it to find many stand-alone applications.

Other Resources

The 2004 *Survey of Energy Resources* also presents the status of other energy resources, namely peat, oil shale, tidal, OTEC (ocean thermal energy conversion), natural bitumen and extra-heavy oil, wave and wood, all of which have a potential to help meet growing demand for energy around the world.

Environmental laws which penalise emissions from power plants or transportation and grant subsidies to selected renewables can create artificial local niches for certain energy sources. This may have an adverse impact on global energy prices and investment.

Conclusions

The overall conclusion drawn from this edition of the *Survey* is that there is no shortage of energy resources around the world. However, the physical concentration of the leading strategic resources in only a few regions is a serious concern for many countries dependent on imported supplies. A diversified fuel mix is a prerequisite for energy security, stability of prices and supply, and should be taken into consideration when developing national energy plans or long-term business strategies, in particular against the background of the growing short term focus of the liberalised energy markets.

Alessandro Clerici
Chairman, Survey of Energy Resources Committee
Chairman, WEC Italian Member Committee

Contents

Chapter 1

Coal (including Lignite)

World Energy Council
CONSEIL MONDIAL DE L'ENERGIE

COMMENTARY

Three years have passed since the last *Survey of Energy Resources* was published, and in that time the coal industry has seen some significant changes. Total world production has increased, yet the number of companies involved in coal mining has reduced, and there has been a notable shift both in demand and production to the Asian market.

That coal can continue to supply the world's energy is not in doubt. The IEA has stated 'World reserves of coal are enormous and, compared with oil and natural gas, widely dispersed.... Proven coal reserves have increased by over 50% in the past 22 years. The correlation of strong growth of proven coal reserves with robust production growth suggests that additions to proven coal reserves will continue to occur in those regions with strong, competitive coal industries'.

Global Recoverable Reserves and Production

Based on figures (data to end-2002) given by WEC Member Committees and from a variety of other sources, total recoverable reserves (i.e. those deposits that are economically viable at today's prices and can be recovered using current technologies) have slightly reduced by 8% since 1999, to just under 910 billion tonnes. This is almost entirely due to economic re-appraisal of the German coal mining industry—worldwide the proven reserve base represents nearly 200 years of production at current rates.

However, production figures show an increase of 11% over 1999 levels. Of this, sub-bituminous coal production remained more or less at the same level, while bituminous coal production increased by 440 million tonnes and lignite by 35 million tonnes.

Africa

The bulk of Africa's 220 mt total coal production in 2002 was bituminous coal, dominated by South Africa. 45 thousand tonnes of sub-bituminous coal was produced, from just two countries—Malawi and Nigeria. Although overall production in Congo and Tanzania is low, there have been significant relative gains since 1999. Egypt has experienced a major decline in coal production, from 200 000 tonnes in 1999 to only 37 000 tonnes in 2002.

The proved recoverable reserves figure for Africa has been downgraded by 9%, owing to a significant reappraisal of the reserves reported for Botswana.

2004 Survey of Energy Resources

North America

The amount of proved recoverable coal reserves in Canada, Mexico and Greenland has remained static, with a slight decrease reported for the USA. Total reserves for North America amount to about 250 billion tonnes. Overall production figures show a similar story—a decrease for the region of around 1%, to just over 1 billion tonnes. The USA accounts for the bulk of this production, with 990 mt. It reports increased production of sub-bituminous coal, with a decrease in bituminous coal production in the order of 50 mt.

South America

Production in South America has increased by roughly 16% over the period—notably due to increased production of bituminous coal in Colombia and Venezuela for the export market. There has been a halving of the already low levels of production in both Argentina and Chile.

In regional terms, South America has the second lowest proved recoverable reserves figure, at just under 20 mt, an 8.5% decrease from the 1999 figures.

Asia

The overall production figures for the Asian region have increased by 26%, reflecting the increasing importance of the region as both a supplier and consumer of coal. The major producers are China (1.4 billion tonnes), India (360 mt) and Indonesia (103 mt). Both Japan and Korea have significantly reduced their coal production, due to the high cost of domestic production compared to the price of imported coal on the Asian market.

Reserve figures for the region have remained stable at nearly 260 billion tonnes, a slight increase of 2% over 1999.

Over recent years there have been some concerns raised over the accuracy of Chinese coal statistics, which as such a large supplier can have significant impacts on global demand and production figures, and affect global environmental issues such as CO_2 emissions. It is interesting to note that the end-2002 reserves figures reported for China are the same as at end-1999.

Europe

Coal production in Europe has slightly increased over the 1999 figures, mainly through higher levels of lignite production. Bituminous coal production has decreased, notably in Germany, Spain, France, Poland and the UK.

Proven reserves are significantly lower than in 1999, with a reduction of over 20% during the period. The bulk of this is due to the very significant decrease in the size of German reserves, which due to the economic reappraisal of the mining industry have reduced by almost 90%, from 66 billion tonnes in 1999 to just under 7 billion by the end of 2002.

Middle East

Iran is the only coal producer in the Middle East region, producing 1.8 mt in 2002, an increase of 17% over 1999. Proved recoverable reserves have fallen by 75%, from 1 710 mt in 1999 to just 419 mt in 2002.

Oceania

Australia is the world's leading exporter of coal, and is ranked 4th worldwide in terms of annual production, with 340 mt in 2002. The majority of the coal produced is bituminous coal for export, although it does have a sizeable lignite industry supplying the domestic power generation market. New Zealand is a small producer for a mainly domestic market, although it does export some specialist coals, e.g. for carbon steel. Reserves in New Zealand have remained static, although a 4% drop in bituminous coal reserves in Australia is reported.

Fig. 1.1 shows proven coal reserves by geographical distribution, highlighting the dominance of three key regions—Europe, North America and Asia. Coal resources are more geographically widespread, while reserves are governed by economic viability—and thus are more likely to be concentrated in countries where coal is a commodity, either for domestic

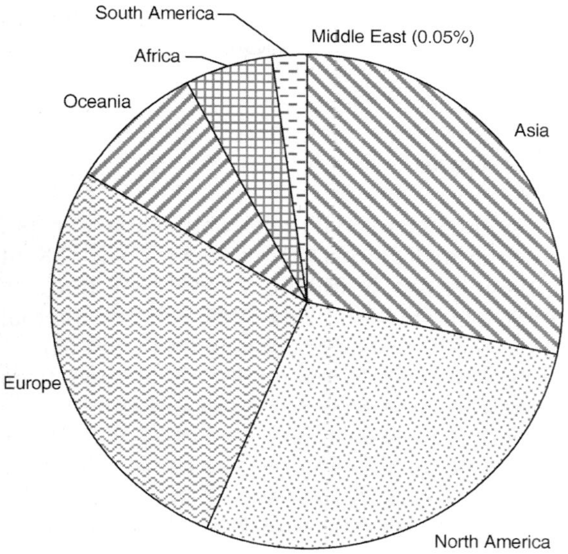

FIGURE 1.1 Proved coal reserves at end-2002—regional distribution.

energy use or as an export product. The top 10 producing countries together make up 85% of total global coal production (Fig. 1.2).

Coal reserves can change significantly and rapidly, as policies change and resources lose

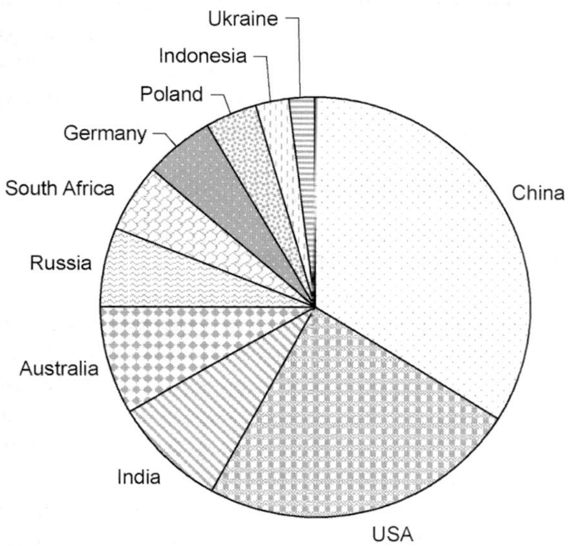

FIGURE 1.2 The top ten coal producing countries in 2002.

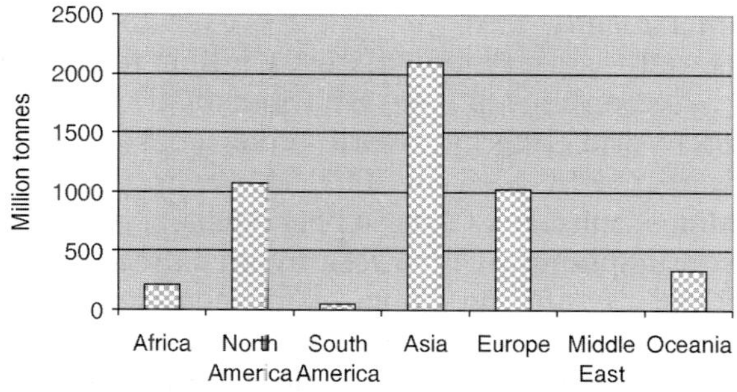

FIGURE 1.3 Coal production, 2002—by region.

or gain viability. This is clearly shown in the European data, where despite maintaining overall production figures (due to increased lignite production in Germany and Greece), European reserves have shifted downwards since 1999, reflecting the decline in the hard coal industry in Western Europe and restructuring in the transitional economies, particularly in Poland.

That Asia is now the focus of global production can clearly be seen in Fig. 1.3. With over 2 billion tonnes of coal produced, an increase of over 25% between 1999 and 2002, the Asian region coal production is double that of the next largest, North America. With global proved recoverable reserves remaining high it would seem there is no practical restraint on the continuing use of coal.

Coal Demand

Significant changes in the location of coal demand have taken place over the last 20 years. In 1980 Europe, the former Soviet Union and North America consumed roughly the same quantities of hard coal for their power generation and steelmaking needs. North America's demand has stayed roughly static, as a percentage of total global consumption (in real terms, an increase of 300 mt over the period). However, by 1990 the trends were of decreasing demand in Europe and the FSU. Demand in the Asia–Pacific region for hard coal, in contrast, has increased dramatically from 34 to 52% over the same period—equivalent to almost 1 billion tonnes.

According to the BP Statistical Review of World Energy, world coal consumption increased by 6.9% in 2002. However, this was almost entirely a Chinese phenomenon: reported consumption in China rose by an extraordinary 27.9%. Excluding China, world coal consumption grew by just 0.6%.

One reason for this is the huge increase in demand for electricity in Asian countries. China's electrification programme, for example, has connected 700 million people over the last 15 years. As a result of the programme, annual electricity production in China has increased by nearly 1 000 TWh: 84% of this is coal-fired. Forecasts indicate that this regional trend will continue, with the bulk of the projected increase in global coal demand coming from the region. According to IEA reports, China has experienced an ongoing decline in coal demand since 1996, but demand increased strongly in 2002, largely owing to continued strong economic growth.

Japan continues to be the largest importer of hard coal—both steam coal and coking coals—and is projected to account for 24% of total world imports by 2020. Other Asia–Pacific countries, such as Malaysia, the Philippines and Thailand, are looking to coal to diversify their energy mix and provide a secure supply of affordable energy to meet their growing electricity needs.

The decline of coal consumption in the EU can be attributed to a number of factors, including more stringent environmental legislation, the availability of gas from the North Sea and Russia, as well as increasingly from North Africa and the Middle East. As older coal-fired plant faced retirement, the capital costs of building combined-cycle gas plant were considerably lower than building a new coal-fired plant with the required environmental controls, and at a time when gas prices were relatively low, were the economic option. However, such long-term decisions can be affected by the vagaries of the gas market—as happened in the UK in 2001 when coal-fired plants were brought back on-line owing to sudden increases in gas prices.

Coal Trade

In 2002, hard coal trade continued to expand, growing to 623 mt (435 mt steam coal, 188 mt coking coal). Worldwide hard coal trade is divided into seaborne trade of 579 mt and internal trade of 44 mt. Steam coal exports from Russia increased by 33% over 2001 levels and from Australia by 6%. Significant reductions in exports were seen from Kazakhstan, China, the USA and Colombia. Coking coal exports generally decreased in 2002, with the exception of China, which achieved a 20% increase in shipments.

In 2002, international hard coal trade in maritime traffic totalled about 16% of the worldwide hard coal output. Almost 85% of hard coal output is thus consumed in the producing country itself—in particular for power generation and by some key industries, such as iron and steel, cement and chemicals. This is especially true for the three biggest hard coal producers—China, the USA and India.

However, between 1998 and 2001 coal exports from China grew from 32 to 90 mt—an increase of 179%, making it the world's second largest coal exporter. Preliminary Chinese data suggest that 2002 exports of 84 mt would maintain China's position in the world export league.

Environmental Issues

Environmental policies are the key factor in determining the *future* role of coal around the world. While coal can and does provide an affordable, reliable, secure and safe source of energy it, along with other fossil fuels, continues to face environmental challenges. The introduction of carbon taxes, emissions trading and other policies to restrict emissions of greenhouse gases to the atmosphere will have an adverse effect on

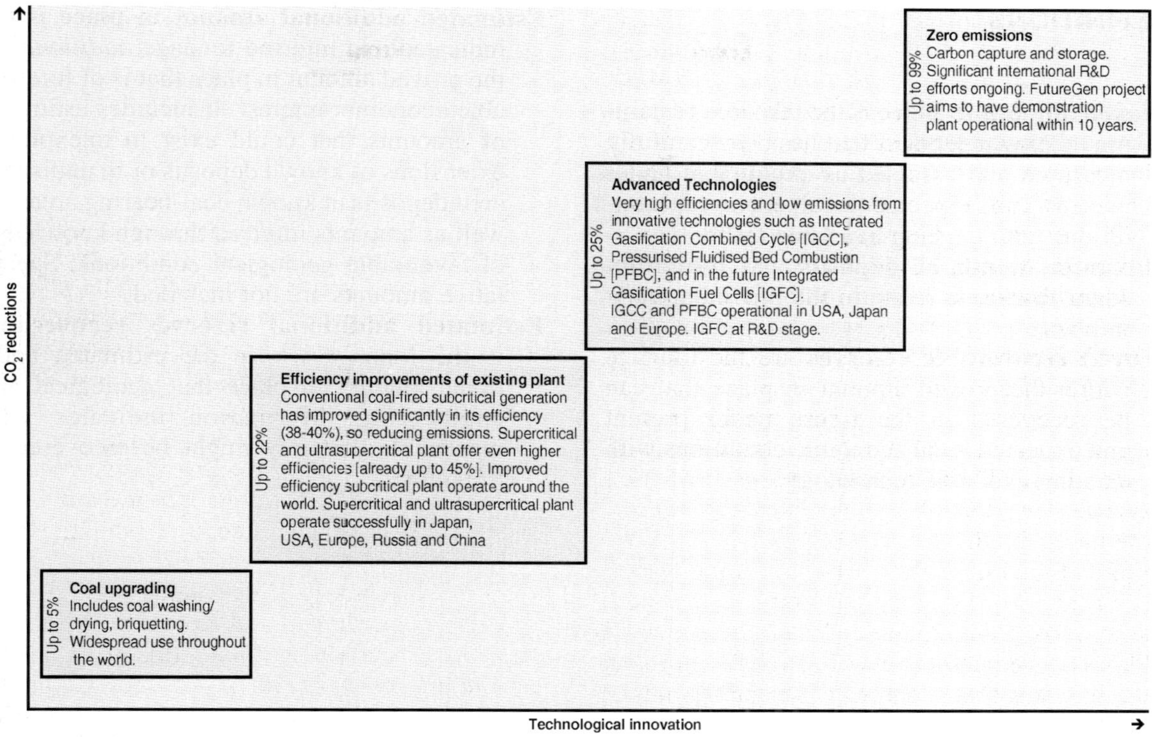

FIGURE 1.4 The coal-fired route to CO_2 reduction (Source: WCI, 2003).

the global coal market. Yet the crucial issue is how coal is used, not the fuel itself (Fig. 1.4).

Technologies have already been developed that are capable of almost entirely eliminating local and regional pollutants from coal-fired power generation, and efficiency gains can significantly reduce carbon dioxide emissions—the thermal efficiency of coal-fired electricity generation underwent an eight-fold improvement during the 20th Century. However, if coal is to maintain its place in the energy mix of the future, the development and deployment of improved coal technologies has to continue.

Ongoing developments in supercritical conventional coal combustion are likely to bring the thermal efficiency of coal burning to over 50%. The gasification of coal in integrated combined cycle (IGCC) systems is becoming increasingly well understood and commercially practical. Near-zero emissions can be realised if such IGCC systems are combined with emerging carbon-capture and storage technology.

Indeed coal, via gasification technology, has the potential to become a mainstay of a future 'hydrogen economy'. It is an abundant potential source of the huge quantities of manufactured hydrogen that would be required for the widespread application of emissions-free hydrogen-based energy systems.

Christine Copley
World Coal Institute

DEFINITIONS

Proved amount in place is the resource remaining in known deposits that has been carefully measured and assessed as exploitable under present and expected local economic conditions with existing available technology.

Maximum depth of deposits and **minimum seam thickness** relate to the proved amount in place.

Proved recoverable reserves are the tonnage *within* the proved amount in place that can be recovered in the future under present and expected local economic conditions with existing available technology.

Estimated additional amount in place is the indicated and inferred tonnage *additional to* the proved amount in place that is of foreseeable economic interest. It includes estimates of amounts that could exist in unexplored extensions of known deposits or in undiscovered deposits in known coal-bearing areas, as well as amounts inferred through knowledge of favourable geological conditions. Speculative amounts are not included.

Estimated additional reserves recoverable is the tonnage *within* the estimated additional amount in place that geological and engineering information indicates with reasonable certainty might be recovered in the future.

Table Notes

The tables cover bituminous coal (including anthracite), sub-bituminous coal and lignite. Data for peat are given in Chapter 8. There is no universally accepted system of demarcation between coals of different rank and what is regarded as sub-bituminous coal tends to vary from one country to another. Moreover, if it is not isolated as such, sub-bituminous is sometimes included with bituminous and sometimes with lignite.

TABLE 1.1

Coal: proved recoverable reserves at end-2002 (million tonnes)

	Bituminous including anthracite	Sub-bituminous	Lignite	Total
Algeria	40			40
Botswana	40			40
Central African Republic			3	3
Congo (Democratic Republic)	88			88
Egypt (Arab Republic)	21			21
Malawi		2		2
Morocco	N			N
Mozambique	212			212
Niger	70			70
Nigeria	21	169		190
South Africa	48 750			48 750
Swaziland	208			208
Tanzania	200			200
Zambia	10			10
Zimbabwe	502			502
Total Africa	**50 162**	**171**	**3**	**50 336**
Canada	3 471	871	2 236	6 578
Greenland		183		183
Mexico	860	300	51	1 211
United States of America	111 338	101 978	33 327	246 643
Total North America	**115 669**	**103 332**	**35 614**	**254 615**
Argentina		424		424
Bolivia	1			1
Brazil		10 113		10 113
Chile	31	1 150		1 181
Colombia	6 230	381		6 611
Ecuador			24	24
Peru	960		100	1 060
Venezuela	479			479
Total South America	**7 701**	**12 068**	**124**	**19 893**

(continued on next page)

TABLE 1.1 (*Continued*)

	Bituminous including anthracite	Sub-bituminous	Lignite	Total
Afghanistan	66			66
China	62 200	33 700	18 600	114 500
India	90 085		2 360	92 445
Indonesia	740	1 322	2 906	4 968
Japan	359			359
Kazakhstan	28 151		3 128	31 279
Korea (DPR)	300	300		600
Korea (Republic)		80		80
Kyrgyzstan			812	812
Malaysia	4			4
Mongolia				
Myanmar (Burma)	2			2
Nepal		1		1
Pakistan		60	2 990	3 050
Philippines	22	144	70	236
Taiwan, China	1			1
Thailand			1 354	1 354
Turkey	278	761	3 147	4 186
Uzbekistan	1 000		3 000	4 000
Vietnam	150			150
Total Asia	**183 358**	**36 368**	**38 367**	**258 093**
Albania			794	794
Austria			20	20
Bulgaria	4	91	2 092	2 187
Croatia	6		33	39
Czech Republic	2 094	3 242	216	5 552
France	15			15
Germany	183		6 556	6 739
Greece			3 900	3 900
Hungary	198	199	2 960	3 357
Ireland	14			14
Italy		27	7	34
Netherlands	497			497
Norway		5		5
Poland	14 000			14 000
Portugal	3		33	36
Romania	22	3	469	494
Russian Federation	49 088	97 472	10 450	157 010
Serbia & Montenegro	9	656	15 926	16 591
Slovakia	N		172	172
Slovenia		40	235	275
Spain	200	300	30	530
Sweden		1		1

TABLE 1.1 (*Continued*)

	Bituminous including anthracite	Sub-bituminous	Lignite	Total
Ukraine	16 274	15 946	1 933	34 153
United Kingdom	220			220
Total Europe	**82 827**	**117 982**	**45 826**	**246 635**
Iran (Islamic Republic)	419			419
Total Middle East	**419**			**419**
Australia	38 600	2 200	37 700	78 500
New Caledonia	2			2
New Zealand	33	205	333	571
Total Oceania	**38 635**	**2 405**	**38 033**	**79 073**
Total World	**478 771**	**272 326**	**157 967**	**909 064**

Notes:
(1) A quantification of proved recoverable reserves for Mongolia is not available.
(2) Sources: WEC Member Committees, 2003; data reported for previous WEC *Surveys of Energy Resources*; national and international published sources.

TABLE 1.2I

Bituminous coal (including anthracite): resources at end-2002

	Proved amount in place			Estimated additional	
	Tonnage (million tonnes)	Maximum depth of deposits (m)	Minimum seam thickness (m)	Amount in place (million tonnes)	Reserves recoverable (million tonnes)
Africa					
Botswana	3 340	85	5.5		
Egypt (Arab Rep.)	27				
South Africa	115 000	350	1.0		
Swaziland	567	550	1.0	450	
North America					
Canada	4 609	1 200	0.6	92 224	62 445

(*continued on next page*)

TABLE 1.2I (Continued)

	Proved amount in place			Estimated additional	
	Tonnage (million tonnes)	Maximum depth of deposits (m)	Minimum seam thickness (m)	Amount in place (million tonnes)	Reserves recoverable (million tonnes)
USA	246 804	671	0.3	445 346	
South America					
Argentina	4				
Chile	64				
Venezuela	1 308			6 955	
Asia					
Indonesia	1 139			5 770	53
Japan	4 772	900	0.6	6 298	
Philippines	36	500	0.6	8	5
Taiwan, China	100	800	0.4		
Turkey	428	1 300	0.3	658	
Europe					
Austria	1				
Bulgaria	4	150	0.6		
Croatia	4				
Czech Republic	7 155	1 600	0.6		7 065
France	160	1 300	1.0		
Germany	383	1 600	0.6	23 000	8 371
Hungary	1 595	1 000	0.4	296	37
Netherlands	1 406	1 500	0.8	2 750	1 375
Poland	46 000	1 000	1.0	59 000–80 000	
Romania	29			2 185	187
Serbia & Montenegro	78		0.5		
Spain	1 300	1 200	0.5	3 000	200
Ukraine	21 699	1 800	0.5	5 423	
Middle East					
Iran	12 700–50 000				
Oceania					
Australia	56 500	600	0.3	125 000	75 000
New Zealand	45			942	313

Notes:

(1) The data on resources are those reported by WEC Member Committees. They thus constitute a sample, reflecting the information available in particular countries: they should not be considered as complete, or necessarily representative of the situation in each region. For this reason, regional and global aggregates have not been computed.

(2) Sources: WEC Member Committees, 2003.

TABLE 1.2II

Sub-bituminous coal: resources at end-2002

	Proved amount in place			Estimated additional	
	Tonnage (million tonnes)	Maximum depth of deposits (m)	Minimum seam thickness (m)	Amount in place (million tonnes)	Reserves recoverable (million tonnes)
North America					
Canada	1 153	300	1.5	48 764	15 165
Greenland	183	550		200	100
USA	165 150	305	1.5	273 593	
South America					
Argentina	697	800	0.5	273	
Brazil	17 033	870	0.5	15 319	7 660
Chile	91				
Asia					
Indonesia	2 034			10 301	95
Japan				5 936	
Korea (Republic)	134	1 000	0.6	377	122
Nepal	5		0.5		
Pakistan	105	2 000	0.2	250	140
Philippines	184	300	0.6	115	80
Turkey	1 526	828	0.4	202	
Europe					
Bulgaria	254	390	1.5		
Czech Republic	1 943	500	2.0		4 403
Hungary	3 204	500	0.8	1 289	78
Italy	60	500	1.4	280	
Norway	52	450	0.6		
Romania	8			223	13
Serbia & Montenegro	949		2.0		
Slovenia	57	190	10.0		
Spain	800	200	3.0	1 600	300
Sweden	4	15	0.5	20	
Ukraine	21 261	1 800	0.6	5 502	
Oceania					
Australia	3 500	200	1.5	27 900	20 700
New Zealand	376			2 085	682

Notes:

(1) The data on resources are those reported by WEC Member Committees. They thus constitute a sample, reflecting the information available in particular countries: they should not be considered as complete, or necessarily representative of the situation in each region. For this reason, regional and global aggregates have not been computed.

(2) Sources: WEC Member Committees, 2003.

TABLE 1.2III

Lignite: resources at end-2002

	Proved amount in place			Estimated additional	
	Tonnage (million tonnes)	Maximum depth of deposits (m)	Minimum seam thickness (m)	Amount in place (million tonnes)	Reserves recoverable (million tonnes)
North America					
Canada	2 961	50	1.5	51 034	42 115
USA	39 558	61	0.8	393 822	
South America					
Argentina	7 350	680			
Asia					
Indonesia	4 474			22 655	209
Japan				1 186	
Pakistan	3 180	280	0.3	67 910	
Philippines	82	100	1.0	53	45
Thailand		700	0.3		
Turkey	5 812	492	0.4	534	
Europe					
Austria	335				
Bulgaria	3 655	120	0.8		
Croatia	41				
Czech Republic	623	130	1.5		190
France	114	1 000	1.0		
Germany	7 136	550	5.0	69 260	34 100
Greece	5 800	400	1.0	2 500	1 000
Hungary	5 828	200	1.0	1 341	
Italy	15	150	3.0	22	20
Poland	13 984	350	3.0	41 000	
Romania	4 053			9 905	1 527
Serbia & Montenegro	20 578		10.0		
Slovakia					389
Slovenia	602	547	8.0		
Spain	30	50	0.5		
Ukraine	2 578	400	2.7	320	
Oceania					
Australia	41 900	300	3.0	175 800	158 200
New Zealand	2 297			9 817	7 078

Notes:

(1) The data on resources are those reported by WEC Member Committees. They thus constitute a sample, reflecting the information available in particular countries: they should not be considered as complete, or necessarily representative of the situation in each region. For this reason, regional and global aggregates have not been computed.

(2) Sources: WEC Member Committees, 2003.

TABLE 1.3

Coal: 2002 production (thousand tonnes)

	Bituminous	Sub-bituminous	Lignite	Total
Algeria	20			20
Botswana	956			956
Congo (Democratic Republic)	100			100
Egypt (Arab Republic)	37			37
Malawi		35		35
Morocco	N			N
Mozambique	20			20
Niger	165			165
Nigeria		10		10
South Africa	214 652			214 652
Swaziland	553			553
Tanzania	75			75
Zambia	70			70
Zimbabwe	4 130			4 130
Total Africa	**220 778**	**45**		**220 823**
Canada	29 600	25 900	11 300	66 800
Mexico	5 168	6 237		11 405
United States of America	520 244	397 668	74 806	992 718
Total North America	**555 012**	**429 805**	**86 106**	**1 070 923**
Argentina	160			160
Brazil		5 100		5 100
Chile	95	265		360
Colombia	39 510			39 510
Peru	25			25
Venezuela	8 000			8 000
Total South America	**47 790**	**5 365**		**53 155**
Afghanistan	2			2
Bhutan	65			65
China	1 343 350		50 000	1 393 350
Georgia	6			6
India	334 970		23 920	358 890
Indonesia	103 372			103 372
Japan	1 369			1 369
Kazakhstan	70 603		2 618	73 221
Korea (Democratic People's Republic)	53 000	15 000		68 000
Korea (Republic)		3 318		3 318
Kyrgyzstan	122		376	498
Laos	100			100
Malaysia		353		353
Mongolia	1 520		3 787	5 307
Myanmar (Burma)	13		27	40

(continued on next page)

TABLE 1.3 (*Continued*)

	Bituminous	Sub-bituminous	Lignite	Total
Nepal		8		8
Pakistan		2 300	1 020	3 320
Philippines	120	1 540		1 660
Taiwan, China				
Tajikistan		33		33
Thailand			19 600	19 600
Turkey	2 200		51 000	53 200
Uzbekistan	75		2 660	2 735
Vietnam	15 000			15 000
Total Asia	**1 925 887**	**22 552**	**155 008**	**2 103 447**
Albania			15	15
Austria			1 210	1 210
Bosnia-Herzogovina			10 000	10 000
Bulgaria	100	3 100	23 200	26 400
Croatia				
Czech Republic	14 467	48 391	501	63 359
FYR Macedonia			8 000	8 000
France	1 480	440	150	2 070
Germany	29 209		181 779	210 988
Greece			70 340	70 340
Hungary	660	4 530	7 570	12 760
Italy		100		100
Norway		2 100		2 100
Poland	102 100		58 200	160 300
Romania	3 700	216	26 530	30 446
Russian Federation	168 420		85 000	253 420
Serbia & Montenegro	22	515	33 536	34 073
Slovakia			3 401	3 401
Slovenia		639	4 048	4 687
Spain	9 751	3 558	8 726	22 035
Ukraine	81 857		1 000	82 857
United Kingdom	29 989			29 989
Total Europe	**441 755**	**63 589**	**523 206**	**1 028 550**
Iran (Islamic Republic)	1 760			1 760
Total Middle East	**1 760**			**1 760**
Australia	256 000	17 000	67 000	340 000
New Zealand	2 269	1 972	218	4 459
Total Oceania	**258 269**	**18 972**	**67 218**	**344 459**
Total World	**3 451 251**	**540 328**	**831 538**	**4 823 117**

Notes:

(1) Sources: WEC Member Committees, 2003; *BP Statistical Review of World Energy*, June 2003; *Energy—Monthly Statistics*, Eurostat; *World Mineral Statistics 1997–2001*, British Geological Survey; national sources; estimates by the editors.

TABLE 1.4

Coal: 2002 consumption (thousand tonnes)

	Bituminous	Sub-bituminous	Lignite	Total
Algeria	928			928
Botswana	974			974
Congo (Democratic Republic)	130			130
Egypt (Arab Republic)	1 630			1 630
Ghana	3			3
Kenya	160			160
Libya/GSPLAJ	5			5
Madagascar	12			12
Malawi	55			55
Mauritania	6			6
Mauritius	250			250
Morocco	5 124			5 124
Mozambique	35			35
Namibia	5			5
Niger	165			165
Nigeria		10		10
South Africa	154 878			154 878
Swaziland	200			200
Tanzania	45			45
Tunisia	1			1
Zambia	65			65
Zimbabwe	4 000			4 000
Total Africa	**168 671**	**10**		**168 681**
Canada	19 000	30 000	11 300	60 300
Cuba	10			10
Dominican Republic	220			220
Guatemala	200			200
Honduras	120			120
Jamaica	60			60
Mexico	2 500	12 500		15 000
Panama	70			70
Puerto Rico	150			150
United States of America	516 650	399 100	75 850	991 600
US Virgin Islands	260			260
Total North America	**539 240**	**441 600**	**87 150**	**1 067 990**
Argentina		850		850
Brazil		17 700		17 700
Chile	3 500			3 500
Colombia	3 100			3 100
Peru	600			600
Uruguay	1			1
Venezuela	60			60

(continued on next page)

TABLE 1.4 (Continued)

	Bituminous	Sub-bituminous	Lignite	Total
Total South America	**7 261**	**18 550**		**25 811**
Afghanistan	2			2
Armenia				
Azerbaijan	5			5
Bangladesh	740			740
Bhutan	70			70
China	1 265 000		50 000	1 315 000
Cyprus	72			72
Georgia	10			10
Hong Kong, China	8 718			8 718
India	355 000		24 000	379 000
Indonesia	28 990			28 990
Japan	159 700			159 700
Kazakhstan	46 900		2 600	49 500
Korea (DPR)	23 500	7 500		31 000
Korea (Republic)	67 144	7 627		74 771
Kyrgyzstan	500		375	875
Malaysia	2 300	3 000		5 300
Mongolia	750		4 400	5 150
Myanmar (Burma)	13		27	40
Nepal		420		420
Pakistan	900	2 300	1 020	4 220
Philippines	7 210	5 670		12 880
Sri Lanka	1			1
Taiwan, China	50 603			50 603
Tajikistan	100	30		130
Thailand			19 570	19 570
Turkey	13 800		51 400	65 200
Uzbekistan	75		2 650	2 725
Vietnam	10 000			10 000
Total Asia	**2 042 103**	**26 547**	**156 042**	**2 224 692**
Albania			90	90
Austria	3 770		1 530	5 300
Belarus	450			450
Belgium	10 300	200	200	10 700
Bosnia-Herzogovina		3 600	5 300	8 900
Bulgaria	3 750	3 000	24 000	30 750
Croatia	900		100	1 000
Czech Republic	10 086	47 193	501	57 780
Denmark	6 708	N	N	6 708
Estonia	120			120
Finland	6 850			6 850
FYR Macedonia	40		8 000	8 040
France	18 720	280	180	19 180
Germany	68 671		172 848	241 519

TABLE 1.4 (Continued)

	Bituminous	Sub-bituminous	Lignite	Total
Greece			68 150	68 150
Hungary	940	5 000	7 570	13 510
Iceland	150			150
Ireland	2 700			2 700
Italy	19 700	100	N	19 800
Latvia	150			150
Lithuania	123		2	125
Luxembourg	150			150
Moldova	100			100
Netherlands	13 300			13 300
Norway		1 200		1 200
Poland	87 700		58 200	145 900
Portugal	5 000			5 000
Romania	7 372	216	26 530	34 118
Russian Federation	139 200		75 000	214 200
Serbia & Montenegro	105	865	33 536	34 506
Slovakia	4 900		4 200	9 100
Slovenia	92	1 198	4 361	5 651
Spain	31 500	4 267	8 726	44 493
Sweden	3 300			3 300
Switzerland	200			200
Ukraine	85 000		1 000	86 000
United Kingdom	58 500			58 500
Total Europe	**590 547**	**67 119**	**500 024**	**1 157 690**
Iran (Islamic Republic)	1 580			1 580
Israel	12 100			12 100
Lebanon	200			200
Total Middle East	**13 880**			**13 880**
Australia	49 000	17 000	67 000	133 000
Fiji	20			20
New Caledonia	150			150
New Zealand	288	1 899	265	2 452
Papua New Guinea	1			1
Total Oceania	**49 459**	**18 899**	**67 265**	**135 623**
Total World	**3 411 161**	**572 725**	**810 481**	**4 794 367**

Notes:

(1) Sources: WEC Member Committees, 2003; *BP Statistical Review of World Energy*, June 2003; *Energy—Monthly Statistics*, Eurostat; national sources; estimates by the editors.

COUNTRY NOTES

The following Country Notes on Coal have been compiled by the editors, drawing upon a wide variety of material, including information received from WEC Member Committees, national and international publications.

Major international published sources consulted included:

- *Energy Balances of OECD Countries 2000–2001*; 2003; International Energy Agency;
- *Energy Balances of Non-OECD Countries 2000-2001*; 2003; International Energy Agency;
- *Major coalfields of the world*; June 2000; IEA Coal Research.

Argentina

Proved amount in place (total coal, million tonnes)	8 051
Proved recoverable reserves (total coal, million tonnes)	424
Production (total coal, million tonnes, 2002)	0.2

The principal coal-mining areas are located in the west of the country along the foothills of the Andes and in the Andes themselves, in the provinces of Catamarca, La Rioja, San Juan, Mendoza, Neuquén, Río Negro, Chubut and Santa Cruz, with smaller coalfields in Córdoba, the centre of Chubut and the Atlantic coast of Santa Cruz.

The biggest coalfield is Río Turbio, located to the west of the city of Río Gallegos in the southern province of Santa Cruz, close to the border with Chile. Río Turbio's coal is a steam coal with low sulphur content (down to 1%), falling into the sub-bituminous rank; it constitutes 99% of the hard coal resources of the country, and supports the only coal extraction activity in the Argentine Republic. The Río Turbio coalfield, including the concession for operating the associated railway and port facilities, was privatised in 1994 but is currently under administration by a Federal auditor.

The Argentinian Member Committee has reported proved amounts in place of 697 million tonnes of sub-bituminous coal and 7 350 million tonnes of lignite, together with a minor quantity (4 million tonnes) of bituminous grade. For sub-bituminous, the maximum deposit depth is 800 m, with seams ranging from 0.5 to 2.0 m in thickness. The lignite resources are at a maximum depth of 680 m. The only proved recoverable reserves reported are 424 million tonnes of sub-bituminous.

Coal output from the Río Turbio mine is quite modest, at around 200 thousand tonnes per annum. The greater part is used for electricity generation, the balance as industrial fuel.

Australia

Proved amount in place (total coal, million tonnes)	101 900
Proved recoverable reserves (total coal, million tonnes)	78 500
Production (total coal, million tonnes, 2002)	340.0

Australia is endowed with very substantial coal resources, with its proved recoverable reserves ranking fifth in the world. The major deposits of black coal (bituminous and sub-bituminous) are located in New South Wales and Queensland; smaller but locally important resources occur in Western Australia, South Australia and Tasmania. The main deposits of brown coal are in Victoria, the only State producing this rank. Other brown coal resources are present in Western Australia, South Australia and Tasmania.

The coal resource data reported for the present *Survey* by the Australian WEC Member Committee have been provided by Geoscience Australia, formerly the Australian Geological Survey Organisation (AGSO). The proved amount of coal in place comprises 56.5 billion tonnes of bituminous coal, 3.5 billion tonnes of sub-bituminous and 41.9 billion tonnes of brown coal/lignite. Within these tonnages, the proportion deemed to be recoverable ranges from 68% of the bituminous coal (with 54% of its reserves surface-mineable) to 90% of the lignite, all of which is suitable for open-cast mining. About 36% of Australia's massive reserves of bituminous coal are of coking quality. The maximum depth of the deposits ranges from 600 m in the case of bituminous coal to 200 m for sub-bituminous and 300 m for lignite. Minimum seam thicknesses are 0.3, 1.5 and 3.0 m, respectively.

Indicated and inferred tonnages, additional to the proved amount in place, are vast: Geoscience Australia's current assessment puts bituminous coal at 125 billion tonnes, sub-bituminous at nearly 28 billion tonnes and lignite at around 175 billion tonnes. In total, more than 250 billion tonnes of this additional coal is considered to be eventually recoverable.

In 2002 Australia produced 273 million tonnes of saleable black coal (bituminous and sub-bituminous) and 67 million tonnes of brown coal. The major domestic market for black coal is electricity generation: in 2001, power stations accounted for 88% of total black coal consumption, with the other large consumers being the iron and steel industry and cement manufacture. Brown coal is used almost entirely for power generation.

Australia has been the world's largest exporter of hard coal since 1984: in 2002, it exported 198 million tonnes. About 53% of 2002 exports were of metallurgical grade (coking coal), destined largely for Japan, the Republic of Korea, India and Europe.

Botswana

Proved recoverable reserves (total coal, million tonnes)	40
Production (total coal, million tonnes, 2002)	1.0

Vast deposits of bituminous coal have been located in Botswana, principally in the eastern part of the country. The only mine to have been developed so far is at Morupule, near the town of Palapye, where Morupule Colliery Limited (controlled by Anglo American Corporation) commenced coal extraction in 1973.

For the present *Survey*, the Botswana WEC Member Committee has reported proved recoverable reserves of 40 million tonnes, of which 50% can be mined by open-cast methods. The reported tonnages relate solely to the economically recoverable reserves that are currently being exploited at the Morupule Mine. With cumulative output to the end of 2002 amounting to nearly 18 million tonnes, Botswana's remaining proved amount of coal in place is reported to be 3 340 million tonnes.

All of Botswana's current coal production (956 thousand tonnes in 2002) is of power generation quality, none of coking quality. The Morupule mine's chief customers are the Botswana Power Corporation, the copper/nickel mine at Selibe-Phikwe and the soda ash plant at Sua Pan. The BPC power station at Morupule (net capacity 118 MW) generates about half of Botswana's electricity supplies, the balance being provided by imports from South Africa.

Brazil

Proved amount in place (total coal, million tonnes)	17 033
Proved recoverable reserves (total coal, million tonnes)	10 113
Production (total coal, million tonnes, 2002)	5.1

Brazil has considerable reserves of sub-bituminous coal, mostly located in the southern states of Rio Grande do Sul, Santa Catarina and Paraná.

For the present *Survey*, the Brazilian WEC Member Committee has reported a proved amount in place (described as covering measured, indicated and inferred reserves) of just over 17 billion tonnes, of which almost 60% is categorised as proved recoverable reserves. The maximum depth of the deposits is 870 m, whilst the minimum seam thickness is 0.5 m. There is estimated to be some 15.3 billion tonnes of additional coal in place, of which 50% is considered to be recoverable.

With respect to the stated level of proved recoverable reserves, it is estimated that 21% could be exploited through surface mining, and that 7% is considered to be of coking quality. In 2002, 63% of Brazilian coal production was obtained by surface mining.

Almost all of Brazil's current coal output is classified as steam coal, of which about 85% is used as power-station fuel and the remainder in industrial plants. Virtually all of Brazil's metallurgical coal is imported: about three-quarters is used as input for coke production.

Canada

Proved amount in place (total coal, million tonnes)	8 723
Proved recoverable reserves (total coal, million tonnes)	6 578
Production (total coal, million tonnes, 2002)	66.8

Canada has considerable coal resources, with proved reserves of more than 6.5 billion tonnes. The levels of resources that have been reported by the Canadian WEC Member Committee are unchanged from those advised for the 2001 *Survey of Energy Resources*. Bituminous coals (including anthracite) are evaluated as 4.6 billion tonnes, based on deposits to a maximum depth of 1 200 m and a minimum seam thickness of 0.6 m; sub-bituminous grades are put at approximately 1.1 billion tonnes (maximum depth 300 m, minimum thickness 1.5 m); and lignite at 3.0 billion tonnes (maximum depth 50 m, minimum thickness 1.5 m). The proved recoverable reserves for each rank have been assessed as approximately 75% of the respective proved amount in place.

Estimates of the tonnages of coal (in-place and recoverable) that are considered to be additional to the 'proved' amounts of each rank all run into tens of billions of tonnes (see Tables 1.2I, II and III). Such numbers can never possess any high degree of accuracy, but they do serve to underline Canada's undoubtedly massive coal endowment.

Canadian coal reserves are mainly located in the western provinces of Saskatchewan, Alberta and British Columbia, with smaller deposits in the eastern provinces of Nova Scotia and New Brunswick. Bituminous deposits are found in the two eastern provinces together with Alberta and British Columbia; Alberta also possesses sub-bituminous grades, while lignite deposits are found only in Saskatchewan.

Western Canada dominates coal production, accounting for 99.5% of the total. Alberta is the largest coal-producing province, mainly of thermal grades. British Columbia is the second largest, producing mainly metallurgical coals. Saskatchewan produces lignite. About 40% of Canadian coal production is exported.

Around 90% of Canadian coal consumption is used for electricity generation, 7% in the steel industry and 3% in other industries. Alberta is the largest coal-consuming province, Ontario the second. Ontario relies on coal imports—there are no coal mines in the province.

The Canadian coal industry is entirely privately owned. Output is mainly from surface mines: only one underground mine is in operation, with a limited production.

China

Proved recoverable reserves (total coal, million tonnes)	114 500
Production (total coal, million tonnes, 2002)	1 393.4

China is a major force in world coal, standing in the front rank in terms of reserves, production and consumption, and recently, as a coal exporter. The levels of proved recoverable reserves (as at end-1990), originally provided by the Chinese WEC Member Committee for the 1992 *Survey*, have been retained for each successive edition; in billions of tonnes, they amount to: bituminous coal and anthracite 62.2; sub-bituminous coal 33.7 and lignite 18.6. The level of proved reserves retained for the present *Survey* implies a coal R/P ratio of 82, on the basis of 2002 production.

It is interesting to note that the same figure (114.5 billion tonnes) for total proved reserves was quoted at the 11th Session of the UN Committee on Sustainable Energy (Geneva, November 2001), in the context of an estimate of 988 billion tonnes for China's coal resources. This reference, in a paper co-authored by Professor Huang Shengchu, a vice-president of the China Coal Information Institute, indicates a degree of continuity in the official assessments of China's coal reserves and supports the retention of the level originally advised by the Chinese WEC Member Committee in 1991.

Coal deposits have been located in most of China's regions but three-quarters of proved recoverable reserves are in the north and north-west, particularly in the provinces of Shanxi, Shaanxi and Inner Mongolia.

After more than 20 years of almost uninterrupted growth, China's coal production peaked at nearly 1.4 billion tonnes in 1996, followed by 4 years during which output fell year by year, largely as a result of the closure of large numbers of small local mining operations. Annual output has regained an upward path in the past two years and reached a new peak in 2002. By far the greater part of output is of bituminous coal: lignite constitutes only about 4%.

The major coal-consuming sectors are power stations (including CHP), which accounted for 50% of total consumption in 2001, the iron and steel industry with a 15% share, and other industrial users with about 17%.

Coal exports more than doubled between 1999 and 2002, when they totalled 84 million tonnes; China is now the world's second largest coal exporter, after Australia.

Colombia

Proved recoverable reserves (total coal, million tonnes)	6 611
Production (total coal, million tonnes, 2002)	39.5

Colombia's vast coal resources are located in the north and west of the country. Data on measured reserves at end-2001, published by the Ministerio de Minas y Energía, indicate a total of some 6.6 billion tonnes, of which Zona Norte and Cerrejón in the department of La Guajira account for 55% and fields in the department of Cesar for 29%. 'Indicated reserves' quoted in the same publication are 2 932 million tonnes. Virtually all Colombia's coal resources fall into the bituminous category: the reserves in the San Jorge field in Córdoba, with an average calorific value in the sub-bituminous/lignite bracket, are shown under sub-bituminous in Table 1.1.

Development of Colombian coal for export has centred on the Cerrejón deposits which are located in the Guajira Peninsula in the far north, about 100 km inland from the Caribbean coast. The coal is found in the northern portion of a basin formed by the Cesar and Rancheria rivers; the deposit has been divided by the Government into the North, Central and South Zones. In October 1975 the Government opened international bidding for the development of El Cerrejón-North Zone reserves and in December 1976 Carbocol (then 100% owned by the Colombian State) and Intercor (an Exxon affiliate) entered into an Association Contract for the development and mining of the North Zone. The contract has three phases and covers a 33-year period with the production phase scheduled to end early in 2009.

Carbocol was privatised in October 2000, the purchasers being a consortium of Anglo-American,

Billiton and Glencore; in early 2002 the three partners acquired the whole of Intercor's interest.

Coal exports from Colombia totalled 39.1 million tonnes in 2001, equivalent to 90% of its coal production. Cerrejón North remains one of the world's largest export mines.

Czech Republic

Proved amount in place (total coal, million tonnes)	9 721
Proved recoverable reserves (total coal, million tonnes)	5 552
Production (total coal, million tonnes, 2002)	63.4

The Czech Republic possesses sizeable coal resources, with a proved amount in place of nearly 10 billion tonnes, of which some 57% is reported to be economically recoverable. In terms of rank, 38% of the proved reserves are classified as bituminous, 58% as sub-bituminous and 4% as lignite. The tonnages reported by the Czech Member Committee for the present *Survey* show little change from those advised in 2000. The maximum depth of deposits varies from 1 600 m in the case of bituminous to 500 m for sub-bituminous and only 130 m for lignite; minimum seam thicknesses range from 0.6 (for bituminous) to 1.5 (lignite) and 2.0 m for sub-bituminous.

Bituminous coal deposits are mainly in the Ostrava-Karviná basin in the east of the country, and lie within the Czech section of the Upper Silesian coalfield. The principal sub-bituminous/lignite basins are located in the regions of North and West Bohemia, close to the Krusne Hory (Ore Mountains), which constitute the republic's north-western border with Germany. Currently all Czech output of bituminous coal and lignite is deep-mined, whereas 99% of sub-bituminous is surface-mined.

Since 1990, Czech output of bituminous coal has fallen by 35%, to 14.5 million tonnes in 2002, whilst sub-bituminous/lignite has declined by 38%, from 79 million tonnes in 1990 to 48.9 million tonnes in 2002. Over half of the republic's bituminous coal production consists of coking coal. In 2002, total exports of coal amounted to 7.7 million tonnes, equivalent to just over 12% of production.

Apart from its coking coal, which is consumed by the iron and steel industry, most of the republic's bituminous coal is used for electricity and heat generation, with industrial and private consumers accounting for relatively modest proportions. This pattern of utilisation also applies to sub-bituminous coal, which is still the main power station fuel.

France

The last lignite mine closed down at the end of January, 2003: all coal extraction will cease in 2004.

Germany

Proved amount in place (total coal, million tonnes)	7 519
Proved recoverable reserves (total coal, million tonnes)	6 739
Production (total coal, million tonnes, 2002)	211.0

The data advised by the German WEC Member Committee for the 2004 Survey reflect a reassessment of coal resources and reserves. The new numbers comply with the recommendations of the UN-ECE, within the context of the definitions specified for the SER. The proved amount in place is now given as approximately 7.5 billion tonnes, the bulk of which consists of lignite. The hard coal component has a maximum deposit depth of 1 600 m below the surface, and a minimum seam thickness of 0.6 m, whilst the corresponding parameters for lignite are 550 and 5 m, respectively.

Proved recoverable reserves comprise 183 million tonnes of hard coal and 6 556 million tonnes of lignite. Earlier assessments of German coal reserves (e.g. end-1996 and end-1999) contained large amounts of speculative resources which are no longer taken into account. Much of the former 'proved amount in place' and 'proved recoverable reserves' has been moved to 'additional amount in place' and 'additional reserves recoverable', respectively.

Germany's output of hard coal has fallen from 76.6 million tonnes in 1990 to 29.2 million tonnes in 2002, whilst lignite production has virtually halved, from 357.5 to 181.8 million tonnes over the same period. Germany is still the world's largest lignite producer.

The Ruhr coalfield produces over three-quarters of German hard coal. The coal qualities range from anthracite to high-volatile, strongly caking bituminous coal. The Saar is the second largest coalfield, with substantial deposits of weakly caking bituminous coal. All German hard coal is deep-mined from seams at depths exceeding 900 m.

The lignite deposit in the Rhine region is the largest single formation in Europe. In the former East Germany there are major deposits of lignite at Halle Leipzig and Lower Lausitz; these have considerable domestic importance.

The principal markets for bituminous coal are electricity generation, iron and steel, and cement manufacture: other industrial and household uses are relatively modest. The bulk of German lignite is consumed in power stations, although a considerable tonnage (11 million tonnes in 2001) is converted into brown coal briquettes for the industrial, residential and commercial markets.

Greece

Proved amount in place (total coal, million tonnes)	5 800
Proved recoverable reserves (total coal, million tonnes)	3 900
Production (total coal, million tonnes, 2002)	70.3

Coal resources are all in the form of lignite. Apart from a very small amount of private mining, all production is carried out by the mining division of the Public Power Corporation (DEI). There are two lignite centres, Ptolemais-Amynteo (LCPA) in the northern region of Western Macedonia, and Megalopolis (LCM) in the southern region of the Peloponnese. These two centres control the operations of seven open-cast mines; LCPA mines account for nearly 80% of DEI's lignite output. In 2002, LCPA produced 55.8 million tonnes of lignite, LCM 14.5 million tonnes.

In the lignite-mining areas, six dedicated power stations (total generating capacity: 4 850 MW) produce more than two-thirds of Greece's electricity supply. Two 330 MW lignite-fired power stations are planned for construction at Florina in Western Macedonia; the first unit came into operation in April 2003.

Greece is the second largest producer of lignite in the European Union and the fourth largest in the world.

India

Proved recoverable reserves (total coal, million tonnes)	92 445
Production (total coal, million tonnes, 2002)	358.9

Coal is the most abundant fossil fuel resource in India, which is the world's third largest coal producer. The principal deposits of hard coal are in the eastern half of the country, ranging from Andhra Pradesh, bordering the Indian Ocean, to Arunachal Pradesh in the extreme north-east: the eastern States of Chhattisgarh, Jharkhand, Orissa and West Bengal together account for about 77% of reserves. The Ministry of Coal (quoting the Geological Survey of India) states that, in addition to 90.1 billion tonnes of proved reserves of bituminous coal, there are 112.6 billion tonnes of indicated reserves and 38.0 billion tonnes of inferred reserves. Coking coals constitute 18% of the tonnage of proved reserves.

Lignite deposits mostly occur in the southern State of Tamil Nadu. India's geological resources of lignite are estimated to be some 35.4 billion tonnes, of which about 2.4 billion tonnes in the Neyveli area are regarded as 'mineable under the presently adopted mining parameters', and taken as proved recoverable reserves in the present *Survey*. Annual production of lignite is currently in the region of 24 million tonnes, almost all of which is used for electricity generation.

Although India's coal reserves cover all ranks from lignite to bituminous, they tend to have a high ash content and a low calorific value. The low quality of much of its coal prevents India from being anything but a small exporter of coal (traditionally to the neighbouring countries of Bangladesh, Nepal and Bhutan) and conversely, is responsible for sizeable imports (around 13 million tonnes/yr of coking coal and 12 million tonnes/yr of steam coal) from Australia, China, Indonesia and South Africa.

Coal is the most important source of energy for electricity generation in India: about three-quarters of electricity is generated by coal-fired power stations. In addition, the steel, cement, fertiliser, chemical, paper and many other medium and small-scale industries are also major coal users.

Indonesia

Proved amount in place (total coal, million tonnes)	7 647
Proved recoverable reserves (total coal, million tonnes)	4 968
Production (total coal, million tonnes, 2002)	103.4

Indonesia possesses very substantial coal resources: according to recent data advised by the Ministry of Energy and Mineral Resources for the purpose of this *Survey*, the proved amount in place totals 7 647 million tonnes, within which Indonesia's proved recoverable reserves amount to 4 968 million tonnes. In each case, lignite has the largest share (58.5%), with sub-bituminous coal accounting for 26.6% and bituminous grades for 14.9%. For all ranks, the recoverable proportion of the coal in place is approximately 65%.

The Ministry reports an estimated additional amount in place (based on 'indicated' and 'inferred' resources in the original data) as some 38.7 billion tonnes, with the same proportionate breakdown by rank as for the proved amount in place and proved reserves.

Indonesian coals in production generally have medium calorific values (5 000–7 000 kcal/kg or 21–29 MJ/kg), with relatively high percentages of volatile matter; they benefit from low ash and sulphur contents, making them some of the cleanest coals in the world.

Competitive quality characteristics have secured substantial export markets for Indonesian coal: in 2002 over 74 million tonnes were shipped overseas, representing nearly 72% of total coal output. Asian customers took 81% of Indonesia's coal exports.

Within Indonesia, coal's main market is power generation, which accounted for 68% of internal consumption in 2002.

Pakistan

Proved amount in place (total coal, million tonnes)	3 285
Proved recoverable reserves (total coal, million tonnes)	3 050
Production (total coal, million tonnes, 2002)	3.3

The republic possesses substantial coal resources. The Geological Survey of Pakistan (GSP) quotes a total coal resource at 30 June 2002 of just over 185 billion tonnes, within which measured reserves are 3 303 million tonnes, and mineable reserves 1 982 million tonnes. The Pakistan WEC Member Committee reports the proved amount in place as 105 million tonnes of sub-bituminous coal and 3 180 million tonnes of lignite; of these quantities, proved recoverable reserves consist of 60 million and 2 990 million tonnes, respectively.

The Pakistan MC has reported all the coal resources of the Sind province as falling within the lignite rank. These lignite resources have been divided into two main categories: (a) non-developed coalfields, and (b) developed coal-fields.

(a) The non-developed coalfields have been further sub-divided into:
   (i) The Thar coalfield: this will be Pakistan's first open-cast coalfield and may come into production within 3–4 years. The proved amount in place is put at 2 700 million tonnes, all of which is considered to be recoverable via open-cast mining. Over and above this tonnage Thar's 'indicated' resources are 9 400 million tonnes, 'inferred' resources 51 billion tonnes and 'hypothetical' resources 113 billion tonnes. Thar accounts for 88.5% of Pakistan's reported recoverable reserves.
   (ii) All other coalfields of Sind Province, such as Thatta, Sonda, Jherruck, Ongar and Indus East: these fields, when developed, will be underground mines: they have little chance of becoming producing fields in the near future. In aggregate, they have 230 million tonnes in place, with a maximum deposit depth of 280 m and seam thicknesses varying from 0.3 to 6.0 m. Proved reserves are 140 million tonnes, equivalent to 60% of the coal in place. 'Indicated' resources are given as 1.5 billion tonnes and 'inferred' as 4.8 billion.

(b) The developed lignite mines are Lakhra and Meting-Jhimpur; the main coal-producing area is Lakhra, where production is expected to rise. The total amount of coal in place is 250 million tonnes, at a maximum deposit depth of 150 m and with seam thicknesses between 0.75 and 2.5 m. Proved recoverable reserves are 150 million tonnes, with additional in-place tonnages of 660 million 'indicated' and 550 million 'inferred'.

The lignite coals are mainly being consumed by the brick kiln industry, with some by the cement industry, although the latter is now mostly converting to imported coal as the local lignite has a low calorific value as well as high sulphur and ash contents.

The coals of the Balochistan, Punjab and NWFP provinces have been grouped in the sub-bituminous rank category; although the rank of such coals ranges from sub-bituminous to bituminous, it is not possible to separate them. The proved amount of sub-bituminous coal in place is 105 million tonnes, to a maximum depth of 2 000 m and with seams from 0.2 to 2.0 m thick. Proved recoverable reserves are 60 million tonnes, with 'indicated' resources put at 35 million and 'inferred' at 215 million. Sub-bituminous coals are, like lignite, mostly used by the brick kiln industry but some coals of Balochistan and NWFP are also being used by the cement industry.

Notwithstanding its massive potential, Pakistan's coal production in recent years has been only some 3–3.5 million tonnes per annum. About half is currently produced in the western province of Balochistan; no Thar coal is produced at present.

Small tonnages of indigenous coal are used for electricity generation and by households but, as seen above, by far the largest portion is used to fire brick kilns. Just over 1 million tonnes of Australian coking coal is imported each year for use in the iron and steel industry.

Poland

Proved amount in place (total coal, million tonnes)	59 984
Proved recoverable reserves (total coal, million tonnes)	14 000
Production (total coal, million tonnes, 2002)	160.3

Most of Poland's substantial tonnage of coal resources is in the form of hard coal, which comprises 77% of the reported proved amount in place. The Polish WEC Member Committee has reported revised resource assessments by comparison with those advised for the 2001 *Survey of Energy Resources* with (in particular) a 10% reduction in the proved amount of bituminous coal in place and a 31% decrease in the corresponding tonnage recoverable.

The latest figures show the proved amount of hard coal in place as 46 billion tonnes, on the basis of a maximum deposit depth of 1 000 m and a minimum seam thickness of 1 m; the corresponding level for lignite is almost 14 billion tonnes, at a maximum deposit depth of 350 m and minimum seam thickness of 3 m. The estimated additional amounts in place are 59–80 billion tonnes of hard coal and 41 billion tonnes of lignite, but assessments of the recoverable portions of these quantities are not available.

The reported 'proved amount in place' comprises the total geological resources of coal (called in Polish terminology 'documented geological resources—category A, B and C'); the 'proved recoverable reserves' comprise the documented reserves of coal, known in the Polish classification as 'industrial reserves'; the 'estimated additional amounts in place' include forecast 'additional resources of coal, which are in unexplored extensions of known deposits below 1 000 m and inferred amounts estimated on the results of geological information'.

Poland's hard coal resources are mainly in the Upper Silesian Basin, which lies in the southwest of the country, straddling the border with the Czech Republic: about 80% of the basin is in Polish territory. Other hard-coal fields are located in the Lower Silesia and Lublin basins. There are a number of lignite deposits in central and western Poland, with four of the larger basins currently being exploited for production.

The quality of the Upper Silesian hard coals is generally quite high, with relatively low levels of sulphur and ash content. One-third of Poland's proved reserves of hard coal are regarded as of coking quality.

Although output of hard coal (and, to a lesser extent, of lignite) has declined during the past 15 years, and especially since 1997, Poland is still among the world's eight largest coal producers (see Table 1.3), with a 2002 output of 102 million tonnes of hard coal and 58 million tonnes of lignite. The decline in hard coal production reflects a deep restructuring of the industry, with the aim of eliminating the non-profitable mines by a reduction in excess production potential, substantially lower employment levels, elimination of government subsidies, etc.

Apart from Russia, Poland is the only world-class coal exporter in Europe: its total exports in 2002 were some 23 million tonnes, of which steam coal accounted for 84% and coking for 16%. Germany, Austria and Denmark are currently Poland's largest export markets for coal.

About 61% of inland consumption of hard coal goes to the production of electricity and bulk heat, industrial uses account for 26% and residential/commercial/agricultural uses 13%. Almost all lignite production is used for base-load electricity generation.

Russian Federation

Proved amount in place (total coal, million tonnes)	200 576
Proved recoverable reserves (total coal, million tonnes)	157 010
Production (total coal, million tonnes, 2002)	253.4

The levels quoted for Russian coal resources and reserves are unchanged from those given for end-1996 in the 1998 *Survey of Energy Resources*, as the WEC Member Committee was unable to obtain any more recent coal data for the present *Survey*.

The proved amount of coal in place at end-1996 comprised 75.8 billion tonnes of bituminous coal, based on a maximum deposit depth of 1 200 m and a minimum seam thickness of 0.6–0.7 m; 113.3 billion tonnes of sub-bituminous grades (at depths of up to 600 m and minimum thickness 1.0–2.0 m); and 11.5 billion tonnes of lignite (at 300 m and 1.5–2.0 m, respectively).

Proved recoverable reserves were reported as just over 49 billion tonnes of bituminous coal, of which 23% was considered to be surface-mineable and 55% was suitable for coking. Of the 97.5 billion tonnes of proved recoverable reserves of sub-bituminous coal, 74% was suitable for surface mining, while all of the 10.5 billion tonnes of recoverable lignite reserves fell into this category. Overall, about 94 billion tonnes of Russia's proved reserves were deemed to be recoverable by opencast or strip mining. Further enormous tonnages of coal, of the order of 20 times the quoted proved reserves, were reported to be recoverable in the future.

Russian coal reserves are widely dispersed and occur in a number of major basins. These range from the Moscow basin in the far west to the eastern end of the Donetsk basin (most of which is within the Ukraine) in the south, the Pechora basin in the far northeast of European Russia, and the Irkutsk, Kuznetsk, Kansk-Achinsk, Lena, South Yakutia and Tunguska basins extending across Siberia to the Far East.

The principal economic hard coal deposits of Russia are found in the Pechora and Kuznetsk basins. The former, which covers an area of some 90 000 km^2, has been extensively developed for underground operations, despite the severe climate and the fact that 85% of the basin is under permafrost. The deposits are in relatively close proximity to markets and much of the coal is of good rank, including coking grades. The Kuznetsk basin, an area of some 26 700 km^2, lies to the east of the city of Novosibirsk and contains a wide range of coals; the ash content is variable and the sulphur is generally low. Coal is produced from both surface and underground mines.

Lying east of the Kuznetsk and astride the trans-Siberian railway, the Kansk-Achinsk basin contains huge deposits of brown (sub-bituminous) coal with medium (in some cases, low) ash content and generally low sulphur; large strip-mines are linked to dedicated power stations and carbo-chemical plants. The vast Siberian coal-bearing areas of the Lena and Tunguska basins constitute largely unexplored resources, the commercial exploitation of which would probably be difficult to establish.

From a peak of around 425 million tonnes in 1988, Russia's total coal production declined dramatically following the disintegration of the USSR, and now stands at about 250–260 million tonnes per annum. In 2001 over 70% of Russian consumption was accounted for by power stations and district heating plants; the iron and steel industry and the residential sector were the other main centres of coal usage.

South Africa

Proved amount in place (total coal, million tonnes)	115 000
Proved recoverable reserves (total coal, million tonnes)	48 750
Production (total coal, million tonnes, 2002)	214.7

The South African WEC Member Committee has reported the proved amount in place on the basis of a maximum deposit depth of 350 m and a minimum seam thickness of 1 m. The coal resources reported for the present *Survey* are based on an assessment published by the Geological Survey of South Africa (now the Council for Geoscience) in 1987, adjusted for cumulative production of coal over the period since its preparation, and thus differ only marginally from those reported for the 2001 *Survey*.

The Council for Geoscience, on behalf of the Department of Minerals and Energy, is currently carrying out a major review of South Africa's coal resources; its report is expected to be completed by the end of 2004. There are some indications that the reassessed reserves could be as low as 38 billion tonnes.

Coal occurs principally in three regions.

- The shaly Volksrust Formation, which covers most of central and northern Mpumalanga province (formerly the Transvaal). The coal is found in isolated basins and troughs which results in the fields being disconnected and widely separated.

- The sandy Vryheid Formation of the northern part of the main Karoo basin (northern Free State, northern Kwazulu-Natal and southern Mpumalanga): this generally continuous area is probably the most important economically.
- The Molteno Formation, which is confined to the north-eastern Cape. It is of minor economic importance compared to other coalfields in South Africa.

Some lignite deposits are known along the Kwazulu-Natal and Cape coasts, but are considered to be of scant economic importance.

Coal occurrences have been divided into 19 separate coalfields, 18 of which are located in an area extending some 600 km from north to south by 500 km from east to west. The Molteno field lies some 300 km south of the main coal-bearing region.

South Africa's coals are generally low in sulphur but high in ash. Beneficiation is essential for export-quality coal. Lower-quality coal is for the local power generation market.

Eskom, the South African electric utility, accounts for nearly 60% of coal consumption. A further large slice is consumed by the Sasol plants in making synthetic fuels and chemicals from coal. The third main user is the industrial sector, including the iron and steel industry. Coal use in residential and commercial premises is relatively small, while demand by the railways has virtually disappeared.

Coal exports are equivalent to about 30% of South African output and are mainly destined for Europe and Asia/Pacific. The main route for exports is via Richards Bay, Kwazulu-Natal, where there is one of the world's largest coal-export terminals.

Thailand

Proved recoverable reserves (total coal, million tonnes)	1 354
Production (total coal, million tonnes, 2002)	19.6

Thailand has sizeable resources of lignite, notably at Mae Moh in the north of the country. The Thai WEC Member Committee reports proved recoverable reserves of 1 354 million tonnes; the maximum deposit depth taken into consideration is approximately 700 m, while the minimum seam thickness is 0.30 m. As an indication of the extent of Thailand's coal resources, the 2002 edition of the annual publication *Thailand Energy Situation*, issued by the Department of Energy Development and Promotion, quotes total lignite reserves as 2 942 million tonnes. In this context, the reserves are defined as including 'the remaining reserve from produced area as well as the measured and indicated reserve from undeveloped area'.

Annual output of lignite increased by almost 90% between 1990 and 1997, but has since levelled out. All of Mae Moh's production is consumed by the Mae Moh power plant (2 625 MW). On the other hand, most of the lignite produced by other Thai mines is used by industry, chiefly in cement manufacture. Imports of bituminous coal are mostly destined for consumption in the iron and steel sector.

United Kingdom

Proved recoverable reserves (total coal, million tonnes)	220
Production (total coal, million tonnes, 2002)	30.0

Coal deposits are widely distributed and for many years the UK was one of the world's largest coal producers, and by far its largest exporter. Production rose to a peak of nearly 300 million tonnes/yr during World War I and thereafter did not fall below 200 million tonnes/yr until 1960. Output began a long-term decline in the mid-1960s, falling to less than 100 million tonnes/yr by 1990. Reflecting continued competition from natural gas and imported coal,

UK coal production sank to only 30 million tonnes in 2002.

The UK coal industry was privatised at the end of 1994, with the principal purchaser being RJB Mining (now UK Coal plc), which acquired 16 deep mines from British Coal. As at 22 April 2003 there were 16 major deep mines, 8 smaller deep mines and 46 open-cast sites in production. Deep-mined coal output in 2002 was 16.4 million tonnes and open-cast sites produced 13.1 million tonnes; production from slurry etc. amounted to 0.5 million tonnes. Most deep-mined coal has a significantly higher content of sulphur and chlorine than that of internationally-traded coal. There is now virtually no UK production of coking coal.

The decline of the British coal industry has been accompanied by a sharp decrease in economically recoverable reserves. For the purpose of the present *Survey*, the level adopted reflects that stated by the Department of Trade and Industry in its publication *UK Energy Sector Indicators 2003*: 'There were estimated to be approximately 220 million tonnes of economically viable coal reserves at mid-November 2001'. This figure is compatible with the view expressed in the Energy White Paper (Version 11, as at 21 February 2003): 'Within 10 years most of our existing deep mines are likely to have exhausted their economic reserves'.

United States of America

Proved amount in place (total coal, million tonnes)	451 512
Proved recoverable reserves (total coal, million tonnes)	246 643
Production (total coal, million tonnes, 2002)	992.7

The United States coal resource base is the largest in the world. The US WEC Member Committee reports a Proved Amount in Place at 1 January 2003 of 451.5 billion tonnes (based on the Energy Information Administration's Demonstrated Reserve Base). This total is comprised of 246.8 billion tonnes of bituminous coal (including anthracite) with a maximum deposit depth of 671 m and minimum seam thickness of 0.25 m; 165.1 billion tonnes of sub-bituminous (at up to 305 m depth and 1.52 m minimum seam thickness) and 39.6 billion tonnes of lignite (at up to 61 m depth and 0.76 m minimum seam thickness).

The reported Proved Recoverable Reserves amount to 246.6 billion tonnes, equivalent to about 27% of the global total. They comprise 111.3 billion tonnes of bituminous coal (including anthracite), 102 billion tonnes of sub-bituminous and 33.3 billion tonnes of lignite. The overall ratio of proved recoverable reserves to the proved amount in place is 0.55. This ratio varies widely from one rank to another, reflecting relative degrees of accessibility and recoverability: bituminous deposits average 0.45, sub-bituminous 0.62 and lignite 0.84. Open-cast or surface mining techniques can be applied to around 28% of bituminous reserves, to nearly 45% of the sub-bituminous and to 100% of the lignite.

Data for Proved Amount in Place and Recoverable Reserves are measured and indicated (proved and probable), in a commingled data base. The data cannot be separated into 'proved only' and 'probable only'.

On top of the tonnages summarised above, the US WEC Member Committee reports enormous quantities of coal as 'estimated additional amounts in place': in total these come to well over a trillion tonnes, composed of 445 billion tonnes of bituminous, 274 billion sub-bituminous and 394 billion lignite. These estimates are derived from a US Department of the Interior study of coal resources as at 1 January 1974, but are regarded as still providing valid indications of the magnitude of the USA's additional coal resources. Data on the Estimated Additional Amount in Place are primarily inferred. These resources extend deeper than the Proved Amount in Place, include thinner beds in some areas, and are based on older source data in many cases. The Estimated Additional Amount in Place has been adjusted only to indicate the arithmetic difference with Proved Amount in Place.

Coal deposits are widely distributed, being found in 38 states and underlying about 13% of the total land area. The Western Region (owing largely to Montana and Wyoming) accounts for about 47% of the EIA's Demonstrated Reserve Base, the Interior Region (chiefly Illinois and western Kentucky) for 32% and the Appalachian Region (chiefly West Virginia, Pennsylvania and Ohio) for 21%. Bituminous coal reserves are recorded for 27 states, whereas only 8 states have sub-bituminous reserves, of which 90% are located in Montana and Wyoming, and 10 have lignite reserves, mostly in Montana and Texas.

US coal output is the second highest in the world, after China, and accounted for about 21% of global production in 2002. Coal is the USA's largest single source of indigenous primary energy; power stations, CHP and heat plants accounted for almost 92% of domestic coal consumption in 2001. Coal exports amounted to 35 million tonnes in 2002: despite a decline in its exports in recent years, the USA remains a leading supplier of coking coal and other bituminous grades.

Chapter 2

Crude Oil and Natural Gas Liquids

World Energy Council
CONSEIL MONDIAL DE L'ENERGIE

COMMENTARY

Introduction

Since the fall of the Berlin Wall we have witnessed the strong influence of three powerful global forces: globalisation, liberalisation and technology. Recently we have seen protest movements take to the streets when related topics have been put on the agenda by nations and NGOs, such as the WTO, the IMF and the World Bank. Reactions matter and they reflect that people are concerned with the lack of meaningful visible results from the past decade, especially in the wake of the UN Summit in Rio in 1992. Unless these three global forces are given a human face, especially in growing economies, we shall have an increasing problem to solve. It is in this setting that we need to address the question of global co-operation being a means to safeguard the interests of host nations, host nation companies and visiting companies.

It is also appropriate to consider possible future developments on the basis of scenarios such as those presented by Shell. Two scenarios published recently catch some alternative developments.

Business Class: Describes a world that is not run by business but that operates like a business, focussing on efficiency and individual freedom of choice. It is shaped by an expanding elite—a global meritocracy in business, government, entertainment, international NGOs—which is interconnected, everywhere.

Prism: Describes a world shaped not by what we have in common but by the interplay of our differences. People find their values in roots, heritage and families. They are still interested in economic well-being and growth, but also in social cohesion, religious faith and national pride. Countries find their own development paths to suit their own circumstances.

Scenarios are not predictions, but coherent and credible stories about the future. They offer an opportunity to see different possible developments and consequently new risks and

2004 Survey of Energy Resources

opportunities that a business may face. When we discuss global co-operation it may be useful to keep these in mind. It is relevant to understand that energy is of fundamental importance for the development of societies and people. The need for affordable and clean energy will continue to increase, especially in growing economies. Both consumption and production of energy is central in all countries, but critical for many. Moreover, it is reasonable to assume that sustainable development will also require cleaner and more effective energy alternatives. Global co-operation is absolutely essential to manage the transition to a sustainable energy world. This is closely linked to the 'war on poverty', or should one rather say crusade for dignified life for present and future generations. The petroleum industry is crucial in this setting, but so also is the work of international organisations and host governments, all co-operating in effective and transparent ways.

Most energy scenarios project that consumption of every primary energy source will increase over the next 20 year forecast horizon. Most of the increment in energy consumption is in the form of fossil fuels (oil, natural gas and coal), because it is expected that fossil fuel prices will remain relatively low and that the cost of generating energy from non-fossil fuels will not be competitive. It is, however, possible that as environmental programmes or government policies—particularly those designed to limit or reduce greenhouse gas emissions—are implemented, the outlook might change, and non-fossil fuels (renewable energy sources such as hydroelectricity, geothermal, biomass, solar and wind power) might become more attractive. But for the future, fossil fuels will dominate (90%) albeit using clean technologies. Overall energy use is closely linked to population growth. Over half the projected growth in energy demand will occur in tandem with strong economic growth in the nations of Asia, including China, India and South Korea, and in Central and South America.

The Start of Globalisation

This commentary is about the petroleum industry, so why not start with a related anecdote. In the 1920s the American entrepreneur, Henry Doherty, became troubled by the 'crude and ridiculous' way that oil producers were operating. Since the mid-19th century the industry had been governed by the 'rule of capture', a principle based on an old English law for hunting migratory animals. This meant that every time a new oil field was struck, there was a scramble among producers to drill into the structure the fastest and to draw off as much oil as possible before their competitors. Mr Doherty recognised that such a haphazard way of drilling was leading to volatile prices as well as damaging the underground pressure needed to bring oil to the surface. To the amazement of the industry he suggested that oil fields should be 'unitised' or drilled as single entities. The number of wells could be limited to preserve the underground pressure for longer and output would be apportioned to the various partners on the basis of their shareholding in the field. This idea—the first time cooperation had been suggested in the ultra-competitive oil industry—was at the time radical and unpopular. It took Mr Doherty several years of hard lobbying until unitisation became an accepted practice. Today, the oil industry remains as competitive as ever but more readily recognises the value of partnerships in some parts of the business. Innovative joint venture agreements are constantly being developed—both between competing oil companies and with specialist organisations outside the industry. Joint ventures and collaborative agreements are a useful way of minimising risks on a number of fronts. Oil companies have to deal with geological risks associated with prospects, political risks in the countries in which they operate and technological risks in various parts of their business. By collaborating with others they potentially lose some of the upside but can pool resources and ideas and spread the risks more thinly.

Joint Ventures

Long-established cooperation between oil companies in the upstream business, refining and marketing is being supplemented by new joint ventures and partnerships, both in traditional and developing areas of the industry. In the mature oil province of the UK North Sea, for example, large oil companies are using collaborative approaches to make their assets work harder while smaller companies entering for the first time are trying to find economic ways of exploiting ever smaller pockets of oil around existing infrastructure. Talisman Energy, the Canadian exploration and production group, entered into a risk-sharing contract with Global Marine Integrated Services that enabled it to access some of the most marginal offshore oil fields in the world. Using advanced drilling Global Marine was able to tap into the tiny Beauly field, which has reserves of only 3 million barrels of oil. A horizontal extension of 1 200 m was drilled from an existing well and this was then tied back through a sub-sea pipeline to a floating production platform.

At the other end of the traditional business, oil companies are seeking ways to minimise risks associated with the construction of their retail outlets. The global nature of the large integrated oil groups means they have retail operations in many countries, each with their own complex planning laws and different suppliers. BP has sought to simplify this process by outsourcing its US$ 3 billion service station construction and maintenance programme in Europe, Japan, Venezuela and the US to Bovis Lend Lease, the Anglo-Australian construction group. To enable the risks to be shared, Bovis is contracted on a long-term partnership agreement and derives its profit purely from the savings it achieves. 'Our aim is to achieve a substantial increase in capital productivity which will enable us to fulfil our future growth operations for the retail network,' says a BP official. 'The partnership with Bovis will be an important component of this, delivering the savings required on all our retail construction and maintenance while continually improving the service we offer to our customers'.

While joint ventures and partnership agreements have helped greatly to streamline traditional activities, the greatest concentration of collaborative working in recent years has been in the non-traditional and developing areas of the business. These fall broadly into two categories—new energy technologies and e-commerce. Major oil companies have realised that they must get involved with the development of renewable and alternative energy forms at an early stage, reasoning that commercially recoverable hydrocarbon reserves will eventually be depleted and that precautions need to be taken against global warming. Some have had in-house research capabilities and have been working in this area for years but, recognising that alternative energy is outside their core areas of expertise, are now forming joint ventures with more specialised companies. The development of hydrogen fuel cells, which could one day replace the internal combustion engine, has been one of the most fertile grounds for this. All the major oil companies have linked up with automotive and fuel cell manufacturers in a bid to share risks and ideas.

Social Concerns

There are increasing calls for corporate responsibility and accountability covering economic, environmental and social issues. Social responsibility is a key issue of concern to all businesses around the world, not just to the oil and gas business. The environment is no longer an isolated subject; it is fundamental to everything we do. Companies are being scrutinised and measured as much for their social performance as for their business performance.

By demonstrating its ethics and integrity, the oil and gas industry safeguards its reputation and credibility. Through this, a respect is shown for people, communities and the environment, safety and compliance with the law. It is

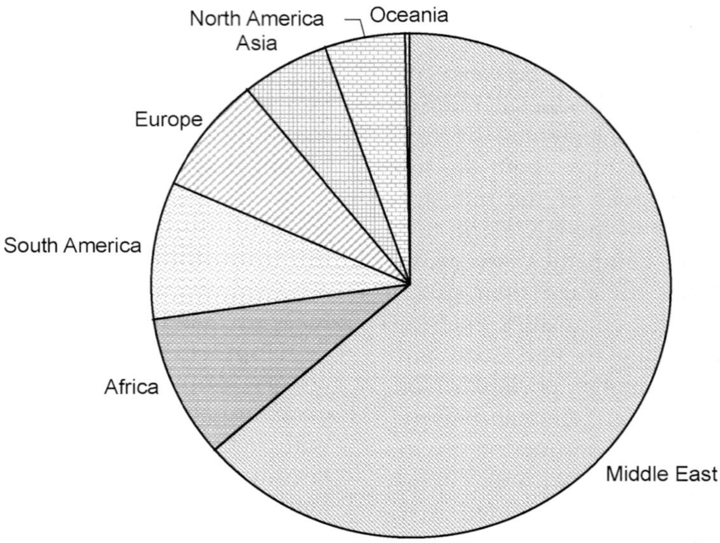

FIGURE 2.1 Proved oil reserves at end-2002: regional distribution.

the right thing to do. The global demand for energy and more specifically 'clean' energy is growing rapidly. There is no doubt that energy is essential to the improvement of human life and economic development. Oil, gas and coal represent around 90% of commercial energy used worldwide, and technological innovation and good stewardship will allow us to go on finding and producing oil and natural gas for many years past current predictions.

To meet this demand, the petroleum industry is called upon not just to provide effective management of oil and gas reserves, but also to marshal its financial resources and technical capacity to further diversify energy sources and delivery options. Energy industries have always evolved to meet people's changing needs. The energy industry of the 21st century will become more diverse and more dynamic. For the longer term, sustainable economic growth, meeting global energy demands in an environmentally sensible and socially acceptable way will require the development of entirely new energy technologies and other advanced fuel sources.

And hence the challenge, how does the industry meet this growing demand for affordable petroleum and natural gas products, while ensuring environmental quality and meeting societal needs? Today, business success, in both developed and developing countries, is increasingly linked to strong performance in financial markets, environmental stewardship and community affairs. Businesses also have to deal with changing sets of expectations from a broader range of stakeholders.

The international petroleum industry takes this challenge very seriously. The actual responses to these environmental and social challenges differ among companies and countries. However, through the World Petroleum Congress (WPC) the international oil industry has a showcase and a forum for communication to define common approaches to global environmental issues and share good environmental and social practices for the benefit of the petroleum industry and the community at large.

Specifically, the international petroleum industry is collectively striving to address

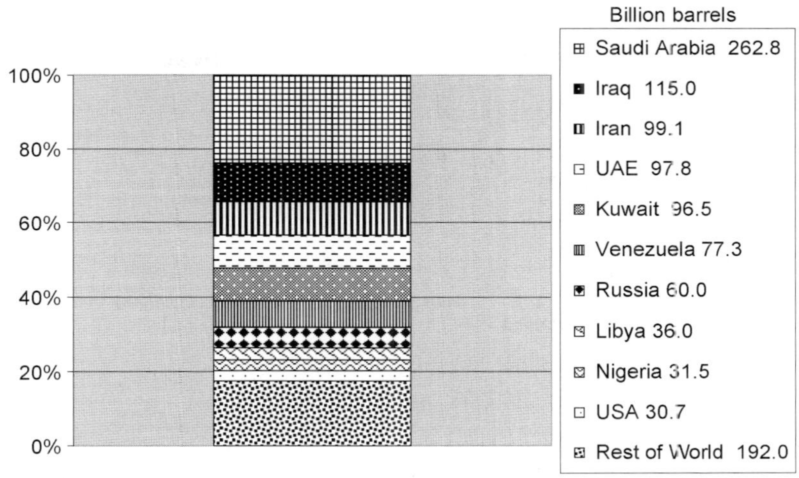

FIGURE 2.2 Proved oil reserves at end-2002: the leading countries.

society's need by helping to develop solutions and responses that are socially, economically and environmentally acceptable, by:

- developing partnership with key UN bodies, global organisations and other industry associations;
- searching for cost-effective solutions based on sound science;
- conducting consultations with all stakeholders.

As members of the global business community, WPC members strive towards maintaining high environmental standards. To accomplish this, companies have developed comprehensive and rigorous 'risk-based' environmental management practices and processes, which they apply to all operations anywhere in the world. This ensures the adequate protection of local values and priorities, while sustaining the economic viability and durability of the business enterprise.

Walking the Talk

There is great similarity among the policy-level environmental commitments and general aspirations of most major oil companies. These commitments include:

- as a minimum, companies comply with local laws and regulations. Responsible approaches are applied where such laws and regulations do not exist and/or where the companies may judge that a more stringent approach is appropriate;
- adoption of a corporate environmental policy that applies to all operations;
- implementation of the policy using a risk-based and a country-specific or site-specific approach;
- setting-up high-level, long-range environmental performance objectives and tracking progress towards them;
- adoption of environmental management systems that are applied consistently throughout the world;
- involvement of appropriate local stakeholders in the decision-making process for issues that affect them;
- establishment of global policy assurance (stewardship) processes.

These common elements provide an umbrella framework with the needed flexibility for each company to implement it according to its core

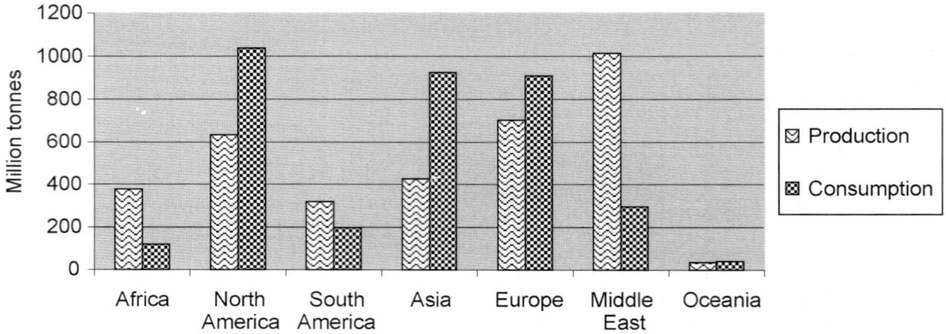

FIGURE 2.3 Oil production and consumption, 2002: regional distribution.

values. Major oil and gas companies are committed to the sharing of good practices and learning from each other so that they can continuously improve their environmental performance.

The petroleum sector has long been at the forefront of the development and implementation of management systems for Environment, Health and Safety and more recently, the emerging issues relating to corporate responsibility. While earlier approaches focussed on end-of-pipe/react and treat controls, more recently proactive, preventative, risk-based management approaches have been adopted as a means of integrating critical elements of risk, then applying quality system approaches to continuously improve on past performance.

The industry produces its own guidelines, designed to encourage operating practices in an environmentally and socially responsible manner and provide an effective mechanism for sharing good practices. There are many examples in the upstream as well as the downstream sector (e.g. International Standards Organisation (ISO) and International Association of Oil & Gas Producers (OGP) management systems, American Petroleum Institute (API) management practices, Canadian Association of Petroleum Producers (CAPP) stewardship process, OGP guidelines for exploration and production in mangrove areas, the International Petroleum Industry Environmental Conservation Association (IPIECA) Oil Spill report series, IMO maritime statutory regulations and many more.

There is a universal need for efficient technologies that will contribute to the sustainable development of the host countries and communities by providing employment, improving quality of life and protecting the environment. The international petroleum industry contributes to economic development and environmental protection through technology cooperation and capacity building. Many hope that technology will provide solutions to meet this challenge. Business and industry, the creators of both wealth and new technology, have a key role to play in these areas. The growing importance of business in promoting a more sustainable future is highlighted by the fact that private financial flows to developing countries have well-outstripped official aid. For the petroleum industry, technology transfer is not a separate activity, but an integral part of the business operation. The establishment of sound management, the availability of a skilled workforce and all the factors in the local supply chain are a vital part of technology transfer.

International Co-operation

In the past 50 years or more, a rapid industrialisation and urbanisation process, while of extreme significance in terms of economic development and personal mobility, has led to the deterioration of urban air quality and strained infrastructure systems. Moreover, in addition to the link between urban air quality deterioration and road transport, it is apparent that a substantial role is also played by the quantity and quality of air emissions from home heating and cooking fuels, manufacturing and small business enterprises.

Within Europe and Scandinavia many petroleum company-led Air Quality Management Programmes are actively promoting policies and practices that are objective, cost-effective and based on sound science, to improve air quality in rapidly developing cities around the world. The industry has developed a framework and a set of tools, derived from the extensive experience gained through the Auto-Oil programmes in OECD countries. These tools enable subjective analysis of options by decision makers and serve as a foundation for new public policies. These are communicated through the use of an air quality 'Toolkit' and regional workshops to promote a scientific basis for the approach to the management of air emissions in order to attain desired levels of improvement of air quality. Over the past few years, the industry has been actively working with the World Bank in its Clean Air Initiative for Latin America, demonstrating the applicability of a number of computerised systems for objective analysis of policy options in cities such as Lima in Peru and Rio de Janeiro in Brazil.

No matter how good preventative measures are, the petroleum industry recognises that accidents can still happen. Particular emphasis has been placed on the developing countries where resources and expertise are scarce. Some initiatives coordinate on-the-ground activity to help the government and the oil industry in specific countries to join forces and develop a credible response and planning strategy. The resources in the oil industry and the specialist expertise available from both the industry and the institutes are used to maximum effect in facilitating national and regional workshops and training programmes attended by both government and industry personnel. It also encourages countries to ratify those IMO Conventions that provide guidance on cooperative response planning and simplify compensation procedures in the aftermath of an incident.

Climate Change

Climate change is a global environmental concern with potentially significant consequences for society and the whole fossil fuel industry. Its significance stems both from possible future impacts of changing world climate as well as potential socio-economic consequences of policies proposed to respond to increased concentrations of greenhouse gases in the atmosphere. It is a unique issue in terms of its scale, time frame, and the complexity and uncertainty of scientific, technical, economic, social and political dimensions.

The principal influences on greenhouse gas emissions are the size of the human population, the amount of energy used by each person and the level of emissions resulting from that use of energy. A variety of technical options are available which could reduce emissions, the most common and easiest to implement quickly being improved energy efficiency. Other options include fuel switching (coal or oil to gas), renewable energy sources, alternative energy economies (methanol, hydrogen) and nuclear power. Yet all of these do not necessarily have lower costs than current systems and some have their own unique environmental and social problems.

Alternatively, greenhouse gases, especially CO_2, may be recovered and stored. CO_2 capture and storage is complementary to the above options, having already demonstrated the potential for deep reductions in emissions whilst allowing continued use of the established fossil fuel infrastructure. It is a real practical option available and demonstrated today. Investigation of options and, especially, application of the most suitable ones will be enhanced by international co-operation. These technical options are most easily applicable to stationary plant, such as power generation and large industrial plants both on and off shore.

The first CO_2 saline reservoir storage project has now been in operation for over 5 years, commissioned offshore in the second half of 1996 by Statoil, the operator of the Sleipner field in the North Sea off the coast of Norway. The company, well known for its socially responsible actions and environmental initiatives, is extracting annually over 1 million tonnes of CO_2 from the natural gas produced by the Sleipner West field and injecting the CO_2 into the nearby Utsira reservoir.

This is also an example of excellent international collaboration, between public and private companies and governments. Statoil on behalf of a group of partners operates the Sleipner field. The additional agreement to carry out a monitoring activity is being supported by Statoil, ExxonMobil, BP, Vattenfall and many geological institutions from Norway, the UK, Denmark, France, the Netherlands and Sweden. In addition there are national ministries involved, including the UK, Norway, Denmark, the Netherlands and France. Thirty countries have been collaborating on practical research in this area since 1990. During the start-up phase, North Sea seismic data was given to the group by all the operators in the North Sea. This saved millions of dollars of work and showed the willingness of many competitors to contribute to this unique project.

Transparency

The petroleum industry has been operating in some of the world's most sensitive environments for more than 100 years. Although the industry's environmental record has not been perfect, it has been a pioneer in developing and using new technologies, systems and practices to minimise the impact of its operations. The industry's continuous efforts to improve have intensified in recent years, reflecting heightened public awareness and interest regarding the way industry manages its interface with the environment, local communities and the public at large. With careful planning, management and consultation, energy resources can be developed without causing lasting damage to the environment, while maximising economic and social benefits to the local communities.

In order to be successful, oil companies need access to resources and a licence to operate. Businesses have to deal with changing sets of expectations from a broader range of stakeholders. In many countries the public is becoming increasingly powerful and plays a growing role in the decision-making processes of both governments and businesses. Establishing and maintaining good relationships with all stakeholders is key to the future of the oil industry.

With the demand for greater openness in corporate behaviour, the success of what the oil industry does increasingly depends upon achieving stakeholder endorsement. This requires that stakeholder views are sought about proposed activities and the industry engages in two-way dialogue. It is essential to effectively engage community stakeholders in decision-making processes that affect them. A trusted company can be part of the process and thus influence the outcome. As trust diminishes, the demand for transparency and assurance increases. There is no hiding place from public scrutiny. Stakeholders want demonstrable assurance that companies are behaving responsibly. They want to be consulted on issues of concern to them.

These key challenges in recent years relate more to the social side than to the environmental aspects of business. The impact of these developments on large companies has been an increasing call for corporate responsibility.

There are inconsistencies between the industry's perception of its responsibilities and what 'others' think is the industry's responsibility. Communities, employees, NGOs and the public all have different expectations of, and concerns about, the industry but there is a general belief outside the sector that the industry is powerful, profitable and insensitive. This is not the case; current approaches utilised by the industry in the area embrace the basic principle: financial performance and responsibility to shareholders as well as acceptable social and environmental performance are of equal key importance to sustained business success.

Cultural Differences

The convergence of oil, human rights, social responsibility and sustainable development has increasingly been in the headlines in recent years, whether it was Nigeria, Myanmar, Indonesia, Colombia, Angola or any number of other countries, and those closer to home where oil and gas projects have served to highlight problems.

The issue of how oil companies interact with the wider world has been brought home by a number of well-publicised incidents. In the summer of 1995 Royal Dutch/Shell found itself the target of a well-organised and coordinated wave of protest across Western Europe over its plan to sink the obsolete Brent Spar oil storage platform in the deep waters of the Atlantic Ocean. Experts and European governments approved the project, but Greenpeace, the environmental pressure group, mounted a big campaign that included occupying the Brent Spar platform. The sophisticated use of video and the media ensured that the Greenpeace protests received maximum exposure on television and radio and in the newspapers.

Sometimes subtle cultural differences between companies from the modern, industrialised world and developing countries can lead to significant friction between the two. In its most basic form, net present value, the means by which oil companies calculate the economics of proposed projects, can be translated simply as 'time is money'. Once a project has been sanctioned, its ultimate value is linked to how quickly it can be completed and become operational. Time may indeed be money in the contemporary commercial world, but in many parts of the Third World time is far from an economic concept, and the rush to complete projects can often raise suspicions about the motives of the oil companies and whether they simply want to extract the oil or gas as quickly as possible.

The rush to complete projects is exacerbated by traditional relationships between oil companies and their main suppliers and contractors. Even in those cases where the oil companies themselves adopt measures to foster local content and employment and mitigate their negative impact on the environment or society, it has often proved difficult to enforce the same attitudes among contractors.

An example of the scale and speed of such operations can be seen in BP's preliminary plans to proceed with full development of the main foreign-operated oil and gas fields in Azerbaijan, on the western shores of the Caspian Sea. The two major BP-operated offshore oil and gas developments along with their associated export pipelines will cost almost US$ 9 billion in the coming years. When translated onto the ground in a country of just 8 million people and a poorly developed and maintained infrastructure, that means up to 400 more heavy trucks a day passing through the potential chokepoint of the Azeri-Georgian border, eight additional trains a day on the Azeri rail network and 10 more ships and three more large barges each month passing through the Volga-Don canal system in Russia into the Caspian. Add to that an influx of thousands of foreign workers and a development timetable that BP acknowledges will be 'aggressive' and the scale of the potential disruption becomes clear.

The Azeri projects and their associated international pipelines promise to transform Azerbaijan's financial condition. In the 4 years leading up to 2005 the Government's share in the present foreign-operated oil project is expected to generate around US$ 150 million a year. But in 2006 that figure is expected to rocket to more than US$ 1 billion and quickly climb to almost US$ 6 billion by 2011. The size of such a windfall can often inflate public expectations, something that the oil industry has to be very careful about, as in many cases, owing to aspects outside their control, such expectations, especially among those more directly affected by such developments, have been dashed.

While the response of oil companies to such situations has not been uniform, there are grounds for optimism that the industry can adapt to new circumstances. Many people working in the industry, especially the younger generation, share some of the goals and aspirations normally espoused by non-governmental organisations and other pressure groups. In many countries oil company employees have gone out of their way to work with local communities and build bridges to the wider society.

Companies such as Statoil, Shell and BP have broken new ground, not only in assessing the social impact of new projects but in continually measuring their impact over the life of such projects. Transparency has also been greatly improved in recent years with the introduction of annual environmental reports which provide unprecedented detail into how companies operate.

Joint Definitions of Petroleum Reserves and Resources

In February 2000 the WPC, the Society of Petroleum Engineers (SPE) and the American Association of Petroleum Geologists (AAPG) agreed on a joint resource classification system. This included definitions of reserves previously developed by the WPC and SPE and marked the culmination of an important international collaborative effort. In 1980 the WPC decided to set up a study group to review the classification and nomenclature of petroleum and petroleum reserves. An interim report was submitted to the 11th Jubilee Congress in London in 1983 and the final document was adopted during the Houston Congress in 1987. Working separately, the SPE had produced strikingly similar sets of petroleum reserves definitions. It made sense to cooperate and at the end of the WPC Stavanger Congress in June 1994 a joint statement of principles was produced by the WPC and SPE. Shortly thereafter a task force was established to develop the ensuing set of definitions. The aim was to provide general guidelines for classification in order to allow for proper comparison of quantities on a worldwide basis, with the stipulation that the nomenclature as proposed should be applied only to discovered accumulations and associated potential deposits. It was recognised that various ways of mathematically handling the data were in use to fix the defining criteria for reasonable certainties of reserves classes. So, no method of calculation was to be excluded. Moreover, it was considered acceptable to evaluate the effectiveness of recovery processes in adjacent reservoirs by use of analogies in reservoir characteristics. The rationale for the cooperation was that if a common nomenclature was presented by the SPE and the WPC it would have a much greater chance of worldwide acceptance and would signify a joint, unique stance on an essential technical issue faced by the international petroleum industry.

The SPE/WPC proposals were adopted in 1997. The need to maintain continuity with past definitions and keep the terminology as close as possible to current common usage was recognised. At the same time, both organisations realised that new technologies had to be accommodated. Over the past several years, the industry had seen the emergence of probabilistic approaches in the evaluation of reserves for all classification categories.

Complete texts of the SPE/WPC/AAPG definitions of petroleum reserves and resources are available via the Internet at www.world-petroleum.org and www.spe.org.

Conclusion

Meeting society's increasing demand for energy and strong environmental and social performance in this century will be a decisive challenge for energy companies and host nations. The petroleum industry recognises the complexities of trying to achieve a balance among economic, environmental and social expectations. There are conflicting demands and contradictory expectations. There are limits to what companies can do, and ought to do. At the international level, the World Petroleum Congress (and other such organisations) encourages dialogue and sharing of experiences between its members, internal discussion and production of case studies, which all help in building on key lessons learned and sharing good practices within the industry, thereby living up to its purpose to promote sound fossil fuel use for the benefit of mankind.

Pierce W Riemer
World Petroleum Congress

DEFINITIONS

Crude oil is a naturally occurring mixture consisting predominantly of hydrocarbons that exists in liquid phase in natural underground reservoirs and is recoverable as liquids at typical atmospheric conditions of pressure and temperature. Crude oil has a viscosity no greater than 10 000 mPa s (centipoises) at original reservoir conditions; oils of greater viscosity are included in Chapter 4—Natural Bitumen and Extra-Heavy Oil.

Natural gas liquids (**NGLs**) are hydrocarbons that exist in the reservoir as constituents of natural gas but which are recovered as liquids in separators, field facilities or gas-processing plants. NGLs include (but are not limited to) ethane, propane, butanes, pentanes, natural gasoline and condensate; they may include small quantities of non-hydrocarbons.

If reserves/resources/production/consumption of NGLs exist but cannot be separately quantified, they are included (as far as possible) under crude oil.

In the tables the following definitions apply to both crude oil and NGLs:

Proved amount in place is the resource remaining in known natural reservoirs that has been carefully measured and assessed as exploitable under present and expected local economic conditions with existing available technology.

Proved recoverable reserves are the quantity *within* the proved amount in place that can be recovered in the future under present and expected local economic conditions with existing available technology.

Estimated additional amount in place is the resource *additional to* the proved amount in place that is of foreseeable economic interest. Speculative amounts are not included.

Estimated additional reserves recoverable is the quantity *within* the estimated additional amount in place that geological and engineering information indicates with reasonable certainty might be recovered in the future.

R/P (**reserves/production**) **ratio** is calculated by dividing the volume of proved recoverable reserves at the end of 2002 by volumetric production in that year. The resulting figure is the time in years that the proved recoverable reserves would last if production were to continue at the 2002 level.

TABLE 2.1

Crude oil and natural gas liquids: proved recoverable reserves at end-2002

	Crude oil (million tonnes)	NGLs (million tonnes)	Total (million tonnes)	Crude oil (million barrels)	NGLs (million barrels)	Total (million barrels)
Algeria	1 347	221	**1 568**	10 693	2 201	**12 894**
Angola			**1 201**			**8 900**
Benin			**1**			**8**
Cameroon			**55**			**400**
Chad			**133**			**900**
Congo (Brazzaville)			**213**			**1 506**
Congo (Dem. Republic)			**26**			**187**
Côte d'Ivoire	38		**38**	277		**277**
Egypt (Arab Republic)	409	83	**492**	2 900	800	**3 700**
Equatorial Guinea			**149**			**1 095**
Ethiopia			**N**			**N**
Gabon			**342**			**2 499**
Ghana			**2**			**17**
Libya/GSPLAJ			**4 688**			**36 000**
Morocco			**N**			**2**
Nigeria			**4 252**			**31 506**
Senegal			**N**			**N**
South Africa	5		**5**	40		**40**
Sudan			**76**			**563**
Tunisia			**40**			**308**
Total Africa			**13 281**			**100 802**
Barbados			**N**			**3**
Canada	614	188	**802**	4 491	1 556	**6 047**
Cuba			**116**			**750**
Guatemala			**80**			**526**
Mexico	2 154	183	**2 337**	15 123	2 072	**17 195**
Trinidad & Tobago			**102**			**716**
USA	3 094	707	**3 801**	22 677	7 994	**30 671**
Total North America			**7 238**			**55 908**
Argentina			**379**			**2 749**
Bolivia			**59**			**486**
Brazil			**1 348**			**9 812**
Chile			**20**			**150**
Colombia			**253**			**1 842**
Ecuador			**650**			**4 630**
Peru			**44**			**323**
Surinam			**26**			**170**
Venezuela			**11 139**			**77 306**
Total South America			**13 918**			**97 468**

(continued on next page)

TABLE 2.1 (Continued)

	Crude oil (million tonnes)	NGLs (million tonnes)	Total (million tonnes)	Crude oil (million barrels)	NGLs (million barrels)	Total (million barrels)
Azerbaijan			**959**			**7 000**
Bangladesh			**8**			**57**
Brunei			**184**			**1 350**
China			**3 300**			**24 156**
Georgia			**5**			**35**
India			**740**			**5 572**
Indonesia			**705**			**5 100**
Japan	9		**9**	68		**68**
Kazakhstan			**1 233**			**9 000**
Kyrgyzstan			**5**			**40**
Malaysia			**393**			**3 000**
Myanmar (Burma)			**7**			**50**
Pakistan	37	2	**39**	276	26	**302**
Philippines	6		**6**	43		**43**
Taiwan, China			**1**			**4**
Tajikistan			**2**			**12**
Thailand	47	32	**79**	364	328	**692**
Turkey	39		**39**	273		**273**
Turkmenistan			**75**			**546**
Uzbekistan			**81**			**594**
Vietnam			**338**			**2 500**
Total Asia			**8 208**			**60 394**
Albania	80		**80**	528		**528**
Austria			**12**			**86**
Belarus			**27**			**198**
Bulgaria			**2**			**15**
Croatia	9		**9**	67		**67**
Czech Republic	12		**12**	81		**81**
Denmark	170		**170**	1 277		**1 277**
France	20	1	**21**	146	7	**153**
Germany	60	**N**	**60**	442	N	**442**
Greece			**1**			**9**
Hungary	22		**22**	164		**164**
Italy			**89**			**622**
Lithuania	64		**64**	467		**467**
Netherlands			**15**			**106**
Norway	1 098	202	**1 300**	8 214	2 234	**10 448**
Poland	16		**16**	118		**118**
Romania	99	1	**100**	742	11	**753**
Russian Federation			**8 219**			**60 000**
Serbia & Montenegro	22		**22**	161		**161**
Slovakia	1		**1**	7		**7**
Slovenia			**N**			**N**
Spain	2		**2**	12		**12**
Ukraine			**197**			**1 436**

TABLE 2.1 (*Continued*)

	Crude oil (million tonnes)	NGLs (million tonnes)	Total (million tonnes)	Crude oil (million barrels)	NGLs (million barrels)	Total (million barrels)
United Kingdom			**591**			**4 580**
Total Europe			**11 032**			**81 730**
Bahrain			**17**			**125**
Iran (Islamic Republic)	10 668	1 981	**12 649**	77 660	21 400	**99 060**
Iraq			**15 520**			**115 000**
Israel			**1**			**4**
Jordan	N		**N**	1		**1**
Kuwait			**13 292**			**96 500**
Oman			**794**			**5 850**
Qatar			**1 996**			**15 207**
Saudi Arabia			**36 098**			**262 790**
Syria (Arab Republic)			**430**			**3 150**
United Arab Emirates			**13 023**			**97 800**
Yemen			**524**			**4 000**
Total Middle East			**94 344**			**699 487**
Australia	130	151	**281**	1 020	1 518	**2 538**
New Zealand			**9**			**71**
Papua New Guinea			**31**			**240**
Total Oceania			**321**			**2 849**
Total World			**148 342**			**1 098 638**

Notes:

(1) The data on the split of total oil reserves between crude and NGLs are those reported by WEC Member Committees in 2003. They thus constitute a sample, reflecting the information available in particular countries: they should not be considered as complete or necessarily representative of the situation in each region. For this reason, regional and global aggregates have not been computed.

(2) Where a split of reserves between crude oil and NGLs is shown, the components have been converted from barrels to tonnes (or vice versa) at specific crude oil and NGL factors for each country; where only total reserves are shown, conversions have been carried out using the crude oil factor for each country.

(3) Sources: WEC Member Committees, 2003; *Oil & Gas Journal*, 22 December, 2003; *Annual Statistical Report 2003*, OAPEC; *Annual Statistical Bulletin 2002*, OPEC; *World Oil*, August 2003; various national sources.

TABLE 2.2I

Crude oil and natural gas liquids: resources at end-2002 (million tonnes)

	Crude oil			Natural gas liquids		
	Proved amount in place	Estimated additional		Proved amount in place	Estimated additional	
		Amount in place	Reserves recoverable		Amount in place	Reserves recoverable
Africa						
Algeria	9 744	222	436	678	16	22
Côte d'Ivoire	84					
North America						
Canada	3 318	22 716	6 815	131	1 476	1 024
Mexico	3 939	1 227		289	99	
USA			709			
South America						
Argentina			156			
Brazil			449			
Asia						
China	22 000	88 000				
Indonesia	1 342					
Philippines	23	180	17			
Thailand			13			25
Turkey	156					
Europe						
Albania	440					
Austria	12					
Croatia	9					
Czech Republic	17		13			
Denmark	1 689	188	73			
France	419			34		
Germany			20			
Hungary	216	33–195	10–58			
Italy	122	116				
Norway	2 130	1 420		155		
Poland	31					
Romania	1 925	204	34	14	4	1
Serbia & Montenegro	60					
Spain	69	1	N			
Oceania						
Australia			72			178

Notes:

(1) The data on resources are predominantly those reported by WEC Member Committees. They thus constitute a sample, reflecting the information available in particular countries: they should not be considered as complete or necessarily representative of the situation in each region. For this reason, regional and global aggregates have not been computed.

(2) Some of the figures above have been converted from data reported in volumetric terms (e.g. barrels), using specific crude oil and NGL factors for each country. The results have generally been left unrounded, although their apparent precision should be disregarded.

(3) Sources: WEC Member Committees, 2003; private communication.

TABLE 2.2II

Crude oil and natural gas liquids: resources at end-2002 (million barrels)

	Crude oil			Natural gas liquids		
	Proved amount in place	Estimated additional		Proved amount in place	Estimated additional	
		Amount in place	Reserves recoverable		Amount in place	Reserves recoverable
Africa						
Algeria	77 365	1 761	3 459	6 730	157	220
Côte d'Ivoire	611					
North America						
Canada	24 258	166 051	49 815	1 082	12 191	8 457
Mexico	27 654	8 611		3 264	1 119	
USA			5 195			
South America						
Argentina			1 132			
Brazil			3 271			
Asia						
China	161 000	644 000				
Indonesia	9 700					
Philippines	168	1 341	125			
Thailand			97			258
Turkey	1 094					
Europe						
Albania	2 904					
Austria	84					
Croatia	67					
Czech Republic	115		88			
Denmark	12 649	1 409	547			
France	3 084			358		
Germany			147			
Hungary	1 620	248–1 463	75–435			
Italy	854	812				
Norway	15 932	10 622		1 702		
Poland	230					
Romania	14 457	1 532	255	146	42	10
Serbia & Montenegro	438					
Spain	510	5	2			
Oceania						
Australia			567			1 789

Notes:

(1) The data on resources are predominantly those reported by WEC Member Committees. They thus constitute a sample, reflecting the information available in particular countries: they should not be considered as complete or necessarily representative of the situation in each region. For this reason, regional and global aggregates have not been computed.

(2) Some of the figures above have been converted from data reported in tonnes, using specific crude oil and NGL factors for each country. The results have generally been left unrounded, although their apparent precision should be disregarded.

(3) Sources: WEC Member Committees, 2003; private communication.

TABLE 2.3

Crude oil and natural gas liquids: 2002 production

	Crude oil (million tonnes)	NGLs (million tonnes)	Total (million tonnes)	Crude oil (thousand barrels per day)	NGLs (thousand barrels per day)	Total (thousand barrels per day)	R/P ratio
Algeria	46.0	24.2	70.2	1 000	658	1 658	21.3
Angola	44.6		44.6	905		905	26.9
Cameroon	3.7		3.7	72		72	15.2
Congo (Brazzaville)	13.3		13.3	258		258	16.0
Congo (Dem. Republic)	1.2		1.2	23		23	22.3
Côte d'Ivoire	0.7		0.7	15		15	50.6
Egypt (Arab Republic)	32.3	4.7	37.0	628	123	751	13.5
Equatorial Guinea	11.7		11.7	237		237	12.7
Gabon	14.7		14.7	295		295	23.2
Ghana	0.3		0.3	6		6	7.5
Libya/GSPLAJ	62.5	2.1	64.6	1 316	60	1 376	71.7
Morocco	N		N	N		N	16.5
Nigeria	96.2	2.4	98.6	1 953	60	2 013	42.9
Senegal	N	N	N	N	N	N	
South Africa	1.0	0.2	1.2	21	6	27	4.1
Sudan	11.5		11.5	233		233	6.6
Tunisia	3.5	0.1	3.6	74	1	75	11.3
Total Africa	**343.2**	**33.7**	**376.9**	**7 036**	**908**	**7 944**	**34.5**
Barbados	0.1		0.1	1		1	6.2
Canada	72.9	22.3	95.2	1 460	678	2 138	7.7
Cuba	3.3		3.3	57		57	36.0
Guatemala	1.4		1.4	25		25	57.6
Mexico	165.2	13.2	178.4	3 177	408	3 585	13.1
Trinidad & Tobago	6.2	1.4	7.6	118	37	155	12.7
United States of America	286.1	60.7	346.8	5 746	1 880	7 626	11.0
Total North America	**535.2**	**97.6**	**632.8**	**10 584**	**3 003**	**13 587**	**11.3**
Argentina	38.0	0.6	38.6	755	19	774	9.7
Bolivia	1.4	0.4	1.8	31	11	42	31.7
Brazil	72.9	1.5	74.4	1 454	45	1 499	17.9
Chile	0.3	0.1	0.4	6	4	10	41.1
Colombia	29.0	0.7	29.7	578	23	601	8.4
Ecuador	20.1	0.3	20.4	393	9	402	31.6
Peru	4.6	0.2	4.8	93	5	98	9.0
Surinam	0.7		0.7	13		13	35.8
Venezuela	143.7	5.2	148.9	2 732	162	2 894	73.2
Total South America	**310.7**	**9.0**	**319.7**	**6 055**	**278**	**6 333**	**42.2**
Azerbaijan	15.1	0.2	15.3	303	6	309	62.1
Bangladesh	N	N	N	N	1	1	>100
Brunei	9.7	0.5	10.2	195	15	210	17.6

TABLE 2.3 (Continued)

	Crude oil (million tonnes)	NGLs (million tonnes)	Total (million tonnes)	Crude oil (thousand barrels per day)	NGLs (thousand barrels per day)	Total (thousand barrels per day)	R/P ratio
China	168.9		168.9	3 387		3 387	19.5
Georgia	0.1	N	0.1	1	N	1	49.5
India	32.9	3.8	36.7	679	114	793	19.3
Indonesia	55.8	6.6	62.4	1 106	172	1 278	10.9
Japan	0.6	0.3	0.9	12	8	20	9.3
Kazakhstan	42.0	5.2	47.2	841	148	989	24.9
Kyrgyzstan	0.1	N	0.1	2	N	2	71.7
Malaysia	28.6	8.3	36.9	599	234	833	9.9
Myanmar (Burma)	0.7	N	0.7	15	N	15	9.1
Pakistan	3.1	0.1	3.2	63	2	65	12.7
Philippines	0.3	0.4	0.7	6	9	15	7.9
Taiwan, China	0.1	N	0.1	1	1	2	6.8
Tajikistan	N	N	N	N	N	N	87.2
Thailand	3.4	4.5	7.9	71	126	197	9.6
Turkey	2.4		2.4	46		46	16.3
Turkmenistan	8.8	0.2	9.0	176	6	182	8.2
Uzbekistan	4.1	3.1	7.2	81	89	170	9.6
Vietnam	16.8	0.5	17.3	341	14	355	19.3
Total Asia	**393.5**	**33.7**	**427.2**	**7 925**	**945**	**8 870**	**18.7**
Albania	0.3		0.3	6		6	>100
Austria	1.0	0.1	1.1	18	3	21	11.2
Belarus	1.8		1.8	37		37	14.7
Bulgaria	N		N	N		N	93.0
Croatia	1.1		1.1	23		23	8.0
Czech Republic	0.3		0.3	5		5	46.3
Denmark	18.1		18.1	370		370	9.5
France	1.3	0.2	1.5	27	5	32	13.1
Germany	3.7		3.7	74		74	16.4
Greece	0.2	N	0.2	4	1	5	4.8
Hungary	1.1	0.4	1.5	22	12	34	13.2
Italy	5.5		5.5	105		105	16.2
Lithuania	0.4		0.4	9		9	>100
Netherlands	2.2	0.9	3.1	44	22	66	4.4
Norway	146.0	11.3	157.3	2 992	338	3 330	8.6
Poland	0.7		0.7	14		14	23.1
Romania	5.7	0.1	5.8	117	3	120	17.2
Russian Federation	367.1	12.5	379.6	7 342	356	7 698	21.4
Serbia & Montenegro	0.7	0.1	0.8	14	2	16	27.6
Slovakia	N		N	N		N	15.2
Slovenia	N		N	N		N	
Spain	0.3		0.3	7		7	4.6
Ukraine	2.6	1.1	3.7	53	32	85	46.3
United Kingdom	107.4	8.5	115.9	2 207	255	2 462	5.1

(*continued on next page*)

TABLE 2.3 (*Continued*)

	Crude oil (million tonnes)	NGLs (million tonnes)	Total (million tonnes)	Crude oil (thousand barrels per day)	NGLs (thousand barrels per day)	Total (thousand barrels per day)	R/P ratio
Total Europe	**667.5**	**35.2**	**702.7**	**13 490**	**1 029**	**14 519**	**15.4**
Bahrain	1.9	0.4	2.3	38	11	49	7.0
Iran (Islamic Republic)	162.8	4.0	166.8	3 248	119	3 367	80.6
Iraq	99.2	0.5	99.7	2 014	16	2 030	>100
Israel	N		N	N		N	>100
Jordan	N		N	N		N	62.4
Kuwait	87.8	4.1	91.9	1 746	125	1 871	>100
Oman	44.4	0.2	44.6	897	5	902	17.8
Qatar	30.8	3.8	34.6	644	111	755	55.2
Saudi Arabia	391.6	26.5	418.1	7 810	870	8 680	82.9
Syria (Arab Republic)	28.2	0.4	28.6	567	10	577	15.0
United Arab Emirates	95.0	10.6	105.6	1 955	315	2 270	>100
Yemen	21.8	0.6	22.4	456	17	473	23.2
Total Middle East	**963.5**	**51.1**	**1 014.6**	**19 375**	**1 599**	**20 974**	**91.4**
Australia	23.0	8.5	31.5	495	235	730	9.5
New Zealand	1.4	0.2	1.6	31	6	37	5.3
Papua New Guinea	2.1		2.1	46		46	14.3
Total Oceania	**26.5**	**8.7**	**35.2**	**572**	**241**	**813**	**9.6**
Total World	**3 240.1**	**269.0**	**3 509.1**	**65 037**	**8 003**	**73 040**	**41.2**

Notes:

(1) Sources: WEC Member Committees, 2003; *BP Statistical Review of World Energy*, June 2003; *Oil & Gas Journal*, 22 December, 2003; other international and national sources.

(2) Conversions from barrels to tonnes (or vice versa) have been carried out using specific crude oil and NGL factors for each country.

TABLE 2.4

Crude oil and natural gas liquids: 2002 consumption

	Crude oil (million tonnes)	NGLs (million tonnes)	Total (million tonnes)	Crude oil (thousand barrels per day)	NGLs (thousand barrels per day)	Total (thousand barrels per day)
Algeria	22.0		22.0	445		445
Angola	2.0		2.0	40		40
Cameroon	1.6		1.6	32		32
Congo (Brazzaville)	0.5		0.5	10		10
Congo (Dem. Republic)	0.1		0.1	2		2
Côte d'Ivoire	3.0		3.0	60		60
Egypt (Arab Republic)	19.6	2.7	22.3	393	80	473
Eritrea						
Gabon	0.6		0.6	12		12
Ghana	1.5		1.5	30		30
Kenya	1.6		1.6	32		32
Libya/GSPLAJ	17.2		17.2	345		345
Madagascar	0.5		0.5	10		10
Mauritania	1.0		1.0	20		20
Morocco	6.3		6.3	127		127
Nigeria	11.0		11.0	220		220
Senegal	1.0		1.0	20		20
South Africa	22.0		22.0	445		445
Sudan	3.0		3.0	60		60
Tanzania						
Tunisia	2.0		2.0	40		40
Zambia	0.5		0.5	10		10
Total Africa	**117.0**	**2.7**	**119.7**	**2 353**	**80**	**2 433**
Aruba	10.0		10.0	200		200
Canada	89.2	8.5	97.7	1 791	253	2 044
Costa Rica	0.3		0.3	6		6
Cuba	5.0		5.0	100		100
Dominican Republic	2.0		2.0	40		40
El Salvador	1.1		1.1	22		22
Guatemala	1.0		1.0	10		10
Jamaica	1.1		1.1	22		22
Martinique	0.8		0.8	15		15
Mexico	69.1		69.1	1 387		1 387
Netherlands Antilles	12.5		12.5	250		250
Nicaragua	1.0		1.0	20		20
Panama	2.2		2.2	45		45
Puerto Rico	2.5		2.5	50		50
Trinidad & Tobago	7.4		7.4	150		150
United States of America	744.3	63.2	807.5	14 947	1 880	16 827
US Virgin Islands	17.5		17.5	350		350
Total North America	**967.0**	**71.7**	**1 038.7**	**19 405**	**2 133**	**21 538**

(continued on next page)

TABLE 2.4 (Continued)

	Crude oil (million tonnes)	NGLs (million tonnes)	Total (million tonnes)	Crude oil (thousand barrels per day)	NGLs (thousand barrels per day)	Total (thousand barrels per day)
Argentina	27.2	0.6	27.8	546	19	565
Bolivia	1.8		1.8	35		35
Brazil	80.1	1.5	81.6	1 609	45	1 654
Chile	10.0	N	10.0	200	N	200
Colombia	14.4		14.4	289		289
Ecuador	8.2		8.2	165		165
Paraguay	0.1		0.1	2		2
Peru	8.0		8.0	160		160
Surinam	0.5		0.5	10		10
Uruguay	1.7		1.7	35		35
Venezuela	43.2		43.2	868		868
Total South America	**195.2**	**2.1**	**197.3**	**3 919**	**64**	**3 983**
Azerbaijan	6.2	0.1	6.3	125	3	128
Bangladesh	1.3	0.1	1.4	26	3	29
Brunei	0.6		0.6	13		13
China	221.0		221.0	4 438		4 438
Cyprus	1.1		1.1	22		22
Georgia	N		N	1		1
India	110.1		110.1	2 212		2 212
Indonesia	49.0		49.0	985		985
Japan	198.0	4.1	202.1	3 976	123	4 099
Kazakhstan	6.0	4.0	10.0	120	120	240
Korea (DPR)	0.7		0.7	14		14
Korea (Republic)	107.6		107.6	2 160		2 160
Kyrgyzstan	0.1		0.1	2		2
Malaysia	22.6		22.6	453		453
Myanmar (Burma)	1.0		1.0	20		20
Pakistan	10.0		10.0	200		200
Philippines	12.9		12.9	259		259
Singapore	41.1		41.1	825		825
Sri Lanka	2.3		2.3	45		45
Taiwan, China	39.3		39.3	789		789
Tajikistan	N		N	N		N
Thailand	39.2	0.4	39.6	787	11	798
Turkey	29.6		29.6	594		594
Turkmenistan	7.5	0.7	8.2	150	20	170
Uzbekistan	4.0	3.2	7.2	80	95	175
Vietnam						
Total Asia	**911.2**	**12.6**	**923.8**	**18 296**	**375**	**18 671**
Albania	1.1		1.1	21		21
Austria	8.8	0.1	8.9	176	2	178
Belarus	13.0		13.0	260		260

TABLE 2.4 (*Continued*)

	Crude oil (million tonnes)	NGLs (million tonnes)	Total (million tonnes)	Crude oil (thousand barrels per day)	NGLs (thousand barrels per day)	Total (thousand barrels per day)
Belgium	33.5		33.5	673		673
Bulgaria	5.5		5.5	112		112
Croatia	4.6		4.6	92		92
Czech Republic	6.2		6.2	125		125
Denmark	7.3		7.3	146		146
Finland	11.1	1.0	12.1	223	30	253
FYR Macedonia	0.7		0.7	14		14
France	81.6	0.2	81.8	1 638	6	1 644
Germany	107.4		107.4	2 157		2 157
Greece	18.5	N	18.5	370	1	371
Hungary	6.4	0.4	6.8	128	13	141
Ireland	3.2		3.2	64		64
Italy	91.6		91.6	1 840		1 840
Latvia						
Lithuania	6.4	0.1	6.5	129	3	132
Netherlands	48.3	9.2	57.5	970	273	1 243
Norway	12.5	1.1	13.6	252	31	283
Poland	17.8		17.8	357		357
Portugal	11.8		11.8	237		237
Romania	11.5		11.5	231		231
Russian Federation	189.0		189.0	3 795		3 795
Serbia & Montenegro	3.2	0.1	3.3	65	3	68
Slovakia	5.3		5.3	106		106
Slovenia						
Spain	71.4		71.4	1 433		1 433
Sweden	18.8		18.8	378		378
Switzerland	4.8		4.8	96		96
Ukraine	16.0	1.2	17.2	320	35	355
United Kingdom	77.8	1.6	79.4	1 562	47	1 609
Total Europe	**895.1**	**15.0**	**910.1**	**17 970**	**444**	**18 414**
Bahrain	12.0		12.0	240		240
Iran (Islamic Republic)	53.6	3.0	56.6	1 076	90	1 166
Iraq	25.5		25.5	512		512
Israel	10.5		10.5	211		211
Jordan	3.9		3.9	78		78
Kuwait	45.5		45.5	914		914
Oman	4.1	0.1	4.2	83	1	84
Qatar	2.5		2.5	50		50
Saudi Arabia	95.0		95.0	1 908		1 908
Syria (Arab Republic)	13.5		13.5	271		271
United Arab Emirates	23.0		23.0	462		462
Yemen	5.0		5.0	100		100
Total Middle East	**294.1**	**3.1**	**297.2**	**5 905**	**91**	**5 996**

(*continued on next page*)

TABLE 2.4 (*Continued*)

	Crude oil (million tonnes)	NGLs (million tonnes)	Total (million tonnes)	Crude oil (thousand barrels per day)	NGLs (thousand barrels per day)	Total (thousand barrels per day)
Australia	33.1	2.5	35.6	664	75	739
New Zealand	4.7	0.1	4.8	94	3	97
Total Oceania	**37.8**	**2.6**	**40.4**	**758**	**78**	**836**
Total World	**3 417.4**	**109.8**	**3 527.2**	**68 606**	**3 265**	**71 871**

Notes:

(1) The data refer to consumption of indigenous and imported crude oil and NGLs, comprising refinery throughput plus direct use of crude oil/NGLs as fuel.

(2) It is often not possible to isolate consumption of NGLs; if details are unavailable they are included with crude oil. This situation makes it impossible to calculate accurate conversions of oil consumption from tonnes to barrels in all cases.

(3) Sources: WEC Member Committees 2003; *Oil, Gas, Coal and Electricity Quarterly Statistics*, third quarter, 2003, IEA/OECD; other international and national sources; estimates by the editors.

COUNTRY NOTES

The following Country Notes on Crude Oil and Natural Gas Liquids provide a brief account of each country with significant oil reserves/ production. They have been compiled by the editors, drawing upon a wide variety of material, including information received from WEC Member Committees, national and international publications[1].

The principal international published sources consulted were:

- *Annual Statistical Bulletin* 2002, OPEC;
- *BP Statistical Review of World Energy, June 2003*;
- *Energy Balances of OECD Countries 2000–2001*, 2003, International Energy Agency;
- *Energy Balances of Non-OECD Countries 2000–2001*, 2003, International Energy Agency;
- *Energy Statistics of OECD Countries 2000–2001*, 2003, International Energy Agency;
- *Energy Statistics of Non-OECD Countries 2000–2001*, 2003, International Energy Agency;
- *Oil & Gas Journal*, various issues, PennWell Publishing Co.;
- *Our Industry Petroleum*, 1977, The British Petroleum Company Ltd;
- *Petroleum Economist*, various issues;
- *Secretary General's Twenty-Ninth Annual Report, A.H. 1422–1423/A.D. 2002, 2003*, OAPEC;
- *World Oil*, August 2003, Gulf Publishing Company.

Brief salient data are shown for each country, including the year of first commercial production (where ascertained).

Please note that Reserves/Production (R/P) ratios have been calculated on volumetric data (barrels): owing to differential conversion factors, R/P ratios based on tonnes would not generally equate to those based on volumes.

[1] For background notes on countries with relatively low levels of reserves/production, please refer to the 2001 edition of the *Survey*.

Algeria

Proved recoverable reserves (crude oil and NGLs, million tonnes)	1 568
Production (crude oil and NGLs, million tonnes, 2002)	70.2
R/P ratio (years)	21.3
Year of first commercial production	1950

Indigenous oil reserves are the third largest in the African region, after Libya and Nigeria. The principal oil provinces are located in the central and south-eastern parts of the country, with the largest oil field being Hassi Messaoud, which was discovered in 1956. Substantial volumes of NGLs (condensate and LPG) are produced at Hassi R'mel and other gas fields. Algerian crudes are of high quality, with a low sulphur content.

A variety of levels (reflecting, inter alia, differences in definitions) have been published for Algeria's proved oil reserves as at end-2002, ranging (in billions of barrels) from *Oil & Gas Journal*'s 9.2 to *World Oil*'s 13.0; intermediate levels are quoted by OPEC (11.314) and OAPEC (12.000). For the present *Survey*, the level adopted is that advised by the Algerian WEC Member Committee: 10 693 million barrels of crude oil plus 2 201 million barrels of NGLs. These proved reserves have been reported in a context of proved amounts in place of 77 365 million barrels of crude plus 6 730 million barrels of NGLs.

Algeria has been a member of OPEC since 1969 and is also a member of OAPEC. It exported more than half its output of crude oil (excluding condensate) in 2002, mainly to Western Europe and North America.

Angola

Proved recoverable reserves (crude oil and NGLs, million tonnes)	1 201

Production (crude oil and NGLs, million tonnes, 2002)	44.6
R/P ratio (years)	26.9
Year of first commercial production	1956

Proved reserves of oil (8.9 billion barrels, as quoted by *World Oil*) are the second largest in sub-Saharan Africa. *Oil & Gas Journal* and OPEC quote a lower level (5 412 million barrels), which does not appear to allow for recent major discoveries in Angola's offshore waters. The early discoveries (from 1955 onwards) were made on land, but the greater part of Angola's oil resources lies in the coastal waters of its enclave of Cabinda and off the north-western mainland. Major discoveries have been made in recent times in deep water locations. Offshore exploration and production activities largely escaped disruption during the long civil war, and output has risen steadily since the early 1980s. By far the greater part of the crude produced is exported.

Argentina

Proved recoverable reserves (crude oil and NGLs, million tonnes)	379
Production (crude oil and NGLs, million tonnes, 2002)	38.6
R/P ratio (years)	9.7
Year of first commercial production	1907

In terms of oil resources, Argentina lies in the middle ranks of South American countries, with a rising level of reserves which now exceeds those of Colombia and Ecuador. The main oil-producing areas are the west-central areas of Neuquén and Cuyo-Mendoza, the Noroeste area near Bolivia in the north, the southern province of Chubut and the Austral area in the far south (including Argentina's portion of Tierra del Fuego). Offshore fields have been discovered in the San Jorge basin off Chubut province and near Tierra del Fuego.

Proved recoverable oil reserves at end-2002 are reported by the WEC Member Committee for Argentina as 437 million m^3 (2 749 million barrels), of which 12 million m^3 (75 million barrels) located offshore. The estimated level of additional recoverable reserves is given as 180 million m^3 (1 132 million barrels). Published assessments of reserves accord quite closely with the reported level.

Oil output rose strongly during most of the 1990s, but a decline set in at the end of the decade: a considerable proportion is exported (37% in 2002). The Golfo San Jorge and Neuquina basins account for 82% of Argentina's oil production.

Australia

Proved recoverable reserves (crude oil and NGLs, million tonnes)	281
Production (crude oil and NGLs, million tonnes, 2002)	31.5
R/P ratio (years)	9.5
Year of first commercial production	1964

Although drilling for oil took place as long ago as 1892, it was not until well after World War II that Australia achieved oil-producer status. Since then, numerous oil fields have been discovered, notably in the following areas: Gippsland Basin (Bass Strait), off Victoria; Cooper Basin, South Australia; Eromanga and Surat Basins, Queensland; Carnarvon Basin (North-west Shelf) off Western Australia; Bonaparte Basin in the Timor Sea.

The Australian WEC Member Committee, quoting Geoscience Australia data as at 1st January 2001, reports proved recoverable reserves (= 'remaining commercial reserves') as 162.2 gigalitres of crude oil and 112.5 gigalitres of condensate. With the inclusion of 128.9 gigalitres of naturally occurring LPG, total proved recoverable oil reserves amount to 403.6 gigalitres, equivalent to 2 538 million barrels or just over 280 million tonnes. Commercially published estimates of Australian oil reserves tend to be considerably higher than the level adopted for the present

Survey: *Oil & Gas Journal* quotes 3 500 million barrels and *World Oil* 3 700.

The estimated additional reserves recoverable, on the basis of Geoscience Australia's 'estimates of reserves that have not yet been declared commercially viable (non-commercial reserves)', are as follows (in gigalitres): crude oil 90.1; condensate 284.5; naturally occurring LPG 230.8.

Production of oil (including condensate and other NGLs) has fluctuated in recent years: in 2002 it averaged 730 000 b/d, of which crude oil accounted for 68%, condensate 17% and LPG/ethane for 15%. About 60% of Australia's total oil output in 2002 was exported, mostly to Japan and other Asian destinations, the USA and New Zealand.

Azerbaijan

Proved recoverable reserves (crude oil and NGLs, million tonnes)	959
Production (crude oil and NGLs, million tonnes, 2002)	15.3
R/P ratio (years)	62.1
Year of first commercial production	1873

This is one of the world's oldest oil-producing areas, large-scale commercial production having started in the 1870s. During World War II the republic was the USSR's major source of crude, but then decreased in importance as the emphasis moved to Siberia. Azerbaijan's proved recoverable reserves (as reported by *Oil & Gas Journal*) have been substantially increased since the 2001 *Survey of Energy Resources*, and are now put at 7 billion barrels.

The development of Azerbaijan's offshore oil resources in the Caspian Sea, currently under way, is re-establishing the republic as a major oil producer and exporter. With new Caspian fields coming into production, oil output has risen year by year since 1998. The bulk of Azerbaijan's production is obtained offshore.

Brazil

Proved recoverable reserves (crude oil and NGLs, million tonnes)	1 348
Production (crude oil and NGLs, million tonnes, 2002)	74.4
R/P ratio (years)	17.9
Year of first commercial production	1940

Brazil's proved reserves feature significantly within the Western Hemisphere—not in the same league as the three largest producers (USA, Venezuela and Mexico), but well exceeding those of Canada and greater than those of Argentina, Colombia and Ecuador combined. Most of the reserves located up until the mid-1970s were in the north-east and central regions, remote from the main centres of oil demand in the south and south-east. Discoveries in offshore areas, in particular the Campos Basin, transformed the reserves picture.

The level of proved recoverable reserves of oil (1 560 million m^3) reported by the Brazilian WEC Member Committee is 16.6% higher than that advised for the 2001 *Survey*. Among published assessments of proved reserves, *World Oil* quotes essentially the same figure as that reported, whilst *Oil & Gas Journal* gives a level some 15% lower. Of the reserves reported by the Member Committee, 90.6% is located offshore.

Oil production has followed a strongly upward trend for more than 10 years; over 90% of Brazil's output is processed in domestic refineries.

Brunei

Proved recoverable reserves (crude oil and NGLs, million tonnes)	184
Production (crude oil and NGLs, million tonnes, 2002)	10.2
R/P ratio (years)	17.6
Year of first commercial production	1929

Although the earliest discoveries (Seria and Rasau fields) were made on land, virtually all subsequent oil fields have been found in offshore waters. Proved reserves have remained at 1 350 million barrels, according to *Oil & Gas Journal*, since 1990, although *World Oil* quotes a somewhat lower figure (1 100 million barrels), which allows for some degree of attrition from production in recent years. There were eight offshore fields in production in 2002, together with three onshore fields: total output (including 15 000 b/d of natural gasoline) was 210 000 b/d, somewhat higher than in recent years. More than 90% of Brunei's oil output is exported, mostly to Japan, Thailand, South Korea and Singapore.

Canada

Proved recoverable reserves (crude oil and NGLs, million tonnes)	802
Production (crude oil and NGLs, million tonnes, 2002)	95.2
R/P ratio (years)	7.7
Year of first commercial production	1862

There is a very substantial oil resource base, albeit not on the scale of the USA and Mexico. The levels of proved recoverable reserves adopted for the present *Survey* correspond with the 'Remaining Reserves as at 2002-12-31' reported by the Reserves Committee of the Canadian Association of Petroleum Producers (CAPP) in the 2002 CAPP Statistical Handbook (November 2003). Reserves are quoted (after rounding to millions of cubic metres) as 714 conventional crude oil, 82 pentanes plus and 165 ethane/propane/butane. Two provinces (Alberta (42.7%) and Saskatchewan (25.8%)) account for more than two-thirds of Canada's conventional crude oil reserves; offshore reserves (Mackenzie Delta/Beaufort Sea and East Coast Offshore) comprise 26.4% of the total. More than three-quarters of the reserves of pentanes plus, and well over 90% of those of LPG's, are located in Alberta.

Based on assessments made in 1999 and regarded as still valid, the Canadian WEC Member Committee reports further huge quantities of crude oil (26.4 billion m^3) and NGLs (1.9 bcm) as additional amounts in place, of which approximately 7.9 bcm of crude and 1.3 bcm of NGLs are considered to be potentially recoverable.

After many years as a comparatively minor producer, Canada's oil output became of real significance only after major discoveries such as the Leduc field in 1947. Output advanced rapidly from around 1950, crude oil production passing the million b/d mark in 1968. In 2002 output of crude was 1.46 million b/d and that of NGLs (including pentanes) 678 000 b/d.

Canada is the world leader in the production of oil from deposits of oil sands—see the chapter on natural bitumen.

Chad

Proved recoverable reserves (crude oil and NGLs, million tonnes)	133
Production (crude oil and NGLs, million tonnes, 2002)	–
R/P ratio (years)	–
Year of first commercial production	2003

The West African republic of Chad has recently joined the ranks of the world's crude oil producers. Oil production began in July 2003, after the construction of a 1 070 km export pipeline from the oil fields in the Doba Basin of southern Chad through Cameroon to a new terminal at Kribi. The development of the Doba Basin fields (Bolobo, Komé and Miandoum) and the pipeline is handled by a consortium consisting of ExxonMobil (40%), Petronas, the Malaysian state oil company (35%), and ChevronTexaco (25%).

Recoverable reserves have been stated by Esso Exploration & Production Chad, Inc. to be 'slightly more than 900 million barrels', mostly of relatively heavy crude (17–24° API) with a sulphur content of less than 0.1%.

For the purpose of the present *Survey*, a (possibly conservative) level of 900 million barrels has been adopted for proved reserves.

Initial output was from the Miandoum field, at around 50 000 b/d, with Komé planned to come on-stream during 2004, when production would reach a plateau rate of 225 000 b/d. The oil offered for export is called Doba Blend. Initial supplies were typically of 24.8° API and 0.14% sulphur; from March 2004, when Komé is due to start production, the blend's characteristics are expected to move to a lower gravity (20.8° API) and a somewhat higher sulphur content (0.17%).

China

Proved recoverable reserves (crude oil and NGLs, million tonnes)	3 300
Production (crude oil and NGLs, million tonnes, 2002)	168.9
R/P ratio (years)	19.5
Year of first commercial production	1939

The first significant oil find was the Lachunmia field in the north-central province of Gansu, which was discovered in 1939. An extensive exploration programme, aimed at self-sufficiency in oil, was launched in the 1950s; two major field complexes were discovered: Daqing (1959) in the north-eastern province of Heilongjiang and Shengli (1961) near the Bo Hai gulf.

Chinese expert opinion provided in a private communication for the present *Survey* puts proved recoverable oil reserves at 3 300 million tonnes at the end of 2002. This assessment (equivalent to just over 24 billion barrels) is close to the level currently quoted by *World Oil*, but well in excess of that given by *Oil & Gas Journal* (18.25 billion barrels), a figure echoed by OAPEC and OPEC. Around 54% (1 790 million tonnes) of the reported proved reserves is located in China's offshore waters. The reported reserves are set in the context of a proved amount in place of 22 billion tonnes, with an additional (non-proved) amount in place assessed as 88 billion tonnes.

China's oil reserves are by far the largest of any country in Asia: oil output is on a commensurate scale, with the 2002 level accounting for about 40% of the regional tonnage. China exported 7.2 million tonnes of its crude oil in 2002.

Colombia

Proved recoverable reserves (crude oil and NGLs, million tonnes)	253
Production (crude oil and NGLs, million tonnes, 2002)	29.7
R/P ratio (years)	8.4
Year of first commercial production	1921

Initially, oil discoveries were made principally in the valley of the Magdalena. Subsequently, other fields were discovered in the north of the country (from the early 1930s), and in 1959 oil was found in the Putamayo area in southern Colombia, near the border with Ecuador. More recently, major discoveries have included the Caño Limón field near the Venezuelan frontier and the Cusiana and Cupiagua fields in the Llanos Basin to the east of the Andes. However, the remaining proved reserves have been shrinking since 1992, and are now at a very low level in relation to production (R/P ratio of only 8.4), on the basis of data published by the state oil company, Ecopetrol.

Colombia's oil production grew strongly between 1994 and 1999, increasing by about 80% over the period: 2000, however, displayed a sharp contraction, and output has continued to fall year by year.

Congo (Brazzaville)

Proved recoverable reserves (crude oil and NGLs, million tonnes)	213
Production (crude oil and NGLs, million tonnes, 2002)	13.3

R/P ratio (years)	16.0
Year of first commercial production	1957

Oil & Gas Journal has reported a constant level of proved oil reserves for 8 years, which is reflected in the tonnage shown above. *World Oil* quotes a very similar figure for end-2002 reserves (1 515 million barrels versus *Oil & Gas Journal*'s 1 506).

After becoming a significant oil producer in the mid-1970s, the country is now the fourth largest in sub-Saharan Africa. Most of the fields in current production were discovered by Elf or Agip between 1969 and 1983, in shallow coastal waters. The average quality of oil output has improved over the years, aided by the coming on-stream of Elf's deep water Nkossa field. The bulk of oil production is exported.

Denmark

Proved recoverable reserves (crude oil and NGLs, million tonnes)	170
Production (crude oil and NGLs, million tonnes, 2002)	18.1
R/P ratio (years)	9.5
Year of first commercial production	1972

The Danish Energy Authority does not use the terms 'proved' and 'additional' reserves, but uses the categories 'ongoing', 'approved', 'planned' and 'possible' recovery. The figure for proved reserves (as reported by the Danish WEC Member Committee) has been calculated as the sum of ongoing and approved reserves, while the figure for additional reserves has been calculated as the sum of planned and possible reserves. All the oil fields discovered so far are located in the North Sea.

Denmark's proved recoverable reserves are the fourth largest in Europe (excluding the Russian Federation). Within a proved amount in place of 2 011 million m^3 (approximately 12.65 billion barrels), 203 million m^3 (1 277 million barrels) is reported to be recoverable. Beyond these quantities are an estimated additional amount in place of 224 million m^3 (over 1.4 billion barrels), of which 87 million m^3 (547 million barrels) is deemed to be recoverable in the future.

The principal fields in production in 2002 were Dan, Halfdan, Gorm, South Arne and Siri, which together accounted for 78% of national oil output. About 75% of Danish crude is exported, chiefly to other countries in Western Europe.

Ecuador

Proved recoverable reserves (crude oil and NGLs, million tonnes)	650
Production (crude oil and NGLs, million tonnes, 2002)	20.4
R/P ratio (years)	31.6
Year of first commercial production	1917

The early discoveries of oil (1913–1921) were made in the Santa Elena peninsula on the south-west coast. From 1967 onwards, numerous oil fields were discovered in the Amazon Basin to the north-east of the country, adjacent to the Putamayo fields in Colombia: these eastern (Oriente) fields are now the major source of Ecuador's oil production.

The level of proved reserves (as reported by *Oil & Gas Journal*) was raised by over 25% in 1993 and again (marginally) in 1995. The level published for end-2002 reflected a further major upgrade, with Ecuador's reserves leaping from 2 115 to 4 630 million barrels; *World Oil* quote an end-2002 figure very close to *Oil & Gas Journal*'s. About 60% of oil production is exported, the rest being refined locally.

Egypt (Arab Republic)

Proved recoverable reserves (crude oil and NGLs, million tonnes)	492
Production (crude oil and NGLs, million tonnes, 2002)	37.0
R/P ratio (years)	13.5
Year of first commercial production	1911

Egypt has the fifth largest proved oil reserves in Africa, with over half located in its offshore waters. According to the Egyptian WEC Member Committee, Egypt's crude oil reserves were 2.9 billion barrels (409 million tonnes) at the end of 2002, together with 0.8 billion barrels (83 million tonnes) of NGLs (condensates and LPG). The main producing regions are in or alongside the Gulf of Suez and in the Western Desert. In 2000 some 64% of crude oil production came from fields in the Gulf of Suez region, 20% from the Western Desert, 9% from Sinai and 7% from the Eastern Desert.

Egypt is a member of OAPEC: crude oil exports account for about 20% of its production. Total oil output (including condensate and gas-plant LPGs) has been gradually declining in recent years. In 2002 crude oil production was 32.3 million tonnes (628 000 b/d), condensate production was 3.5 million tonnes (87 000 b/d), and LPGs from gas-processing plants amounted to just over 1.1 million tonnes (36 000 b/d).

Equatorial Guinea

Proved recoverable reserves (crude oil and NGLs, million tonnes)	149
Production (crude oil and NGLs, million tonnes, 2002)	11.7
R/P ratio (years)	12.7
Year of first commercial production	1992

The Alba offshore condensate field was discovered in 1984 near the island of Bioko, a province of Equatorial Guinea, by the American company Walter International. In 1996, 4 years after Alba was brought into production, Mobil and its US partner United Meridian began producing from Zafiro, another offshore field. Output built up rapidly in subsequent years: crude oil production in Equatorial Guinea averaged nearly 240 000 b/d in 2002.

For the purposes of the present *Survey*, the level of proved reserves published by *World Oil* (1 095 million barrels) has been adopted; *Oil & Gas Journal* continues to quote a figure of 12 million barrels, unchanged since 1994.

Gabon

Proved recoverable reserves (crude oil and NGLs, million tonnes)	342
Production (crude oil and NGLs, million tonnes, 2002)	14.7
R/P ratio (years)	23.2
Year of first commercial production	1961

Extensive oil resources have been located, both on land and offshore. In terms of proved recoverable reserves, Gabon ranks third largest in sub-Saharan Africa. The published volume of proved reserves (as reported in *Oil & Gas Journal*) was raised from 1 340 to 2 499 million barrels in 1997, at which level it has remained.

Gabon was a member of OPEC from 1975 to 1995, when it withdrew on the grounds that it was unfair for it to be charged the same membership fee as the larger producers but not to have equivalent voting rights.

In recent years over 95% of Gabon's oil output has been exported, mainly to the USA.

India

Proved recoverable reserves (crude oil and NGLs, million tonnes)	740
Production (crude oil and NGLs, million tonnes, 2002)	36.7
R/P ratio (years)	19.3
Year of first commercial production	1890

The amount of proved recoverable reserves reported for this *Survey* is 740 million tonnes, of which 406 million tonnes is located offshore.

For more than 60 years after its discovery in 1890, the Digboi oil field in Assam, in the north-east of the country, provided India with its only commercial oil production: this field was still producing in 2002, albeit at less than 300 b/d. Since 1960 numerous onshore discoveries have been made in the western, eastern and southern parts of India; the outstanding find was, however, made in offshore waters in 1974, when the Bombay High oil and gas field was discovered. In 2001 this one field provided almost 35% of national oil output.

Total production of oil (including gas-plant liquids) has fluctuated in recent years within a range of 36–37 million tonnes per annum. In 2002, India produced (in million tonnes) 32.9 of crude oil, 1.7 of natural gasoline and an estimated 2.1 of gas-plant LPG, all of which was used internally.

Indonesia

Proved recoverable reserves (crude oil and NGLs, million tonnes)	705
Production (crude oil and NGLs, million tonnes, 2002)	62.4
R/P ratio (years)	10.9
Year of first commercial production	1893

The first commercial discovery of oil was made in north Sumatra in 1885; subsequent exploration led to the finding of many more fields, especially in southern Sumatra, Java and Kalimantan.

The Indonesian WEC Member Committee has reported that proved recoverable oil reserves were 5.1 billion barrels at the end of 2002, within a proved amount in place of 9.7 billion barrels. The reported level of reserves is situated towards the middle of a band of published figures, ranging (in millions of barrels) from OPEC's 4 722 and *Oil & Gas Journal*'s 5 000 to OAPEC's 5 120 and *World Oil*'s 5 945.

In 2001 Indonesia exported about 49% of its output of crude oil and condensate, as well as a large part of its production of gas-plant LPGs. The bulk of its oil exports are consigned to Japan, the Republic of Korea, Australia and China. It has been a member of OPEC since 1962.

Iran (Islamic Republic)

Proved recoverable reserves (crude oil and NGLs, million tonnes)	12 649
Production (crude oil and NGLs, million tonnes, 2002)	166.8
R/P ratio (years)	80.6
Year of first commercial production	1913

The first commercial crude oil discovered in Iran was at Masjid-i-Sulaiman in 1908. Further exploration in the next two decades resulted in the discovery of a number of major oil fields, including Agha Jari and Gach Saran. Fields such as these confirmed Iran in its role as a global player in the oil industry. After many years as a major oil producer, the country's oil resources are still enormous: proved reserves, as reported for the present *Survey* by the Iranian WEC Member Committee, comprise 77.66 billion barrels of crude oil plus 21.40 billion barrels of NGLs. In total, the reported reserves are almost identical to those quoted by OAPEC and OPEC (99.08 billion barrels) and *World Oil* (100.06 billion barrels); *Oil & Gas Journal* has a lower figure (89.7 billion barrels). Approximately 17% of the proved reserves of crude and 74% of the NGLs is located offshore.

Iran was a founder member of OPEC in 1960. In 2002, 64% of Iran's crude oil output of 3.25 million b/d was exported, mostly to Europe and Asia.

Iraq

Proved recoverable reserves (crude oil and NGLs, million tonnes)	15 520
Production (crude oil and NGLs, million tonnes, 2002)	99.7
R/P ratio (years)	> 100
Year of first commercial production	1927

Crude oil deposits were discovered near Kirkuk in northern Iraq in 1927, with large-scale production getting under way in 1934–1935 following the construction of export pipelines to the Mediterranean. After World War II more oil fields were discovered and further export lines built. Proved reserves, as quoted by OAPEC, OPEC and *World Oil*, now stand at 115 billion barrels, second only to Saudi Arabia in the whole of the Middle East, and indeed in the world.

Iraq was a founder member of OPEC in 1960 and it is also a member of OAPEC. According to provisional data published by OPEC, crude oil exports amounted to almost 1.5 million b/d in 2002, with 59% destined for the USA and 28% for Western Europe.

Italy

Proved recoverable reserves (crude oil and NGLs, million tonnes)	89
Production (crude oil and NGLs, million tonnes, 2002)	5.5
R/P ratio (years)	16.2
Year of first commercial production	1861

Like France and Germany, Italy has a long history of oil production, albeit on a very small scale until the discovery of the Ragusa and Gela fields in Sicily in the mid-1950s. Subsequent exploration led to the discovery of a number of fields offshore Sicily, several in Adriatic waters and others onshore in the Po Valley Basin. In the absence of reported reserves data from WEC sources, the level of proved recoverable reserves quoted by *Oil & Gas Journal* has been adopted for the present *Survey*. Total oil output (including minor quantities of NGLs) recovered somewhat in 2002, after falling for several years.

Kazakhstan

Proved recoverable reserves (crude oil and NGLs, million tonnes)	1 233
Production (crude oil and NGLs, million tonnes, 2002)	47.2
R/P ratio (years)	24.9
Year of first commercial production	1911

Kazakhstan's oil resources are the largest of all the former Soviet republics (apart from the Russian Federation): *Oil & Gas Journal* puts end-2002 proved reserves (as adopted above) at 9 billion barrels, a 66% leap from the level reported by *OGJ* at end-2001. Most of the republic's oil fields are in the north and west of the country. Output of oil more than doubled between 1996 and 2002 to just over 47 million tonnes (989 000 b/d), including 5.2 million tonnes (148 000 b/d) of NGLs. In 2001, exports accounted for 80% of the republic's oil production.

Kuwait

Proved recoverable reserves (crude oil and NGLs, million tonnes)	13 292
Production (crude oil and NGLs, million tonnes, 2002)	91.9
R/P ratio (years)	> 100
Year of first commercial production	1946

Note: Kuwait data include its share of Neutral Zone.

The State of Kuwait is one of the most oil-rich countries in the world: it currently ranks fifth in terms of the volume of proved reserves. Oil was discovered at Burgan in 1938 and commercial production commenced after World War II. Seven other oil fields were discovered during the next 15 years and output rose rapidly. Kuwait was one of the founder members of OPEC in 1960 and is also a member of OAPEC.

The level of proved recoverable reserves adopted for the present *Survey* is 96.5 billion barrels (as quoted by OAPEC, *Oil & Gas Journal* and OPEC)—a figure unchanged since 1991. *World Oil* gives a slightly higher figure: 98.85 billion barrels.

Kuwait's crude production in 2002 averaged 1.75 million b/d, of which 1.14 million b/d, or 65%, was exported. The main markets for Kuwaiti crude were Japan, other Asian countries, North America and Western Europe.

Libya/GSPLAJ

Proved recoverable reserves (crude oil and NGLs, million tonnes)	4 688
Production (crude oil and NGLs, million tonnes, 2002)	64.6
R/P ratio (years)	71.7
Year of first commercial production	1961

With proved oil reserves of 36 billion barrels, Libya accounts for 36% of the total for Africa. The majority of the known oil reservoirs lie in the northern part of the country; there are a few offshore fields in western waters near the Tunisian border. The crudes produced are generally light (over 35° API) and very low in sulphur.

The level of proved reserves adopted for the present *Survey* is based upon data published by OPEC in its *Annual Statistical Bulletin 2002*, and is in line with the level quoted by OAPEC in its *Annual Report 2002*. *Oil & Gas Journal* and *World Oil* give somewhat lower figures (29 500 and 30 000, respectively).

Libya joined OPEC in 1962 and is also a member of OAPEC. It exported about 70% of its crude oil output in 2002, almost all to Western Europe.

Malaysia

Proved recoverable reserves (crude oil and NGLs, million tonnes)	393
Production (crude oil and NGLs, million tonnes, 2002)	36.9
R/P ratio (years)	9.9
Year of first commercial production	1913

Oil was discovered at Miri in northern Sarawak in 1910, thus ushering in Malaysia's long history as an oil producer. However, it was not until after successful exploration in offshore areas of Sarawak, Sabah and peninsular Malaysia in the 1960s and 1970s that the republic really emerged as a major producer.

Proved reserves, as reported by *Oil & Gas Journal,* remained in the vicinity of 4 billion barrels from the early 1990s to end-2001, when they were reduced to 3 billion barrels, a level retained for end-2002. After following a rising trend for many years, oil production has levelled off since 1998, in line with the Government's National Depletion Policy. In 2001, over 50% of Malaysian crude oil production was exported, chiefly to Thailand, the Republic of Korea, Indonesia, Japan and India.

Mexico

Proved recoverable reserves (crude oil and NGLs, million tonnes)	2 337
Production (crude oil and NGLs, million tonnes, 2002)	178.4
R/P ratio (years)	13.1
Year of first commercial production	1901

Mexico's massive oil resource base has given rise to one of the world's largest oil industries, centred on the state enterprise Petróleos Mexicanos (Pemex), founded in 1938.

In 2002, Mexico's proven reserves were radically reduced in order to conform to the definitions laid down by the Securities and Exchange Commission of the USA. The Mexican WEC Member Committee has reported proved recoverable reserves (on the new basis) of 15 123 million barrels of crude oil and 2 072 million barrels of NGLs, which correspond with the proved reserves given by Pemex in its *Informe Estadístico de Labores 2002*. Pemex quotes its proved reserves (in millions of barrels) as: crude oil 15 123.6, condensate 550.5 and gas-plant liquids 1 521.9. In addition to these *Proved* oil reserves (totalling 17 196 million barrels), *Probable* reserves are given as 13 723 million barrels and *Possible* reserves as 9 731 million barrels: a total of 40.65 billion barrels.

Commercial oil production began in 1901 and by 1918 the republic was the second largest producer in the world. The discovery and development of oil fields along the eastern side of the country—in particular, offshore fields in the Gulf of Campeche—have brought annual production up to its present level. In 2002 oil output comprised 3 177 tb/d crude oil, 80 tb/d condensate and 328 tb/d gas-plant liquids; exports of crude totalled 1 664 tb/d, of which some 78% was consigned to the USA.

Nigeria

Proved recoverable reserves (crude oil and NGLs, million tonnes)	4 252
Production (crude oil and NGLs, million tonnes, 2002)	98.6
R/P ratio (years)	42.9
Year of first commercial production	1957

Nigeria's proved oil reserves are the second largest in Africa, after those of Libya. The country's oil fields are located in the south, mainly in the Niger delta and offshore in the Gulf of Guinea. Nigeria has been a member of OPEC since 1971.

Published assessments of Nigeria's proved recoverable reserves (as at end-2002) are closer to consensus than those for earlier years. For the purposes of the present *Survey*, the level of 31 506 million barrels reported by OPEC (*Annual Statistical Bulletin 2002*) has been adopted (OAPEC and *World Oil* quote very similar figures). *Oil & Gas Journal* provides the exception, by retaining a significantly lower level—24 000 million barrels. The latest OPEC level for Nigerian reserves represents a substantial advance on previous assessments from this source.

Nigeria exported about 90% of its crude oil output in 2001, to a wide spread of regions throughout the world.

Norway

Proved recoverable reserves (crude oil and NGLs, million tonnes)	1 300
Production (crude oil and NGLs, million tonnes, 2002)	157.3
R/P ratio (years)	8.6
Year of first commercial production	1971

Starting with the discovery of the Ekofisk oil field in 1970, successful exploration in Norway's North Sea waters has brought the country into No. 1 position in Europe (excluding the Russian Federation), in terms of oil in place, proved reserves and production. On the basis of data published by the Norwegian Petroleum Directorate, total oil reserves at end-2002 amounted to 1 306 million m^3 (approximately 1 098 million tonnes) of crude oil, 118 million tonnes of NGLs and 130 million m^3 (about 84 million tonnes) of condensate. In addition to the quoted proved amount, there is estimated to be a further 1 420 million m^3 (1 194 million tonnes) of crude oil in place: the proportion of these 'undiscovered resources' that might be ultimately recoverable is not specified.

Norway's recoverable reserves are over twice those of the UK; its oil output in 2002 was, however, only about 35% higher than that of the UK. Norwegian oil production levelled off after 1997, following 16 years of unremitting growth. The groups of fields with the largest output of crude oil in 2002 were Troll, Ekofisk, Snorre, Oseberg, Draugen, Norne and Heidrun. Over 90% of Norwegian oil production was exported in 2002, mostly to Western European countries, the USA and Canada.

Oman

Proved recoverable reserves (crude oil and NGLs, million tonnes)	794
Production (crude oil and NGLs, million tonnes, 2002)	44.6
R/P ratio (years)	17.8
Year of first commercial production	1967

In a regional context, this is one of the less well-endowed Middle East countries but its proved reserves are, nevertheless, quite substantial (5.85 billion barrels at end-2002, according to OAPEC and OPEC). Three oil fields were discovered in the north-west central part of Oman in the early 1960s; commercial production began after the construction of an export pipeline. Many other fields have subsequently been located and brought into production, making the country a significant oil producer and exporter; it has, however, never joined OPEC or OAPEC. Production levels have steadily increased over the years and averaged 902 000 b/d in 2002: a high proportion of crude oil output is exported, mainly to Japan, South-east Asia and China.

Papua New Guinea

Proved recoverable reserves (crude oil and NGLs, million tonnes)	31
Production (crude oil and NGLs, million tonnes, 2002)	2.1
R/P ratio (years)	14.3
Year of first commercial production	1992

Five sedimentary basins are known to exist in PNG. Most exploration activity, and all hydrocarbon discoveries to date, have been made in the Papuan Basin in the southern part of the mainland. After many campaigns of exploration (starting in 1911), the first commercial discoveries were eventually made during the second half of the 1980s. Commercial production began in 1992 after an export pipeline had been built. The oil exported is a blend called Kutubu Light (45° API). Output in 2002 averaged 46 000 b/d.

Peru

Proved recoverable reserves (crude oil and NGLs, million tonnes)	44
Production (crude oil and NGLs, million tonnes, 2002)	4.8
R/P ratio (years)	9.0
Year of first commercial production	1883

Peru is probably the oldest commercial producer of oil in South America, but its remaining recoverable reserves are comparatively slender, at less than 45 million tonnes (based on *Oil & Gas Journal*'s figure of 323 million barrels). However, it is of note that *World Oil*, which habitually quotes substantially higher figures for Peru's oil reserves than does *Oil & Gas Journal*, gives 963 million barrels (equivalent to about 130 million tonnes) at end-2002.

For many years oil production was centred on the fields in the Costa (coastal) area in the north-west; from about 1960 onwards the Zocalo (continental shelf) off the north-west coast and the Oriente area east of the Andes came into the picture. In 2001 the Oriente fields accounted for 67% of total oil output, the Costa fields for 19% and the Zocalo for nearly 14%. Production overall has followed a gently downward slope in recent years.

Qatar

Proved recoverable reserves (crude oil and NGLs, million tonnes)	1 996
Production (crude oil and NGLs, million tonnes, 2002)	34.6
R/P ratio (years)	55.2
Year of first commercial production	1949

In regional terms, oil resources are relatively small, Qatar's strength being much more in natural gas. In the 1930s interest in its prospects was aroused by the discovery of oil in neighbouring Bahrain. The Dukhan field was discovered in 1939 but commercialisation was deferred until after World War II. During the period 1960–1970, several offshore fields were found, and Qatar's oil output grew steadily. It joined OPEC in 1961 and also became a member of OAPEC.

The level of proved recoverable oil reserves (15 207 million barrels) adopted for the present *Survey* is that quoted by OPEC in its *Annual Statistical Bulletin 2002*, and is one that represents a more than three-fold increase over the end-1999 level quoted in the 2001 *Survey*. Currently OAPEC (in slightly rounded form) and *Oil & Gas Journal* both concur with OPEC's assessment, but *World Oil* differs considerably at 19 559 million barrels.

Qatar is a major producer of NGLs: its output was around 90 000 b/d in 2002, almost all of which was exported, very largely to Japan, the Republic of Korea and other Asia/Pacific countries. Crude oil exports follow a similar pattern of destinations.

Romania

Proved recoverable reserves (crude oil and NGLs, million tonnes)	100
Production (crude oil and NGLs, million tonnes, 2002)	5.8
R/P ratio (years)	17.2
Year of first commercial production	1857

Besides being one of Europe's oldest oil producers, Romania still possesses substantial reserves of recoverable oil. Within a proved amount of crude oil in place of 1 925 million tonnes, plus a corresponding figure of 14 million tonnes for NGLs, recoverable reserves are reported as 99 million tonnes of crude plus 1 million tonnes of NGLs. The estimated additional amounts in place (in millions of tonnes) are given as approximately 204 and 4, respectively, with recoverable amounts of 34 and 1.

The principal region of production has long been the Ploesti area in the Carpathian Basin to the north-west of Bucharest, but a new oil province has come on the scene in recent years with the start-up of production from two offshore fields (West and East Lebada) in the Black Sea. Within the figures of proved recoverable reserves given above, the amounts located in offshore waters are 2.6 million tonnes of crude oil and 0.1 million tonnes of NGLs. In national terms, oil output (including NGLs) has been slowly contracting since around 1995.

Russian Federation

Proved recoverable reserves (crude oil and NGLs, million tonnes)	8 219
Production (crude oil and NGLs, million tonnes, 2002)	379.6
R/P ratio (years)	21.4

The Russian oil industry has been developing for well over a century, much of that time under the Soviet centrally planned and state-owned system, in which the achievement of physical production targets was of prime importance. After World War II, hydrocarbons' exploration and production development shifted from European Russia to the east, with the opening-up of the Volga-Urals and West Siberia regions.

The level of proved recoverable reserves adopted for the present *Survey* is based on the figure of 60 000 million barrels quoted by *Oil & Gas Journal*, and is 23.5% higher than the end-1999 level quoted in the 2001 *Survey*, based on the same source. *World Oil* gives a figure that is similar to *Oil & Gas Journal*'s, but less rounded: 58 765 million barrels.

Production levels in Russia advanced strongly from the mid-1950s to around 1980 when output levelled off for a decade. After a sharp decline in the first half of the 1990s, oil production levelled off again, at around 305 million tonnes/yr, until an upturn in 2000–2002 brought the total up to nearly 380 million tonnes in 2002. In that year, Russia exported nearly 50% of its oil production.

Saudi Arabia

Proved recoverable reserves (crude oil and NGLs, million tonnes)	36 098
Production (crude oil and NGLs, million tonnes, 2002)	418.1
R/P ratio (years)	82.9
Year of first commercial production	1936

Note: Saudi Arabia data include its share of Neutral Zone.

The Kingdom has been a leading oil producer for more than 40 years and currently has by far the world's largest proven reserves of oil: at end-2002 these represented about 24% of the global total. The first major commercial discovery of oil in Saudi Arabia was the Dammam field, located by Aramco in 1938; in subsequent years the company discovered many giant fields, including Ghawar (1948), generally regarded as the world's largest oil field, and Safaniyah (1951), the world's largest offshore field.

Whilst not displaying an exact consensus, current published assessments of Saudi Arabia's proved oil reserves fall within a narrow bracket: namely (in billions of barrels), *World Oil* 261.75, *Oil & Gas Journal* 261.8, OPEC (as used in this *Survey*) 262.79, OAPEC 262.80.

Saudi Arabia was a founder member of OPEC and also of OAPEC. It exports about three-quarters of its crude oil output; major destination regions are Asia, North America and Western Europe.

Sudan

Proved recoverable reserves (crude oil and NGLs, million tonnes)	76
Production (crude oil and NGLs, million tonnes, 2002)	11.5
R/P ratio (years)	6.6
Year of first commercial production	1992

Several oil fields, including Heglig and Unity, were discovered in south-central Sudan in the early 1980s but terrorist action forced the companies concerned to withdraw. Other foreign companies started to undertake exploration and development activities some 10 years later. According to *Oil & Gas Journal*, Sudan's proved recoverable reserves have risen from 262 million barrels (35 million tonnes) at end-1999 to 563 million barrels (76 million tonnes) at end-2002. Other published sources tend to quote higher figures—in million barrels, *World Oil* gives 700 and OAPEC 810.

Commercial production from the Heglig field began in 1996, since when Sudan has developed into an oil producer and exporter of some significance, a key factor being the construction of a 250 000 b/d export pipeline to the Red Sea. By the end of 2002, Sudan's oil production was approaching the capacity of the pipeline, having averaged about 230 000 b/d over the year.

Syria (Arab Republic)

Proved recoverable reserves (crude oil and NGLs, million tonnes)	430
Production (crude oil and NGLs, million tonnes, 2002)	28.6
R/P ratio (years)	15.0
Year of first commercial production	1968

After many years (1930–1951) of unsuccessful exploration, oil was eventually found in 1956 at Karachuk. This and other early discoveries mostly consisted of heavy, high-sulphur crudes. Subsequent finds, in particular in the Deir al-Zor area in the valley of the Euphrates, have tended to be of much lighter oil.

Proved recoverable reserves are taken as 3.15 billion barrels, the level quoted by OAPEC: other published sources quote lower levels—*Oil & Gas Journal* and OPEC 2.5 billion barrels, *World Oil* 2.28.

National oil output has plateaued at just under 600 000 b/d in recent years, with nearly 60% being exported. Syria's principal customers for its crude oil are Italy, France and Spain: it is a member of OAPEC.

Thailand

Proved recoverable reserves (crude oil and NGLs, million tonnes)	79
Production (crude oil and NGLs, million tonnes, 2002)	7.9

R/P ratio (years)	9.6
Year of first commercial production	1959

Resources of crude oil and condensate are not very large in comparison with other countries in the region. The data reported by the Thai WEC Member Committee for the present *Survey* show proved reserves of oil as 364 million barrels of crude oil and 328 million barrels of NGLs. Just over three-quarters of the crude reserves and 100% of the NGL reserves are located in Thailand's offshore waters. The estimated additional reserves recoverable are reported as 97 million barrels of crude and 258 million barrels of NGLs.

Total output of oil (crude oil, condensate and other NGLs) has doubled since 1996. Exports of crude oil have risen sharply in recent years, averaging nearly 49 000 b/d in 2002.

Trinidad & Tobago

Proved recoverable reserves (crude oil and NGLs, million tonnes)	102
Production (crude oil and NGLs, million tonnes, 2002)	7.6
R/P ratio (years)	12.7
Year of first commercial production	1908

The petroleum industry of Trinidad is approaching its centenary, several oil fields that are still in production having been discovered in the first decade of the 20th century. The remaining recoverable reserves are small in regional terms, at just over 700 million barrels, according to *Oil & Gas Journal*; *World Oil*, however, quotes a somewhat higher figure (990 million barrels). The oil fields that have been discovered are all virtually in the southern part of the island or in the corresponding offshore areas (in the Gulf of Paria to the west and off Galeota Point at the south-east tip of the island).

After several years in decline, output of crude oil recovered in 2002 to 118 000 b/d; condensate output was little changed at 13 000 b/d. Production of gas-plant NGLs began in 1991 and averaged about 24 000 b/d in 2002. Over 40% of Trinidad's crude output is exported.

Turkmenistan

Proved recoverable reserves (crude oil and NGLs, million tonnes)	75
Production (crude oil and NGLs, million tonnes, 2002)	9.0
R/P ratio (years)	8.2
Year of first commercial production	1911

This republic has been an oil producer for over 90 years, with a cumulative output of more that 5 billion barrels. According to *Oil & Gas Journal*, proved reserves are 546 million barrels. Known hydrocarbon resources are located in two main areas: the South Caspian Basin to the west and the Amu-Daria Basin in the eastern half of the country. After a period of decline, oil output (including NGLs) has grown quite strongly since 1995.

United Arab Emirates

Proved recoverable reserves (crude oil and NGLs, million tonnes)	13 023
Production (crude oil and NGLs, million tonnes, 2002)	105.6
R/P ratio (years)	> 100
Year of first commercial production	1962

The United Arab Emirates comprises Abu Dhabi, Dubai, Sharjah, Ras al-Khaimah, Umm al-Qaiwain, Ajman and Fujairah. Exploration work in the three last-named has not found any evidence of oil deposits on a commercial scale. On the other hand, the four emirates endowed with oil resources have, in aggregate, proved reserves on a massive scale, in the same bracket as those of Iran, Iraq and Kuwait. Abu Dhabi has by far the largest share of UAE reserves and production, followed at some distance by Dubai. The other two oil-producing emirates are relatively minor operators.

At end-2002, the UAE's proved oil reserves were quoted by OAPEC as 97.8 billion barrels, a level virtually unchanged since 1990. According to OPEC, quoting the same total figure, Abu Dhabi accounts for 94.3% of proved reserves, Dubai for 4.1%, Sharjah for 1.5% and Ras al-Khaimah for 0.1%. Total crude output (including a considerable amount of production offshore) amounted to about 2.3 million b/d in 2002, of which some 85% was exported, almost all to Japan and other Asia/Pacific destinations. The UAE has been a member of OPEC since 1967 and is also a member of OAPEC.

United Kingdom

Proved recoverable reserves (crude oil and NGLs, million tonnes)	591
Production (crude oil and NGLs, million tonnes, 2002)	115.9
R/P ratio (years)	5.1
Year of first commercial production	1919

Proved recoverable reserves are indicated by the UK Department of Trade and Industry in the *Digest of United Kingdom Energy Statistics 2003* as 591 million tonnes at end-2002. This figure represents the difference between the proven amount of 'initial recoverable oil reserves in present discoveries' (3 390 million tonnes) and cumulative production to end-2002 (2 799 million tonnes).

In addition, there are estimated to be 325 million tonnes of 'probable' reserves, with 'a better than 50% chance of being technically and economically producible', and a further 425 million tonnes of 'possible' reserves, with 'a significant but less than 50% chance of being technically and economically producible'.

'Reserves in potential future discoveries' are quoted in the 2003 *DUKES* as ranging from 270 to 1 770 million tonnes, whilst 'potential additional reserves' (which exist in discoveries that do not meet the criteria for inclusion as possible reserves) are assessed as ranging from 90 to 490 million tonnes.

Total output of crude oil and NGLs increased from about 92 million tonnes/yr in 1989–1991 to an all-time high of 137 million tonnes in 1999, since when production has tended to decline. The UK exported 76% of its crude oil output in 2002; 58% of such exports were consigned to EU countries and 32% to the USA.

United States of America

Proved recoverable reserves (crude oil and NGLs, million tonnes)	3 801
Production (crude oil and NGLs, million tonnes, 2002)	346.8
R/P ratio (years)	11.0
Year of first commercial production	1859

The United States has one of the largest and oldest oil industries in the world. Although its remaining recoverable reserves are dwarfed by some of the Middle East producers, it is the third largest oil producer, after Saudi Arabia and the Russian Federation.

Proved reserves at end-2002, as published by the Energy Information Administration of the US Department of Energy in December 2003, were 22 677 million barrels of crude oil and 7 994 million barrels of NGLs. Compared with the levels at end-1999, crude reserves are 4.2% higher and those of NGLs up by 1.1%. The 912 million barrel net increase in crude reserves was the result of 4 802 from extensions and discoveries in old and new fields, plus revisions and adjustments of 1 780, minus crude production of 5 670. The comparable figures for NGLs (also in millions of barrels) are 2 574 from extensions and discoveries, plus 209 net revisions, etc. less 2 695 production, giving a net increase of 88 in proved reserves.

Crude oil production in 2002 was 5 746 000 b/d and that of NGLs (including 'pentanes plus') was 1 880 000 b/d. The USA exported only 9 000 b/d of crude oil in 2002, almost all to Canada.

Uzbekistan

Proved recoverable reserves (crude oil and NGLs, million tonnes)	81
Production (crude oil and NGLs, million tonnes, 2002)	7.2
R/P ratio (years)	9.6

Although an oil producer for more than a century, large-scale developments in the republic mostly date from after 1950. The current assessment published by *Oil & Gas Journal* shows proved reserves as 594 million barrels, a level unchanged since 1996. Oil fields discovered so far are located in the south-west of the country (Amu-Daria Basin) and in the Tadzhik-Fergana Basin in the east.

Total oil output (including NGLs) followed a rising trend for about 10 years from 1988, since when the trend has been moderately negative. In 2001, all of Uzbekistan's production of crude and condensate was processed in domestic refineries or used directly as feedstock for petrochemicals.

Venezuela

Proved recoverable reserves (crude oil and NGLs, million tonnes)	11 139
Production (crude oil and NGLs, million tonnes, 2002)	148.9
R/P ratio (years)	73.2
Year of first commercial production	1917

The oil resource base is truly massive, and proved recoverable reserves are easily the largest of any country in the Western Hemisphere. Starting in 1910, hydrocarbons exploration established the existence of four petroliferous basins: Maracaibo (in and around the lake), Apure to the south of the lake, Falcón to the north-east and Oriental in eastern Venezuela. The republic has been a global-scale oil producer and exporter ever since the 1920s, and was a founder member of OPEC in 1960.

The Venezuelan WEC Member Committee reports that end-2002 proved reserves were 77 306 million barrels (equivalent to just over 11 billion tonnes): these figures include reserves of extra-heavy oil (less than 8° API). The principal published sources all quote similar levels of reserves, with the exception of *World Oil*, which gives approximately 53 billion barrels, possibly reflecting differing definitions or coverage.

In 2000 about 50% of national oil output came from the Oriental Basin, 47% from the Maracaibo, 3% from the Apure and a minimal proportion from the Falcón Basin. Of total crude oil output of 3 146 000 b/d in 2000 (including condensate and bitumen for Orimulsion® (registered trade mark belonging to Bitúmenes Orinoco S.A.)), 1 998 000 b/d (63.5%) was exported, the bulk of which being consigned to North and South America: the United States took nearly 59% of Venezuela's crude exports.

Vietnam

Proved recoverable reserves (crude oil and NGLs, million tonnes)	338
Production (crude oil and NGLs, million tonnes, 2002)	17.3
R/P ratio (years)	19.3
Year of first commercial production	1986

During the first half of the 1980s oil was discovered offshore in three fields (Bach Ho, Rong and Dai Hung), and further discoveries have since been made. For the present *Survey*, proved recoverable reserves (2 500 million barrels) have been derived from *World Oil*, as the alternative source (*Oil & Gas Journal*) continues to quote a level of only 600 million barrels, which would imply an unrealistically low R/P ratio of 4.6. Production of crude oil (averaging 34° API) began in 1986 and has risen steadily: at present all output is exported.

Yemen

Proved recoverable reserves (crude oil and NGLs, million tonnes)	524
Production (crude oil and NGLs, million tonnes, 2002)	22.4
R/P ratio (years)	23.2
Year of first commercial production	1986

After many years of fruitless searching, exploration in the 1980s and 1990s brought a degree of success, with the discovery of a number of fields in the Marib area, many yielding very light crudes. Oil discoveries have been made in two other areas of the country (Shabwa and Masila) and Yemen has evolved into a fairly substantial producer and exporter of crude. The level of proved recoverable reserves, as quoted by OAPEC and *Oil & Gas Journal*, has been unchanged at 4 billion barrels for the past 10 years. *World Oil*, however, reports a total of only 2 855 million barrels, albeit an increase of 19% over its end-2001 figure. Estimated total output (including about 17 000 b/d of NGLs) in 2002 was nearly 475 000 b/d, of which about three-quarters was exported, largely to Singapore, Japan, the Republic of Korea and other Asia/Pacific destinations.

Chapter 3

Oil Shale

World Energy Council
CONSEIL MONDIAL DE L'ENERGIE

COMMENTARY

Introduction

Oil shales ranging from Cambrian to Tertiary in age occur in many parts of the world. Deposits range from small occurrences of little or no economic value to those of enormous size that occupy thousands of square miles and contain many billions of barrels of potentially extractable shale oil. Total world resources of shale oil are conservatively estimated at 3.3 trillion barrels (see Table 3.1). However, petroleum-based crude oil is cheaper to produce today than shale oil because of the additional costs of mining and extracting the energy from oil shale. Because of these higher costs, only a few deposits of oil shale are currently being exploited in Australia, China, Brazil and Estonia. However, with the continuing decline of petroleum supplies, accompanied by increasing costs of petroleum-based products, oil shale presents opportunities for supplying some of the fossil energy needs of the world in the years ahead.

Definition of Oil Shale

Most oil shales are fine-grained sedimentary rocks containing relatively large amounts of organic matter from which significant amounts of shale oil and combustible gas can be extracted by destructive distillation. Included in most definitions of 'oil shale', either stated or implied, is the potential for the profitable extraction of shale oil and combustible gas or for burning as a fuel.

The organic matter in oil shale is composed chiefly of carbon, hydrogen, oxygen, and small amounts of sulphur and nitrogen. It forms a complex macromolecular structure that is insoluble in common organic solvents (e.g. carbon disulphide). The organic matter (OM) is mixed with varied amounts of mineral matter (MM) consisting of fine-grained silicate and carbonate minerals. The ratio of OM:MM for commercial grades of oil shale is about 0.75:5–1.5:5. Small amounts of bitumen that are soluble in organic solvents are present in some oil shales. Because of its insolubility, the organic matter must be retorted at temperatures of about 500 °C to decompose it into shale oil and gas. Some organic carbon remains with the shale residue after retorting

[☆] This paper was first published by the Energy Minerals Division of the American Association of Petroleum Geologists, 27 February 2000. It has been edited for inclusion in this *Survey*.

2004 Survey of Energy Resources

but can be burned to obtain additional energy. Oil shale differs from coal whereby the organic matter in coal has a lower atomic H:C ratio, and the OM:MM ratio of coal is usually greater than 4.75:5.

Origin of Oil Shale

Oil shales were deposited in a wide variety of environments, including freshwater to saline ponds and lakes, epicontinental marine basins and related subtidal shelves. They were also deposited in shallow ponds or lakes associated with coal-forming peat in limnic and coastal swamp depositional environments. It is not surprising, therefore, that oil shales exhibit a wide range in organic and mineral composition. Most oil shales were formed under dysaerobic or anaerobic conditions that precluded the presence of burrowing organisms that could have fed on the organic matter. Many oil shales show well-laminated bedding attesting to a low-energy environment free of strong currents and wave action. In the oil shale deposits of the Green River Formation in Colorado and Utah, numerous beds, and even individual laminae, can be traced laterally for many kilometres. Turbiditic sedimentation is evidenced in some deposits as well as contorted bedding, microfractures, and faults.

Most oil shales contain organic matter derived from varied types of marine and lacustrine algae, with some debris of land plants, depending upon the depositional environment and sediment sources. Bacterial processes were probably important during the deposition and early diagenesis of most oil shales. Such processes could produce significant quantities of biogenic methane, carbon dioxide, hydrogen sulphide, and ammonia. These gases in turn could react with dissolved ions in the sediment waters to form authigenic carbonate and sulphide minerals such as calcite, dolomite, pyrite, and even such rare authigenic minerals as buddingtonite, an ammonium feldspar.

Classification of Oil Shales

Oil shales, until recent years, have been an enigmatic group of rocks. Many were named after a locality, mineral or algal content, or type of product the shale yielded. The following are some names applied to oil shales, a few of which are still in use today:

- algal coal
- alum shale
- bituminite
- boghead coal
- cannel coal
- gas coal
- kerosene shale
- kukersite
- schistes bitumineux
- stellarite
- tasmanite
- torbanite
- wollongongite

A.C. Hutton (1987) developed a workable scheme of classifying oil shales on the basis of their depositional environments and by differentiating components of the organic matter with the aid of ultraviolet/blue fluorescent microscopy (Fig. 3.1). His classification has proved useful in correlating components of the organic matter with the yields and chemistry of the oil obtained by retorting.

Hutton divided the organic-rich sedimentary rocks into three groups. These groups are (1) humic coals and carbonaceous shales, (2) bitumen-impregnated rock (tar sands and petroleum reservoir rocks), and (3) oil shale. On the basis of the depositional environment, three basic groups of oil shales were recognised: terrestrial, lacustrine, and marine. Terrestrial oil shales include those composed of lipid-rich organic matter such as resins, spores, waxy cuticles, and corky tissue of roots and stems of vascular terrestrial plants commonly found in coal-forming swamps and bogs. Lacustrine oil shales are those containing lipid-rich organic matter derived from algae that lived in freshwater, brackish, or saline lakes. Marine oil shales are composed of lipid-rich

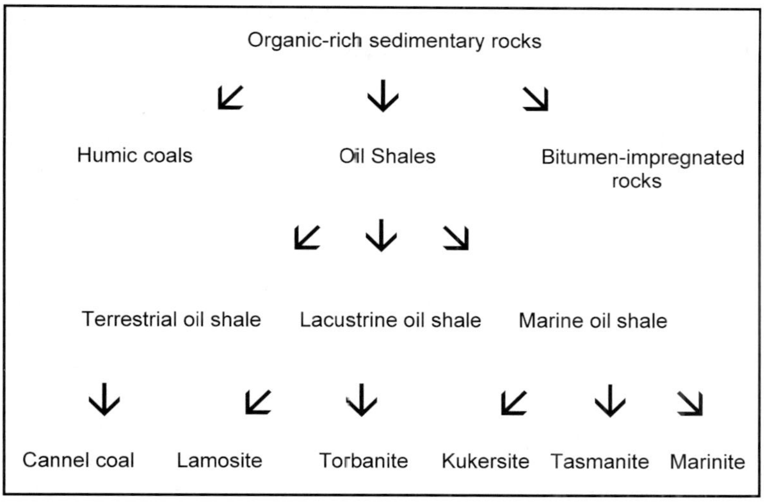

FIGURE 3.1 Classification of organic-rich rocks (from Hutton, 1987).

organic matter derived from marine algae, acritarchs (unicellular microorganisms of questionable origin), and marine dinoflagellates (one-celled organisms with a flagellum).

Hutton (1987) recognised three major macerals in oil shale: telalginite, lamalginite, and bituminite. Telalginite is defined as structured organic matter composed of large colonial or thick-walled unicellular algae such as Botryococcus and Tasmanites. Lamalginite includes thin-walled colonial or unicellular algae that occurs as distinct laminae, but displays little or no recognisable biologic structures. Under the microscope, telalginite and lamalginite are easily recognised by their bright shades of yellow under ultraviolet/blue fluorescent light. The third maceral, bituminite, is another important component in many oil shales. It is largely amorphous, lacks recognisable biologic structures, and displays relatively low fluorescence under the microscope. This material has not been fully characterised with respect to its composition or origin although it is often a quantitatively important component of the organic matter in many marine oil shales. Other organic constituents include vitrinite and inertinite, which are macerals derived from humic matter of land plants. These macerals are usually found in relatively small amounts in most oil shales.

History of the Oil Shale Industry

The use of oil shale can be traced back to ancient times. By the 17th century, oil shales were being exploited in several countries. One of the interesting oil shales is the Swedish alum shale of Cambrian and Ordovician age that is noted for its alum content and high concentrations of metals including uranium and vanadium. As early as 1637, the alum shales were roasted over wood fires to extract potassium aluminium sulphate, a salt used in tanning leather and for fixing colours in fabrics. Late in the 1800s, the alum shales were retorted on a small scale for hydrocarbons. Production continued through World War II but ceased in 1966 because of the availability of cheaper supplies of petroleum crude oil. In addition to hydrocarbons, some hundreds of tonnes of uranium and small amounts of vanadium were extracted from the Swedish alum shales in the 1960s (Andersson et al., 1985, p. 8).

An oil shale deposit at Autun, France, was exploited commercially as early as 1839. The Scottish oil shale industry began about 1859, the year that Colonel Drake drilled his pioneer well at Titusville, Pennsylvania. As many as 20 beds of oil shale were mined at different times. Mining continued during the 1800s and by 1881 oil shale production had reached 1 million tonnes per year. With the exception of the World War II years, between 1 and 4 million tonnes of oil shale were mined yearly in Scotland from 1881 to 1955 when production began to decline, then ceased in 1962. Canada produced some shale oil from deposits in New Brunswick and Ontario in the mid-1800s.

Common products made from oil shale from these early operations were kerosine and lamp oil, paraffin wax, fuel oil, lubricating oil and grease, naphtha, illuminating gas, and the fertiliser chemical, ammonium sulphate. With the introduction of the mass production of automobiles and trucks in the early 1900s, the supposed shortage of gasoline encouraged the exploitation of oil shale deposits for transportation fuels. Many companies were formed to develop the oil shale deposits of the Green River Formation in the western United States, especially in Colorado. Oil placer claims were filed by the thousand on public lands in western United States. The Mineral Leasing Act of 1920 removed oil shale and certain other fossil fuels and minerals on public lands administered by the Federal Government from the status of locatable to leaseable minerals. Under this Act, the ownership of the public mineral lands is retained by the Federal Government and the mineral, e.g. oil shale, is made available for development by private industry under the terms of a mineral lease.

Several oil shale leases on Federal lands in Colorado and Utah were issued to private companies in the 1970s. Large-scale mine facilities were developed on the properties and experimental underground 'modified in situ' retorting was carried out on one of the lease tracts. However, all work has ceased and the leases have been relinquished to the Federal Government. Unocal operated the last large-scale experimental mining and retorting facility in western United States from 1980 until its closure in 1991. The company produced 4.5 million barrels of oil from oil shale averaging 34 gal of shale oil per ton of rock over the life of the project.

The tonnages mined in six oil shale producing countries for the period 1880 to 2000 are shown in Fig. 3.2. By the late 1930s, total yearly production of oil shale for these six countries had risen to over 5 million tonnes. Although production fell in the 1940s during World War II, it continued to rise for the next 35 years, peaking in 1979–1980 when in excess of 46 million tonnes of oil shale per year was mined,

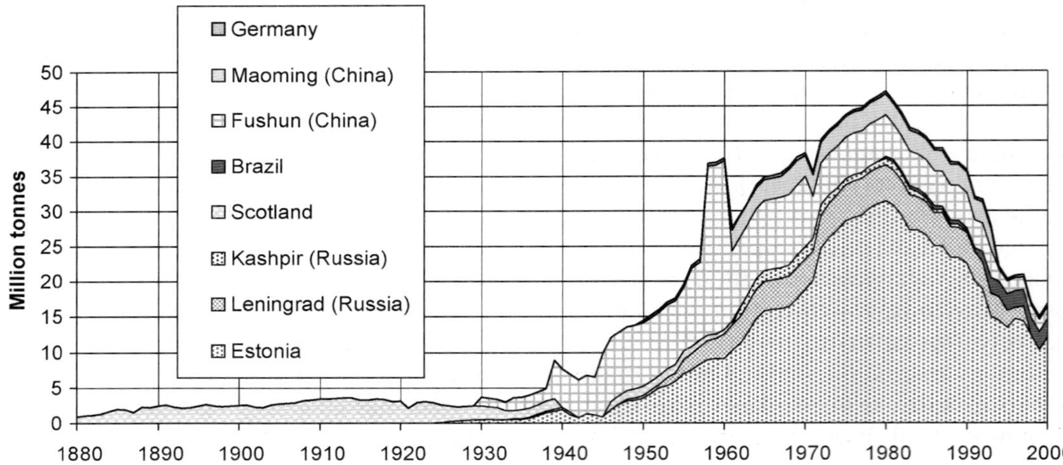

FIGURE 3.2 Oil shale, mined from deposits in Brazil, China, Estonia, Germany, Russia and Scotland, 1880–2000.

two-thirds of which was in Estonia. Assuming an average shale oil content of 100 l/tonne, 46 million tonnes of oil shale would be equivalent to 4.3 million tonnes of shale oil. Of interest is a secondary period of high production reached by China in 1958–1960 when as much as 24 million tonnes of oil shale per year were mined at Fushun.

The oil shale industry as represented by the six countries in Fig. 3.2 maintained a combined yearly production of oil shale in excess of 30 million tonnes from 1963 to 1992. From the peak year of 1981, yearly production of oil shale steadily declined to a low of about 15 million tonnes in 1999. Most of this decline is due to the gradual downsizing of the Estonian oil shale industry. This decline was not due to diminishing supplies of oil shale but to the fact that oil shale could not compete economically with petroleum as a fossil energy resource. On the contrary, the potential oil shale resources of the world have barely been touched.

Oil Shale Resources

Although information about many oil shale deposits is rudimentary and much exploratory drilling and analytical work needs to be done, the potential resources of oil shale in the world are enormous. An evaluation of world oil shale resources is made difficult because of the numerous ways by which the resources are assessed. Gravimetric, volumetric, and heating values have all been used to determine the oil shale grade. For example, oil shale grade is expressed in litres per tonne or gallons per short ton, weight percent shale oil, kilocalories of energy per kilogram of oil shale or Btu, and others. If the grade of oil shale is given in volumetric measure (litres of shale oil per tonne), the density of the oil must be known to convert litres to tonnes of shale oil.

By-products can add considerable value to some oil shale deposits. Uranium, vanadium, zinc, alumina, phosphate, sodium carbonate minerals, ammonium sulphate, and sulphur add potential value to some deposits. The spent shale obtained from retorting may also find use in the construction industry as cement. Germany and China have used oil shale as a source of cement. Other potential by-products from oil shale include specialty carbon fibres, adsorbent carbons, carbon black, bricks, construction and decorative building blocks, soil additives, fertilisers, rock wool insulating materials, and glass. Many of these by-products are still in the experimental stage, but the economic potential for their manufacture seems large.

Many oil shale resources have been little explored and much exploratory drilling needs to be done to determine their potential. Some deposits have been fairly well explored by drilling and analyses. These include the Green River oil shale in western United States, the Tertiary deposits in Queensland, Australia, the deposits in Sweden and Estonia, the El-Lajjun deposit in Jordan, perhaps those in France, Germany and Brazil, and possibly several in Russia. It can be assumed that the deposits will yield at least 40 litres of shale oil per tonne of shale by Fischer assay. The remaining deposits are poorly known and further study and analysis are needed to adequately determine their resource potential.

By far the largest known deposit is the Green River oil shale in the western United States, which contains a total estimated resource of over 1.7 trillion barrels. In Colorado alone, the total resource reaches 1 trillion barrels of oil. The Devonian black shales of the eastern United States are estimated at 423 billion barrels. Other important deposits include those of Estonia, Brazil, Australia, Jordan, and Morocco.

The total world resource of shale oil is estimated at 3.3 trillion barrels. This figure is considered to be conservative in view of the fact that oil shale resources of some countries are not reported and other deposits have not been fully investigated. On the other hand, several deposits, such as those of the Heath and Phosphoria Formations and portions of the Swedish alum oil shale, have been degraded by geothermal heating. Therefore, the resources reported for such deposits are probably too high and somewhat misleading.

Recoverable Resources

The amount of shale oil that can be recovered from a given deposit depends upon many factors. As alluded to above, geothermal heating, or other factors, may have degraded some or all of a deposit, so that the amount of recoverable energy may be significantly decreased. Some deposits or portions thereof, such as large areas of the Devonian black shales in the eastern United States, may be too deeply buried to mine economically in the foreseeable future. Surface land uses may greatly restrict the availability of some oil shale deposits for development, especially those in the industrial western countries. The obvious need today is new and improved methods for the economic recovery of energy and by-products from oil shale. The bottom line in developing a large oil shale industry will be governed by the price of petroleum-based crude oil. When the price of shale oil is comparable to that of crude oil because of diminishing resources of crude, then shale oil may find a place in the world fossil energy mix.

John R Dyni
US Geological Survey

References

Andersson, A., et al., 1985. The Scandinavian Alum Shales: Sveriges Geologiska Undersökning, Avhandlingar Och Uppsatser I A4, Ser. Ca, nr. 56, 50 p.

Batista, A.R.D., Terabe, K., 1988. Proceedings International Conference on Oil Shale and Shale Oil, May 16–19, 1988, Chemical Industry Press, Beijing, China, 635–642.

Bauert, H., 1994. The Baltic oil shale basin: an overview, Proceedings, 1993 Eastern Oil Shale Symposium, University of Kentucky Institute for Mining and Minerals Research, 411–421.

Crisp, P.T., et al., 1987. Australian Oil Shale: A Compendium of Geological and Chemical Data, CSIRO Inst. Energy and Earth Sciences, Division of Fossil Fuels, North Ryde, NSW, Australia, 109 p.

Duncan, D.C., Swanson, V.E. Organic-rich shale of the United States and world land areas, US Geological Survey Circular 523, 30 p.

Dyni, J.R., et al., 1989. Comparison of hydroretorting, Fischer assay, and Rock-Eval analyses of some world oil shales, Proceedings 1989 Eastern Oil Shale Symposium, University of Kentucky Institute for Mining and Minerals Research, 270–286.

Hutton, A.C., 1987. Petrographic classification of oil shales. International Journal of Coal Geology 8, 203–231.

Johnson, E.A., 1990. Geology of the Fushun coalfield, Liaoning Province, People's Republic of China. International Journal of Coal Geology 14, 217–236.

Macauley, G., 1981. Geology of the oil shale deposits of Canada, Geological Survey of Canada Open File Report OF 754, 155 p.

Matthews, R.D., 1983. The Devonian–Mississippian oil shale resource of the United States. In: Gary, J.H., (Ed.), Sixteenth Oil Shale Symposium Proceedings, Colorado School of Mines Press, 14–25.

Novicki, R., 2000. Personal communication.

Padula, V.T., 1969. Oil shale of Permian Iratí Formation. Bulletin American Association of Petroleum Geologists 53, 591–602.

Pitman, J.K., et al., 1989. Thickness, oil-yield, and kriged resource estimates for the Eocene Green River Formation, Piceance Creek Basin, Colorado, US Geological Survey Oil and Gas Investigations Chart OC-132.

Qian, J., 2000. Personal communication.

Reinsalu, E., 2000. An Acceptable Scenario for Oil Shale Industry: Oil Shale, 16, 289. Personal communication.

Russell, P.L., 1990. Oil Shales of the World, Their Origin, Occurrence, and Exploitation, Pergamon Press, New York, 753 p.

Salvador, A., 2000. Demand and supply of energy in the 21st century.

Smith, J.W., 1980. Oil Shale Resources of the United States, Colorado School of Mines Mineral and Energy Resources, vol. 23, no. 6, 30 p.

DEFINITIONS

In Table 3.1 the following definition applies:

Shale oil refers to synthetic oil obtained by heating organic material (kerogen) contained in oil shale rock to a temperature which will convert it to oil, combustible gas, and residual carbon that remains in the spent shale. Properly processed, kerogen can be thermally decomposed (pyrolysis) into a substance somewhat similar to petroleum. Kerogen has not gone through the 'oil window' of heat and therefore, to be changed into an oil-like substance, it must be heated to temperatures as high as 450–500 °C. By this process the organic material is converted into a liquid, which, like crude oil, must be refined to produce transportation fuels (gasoline, jet kerosine and diesel fuel) and other useful petrochemicals.

TABLE 3.1

Shale oil: resources and production at end-2002

	In-place shale oil resources (million barrels)	In-place shale oil resources (million tonnes)	Production in 2002 (thousand tonnes (oil))
Egypt (Arab Rep.)	5 700	816	
Congo (Dem. Rep.)	100 000	14 310	
Madagascar	32	5	
Morocco	53 381	8 167	
South Africa	130	19	
Total Africa	**159 243**	**23 317**	
Canada	15 241	2 192	
USA	2 587 228	380 566	
Total North America	**2 602 469**	**382 758**	
Argentina	400	57	
Brazil	82 000	11 734	157
Chile	21	3	
Total South America	**82 421**	**11 794**	**157**
Armenia	305	44	
China	16 000	2 290	100
Kazakhstan	2 837	400	
Mongolia	294	42	
Myanmar	2 000	286	
Thailand	6 400	916	
Turkey	1 985	284	
Turkmenistan	7 687	1 100	
Uzbekistan	8 386	1 200	
Total Asia	**45 894**	**6 562**	**100**
Austria	8	1	
Belarus	6 988	1 000	
Bulgaria	125	18	
Estonia	16 286	2 494	275
France	7 000	1 002	
Germany	2 000	286	
Hungary	56	8	
Italy	73 000	10 446	
Luxembourg	675	97	
Poland	48	7	
Russian Federation	247 883	35 470	
Spain	280	40	

(continued on next page)

TABLE 3.1 (*Continued*)

	In-place shale oil resources (million barrels)	In-place shale oil resources (million tonnes)	Production in 2002 (thousand tonnes (oil))
Sweden	6 114	875	
Ukraine	4 193	600	
United Kingdom	3 500	501	
Total Europe	**368 156**	**52 845**	**275**
Israel	4 000	550	
Jordan	34 172	5 242	
Total Middle East	**38 172**	**5 792**	
Australia	31 729	4 531	46
New Zealand	19	3	
Total Oceania	**31 748**	**4 534**	**46**
Total world	**3 328 103**	**487 602**	**578**

Notes:
(1) The figures for Turkmenistan refer to the Amu-Daria Basin, which also extends into Uzbekistan.
(2) Source: J.R. Dyni, US Geological Survey.

COUNTRY NOTES

The following Country Notes on Oil Shale have been compiled by the editors, drawing upon a wide variety of material, including papers authored by J.R. Dyni of the USGS, papers presented at the Symposium on Oil Shale in Tallinn, Estonia, November 2002, national and international publications, and direct communications with oil shale experts.

Australia

The total demonstrated oil shale resource is estimated to be in the region of 58 billion tonnes, of which about 24 billion barrels of oil is recoverable. The deposits are spread through the eastern and southern states of the country (Queensland, New South Wales, South Australia, Victoria and Tasmania), although it is the eastern Queensland deposits that have the best potential for economic development.

Production from oil shale deposits in south-eastern Australia began in the 1860s, coming to an end in 1952 when government funding ceased. Between 1865 and 1952 some 4 million tonnes of oil shale were processed.

During the 1970s and early 1980s a modern exploration programme was undertaken by two Australian companies, Southern Pacific Petroleum N.L. and Central Pacific Minerals N.L. (SPP/CPM). The aim was to find high-quality oil shale deposits amenable to open-pit mining operations in areas near infrastructure and deepwater ports. The programme was successful in finding a number of silica-based oil shale deposits of commercial significance along the coast of Queensland. Ten deposits clustered in an area north of Brisbane were investigated and found to have an oil shale resource in excess of 20 billion barrels (based on a cutoff grade of 50 l/t at 0% moisture), which could support production of more than 1 million barrels a day.

In 1995 SPP/CPM signed a joint venture agreement with the Canadian company Suncor Energy Inc. to commence development of one of the oil shale deposits, the Stuart Deposit. Located near Gladstone, it has a total in situ shale oil resource of 2.6 billion barrels and the capacity to produce more than 200 000 b/d. Suncor had had the role of operator of the Stuart project, but in April 2001, SPP/CPM purchased Suncor's interest. In February 2002, a corporate restructuring was undertaken so that SPP became the holding company for the group's interests and CPM was delisted from the Australian stock exchange.

The Stuart project, incorporating the Alberta-Taciuk Processor (ATP) retort technology (initially developed for potential application to the Alberta oil sands), has three stages. The Stage 1 demonstration plant was constructed between 1997 and 1999 and to date has produced over 500 000 barrels. The plant is designed to process 6 000 tonnes per stream day of run-of-mine (wet shale) to produce 4 500 bpsd of shale oil products. Technical and economic feasibility having been proved, it is planned that the ATP in Stage 2 will be scaled up by a factor of 4 to a commercial-sized module processing 23 500 tpsd and producing 15 500 bpsd oil products. Once regulatory approvals have been granted and financing is in place, it is planned that Stage 2 will come on stream during 2006. It is envisaged that multiple commercial ATP units will come on stream during 2010–2013 processing up to 380 000 tpsd and producing up to 200 000 bpsd of oil products for a period in excess of 30 years.

The Stage 1 raw shale oil produced is a relatively light crude with a 42° API gravity, 0.4 wt% sulphur and 1.0 wt% nitrogen. To meet the needs of the market, the raw oil requires further processing, resulting in low-sulphur naphtha and light fuel oil. The ultra low-sulphur 'water-white' product can be readily used to make gasoline and jet fuel. Shale oil has been certified as a feedstock for jet fuel production by the world's leading accreditation agencies. A long-term contract has been signed for the sale of naphtha to Mobil Oil Australia. The light fuel oil is being shipped to Singapore and sold into the fuel oil blending market.

Having committed itself to ensuring that the Stuart oil shale project has a sustainable development, SPP has put various schemes into operation to achieve its stated environmental goals. One in particular launched in 1998 is a reforestation carbon dioxide sink. To date, 250 000 trees have been planted on deforested lands in Central Queensland, with data on carbon dioxide sequestration rates in both the trees and surrounding soils being collected. In September 2000, the first carbon trade in Queensland was announced. It is between SPP and the state government and is based on the reforestation trials.

SPP announced at the beginning of December 2003 that receivers had been appointed to the company. In February 2004 it was reported that a new company, Queensland Energy Resources, had been formed to acquire most of the assets of SPP.

Brazil

The oil shale resource base is one of the largest in the world and was first exploited in 1884 in the State of Bahia. In 1935 shale oil was produced at a small plant in São Mateus do Sul in the State of Paraná and in 1950, following government support, a plant capable of producing 10 000 b/d shale oil was proposed for Tremembé, São Paulo.

Following the formation of Petrobras in 1953, the company developed the Petrosix process for shale transformation. Operations are concentrated on the reservoir of São Mateus do Sul, where the ore is found in two layers: the upper layer of shale (6.4 m thick), with an oil content of 6.4%, and the lower 3.2 m layer with an oil content of 9.1%. The company brought a pilot plant (8 in. internal diameter retort) into operation in 1982, its purpose being for oil shale characterisation, retorting tests and developing data for economic evaluation of new commercial plants. A 6 ft (internal diameter) retort demonstration plant followed in 1984 and is used for the optimisation of the Petrosix technology.

A 2 200 (nominal) tonnes per day, 18 ft (internal diameter) semi-works retort (the Iratí Profile Plant), originally brought on line in 1972, began operating on a limited commercial scale in 1981 and a further commercial plant—the 36 ft (internal diameter) Industrial Module retort—was brought into service in December 1991. Together the two commercial plants process some 7 800 tonnes of bituminous shale daily. The retort process (Petrosix) where the shale undergoes pyrolysis yields a nominal daily output of 3 870 barrels of shale oil, 120 tonnes of fuel gas, 45 tonnes of liquefied shale gas and 75 tonnes of sulphur. Output of shale oil in 2002 was an estimated 3 000 b/d (157 000 tpa).

The Ministry of Mines and Energy quotes end-1999 shale oil reserves as 445.1 million m^3 measured/indicated/inventoried and 9 402 million m^3 inferred/estimated, with shale gas reserves as 111 billion m^3 measured/indicated/inventoried and 2 353 billion m^3 inferred/estimated.

Canada

Oil shales occur throughout the country, with as many as 19 deposits having been identified. However, the majority of the in-place shale oil resources remain poorly known. The most explored deposits are those in the provinces of Nova Scotia and New Brunswick. Of the areas in Nova Scotia known to contain oil shales, development has been attempted at two—Stellarton and Antigonish. Mining took place at Stellarton from 1852 to 1859 and 1929 to 1930 and at Antigonish around 1865. The Stellarton Basin is estimated to hold some 825 million tonnes of oil shale, with an in situ oil content of 168 million barrels. The Antigonish Basin has the second largest oil shale resource in Nova Scotia, with an estimated 738 million tonnes of shale and 76 million barrels of oil in situ.

Investigations into retorting and co-combustion (with coal for power generation) of Albert Mines shale (New Brunswick) have been conducted, including some experimental processing in 1988 at the Petrobras plant in Brazil. Interest has been shown in the New Brunswick deposits for the potential they might offer to reduce sulphur emissions by co-combustion of carbonate-rich shale residue with high-sulphur coal in power stations.

China

Oil shale deposits are widespread, with proved reserves estimated to be in the region of 32 billion tonnes.

Fushun, a city in the north-eastern province of Liaoning, is known as the Chinese 'capital of coal'. Within the Fushun coalfield the West Open Pit mine is the largest operation of coal mining and is where, above the coal layer, oil shale from the Tertiary Formation is mined as a by-product. It was estimated that the entire Fushun area had a resource of approximately 2 billion tonnes with the oil content greater than 4.7%.

It has been reported that the average oil content is 7–8%, which would produce in the region of 78–89 l of oil per tonne of oil shale (assuming a 0.9 specific gravity).

The commercial extraction of oil shale and the operation of heating retorts for processing the oil shale were developed in Fushun between 1920 and 1930. After World War II, Refinery No. 1 had 200 retorts, each with a daily throughput of 100–200 tonnes of oil shale. It continued to operate and was joined by Refinery No. 2, restored in 1954. In Refinery No. 3 shale oil was hydrotreated for producing light liquid fuels. Shale oil was also open-pit mined in Maoming, Guangdong Province and 64 retorts were put into operation there in the 1960s.

At the beginning of the 1960s, 266 retorts were operating in Fushun's Refinery Nos. 1 and 2 and production peaked at about 23 million tonnes of oil shale (about 780 000 tonnes of shale oil). However, during the 1980s production had dropped to about 300 000 tonnes of shale oil and at the beginning of the 1990s the availability of much cheaper crude oil had led to the Maoming operation and Fushun Refinery Nos. 1 and 2 being shut down.

A new facility—the Fushun Oil Shale Retorting Plant—came into operation under the management of the Fushun Bureau of Mines. It at first consisted of 60 retorts producing 60 000 tonnes per year of shale oil to be sold as fuel oil, with carbon black as a by-product. Subsequently 20 further retorts were constructed and by 2001 the annual production of shale oil had increased to 80 000 tonnes.

In addition to Fushun, small quantities of oil shale have been extracted from underground mines in Huadian, Jilin Province in recent years. This has been burnt in a fluidised bed boiler for power generation.

In 2002 the Fushun shale oil plants produced 100 000 tonnes of shale oil and owing to favourable economic factors it is planned to triple the production capacity. This will be achieved with improved technology and the installation of larger retorts.

Future plans involve a plant with a processing capacity of 2 million tonnes of oil shale yearly (6 000 tonnes oil shale daily) being built in Huadian, and the following plants being constructed as a result of the oil shale being a by-product of coal mining:

- Harbin Gas and Chemical Company in Heilongjang Province—1 000 tonnes oil shale daily;
- Song Ya San Coal Mining Company in Heilongjang Province—1 000 tonnes oil shale daily;
- Longkow Coal Mining Company in Shandong Province—2 500 tonnes oil shale daily.

Egypt (Arab Republic)

Oil shale was discovered during the 1940s as a result of oil rocks self-igniting whilst phosphate mining was taking place. The phosphate beds in question lie adjacent to the Red Sea in the Safaga-Quseir area of the Eastern Desert. Analysis was at first undertaken in the Soviet Union in 1958 and was followed by further research in Berlin in the late 1970s. This latter work concentrated on the phosphate belt in the Eastern Desert, the Nile Valley and the southern Western Desert. The results showed that the Red Sea area was estimated to have about 4.5 billion barrels of in-place shale oil and that in the Western Desert, the Abu Tartour area contained about 1.2 billion barrels.

The studies concluded that the oil shale rocks in the Red Sea area were only accessible by underground mining methods and would be uneconomic for oil and gas extraction. However, the Abu Tartour rocks could be extracted whilst mining for phosphates and then utilised for power production for use in the mines. Additionally, although in both areas power could be generated for the in-place cement industry, the nature of the shale as a raw material would not be conducive to the manufacture of high-quality cement.

In view of the depletion of Egyptian fossil fuel reserves, a research project was implemented during 1994–1998 on the 'Availability of Oil Shale in Egypt and its Potential Use in Power Generation'. The project concluded that the burning of oil shale and its use as fuel for power production was feasible, but only became economic when heavy fuel oil and coal prices rose to significantly higher levels. Many recommendations for a technological and environmental nature were made and economic studies continue. A 20 MW oil shale pilot plant for power generation in Quseir was recommended as part of a first step towards the exploitation of Egyptian oil shale.

Estonia

Oil shale was first scientifically researched in the 18th century. In 1838 work was undertaken to establish an open-cast pit near the town of Rakvere and an attempt was made to obtain oil by distillation. Although it was concluded that the rock could be used as solid fuel and, after processing, as liquid or gaseous fuel, the 'kukersite' (derived from the name of the locality) was not exploited until the fuel shortages created by World War I began to impact.

The Baltic Oil Shale Basin is situated near the north-western boundary of the East European Platform. The Estonia and Tapa deposits are both situated in the west of the Basin, the former being the largest and highest-quality deposit within the Basin.

Since 1916 oil shale has had an enormous influence on the energy economy, particularly during the period of Soviet rule and then under the re-established Estonian Republic. At a very early stage, an oil shale development programme declared that kukersite could be used directly as a fuel in the domestic, industrial or transport sectors. Moreover, it was easily mined and could be even more effective as a combustible fuel in power plants or for oil distillation. Additionally kukersite ash could be used in the cement and brick-making industries.

Permanent mining began in 1918 and has continued until the present day, with capacity (both underground mining and open-cast) increasing as demand rose. By 1955 oil shale output had reached 7 million tonnes and was mainly used as power station/chemical plant fuel and in the production of cement. The opening of the 1 400 MW Baltic Thermal Power Station in 1965 followed, in 1973, by the 1 600 MW Estonian Thermal Power Station again boosted production and by 1980 (the year of maximum output) the figure had risen to 31.35 million tonnes.

In 1981, the opening of a nuclear power station in the Leningrad district of Russia signalled the beginning of the decline in Estonian oil shale production. No longer were vast quantities required for power generation and the export of electricity. The decline lasted until 1995, with some small annual increases thereafter.

The Estonian government has taken the first steps towards privatisation of the oil shale industry and is beginning to tackle the air and water pollution problems that nearly a century of oil shale processing has brought.

The total Estonian in-place shale oil resource is currently estimated to be in the region of

16 billion barrels and at the present time continues to play a dominant role in the country's energy balance.

In 2002 12.4 million tonnes of oil shale were produced. Imports amounted to 0.7 million tonnes, 9.4 million tonnes were used for electricity generation, 1.0 for heat generation and 2.3 million tonnes were processed for shale oil and coke production. Production of shale oil was 275 000 tonnes, 160 000 tonnes were exported, 7 000 tonnes were utilised for electricity generation and 114 000 for heat generation.

The historical ratio of underground mining to opencast (approximately 50:50) is tending to move away from opencast production as the bed depths increase. The exhausted opencast areas are gradually being recultivated and reforested but the 'Restructuring Plan of the Estonian oil-shale sector' published by the Ministry of Economic Affairs in 2001 sets out the many problems associated with both the mining of oil shale and its use for electricity production. Nevertheless the Government and Parliament have taken the strategic decision to continue energy production based on oil shale until 2015, albeit by then with a leaner and fitter industry.

The task of the government in restructuring the oil shale industry is to achieve the following strategic goals, not least to meet the environmental conditions of membership of the European Union:

- increase the efficiency in oil shale based energy production and significantly reduce the harmful environmental impact via renovation of combustion technologies;
- renovate the power plants and oil processing plants in order to ensure the rational use of the oil shale resource in the existing underground and opencast mines;
- develop the Estonian oil shale based energy production in conformity with EU legislation and general trends.

Furthermore, the 2002 legislation, which permitted the liberalisation of the electricity and gas market, is having a significant social effect on Ida-Virumaa County, the area in north-eastern Estonia with the greatest concentration of oil shale industry in the country. The oil shale sector is the main employer in the region and the reorganisation of the industry has led to high unemployment. The once high dependence on oil shale processing has yet to be replaced by a culture of private enterprise—it is hoped that international co-operation and the injection of foreign investment may help this situation in due course.

France

Oil shale was irregularly exploited in France between 1840 and 1957 but at its highest (1950), output only reached 0.5 million tonnes per year of shale. During its 118 year life, the Government imposed taxes and duties on foreign oil, thus preserving the indigenous industry.

In 1978 it was estimated that the in-place shale oil resources amounted to 7 billion barrels.

Germany

The German oil shale industry was developed in the middle of the 19th century and during the 1930s and 1940s the development of retorted oil contributed to the depleted fuel supplies during World War II.

Today the only active plant is located in Dotternhausen in southern Germany, where Rohrbach Zement has been using oil shale since the 1930s. The oil shale from this area has a low energy content, a low oil yield and a high ash content but by using a complex process the complete utilisation of both the oil shale energy and all its minerals can be accomplished and incorporated into the manufacture of cement and other hydraulic binding agents. Part of the oil shale is directly used for cement clinker production. Finely ground oil shale supplies 20% of the thermal energy and 10% of the raw materials needed for the clinker burning process. Most of the oil shale, however, is burnt in fluidised bed units, the heat of which is used simultaneously to produce electricity.

A minimal quantity of oil shale is produced for use at the Rohrbach works. In 2002 production amounted to 367 000 tonnes.

In 1965 it was estimated that in-place shale oil resources amounted to 2 billion barrels.

Israel

Sizeable deposits of oil shale have been discovered in various parts of Israel, with the principal resources located in the north of the Negev desert. About 12 billion tonnes of oil shale reserves have been identified. The largest deposit (Rotem Yamin) has shale beds with a thickness of 35–80 m, yielding 60–71 l of oil per tonne. Generally speaking, Israeli oil shales are relatively low in heating value and oil yield, and high in moisture, carbonate, and sulphur content, compared with other major deposits.

A 25 MW experimental electric power plant fuelled by oil shale from the Rotem Yamin deposit began operation in 1989 but is now closed.

Jordan

There are about 26 known occurrences which result in Jordan having an extremely large proven and exploitable oil shale resource. Geological surveys estimate that the existing shale reserves cover more than 60% of the country and amount to in excess of 65 billion tonnes.

The eight most important deposits are located in west-central Jordan and of these, El Lajjun, Sultani, and the Juref ed Darawish have been most extensively explored. They are all classified as shallow and are suitable for open-cast mining, albeit some are underlain by phosphate beds. One more deposit, Yarmouk, located close to the northern border is thought to extend into Syria and may prove to be exceptionally large, both in area and thickness. Reaching some 400 m in thickness, it would only be exploitable by underground mining.

Jordanian shales are generally of quite good quality, having relatively low ash and moisture content (2–5.5 wt%), a gross calorific value of 1 200–2 000 kcal/kg and an average oil yield of 10% (by weight). The sulphur content of the oil shale ranges from 0.3 to 4.3%, whilst the range for the shale oil from the Juref ed Darawish and the Sultani deposits is very high at 8 and 10%, respectively. Also of relevance is the relatively high trace metal content from the major deposits (copper, nickel, zinc, chromium, vanadium).

During the past two decades the Government has undertaken a number of feasibility studies and test programmes. These have been carried out in co-operation with companies from the former Soviet Union, Germany, China, Russia, Canada and Finland. They were all intended to demonstrate utilisation through either direct burning or retorting. All tests proved that burning Jordanian oil shale is very stable, emission levels are low and carbon burn-out is high. Furthermore, research on catalytic gasification was undertaken in the FSU, with positive results. Solvent extraction of organic matter was the subject of a joint study by the Jordanian Natural Resources Authority and the National Energy Research Center.

Following combustion, the resultant oil shale ash appears to be suitable for using in a wide range of products, ranging from construction materials to being used as a supplement to animal food.

The eventual exploitation of Jordan's fuel resource to produce liquid fuels and/or electricity, together with chemicals and building materials, would be favoured by three factors—the high organic matter content of Jordanian oil shale, the suitability of the deposits for surface-mining and their location near potential consumers (i.e. phosphate mines, potash and cement works).

The Government is keen for private investment to develop the industry through BOO and BOT schemes and to this end passed a new electricity law in 1996 which allowed such co-operation to occur. The Ministry of Energy and Mineral Resources (MEMR) has invited qualified companies with proven technology and experience in oil shale to submit their proposals. In particular, Suncor of Canada has developed a technology for extracting and processing oil shale at about US$ 10 per barrel. The company has proposed a project, the first stage of which would result in the production of 17 000 b/d oil. The success of this initial stage, together with that of the first operational oil shale project in Australia, would permit the Government to grant continued expansion of other projects on a BOO basis.

Kazakhstan

At the beginning of the 1960s successful experimentation was carried out on a sample of Kazakhstan's oil shale in the former Soviet Republic of Estonia. Both domestic gas and shale oil were produced. It was found that the resultant shale oil had a low-enough sulphur content for the production of high-quality liquid fuels.

Beginning in early 1998 and lasting until end-2001, a team funded by INTAS (an independent, international association formed by the European Community to preserve and promote scientific co-operation with the newly independent states) undertook a project aimed at completely re-evaluating Kazakhstan's oil shales. The resultant report testified that Kazakhstan's oil shale resources could sustain the production of various chemical and power-generating fuel products.

The research undertaken concluded that the occurrence of oil shale is widespread, the most important deposits having been identified in western (the Cis-Urals group of deposits) and eastern (the Kenderlyk deposit) Kazakhstan. Further deposits have been discovered in both the southern region (Baikhozha and the lower Ili river basin) and the central region (the Shubarkol deposit).

In excess of 10 deposits have been studied: the Kenderlyk Field has been revealed as the largest (in the region of 4 billion tonnes) and has undergone the greatest investigation. However, studies on the Cis-Urals group and the Baikhozha deposit have shown that they have important concentrations of rare elements (rhenium and selenium), providing all these deposits with promising prospects for future industrial exploitation.

The in-place shale oil resources in Kazakhstan have been estimated to be in the region of 2.8 billion barrels. Moreover, many of the deposits occur in conjunction with hard and brown coal accumulations which, if simultaneously mined, could increase the profitability of the coal production industry whilst helping to establish a shale-processing industry.

The recommendations made to INTAS were that collaboration between the project's participants should continue and further research should be undertaken on a commercial basis with interested parties, as a precursor to the establishment of such an industry.

Morocco

Morocco has very substantial oil shale reserves but to date they have not been exploited. During the early 1980s, Shell and the Moroccan state entity ONAREP conducted research into the exploitation of the oil shale reserves at Tarfaya, and an experimental shale-processing plant was constructed at another major deposit (Timahdit). At the beginning of 1986, however, it was decided to postpone shale exploitation at both sites and to undertake a limited programme of laboratory and pilot-plant research.

Although deposits have been identified at 10 localities, the resources of the two largest deposits, Tarfaya and Timahdit, with over 50 billion barrels, rank the country sixth in terms of worldwide in-place shale oil.

Russian Federation

In excess of 80 oil shale deposits have been identified in Russia.

The deposits in the Volga-Petchyorsk province, although of reasonable thickness (ranging from 0.8 to 2.6 m), contain high levels of sulphur.

Extraction began in this area in the 1930s with the oil shale being used to fuel two power plants, but the operation was abandoned owing to environmental pollution. However, most activity has centred on the Baltic Basin where the kukersite oil shale has been exploited for many years. In 2002 the Leningradslanets Oil Shale Mining Public Company produced 1.12 million tonnes (an output of 1.2 million tonnes is expected in 2003). Since June 2003 all shale mined has been delivered to the Estonian Baltic power station with the resultant electricity delivered to UES (Unified Energy System of Russia).

Until 1998, the Slantsy electric power plant (located close to the Estonian border, 145 km from St Petersburg) was equipped with oil shale fired furnaces but in 1999 its 75 MW plant was converted to use natural gas. It continued to process oil shale for oil until June 2003, since when its main activities have been electrode coke annealing and the processing of coals and natural gas oil components.

In 1995 a small processing plant operated at Syzran with an input of less than 50 000 tonnes of shale per annum. Although the accompanying mine has now closed it has been said that a group of about 10 miners are producing in the region of 10 000 tonnes per year. Using the Syzran plant the oil shale is being processed for the manufacture of a pharmaceutical product.

It is alleged that the administrators of the Syzran plant are seeking investment for a new plant capable of processing 500 tonnes per day. The mine would be re-opened with the intention of perpetuating the production of pharmaceutical products. To this end a business plan has been issued.

Sweden

The huge shale resources underlying mainland Sweden are more correctly referred to as alum shale; black shale is found on two islands lying off the coast of south-eastern Sweden. The in-place shale oil resource is estimated to be 6.1 billion barrels.

The exploitation of alum shale began as early as 1637 when potassium aluminium sulphate (alum) was extracted for industrial purposes. By the end of the 19th century the alum shale was also being retorted in an effort to produce a hydrocarbon oil. Before and during World War II, Sweden derived oil from its alum shale, but this process had ceased by 1966, when alternative supplies of lower-priced petroleum were available; during the period 50 million tonnes of shale had been mined.

The Swedish alum shale has a high content of various metals including uranium, which was mined between 1950 and 1961. At that time the available uranium ore was of low grade but later higher-grade ore was found and 50 tonnes of uranium were produced per year between 1965 and 1969. Although the uranium resource is substantial, production ceased in 1989 when world prices fell and made the exploitation uneconomic.

Thailand

Some exploratory drilling by the Government was made as early as 1935 near Mae Sot in Tak Province on the Thai–Burmese border. The oil shale beds are relatively thin, underlying about 53 km^2 in the Mae Sot basin and structurally complicated by folding and faulting.

Another deposit at Li, Lampoon Province is small, estimated at 15 million tonnes of oil shale and yielding 50–171 l of oil per tonne.

Some 18.7 billion tonnes of oil shale yielding an estimated 6.4 billion barrels of shale oil have been identified in the Mae Sot Basin, but to date it has not been economic to exploit the deposits. In 2000 the Thai Government estimated that total proved recoverable reserves of shale oil were 810 million tonnes.

Turkey

Although oil shale deposits are known to exist over a wide area in middle and western Anatolia, they have received relatively little investigation. Between 1993 and 1995 it was estimated that in-place resources of shale oil were 1 985 million barrels, but further research is needed to establish a body of data.

United States of America

It is estimated that nearly 78% of the world's potentially recoverable shale oil resources are concentrated in the USA. The largest of the deposits is found in the 42 700 km^2 Eocene Green River formation in north-western Colorado, north-eastern Utah and south-western Wyoming. The richest and most easily recoverable deposits are located in the Piceance Creek Basin in western Colorado and the Uinta Basin in eastern Utah. The shale oil can be extracted by surface and in situ methods of retorting: depending upon the methods of mining and processing used, as much as one-third or more of this resource might be recoverable. There are also the Devonian-Mississippian black shales in the eastern United States. The Green River deposits account for 67% of US shale oil resources, the eastern black shales for 16%.

Oil distilled from shale was burnt and used horticulturally in the second half of the 19th century in Utah and Colorado but very little development occurred at that time. It was not until the early 1900s that the deposits were first studied in detail by the US Geological Survey and the Government established the Naval Petroleum and Oil Shale Reserves, which for much of the 20th century served as a contingency source of fuel for the nation's military. These properties were originally envisioned as a way to provide a reserve supply of oil to fuel US naval vessels.

Oil shale development had always been on a small scale but the project that was to represent the greatest development of the shale deposits was begun immediately after World War II in 1946—the US Bureau of Mines established the Anvils Point oil shale demonstration project in Colorado. However, processing plants had been small and the cost of production high. It was not until the USA had become a net oil importer, together with the oil crises of 1973 and 1979, that interest in oil shale was reawakened. In the latter part of the 20th century military fuel needs changed and the strategic value of the shale reserves began to diminish.

In the 1970s ways to maximise domestic oil supplies were devised and the oil shale fields were opened up for commercial production. Oil companies led the investigations: leases were obtained and consolidated but one by one these organisations gave up their oil shale interests. Unocal was the last to do so in 1991.

Recoverable resources of shale oil from the marine black shales in the eastern United States were estimated in 1980 to exceed 400 billion barrels. These deposits differ significantly in chemical and mineralogical composition from Green River oil shale. Owing to its lower H:C ratio, the organic matter in eastern oil shale yields only about one-third as much oil as Green River oil shale, as determined by conventional Fischer assay analyses. However, when retorted in a hydrogen atmosphere, the oil yield of eastern oil shale increases by as much as 2.0–2.5 times the Fischer assay yield.

Green River oil shale contains abundant carbonate minerals including dolomite, nahcolite, and dawsonite. The latter two minerals have potential by-product value for their soda ash and alumina content, respectively. The eastern oil shales are low in carbonate content but contain notable quantities of metals, including uranium, vanadium, molybdenum, and others which could add significant by-product value to these deposits.

All field operations have ceased and at the present time shale oil is not being produced in the USA. Large-scale commercial production of oil shale is not anticipated before the second or third decade of the 21st century.

Chapter 4

Natural Bitumen and Extra-Heavy Oil

World Energy Council
CONSEIL MONDIAL DE L'ENERGIE

COMMENTARY

Introduction

Crude oil is found in sedimentary rocks throughout the world, except, thus far, in Antarctica. In many places the oil has been degraded, so that it is represented by viscous black oil that is difficult to recover, transport, and refine. Depending upon the degree of degradation the result is extra-heavy oil or, in the extreme case, natural bitumen. Except in Canada, precise quantitative reserves and oil-in-place data on a reservoir basis are seldom available because most countries and companies consider such information to be proprietary.

Natural bitumen is the oil contained in clastic and carbonate reservoir rocks, most frequently in small deposits at, or near, the earth's surface. These rocks are commonly referred to as tar sands or oil sands and have been mined since antiquity for use as paving. Occasionally such deposits are extremely large in areal extent and in contained resources, most notably those in northern Alberta, Canada. In 2003 only the Alberta bitumen deposits were being exploited as a source of crude oil.

Similarly, reservoirs containing extra-heavy oil are geographically widespread but only one such deposit is sufficiently large to have a major supply and economic impact. That deposit is the Orinoco Oil Belt in Eastern Venezuela. Nowhere else in the world is such a concentration of extra-heavy oil known or likely to exist.

Definitions of terms used in this commentary may be found immediately prior to Table 4.1. The resource definitions are those of the World Petroleum Congress-Society of Petroleum Engineers-American Association of Petroleum Geologists, with minor additions. One such addition, e.g. is the term Original Reserves, comprised of Proved Reserves plus Cumulative Production, which tends to place new and mature reservoirs on a more nearly comparable basis than either term alone.

Chemistry

Extra-heavy oil and natural bitumen represent crude oils which have been severely degraded by microbial action, as evidenced by their paucity of low-molecular-weight saturated hydrocarbons. Fig. 4.1 provides a comparison of salient attributes of crude oil and natural bitumen.

2004 Survey of Energy Resources

	Conventional oil	Medium oil	Heavy oil	Extra-Heavy oil	Natural bitumen
Reservoirs (number)	8 102	816	1 375	57	324
Depth to top of reservoir (ft)	5 140	3 284	3 200	3 628	484
Coke in crude oil (wt%)	10.4	17.6	21.8	28.1	
Asphalt in crude oil (wt%)	8.8	25.0	38.4	61.9	69.6
Gasoline yield (vol%)	9.2	2.8	2.0	1.3	1.4
Gas Oil yield (vol%)	17.4	21.9	15.9	16.9	7.2
Residuum yield (vol%)	21.9	39.5	52.6	62.6	18.1
Pour point of crude oil (°F)	16	9	20	66	89
Crude oil density (g/cm³)	0.836	0.920	0.958	1.018	1.041
Crude oil gravity (°API)	38.1	22.3	16.3	7.5	5.0
Crude oil dynamic viscosity (cP, 100°F)	9	63	593	7 936	292 991
Resins (wt%)	6.1	19.3	24.2	21.2	25.2
Asphaltenes (wt%)	2.1	6.6	12.4	13.2	30.6
Total BTEX volatiles (ppm)	10 157.4	4 909.0	2 487.1	N/A	N/A
Total VOC volatiles (ppm)	16 736.1	8 018.3	4 518.2	N/A	N/A
Nickel (ppm)	8.0	33.4	54.0	129.9	78.2
Vanadium (ppm)	18.2	88.2	170.9	777.7	183.0
Nitrogen (wt%)	0.1	0.2	0.5	0.6	0.7
Sulphur (wt%)	0.4	1.5	2.9	4.9	3.3

FIGURE 4.1 Comparative chemical analysis of world oils and natural bitumen.

The numbers of reservoirs involved show that many fewer analyses are available for extra-heavy oil and natural bitumen, which causes the averages to be weighted by a few large deposits with numerous analyses. Nevertheless, the chemical and physical differences among the oil types are clear. From conventional oil to natural bitumen there are increases in density; coke, asphalt, asphaltene, and resin contents; residuum yield; pour point; dynamic viscosity; and in the content of the metals nickel and vanadium and non-metals nitrogen and sulphur. Conversely, the API gravity, gasoline and gas oil yields, and volatile organic compounds (Benzene, Toluene, Ethylbenzene and Xylenes—BTEX and volatile organic compounds—VOC) all decrease. There is also a decrease in average reservoir depth. Very little of the extra-heavy oil and natural bitumen originated with these chemical attributes, which are rather the result of the degradation of originally conventional crude oils, with the consequent loss of most of their low-molecular-weight volatiles.

The degradation has resulted in crude oils which are very dense, highly viscous, and black. The degradation, principally bacterial, requires an active water supply to carry the bacteria, inorganic nutrients and oxygen, and to remove toxic by-products, such as hydrogen sulphide; contact with the reservoir containing the low-molecular-weight hydrocarbon food; and temperatures generally below about 200 °F (Barker, 1979). Other low-molecular-weight components are lost through water washing in the reservoir, thermal fractionation, and evaporation when the reservoir is breached at the earth's surface.

Resources

World summaries of natural bitumen and extra-heavy oil resources are given in Tables 4.1 and 4.2. Although natural bitumen and extra-heavy oil are worldwide in occurrence, a single extraordinary deposit in each category is dominant. The three Canadian oil sands deposits in Alberta together contain at least 63% of the discovered world total bitumen in place and constitute the only bitumen deposits that are economically recoverable as sources of synthetic oil. Additionally, Alberta has about 90% of the world's undiscovered or poorly known natural bitumen. Canada's known bitumen in place amounts to about 1 700 billion barrels. Similarly, the extra-heavy crude oil deposit of the Orinoco Oil Belt, a part of the Eastern Venezuela basin, represents about 98% of that known to be in place or some 2 000 billion barrels. Between them, these two deposits contain about 3 700 billion barrels of oil in place. These are only the remaining, degraded remnants of petroleum deposits that must have originally totalled as much as 18 000 billion barrels of oil in place.

Extra-heavy oil is recorded in 91 deposits. Some of these represent separate reservoirs in a single field, of which some are producing and others abandoned. The deposits are found in 21 countries, with 11 of the deposits being offshore or partially offshore.

Natural bitumen is found in 183 identified deposits in 21 countries. These are generally reported as tar sands or, in Canada, oil sands. Clearly, many more such deposits are identified but, as in the case of oil seepages, no resource estimate is possible. Very large resource deposits are known in eastern Siberia in the Russian Federation but insufficient data are available to make more than conservative-size estimates.

Two types of basins contain, respectively, most of the world's natural bitumen and extra-heavy oil and, indeed, contain about three-quarters of all the oil reserves in the world. These basins are architecturally similar, either lying within or accreted to continental cratons. In profile, the sediments are thick and strongly folded or rift-faulted in the seaward direction and become thinner and structurally higher as they encroach upon the craton. Oil is generated in the deeply buried, thick seaward sediments and migrates upward to be trapped adjacent to the craton. Biodegradation is promoted at the cratonic edge, where the sediments have been brought near to the earth's surface. This permits an influx of fresh water, providing oxidising conditions, and both evaporation and washing out of light, high API gravity oil components. Complete degradation results in highly viscous, very low-API gravity bitumen exemplified by the Alberta deposits. If the edge is reburied before the oil is completely degraded, the result is likely to be extra-heavy oil like that in the Venezuelan Orinoco Oil Belt, which is somewhat less viscous and of higher API gravity than the natural bitumen.

Production Methods

The chemical and physical attributes of extra-heavy oil lead to an array of problems with respect to exploitation, transportation, storage, and refining. This, of course, is reflected in the increased cost of extraction and processing and physical limitations on production capacity. Due to the high viscosity of the crude, some form of improved recovery is usually required for production. Steam injection has been common practice, in both vertical and lateral wells. A notable addition to recovery technology has been SAGD, or steam-assisted gravity drainage, combined with horizontal drilling. In this method a horizontal steam-injection well is drilled a few metres above a production well. A similar technology involves the injection of solvent rather than steam in the superjacent well. It is also common practice to inject a low-API gravity hydrocarbon fluid (frequently gas condensate) as a diluent into the reservoir to improve mobility. An important production improvement is recovery of cold heavy oil with sand (CHOPS). Cold production is achieved in Venezuela through horizontal lateral wells in combination with electric submersible pumps and progressing cavity pumps. Finally, efforts are continuing to improve production of viscous oil through down-hole electrical resistance heating.

Natural bitumen is immobile in near-surface reservoirs, where it can be recovered only by mining and surface separation of the bitumen from the rock. Where the bitumen is buried deeply enough to prevent severe heat loss, it may be produced from wells by the use of

FIGURE 4.2 Canadian Oil Sands Mine Truck & Shovel (Source: Suncor Energy Inc.).

steam injection from vertical wells, by taking advantage of horizontal well technology, or by utilising SAGD. In these cases the bitumen is actually extra-heavy oil. For the bitumen in reservoirs too deep to strip-mine economically but too shallow for steam injection from wells to be effective, a combination of mining and steam injection has been developed, with injection wells emplaced from within the mine tunnel, the oil being recovered by gravity drainage.

Most production schemes for both extra-heavy oil and natural bitumen entail the incorporation of upgrading facilities at or near the production sites. The benefit is the simplification of pipeline movement of the upgraded oil.

Upgrading and Refining

Two fundamental upgrading processes are presently employed to prepare heavy oil and natural bitumen for transportation and refining to finished products. These processes are carbon rejection and hydrogen addition. Each process improves the hydrogen-to-carbon ratio but by following different paths. Carbon rejection, such as Flexicoking, yields a large quantity of low-Btu gas at the expense of produced liquids, a large amount of petroleum coke, and therefore moderate conversion at low pressure. A hydrogen addition process, such as VEBA-Combi-Cracking (VCC) heats the raw material under pressure, the resulting gas being combined with added hydrogen to maximise liquids yield through high conversion. High conversion carries an economic penalty because of the cost of the added hydrogen and the high pressures required. The choice, therefore, is economic, being related to demand for the resulting products.

The yield of upgraded oil (synthetic crude oil) from the natural bitumen, based on the Alberta experience, varies with the technology employed, the consumption of product for fuel in the upgrader, the extent of natural gas liquids recovery, and the degree of residue upgrading. The Canadian company Suncor uses delayed coking for a yield of 0.81, whereas Syncrude (another Canadian company) obtains a yield of 0.85 through fluid coking combined with hydrocracking. The expected yield for the Albian Sands sub-project of the Shell/Chevron/Western Athabasca Oil Sands Project is 0.90, using hydrocracking.

Company	Project	Location	Recovery method	Company production 2002 (b/d)	Company production target (b/d)	Company production target (year)	Targets
Canadian Natural Resources	Brintnell, Pelican Lake	Athabasca In situ	Primary	7 791			
Canadian Natural Resources	Wabasca	Athabasca In Situ	Primary	21 270			
EnCana	Pelican Lake, Woodenhouse	Athabasca In Situ	Primary	13 656	35 000	2004	w/o Syncrude
Koch Exploration Canada	North Wabasca	Athabasca In Situ	Primary	185			
Canadian Natural Resources	Kirby Thermal	Athabasca In Situ	SAGD	223			
ConocoPhillips	Surmont (43.5%)	Athabasca In Situ	SAGD		100 000	2012	Total Surmont
Deer Creek	Jocelyn	Athabasca In Situ	SAGD		30 000	2012	Total Jocelyn
Devon	Dover	Athabasca In Situ	SAGD	1 470			
Devon	Surmont (13%)	Athabasca In Situ	SAGD				[Surmont]
Devon	Jackfish	Athabasca In Situ	SAGD		35 000	2003	Total Jackfish
EnCana	Foster Creek	Athabasca In Situ	SAGD	14 563			
EnCana	Christina Lake	Athabasca In Situ	SAGD	1 087			
Husky Oil	Kearl Lake	Athabasca In Situ	SAGD		100 000	2012	Total Kearl Lake
Japan Canada Oilsands	Hangingstone	Athabasca In Situ	SAGD	4 932	60 000	2012	Total Hangingstone
Nexen	Long Lake (50%)	Athabasca In Situ	SAGD/Upgrader		70 000	2007	Total Long Lake
Opti	Long Lake (50%)	Athabasca In Situ	SAGD/Upgrader				[Long Lake]
Petro-Canada	MacKay River	Athabasca In Situ	SAGD	6 672	30 000	2004	Total MacKay River
Petro-Canada	Meadow Creek	Athabasca In Situ	SAGD		80 000	2008	Total Meadow Creek
Suncor	Firebag	Athabasca In Situ	SAGD				[Suncor]
TotalFinaElf	Surmont (43.5%)	Athabasca In Situ	SAGD				[Surmont]
Canadian Natural Resources	Horizon	Athabasca Mining	Mining/Upgrader		232 000	2012	Total Horizon
Canadian Oil Sands Trust	Syncrude (21.7%)	Athabasca Mining	Mining/Upgrader	58 601			[Syncrude]
ChevronTexaco	Athabasca Oil Sands Project (20%)	Athabasca Mining	Mining/Upgrader	<100			[AOSP]
ConocoPhillips	Syncrude (9%)	Athabasca Mining	Mining/Upgrader	24 305			[Syncrude]
EnCana	Syncrude (15%)	Athabasca Mining	Mining/Upgrader	40 508			[Syncrude]
ExxonMobil	Kearl Mine	Athabasca Mining	Mining				[Kearl Mine]
Imperial Oil	Syncrude (25%)	Athabasca Mining	Mining/Upgrader	67 513	550 000	2012	Total Syncrude
Imperial Oil	Kearl Mine	Athabasca Mining	Mining		200 000	2012	Total Kearl Mine
Mocal	Syncrude (5%)	Athabasca Mining	Mining/Upgrader	13 530			[Syncrude]
Murphy Oil	Syncrude (5%)	Athabasca Mining	Mining/Upgrader	13 503			[Syncrude]
Nexen	Syncrude (7.2%)	Athabasca Mining	Mining/Upgrader	19 444			[Syncrude]
Petro-Canada	Syncrude (12%)	Athabasca Mining	Mining/Upgrader	32 406			[Syncrude]
Shell Canada	Athabasca Oil Sands (60%)	Athabasca Mining	Mining/Upgrader	<100	525 000	2012	Total AOSP
Suncor	Suncor Mine (Steepbank & Millenium)	Athabasca Mining	Mining/Upgrader	269 781	550 000	2010	Total Suncor
Synenco	Northern Lights	Athabasca Mining	Mining		80 000	2008	Total Northern Lights
TrueNorth Energy	Fort Hills (78%)	Athabasca Mining	Mining		[190 000]	2008	Total Fort Hills
UTS	Fort Hills (22%)	Athabasca Mining	Mining				[Fort Hills]
Western Oil Sands	Athabasca Oil Sands Project (20%)	Athabasca Mining	Mining/Upgrader	<100			[AOSP]
Baytex	Reita Lake	Cold Lake	Primary	2 300			
Bonavista Petroleum	Frog, Swimming, Irish, Elk Pt, Fishing	Cold Lake	Primary	4 320			
Canadian Natural Resources	16 projects	Cold Lake	Primary	32 119			
Crispin Energy	Mann Lake, Mann North	Cold Lake	Primary	222			
Devon	Manatokan, Tulabi John Lake	Cold Lake	Primary	4 896			
Husky Oil	Frog Lake, Cold Lake	Cold Lake	Primary	3 286			
Krang Energy	Lindbergh South	Cold Lake	Primary	100			
Murphy Oil	Lindbergh	Cold Lake	Primary	296			
Petrovera	Elk Point/Frog Lake/ Lindbergh	Cold Lake	Primary	4 391			
Ricks Nova Scotia	Beaverdam	Cold Lake	Primary	94			
Risa Energy	John Lake	Cold Lake	Primary	94			
Blackrock Ventures	Hilda Lake	Cold Lake	SAGD	500			
Blackrock Ventures	Orion	Cold Lake	SAGD		20 000	2007	Total Orion
Canadian Natural Resources	Primrose	Cold Lake	CSS/SAGD	32 007			
Canadian Natural Resources	Burnt Lake Crown Agreement	Cold Lake	SAGD	1 145			
Canadian Natural Resources	Wolf Lake Crown Agreement	Cold Lake	CSS/SAGD	4 615			
Husky Oil	Tucker	Cold Lake	SAGD		30 000	2006	Total Tucker
Imperial Oil	Cold Lake	Cold Lake	CSS	111 423	180 000	2008	Total Cold Lake
Blackrock Ventures	Seal I, II	Peace River	Primary	1 111	15 000	2005	Total Seal
Shell Canada	Peace River	Peace River	Pressure Pulse/SAGD	8 854	17 000	2005	Total Peace River

FIGURE 4.3 Canadian Oil Sands Projects, 2002.

Type	Company	Project	Sub-project	Initial production (b/d)	Supply (tonnes/year)
Orimulsion	PDVSA Bitor	South Korea	Orimulsion (2003 actual)		300 000
	PDVSA Bitor	South Korea	Orimulsion (design)		2 000 000
	PDVSA Bitor	Singapore	Orimulsion (signed)		
	PDVSA Bitor	Italy	Orimulsion (signed)		
	PDVSA Bitor	China	Orimulsion (signed)		
	PDVSA Bitor	Canada	Orimulsion (agreed)		
	PDVSA Bitor	Thailand	Orimulsion (proposed)		
	PDVSA Bitor	Philippines	Orimulsion (proposed)		
Extra-heavy oil	ConocoPhillips/PDVSA	Petrozuata	Produce, transport, upgrade (2003)	120 000	
	ExxonMobil/PDVSA/Veba	Cerro Negro	Produce, transport, upgrade (2003)	121 000	
	TotalFinaElf/PDVSA/Statoil	Sincor	Produce, transport, upgrade (design)	200 000	
	ConocoPhillips/ChevronTexaco / PDVSA	Hamaca	Produce, transport, upgrade (2003)	85 000	
	Ameriven (ARCO/PDVSA/ ConocoPhiliips/ChevronTexaco)	Hamaca	Produce, transport, upgrade (2003)	41 000	
	Ameriven	Hamaca	Produce, transport, upgrade (design)	165 000	

FIGURE 4.4 Venezuelan Extra-Heavy Oil Projects.

Conclusions

The recoverable volumes of oil contained in deposits of extra-heavy oil and natural bitumen are immense. If oil in place for conventional oil is estimated to be about three times original reserves, then remaining oil in place, after deduction of cumulative production, is about 4 925 billion barrels. On the same basis, the remaining oil in place in Venezuela's Orinoco Oil Belt is about 1 968 billion barrels, plus 235 billion barrels of contingent resources, and, in northern Alberta, Canada, an additional 1 455 billion barrels of natural bitumen plus 917 billion barrels of bitumen in place in the less well known carbonate deposits. Future deposits on the scale of the Orinoco Oil Belt are not expected but additional exploitable natural bitumen deposits are known in the Russian Federation, the United States, and elsewhere.

Relatively small portions of the two major deposits are currently being produced, with this production increasing annually. New technologies have allowed production rates comparable to those of conventional oil reservoirs. Major cost-reduction breakthroughs in upgrading, transportation, and refining are enhancing the movement of these hydrocarbons into mainstream world oil supply.

Richard F. Meyer
Emil Attanasi
US Geological Survey

References

Alberta Department of Energy, 2003. Alberta's oil sands: www.energy.gov.ab.ca

Barker, C., 1979. Organic geochemistry in petroleum exploration: American Association of Petroleum Geologists Education Course Note Series #10, 159p.

Harrison, R.S., 1984. Geology and production history of the Grosmont carbonate pilot project, Alberta, Canada. In: Meyer, R.F., Wynn, J.C., Olson, J.C. (Eds.), The Second UNITAR International Conference on Heavy Crude and Tar Sands, Caracas, February 7–17, 1982, McGraw-Hill, New York, pp. 199–204.

Luhning, R.W. et al., 2002. Pipeline transportation of emerging partially upgraded bitumen: CIPF Conference, Calgary, AB, Canada, June 11–13, 2002, pp. 1–18.

Marchant, L.C., 1988. US tar sand oil recovery projects–1985. In: Meyer, R.F. (Ed.), The Third UNITAR/UNDP International Conference on Heavy Crude and Tar Sands, Long Beach, CA, July 22–31, 1985: Edmonton, Canada, Alberta Oil Sands Technology and Research Authority, pp. 931–940.

McPhee, D. and Ranger, M.J., 1998. The geological challenge for development of heavy crude and oil sands of western Canada: UNITAR International Conference on Heavy Crude and Tar Sands, 7th, Beijing, October 27–30, proceedings, 10p.

Oil & Gas Journal Online, 2003a. Athabasca oil sands project opens in Canada, June 23.

Oil & Gas Journal Online, 2003b. US House passes tar sands leasing bill, October 9.

Oil & Gas Journal Online, 2003c. PDVSA-BITOR and KOSPO of South Korea signed Orimulsion® supply contract, April 16.

Oil & Gas Journal Online, 2003d. New techniques improve heavy oil production feasibility, October 19.

Walters, E.J., 1974. Review of the world's major oil sands deposits, In: Hills, L.V. (Ed.), Oil sands: fuel of the future: Canadian Society of Petroleum Geologists Memoir 3, pp. 240–263.

DEFINITIONS

In Tables 4.1 and 4.2 the following definitions apply:

Original oil in place: Discovered original oil in place.

Reserves: Those amounts of oil commonly reported as reserves or probable reserves, generally with no further distinction; only in Canada are reserves reported separately as recoverable by primary or enhanced methods. Russian A, B, and C_1 reserves are included here. The term reserve generally refers to quantities of petroleum that are anticipated to be recoverable from known accumulations.

Contingent resources: Quantities of petroleum estimated to be potentially recoverable from known accumulations but not commercially recoverable at the time of reporting, including, in Russia, C_2 deposits.

Undiscovered original oil in place: The original oil in place in undiscovered deposits. This category also includes material that is identified but is too poorly known to be considered as discovered.

Prospective resources: Quantities of petroleum potentially recoverable from undiscovered deposits. This category includes some oil categorised by the authors as possible, speculative, undiscovered recoverable or, in Russia, C_3, D_1, and D_2.

Original reserves: Reserves plus cumulative production. This category includes oil that is frequently reported as estimated ultimate recovery, particularly in the case of new discoveries.

Cumulative production: Total of production to latest date.

Annual production: Production for latest year reported.

Conventional oil: API gravity above 25° (density below 0.904 g/cm^3).

Medium oil: API gravity 20–25° (density 0.934–0.904 g/cm^3).

Heavy oil: API gravity 10–20° (density 0.934–1.000 g/cm^3).

Extra-heavy oil: API gravity below 10° (density above 1.000 g/cm^3).

Natural bitumen: Dynamic viscosity above 10 000 mPa s. (Natural bitumen is immobile in the reservoir. Because of lateral variations in chemistry as well as in depth, and therefore temperature, many reservoirs contain both extra-heavy oil, and occasionally heavy oil, in addition to natural bitumen).

Oil Field: A geographic area below which are one or more discrete reservoirs from which petroleum is produced. Each reservoir may be comprised of one or more zones, the production from which is commingled. The production of the reservoirs themselves may be commingled, in which case production and related data cannot be distinguished. This is a vexing problem if one or more of the reservoirs contains heavy or extra-heavy oil and others, medium or light oil.

Table Notes

The data in the tables are largely estimates by Richard Meyer of the US Geological Survey. They have been based upon a detailed review of the literature combined with available databases, and suggest (but do not define) the resource volumes that could someday be of commercial interest.

TABLE 4.1

Natural Bitumen: resources, reserves and production at end-2002

	Fields (number)	Reservoir depth (average, m)	Original oil in place—undiscovered (million barrels)	Prospective resources (million barrels)	Original oil in place—discovered (million barrels)	Cumulative oil production (million barrels)	Reserves (million barrels)	Original reserves (million barrels)	Contingent resources (million barrels)	Annual oil production (million barrels)
Angola	2				4 511		465	465		
Congo (Brazzaville)	1				214		6	6		
Congo (Dem. Rep.)	1				600		30	30		
Madagascar	1	10			1 750		221	221	192	
Nigeria	1		32 580		421 124		574	574	41 538	
Total Africa	**6**		**32 580**		**428 199**		**1 296**	**1 296**	**41 730**	
Canada	18	329	917 141		1 633 307	3 591	174 951	178 542	258 220	271
USA	85	89	19 775		42 119	64		64	148	
Total N. America	**103**		**936 916**		**1 675 426**	**3 655**	**174 951**	**178 606**	**258 368**	**271**
Peru	1				132					
Trinidad & Tobago	16				1 127				67	N
Venezuela	1				62					
Total S. America	**18**				**1 321**				**67**	**N**
Azerbaijan	3				90		<1	<1		
China	4	2			1 593					
Georgia	1				630		3	3		
Indonesia	1				8 912	24	422	446		3
Kazakhstan	6	36			252 922		42 009	42 009		
Total Asia	**15**				**264 147**	**24**	**42 434**	**42 458**		**3**
Germany	1	250			220	<1		<1		
Italy	1				1 260		210	210		
Russian Federation	36	116	51 345	10 650	202 087	14	28 386	28 400		
Switzerland	1				10					
Total Europe	**39**		**51 345**	**10 650**	**203 577**	**14**	**28 596**	**28 610**		

(continued on next page)

TABLE 4.1 (*Continued*)

	Fields (number)	Reservoir depth (average, m)	Original oil in place—undiscovered (million barrels)	Prospective resources (million barrels)	Original oil in place—discovered (million barrels)	Cumulative oil production (million barrels)	Reserves (million barrels)	Original reserves (million barrels)	Contingent resources (million barrels)	Annual oil production (million barrels)
Bahrain	1				320					
Syria (Arab Republic)	1				13				1	1
Total Middle East	**2**				**333**				**1**	**1**
Total World	**183**		**1 020 841**	**10 650**	**2 573 003**	**3 693**	**247 277**	**250 970**	**300 166**	**275**

Source: R.F. Meyer, US Geological Survey.

TABLE 4.2

Extra-Heavy Oil: resources, reserves and production at end-2002

	Fields (number)	Of which: fields offshore (number)	Reservoir depth (average, m)	Original oil in place—discovered (million barrels)	Cumulative oil production (million barrels)	Reserves (million barrels)	Original reserves (million barrels)	Contingent resources (million barrels)	Annual oil production (million barrels)
Egypt (Arab Rep.)	1		594	500	<1	50	50		<1
Total Africa	**1**			**500**	**<1**	**50**	**50**		**<1**
Canada	5		916						
Mexico	2		2 499	60	5	1	6		<1
USA	41	1	2 101	2 801	175	16	191		1
Total N. America	**48**	**1**		**2 861**	**180**	**17**	**197**		**1**

	Fields (number)	Of which: fields offshore (number)	Reservoir depth (average, m)	Original oil in place—discovered (million barrels)	Cumulative oil production (million barrels)	Reserves (million barrels)	Original reserves (million barrels)	Contingent resources (million barrels)	Annual oil production (million barrels)
Colombia	1		2 335	145	9	16	25		2
Cuba	1	1	1 500	960	27	50	77		2
Ecuador	3		2 462	438	30	25	54		
Peru	2		2 956	66	15	5	20		2
Trinidad & Tob.	1		76	300					
Venezuela			1 341	2 027 271	12 026	47 218	59 244	235 440	181
Total S. America	**8**	**1**		**2 029 180**	**12 107**	**47 314**	**59 420**	**235 440**	**187**
China	1		600	1 500	137	463	600		
Indonesia	1		169	<1	<1		<1		
Uzbekistan	1								
Total Asia	**3**			**1 500**	**137**	**463**	**600**		
Albania	3		1 176	2 174	179	83	261		4
Italy	14	6	2 219	2 371	152	81	234		4
Poland	2		767	12					
Russian Federation	6		1 716						
United Kingdom	2	2	1 562	2 428	918	96	1 014		9
Total Europe	**27**	**8**		**6 985**	**1 249**	**260**	**1 509**		**16**
Iran (Islamic Republic)	2	1	871	23 030		250	250		
Iraq	1								
Israel	1		1 551	<1	<1	<1	<1		<1
Total Middle East	**4**	**1**		**23 030**	**< 1**	**250**	**250**		**< 1**
Total World	**91**	**11**		**2 064 056**	**13 674**	**48 354**	**62 026**	**235 440**	**205**

Source: R.F. Meyer, US Geological Survey.

COUNTRY NOTES

The Country Notes on Natural Bitumen and Extra-Heavy Oil have been compiled by the commentary authors and the editors. Since 2001 there has been considerable activity on both the Canadian and Venezuelan fronts.

In addition to material provided by the commentary authors, information has been drawn from the companies directly involved with the resource extraction and from national and governmental organisations. Recourse has also been made to the papers given at the 7th UNITAR International Conference on Heavy Crude and Tar Sands (1998).

Albania

Three of Albania's oil fields contain extra-heavy oil, with perhaps 4 million barrels of annual production. The fields lie in the Durres Basin, a continental interior basin. In addition, the Selenzza natural bitumen deposit lies in the immediate area of the oil fields. This deposit, the most extensive of European bitumen deposits, contains an estimated 371 million barrels of bitumen in place (Walters, 1974).

Angola

Two natural bitumen deposits are located in the Cuanza Basin, Bengo Province. They contain about 4.5 billion barrels of bitumen in place but have not been worked as an energy source and are not likely to be. When conventional oil resources have been exhausted and the political stability required for mining facilities is established, exploitation could be an option.

Azerbaijan

The natural bitumen deposits are small and will probably never serve as sources for energy. They fall within the South Caspian Basin. The best known of the three is Cheildag, near Baku oil field, which has been reported frequently and contains an estimated 24 million barrels of oil in place (Walters, 1974).

Bahrain

One small natural bitumen deposit is found in Bahrain. It lies within the enormously productive Arabian Basin.

Canada

Resource information for Alberta bitumen deposits is derived from Alberta Energy and Utilities Board (2002), supplemented by estimates of undiscovered resources for Peace River (Harrison, 1984) and Athabasca (McPhee and Ranger, 1998 and Harrison, 1984). The deposits are found in Lower Cretaceous sandstones and in the Mississippian and Devonian carbonates unconformably overlain by the Lower Cretaceous. The oil sands occur along the up-dip edge of the Western Canada Sedimentary Basin.

East of the Athabasca and Cold Lake deposits, in Alberta and Saskatchewan, large quantities of heavy and medium oil are found in the Lower Cretaceous sandstones but occurrences of extra-heavy oil are few and of little economic importance.

The data in Fig. 4.3, derived from the Alberta Department of Energy (2003), represent oil sands production, by company, for 2002. Many of the companies are participants in major oil sands projects, as shown by their percentage shares in the projects. The projected future production from the projects appears as the target production for the target year.

Substantial production in the Athabasca in-situ and Cold Lake areas is primary. This bitumen is of sufficiently low viscosity to permit it to flow, albeit with two to three times the amount of sand accompanying heavy oil production. Most primary production is outside the oil sands projects and much of it falls within the conventional royalty regime. On the other hand, nearly all the oil sands project production falls within the oil sands royalty regime, permitting low royalties until capital costs are recovered. Of the oil sands royalty production, 65.5% is from mining projects, 22.8% from in situ, thermally assisted projects, and 11.7% is primary.

If all the currently proposed and approved projects move forward, production from oil sands would be about 3 million barrels per day in 2012. This amount would be supplemented by projects presently on hold, by new projects, and by continued primary production.

The National Energy Board (NEB) distinguishes between two types of non-conventional oil obtained from deposits of oil sands, defining them as follows:

- Bitumen (also known as crude bitumen)—'a naturally occurring viscous mixture, mainly of hydrocarbons heavier than pentanes that may contain sulphur compounds and other minerals, and that in its natural viscous state is not recoverable at a commercial rate through a well'.
- Upgraded Crude Oil (also known as synthetic crude)—'a mixture of hydrocarbons similar to light crude oil derived by upgrading oil sands bitumen'.

Canada's 'original reserves' (corresponding to NEB's 'initial established reserves') of oil sands bitumen are given in Table 4.1 as 178.5 billion barrels, of which 'cumulative production' is about 3 600 million barrels. Only some 11% of the 'original oil in place discovered' is regarded as recoverable.

Within these huge resources, the 'remaining established reserves' (representing only 'those recoverable from developed experimental/demonstration and commercial projects') at end-2002 have been assessed by the Canadian Association of Petroleum Producers (CAPP) as 775.6 million m^3 of mining-integrated SCO and 321.7 million m^3 of in situ bitumen.

Although the existence of oil sands deposits was noted in the 18th century, it was not until 1875 that a complete survey was undertaken and it was the 20th century before exploitation was embarked upon. The deposits range from being several hundred metres below ground to surface outcroppings. The extraction of bitumen from the oil sands was initially based on surface-mining but in situ techniques became necessary in order to reach the deeper deposits.

There was much experimentation with oil sands technology in the first half of the 20th century but it was not until the effects of the economic climate of the 1950s and early 1960s began to be felt that commercial development became viable. The Government of Alberta's oil sands development policy was announced in 1962.

There are now many different oil sands projects, both mining and in situ, at different stages of development. The following are the main ventures:

Suncor: control of the Great Canadian Oil Sands (GCOS) project passed to Sun Oil Company in 1963 and in 1967 the world's first integrated oil sands production and upgrading plant was started up by Suncor (formerly Sun Oil). Suncor's area of operation, 40 km north of Fort McMurray, is within the Athabasca deposits. The processing capability of the original Oil Sands Plant has been steadily increased so that by 2002 production had reached a record average of 205 800 b/d. Sales in 2002 amounted to 104.7 tb/d light sweet crude oil, 23.0 tb/d diesel, 68.3 tb/d of light sour crude oil and 9.3 tb/d bitumen.

At the beginning of 1999 the company announced its 'Project Millennium', a phased series of expansions to the Steepbank mine (on the opposite side of the Athabasca River), adding bitumen extraction plants and increasing upgrader capacity. December 2001 saw the project, with a design capacity of 225 000 b/d, move from construction to production.

In early 2000, the establishment of a four-stage in-situ project at Firebag (40 km north-east of the Oil Sands Plant) was announced. The first stage is on schedule and is expected to begin commercial bitumen production in 2004. By mid-2005 it is intended that full production from the first stage will be 35 000 b/d of bitumen. It is planned that the first stage of Firebag, in conjunction with the further development of the upgrading facility, will result in Suncor production capacity reaching 260 000 b/d in 2005. In February 2003 it was reported that development of the second stage was underway and, on completion of all four stages, Firebag's combined bitumen production potential will be 140 000 b/d.

Suncor's plan is to increase overall production capacity to 500 000–550 000 b/d in 2010–2012. To this end the company announced (in late 2001) its multi-phased Voyageur growth strategy encompassing the Firebag project. Details of the first phase, released in April 2003, stated that the increased production capacity would reach 330 000 b/d by late-2007.

To achieve greater sustainability whilst vastly reducing the impact on the environment, Suncor is employing Steam Assisted Gravity Drainage (SAGD) technology. SAGD uses underground wells to inject steam into the oil sands deposits thereby collecting the bitumen released by heat. Additionally, all water used in the Firebag project will be recycled. To enable the impact on the environment to be reduced even further, Suncor is investigating ways in which CO_2 and light hydrocarbons could be injected into the deposits in order to decrease emissions and achieve a reduction in the quantities of natural gas used to generate steam. Another possibility being studied is the use of the in situ reservoir as a permanent repository for GHG.

Syncrude: is a joint venture with eight participants: Imperial (a subsidiary of ExxonMobil, with a 25% holding), Mocal, Murphy, Nexen, ConocoPhillips, Petro-Canada, EnCana and Canadian Oil Sands Trust (COS-Trust—which became the majority shareholder in early 2003). As at October 2003, COS-Trust had a 35.49% share. Imperial Oil operates the Lake Mildred plant, also 40 km north of Fort McMurray. Production began in 1978 and, using open-pit mining methods, the shallow deposits are recovered for bitumen extraction and for the production of an upgraded light sweet crude oil (Syncrude Sweet Blend® or SSB). SSB possesses no residual bottoms and at the present time the crude is rated with the lowest sulphur content in North America. In 2002 Syncrude shipped 83.8 million barrels of SSB.

Syncrude 21, an expansion project launched in 1996, is designed to take production in five stages to 200 million barrels per year by 2013–2015.

Stages 1 and 2 have been completed, bringing production from 73.5 million barrels/yr in 1996 to 81.4 million barrels/yr in 1999 and then to 94 million barrels/yr by 2001. Stage 2 included the start-up of the Aurora mine—a 35 km extension from Lake Mildred that began operating in July 2000. The mine's output is partially processed on-site and then pipelined (using hydrotransport technology) to the upgrader for further treatment.

Stage 3 (2001–2005) is currently underway. It is planned that the design capacity will be increased to 128 million barrels/yr by 2005. A second bitumen production train at the Aurora mine, scheduled for completion in the fourth quarter 2003, will feed the first phase of expansion of the Lake Mildred upgrader (completion expected first quarter 2005).

Stage 4 (2006–2010) is planned first to increase production to 150 million barrels/yr, secondly to bring into operation a new production train at Aurora South and thirdly to complete the second phase of expansion to the Lake Mildred upgrader. By 2010 processing facilities will also have been upgraded to increase the yield and quality obtained from the bitumen.

At end-2003 Stage 5 (2011–2015) was still in a planning mode. It would involve the addition of a new bitumen production train at Aurora South and further upgrading of the Lake Mildred plant.

Imperial: the Cold Lake oil sands deposits area is operated by Imperial Oil. The company began commercial development in 1983 and has since gradually expanded facilities in a series of phases. The first 10 phases were completed by 1996 and in 2002 production of bitumen and syncrude were 112 000 and 57 000 b/d, respectively.

Regulatory approval for phases 11–13 was sought in 1997, granted in 1999 and a decision made to proceed taken in 2000. Production began in late-2002, enabling an estimated additional 30 000 b/d of crude bitumen to be brought on line. In May 2003 it was reported that production was averaging 140 000 b/d.

In 2002, Imperial sought approval for phases 14–16. If the necessary approval is granted and market conditions are favourable enough to allow development to occur, bitumen production would again increase by 30 000 b/d in 2007. By 2010 production could reach 180 000 b/d.

The majority of Cold Lake production is taken by US refineries, with some sent to the heavy oil upgrader in Lloydminster (Saskatchewan) and also to Imperial's own refineries.

Imperial also plans to proceed with its Kearl Oil Sands Project (some 70 km north of Fort McMurray). Environmental surveys are being undertaken and if the mine and upgrading facilities are proceeded with, production could reach 200 000 b/d.

Shell: Commercial production of Shell Canada's Peace River in situ deposits (Northwestern Alberta) began in 1986. Bitumen production capacity is set at approximately 12 000 b/d but during first half 2000, output was running at only 4 300 b/d. Over the period August 2001–December 2002 Shell drilled 33 new production wells on four drilling pads using horizontal cyclic steam technology and over the first 10 months of 2003 was producing an average of 9 700 b/d. Although the company is continuing to assess the potential for future development, it has not yet made a decision to enlarge the project.

The Athabasca Oil Sands Project (AOSP), a joint venture between Shell Canada (the majority shareholder, with 60%), Chevron Canada and Western Oil Sands will, on completion, be capable of supplying 10% of Canada's oil needs.

The project consists of two main components: the Muskeg River Mine (75 km north of Fort McMurray), operated by Albian Sands Energy Inc. (a new company formed by the joint venture) and the Scotford Upgrader (some 500 km south and adjacent to Shell's Scotford Refinery), operated by Shell.

The oil sands deposit at the Muskeg River Mine is both close to the surface and contains a high concentration of oil. The plant is designed to produce 155 000 b/d of bitumen. Using hydrogen-addition technology to upgrade the high-viscosity bitumen, the Scotford Upgrader will upgrade the extra-heavy oil into a range of premium quality low-sulphur and low-viscosity SCOs. These will be refined to make high quality transport fuels and other products.

The Corridor Pipeline system connecting the Mine to the Upgrader consists of 2 pipelines, one capable of transporting 215 000 b/d of diluted bitumen and one capable of transporting 65 000 b/d of diluent back to the Mine for re-use.

Construction of the Mine was completed in mid-2002; the Pipeline followed during third quarter 2002, the Upgrader being completed during fourth quarter 2002. Production of bitumen began in December 2002. Development of the project continued throughout 2003 as Trains 1 and 2 of the Mine were brought into service during the first half of the year. Trains 1 and 2 of the Scotford Upgrader took delivery of the bitumen and production of SCO began. By June 2003 all facilities at both Mine and Upgrader were fully operational. Production during third quarter 2003 averaged 115 000 b/d as the project moved towards achieving the design capacity of 155 000 b/d.

It has been reported that a possible expansion of the Muskeg River plant to 250 000 b/d could occur in 2006–2007. The application for regulatory approval of the 200 000 b/d Jackpine Mine submitted by Shell Canada on behalf of the joint venture owners was completed in October 2003. It has been reported that the Mine could ultimately be expanded to 300 000 b/d.

Phase 1 of Jackpine, east of the Muskeg River Mine, would have the capacity to produce in the region of 200 000 b/d of bitumen. Phase 2 would access further reserves enabling capacity to increase by 100 000 b/d. If the plan is adhered to, the timescale for startup production from the Jackpine Mine is estimated as 2008–2010.

Petro-Canada: the first oil from Petro-Canada's in-situ MacKay River development was produced in fourth quarter 2002. It is expected that full production will be attained by end-2003 at a rate of 30 000 b/d. There are further development opportunities at Meadow Creek (45 km south of Fort McMurray), which could be commissioned and started up during 2006. A decision on whether to proceed, together with a decision regarding the upgrading of the company's Strathcona refinery in Edmonton, are expected by end-2003.

EnCana: currently has two oil sands projects: the in-situ Foster Creek Thermal Project and a pilot plant at the Christina Lake Thermal Project.

Foster Creek was producing an average of about 22 000 b/d in third quarter 2003. It is planned that within Phase 1 production will be increased to about 30 000 b/d in 2004. Following construction of Phase 2, production would be further increased by 2006 resulting in a capacity of 100 000 b/d. In May 2003 the Foster Creek co-generation plant began delivering electricity into the Albertan power system.

The first phase of the Christina Lake pilot project is producing about 3 500 b/d using SAGD technology. The design capacity of the project is 10 000 b/d. EnCana carries out evaluation and development methods to aid production and to this end has recently applied to use solvent aided process (SAP), an enhancement to SAGD, in which viscosity can be lowered.

ConocoPhillips: regulatory approval has been granted for ConocoPhillips to develop and operate its Surmont in situ oil sands deposit. If the project proceeds, it is planned that production could begin in 2006 and ultimately reach 100 000 b/d.

Jacos: approximately 50 km southwest of Fort McMurray lies the Japan Canada Oil Sands (JACOS)-operated SAGD pilot plant at Hangingstone lease. Phase 1 of the demonstration project started producing in 1999 with Phases 2 and 3 following in 2000 and 2002 respectively. By end-2002 production had reached 10 000 b/d. If it is decided to develop a commercial SAGD plant, production could be expected to be up to 50 000 b/d and to begin by late 2005.

Opti Canada (a member of the ORMAT group) and Nexen, its joint-venture partner (50/50), are developing the Long Lake Project (about 40 km southeast of Fort McMurray). The initial application for approval was for a 70 000 b/d bitumen extraction and upgrader plant. This was subsequently amended to include an additional 70 000 b/d of upgrading capacity. Since April 2001 a 500 b/d demonstration upgrader (near Cold Lake) has been successfully operating. However, the plan is for commercial production to begin in 2006 and the upgrader to come into operation the following year. This will then represent the first integrated SAGD and field upgrading operation: the bitumen production will be sent to Opti Canada's proprietary OrCrude™ process where it will be partially upgraded. In addition, the liquid asphaltenes concurrently produced will be sent to a gasifier and converted into a synthetic hydrogen-rich gas. The separated hydrogen will be utilised in a hydrocracker, completing the upgrading of the bitumen into a high-grade synthetic oil. The remaining syngas will be used for power generation, thus ensuring almost complete power self-sufficiency.

TrueNorth: as operator, TrueNorth was granted approval to develop the Fort Hills Project located about 90 km north of Fort McMurray. The application was for bitumen production of up to 235 000 b/d. However, as at end-2003 development has been deferred and it has been reported that an amended application for a smaller plant was being considered.

Canadian Natural Resources (CNRL): Primrose/Wolf Lake in-situ Oil Expansion Project was granted permission during 2002 to expand the current production of 40 000 b/d to more than 120 000 b/d. The first stage of expansion, beginning in 2003, will take place over a period of 15–20 years, as deemed economically viable.

An application has been made for CNRL's Horizon Oil Sands Project. It is a multi-phase 5-year project that includes a mine and integrated upgrader. The first synthetic oil production (at a rate of 110 000 b/d) is planned for first-half 2008. By 2010 an additional 45 000 b/d SCO would be added and by 2012, production would be in excess of 230 000 b/d. CNRL is currently developing ancillary facilities.

Fort McKay First Nation: although the 35 000 b/d Fort McKay First Nation oil sands project has been publicly debated, no firm details have emerged for its development.

Devon Energy: if Devon Energy's application for its Jackfish SAGD Project (October 2003) is approved, then the 35 000 b/d (design capacity) plant (located near Conklin) could produce its first oil in 2006/2007. As with many other oil sands operators, Devon is undertaking technological research in order to reduce the quantity of natural gas (and thus CO_2) required during conventional SAGD operations. In this instance Vapex technology, using vaporised solvents in place of steam to extract the in-situ oil, is being employed.

Synenco's Northern Lights Project: located northeast of Fort McMurray, is planned to be an integrated plant incorporating mining, bitumen extraction and upgrading. During 2003 development of plans for the 100 000 b/d, 41° API SCO plant continued. It has been reported that the anticipated start-up date is 2008.

Husky Energy is involved with two in-situ oil sands projects: the Kearl Project and the Tucker Thermal Project. A third property, Caribou, is not currently being developed.

It is estimated that the Kearl Project will first produce 50 000 b/d during 2006/2007, followed by staged, incremental expansions of 50 000 b/d. By 2012 total production is expected to be between 100 000 and 150 000 b/d.

If the Tucker Thermal Project is approved, construction could begin in 2004, with commercial production of around 30 000 b/d by 2005/2006.

Deer Creek: Phase 1 of the Joslyn Oil Sands Project is currently being constructed by Deer Creek Energy. It is anticipated that start-up of Phase 1 of the SAGD plant will occur in 2004, with full production of 600 b/d in 2005. If Phase 2 receives approval, it is expected that construction to raise production to 10 000 b/d will begin in 2004, start up in 2006 and full production in 2007. Further phases could add up to 60 000 b/d by 2012.

BlackRock is the operator of the Hilda Lake (Orion) Project. The company has successfully operated a 5-year SAGD pilot project and has now applied for approval of the commercial development of the deposit. This would involve an expansion in production to 20 000 b/d.

Of the Canadian total synthetic and bitumen production, the former contributes approximately 59% and the latter 41%. Together they represent some 26% of Canada's total production of crude oil and NGL (as at end-2002).

China

A small amount of extra-heavy oil is present in one field, Liaohe Shuguang, in the Huabei Basin. Four natural bitumen deposits have been identified in the Junggar Basin with resources of about 1.6 billion barrels of bitumen in place.

Colombia

The basins of Colombia are rich in heavy oil but extra-heavy oil is identified in only a single field. Numerous oil seepages and small deposits of natural bitumen, especially in the Middle and Upper Magdalena Basins, also characterise the country. None of these deposits appears to be sufficiently large to be of economic importance as a source for synthetic oil.

Congo (Brazzaville)

Heavy oil is found in reservoirs offshore Congo but no extra-heavy oil is known. The natural bitumen deposit at Lake Kitina in the Cabinda Basin has been exploited as road material.

Congo (Democratic Republic)

A natural bitumen deposit occurs in the Democratic Republic of Congo, in the Cabinda Basin near the border with Cabinda. It has served as a source of road material, with nearly 4 000 metric tons having been extracted in 1958. This deposit is not likely to become a source of synthetic oil.

Ecuador

Ecuador is endowed with large amounts of heavy oil but only a small amount, all in the Oriente Basin, is extra-heavy. Natural bitumen is restricted to scattered oil seepages.

Egypt (Arab Republic)

Many fields containing heavy oil are found in Egypt but very little of this is extra-heavy.

Georgia

The only significant natural bitumen deposit in Georgia is in the South Caspian basin, at Natanebi. Neither heavy nor extra-heavy oil are known in Georgia, although conventional oil has been produced there for more than a century.

Germany

Heavy oil is produced from many fields in Germany but extra-heavy oil has not been reported. Extremely viscous natural bitumen is present in the Nordhorn deposit, in the North-west German basin.

Indonesia

In Indonesia a very small amount of extra-heavy oil is reported from a single reservoir in the east Java Basin, although many fields produce heavy oil. Natural bitumen occurs in the well-known Buton Island deposit. This has long been utilised as a source of road asphalt.

Iran (Islamic Republic)

Two fields in Iran contain extra-heavy oil, one (F-Structure) being an offshore discovery. The other, Kuh-e-Mund, was discovered in 1931 and was still producing fifty years later. A number of Iranian fields produce heavy oil.

Iraq

Oil seepages have been known and utilised in Iraq throughout historical times, but are insufficient to serve as sources of synthetic oil. Although heavy oil fields are productive in the country, very little extra-heavy oil is present.

Israel

Little more than a trace amount of extra-heavy oil is known from Israel. Natural bitumen occurs only as the Dead Sea asphalt, blocks of which occasionally rise to the surface.

Italy

The 234 million barrels of original reserves of extra-heavy oil in Italy have evolved in four separate basins, similar geologically to the Durres basin of Albania. Most important of these is the Caltanisetta Basin, mostly offshore and including the Gela field. These basins are all found in the foredeep portion of the basins, where the sediments are thickest and most structurally disturbed. The viscous nature of the oil, the offshore situation, and the limited resources create enormous problems in the exploitation of these fields.

Kazakhstan

Although Kazakhstan possesses large resources of conventional and heavy oil, it contains little if any extra-heavy oil. It does have significant resources of natural bitumen (Table 4.1) in the North Caspian Basin. As with nearly all the large natural bitumen deposits, the geological setting, like that of the Western Canada Sedimentary Basin, is conducive to the development of natural bitumen. In the light of the very large resources of conventional oil and natural gas in this country, exploitation of the bitumen as a source of synthetic oil is unlikely in the foreseeable future.

Madagascar

Bemolanga is the only natural bitumen deposit in Madagascar. It is large, but attempts to produce it for synthetic oil have thus far failed. A large heavy-oil deposit, Tsimiroro, has similarly been the subject of a number of unsuccessful production tests, but no extra-heavy oil is present in the country.

Mexico

Mexico, with numerous heavy oil fields, includes very few extra-heavy oil reservoirs. The latter are small in resources and production. Oil seepages are common in the country but no large natural bitumen deposits are present.

Nigeria

Natural bitumen in place possibly totalling as much as 450 billion barrels is located in southwestern Nigeria, in the Ghana Basin. This extensive deposit has not yet been evaluated as a source of synthetic oil and its exploitation will no doubt be delayed as long as Nigeria is a leading producer of conventional oil.

Peru

Peru contains numerous heavy oil deposits, mostly in the Oriente Basin. However, its extra-heavy oil and natural bitumen deposits are small and of little economic impact.

Poland

The two extra-heavy oil reservoirs of Poland are of little interest. They are very marginal quantitatively.

Russian Federation

Extra-heavy oil has been identified in the Russian Federation in small amounts in the Volga-Urals and North Caucasus-Mangyshlak Basins. As is the case with many countries, accurate and timely data are insufficient for making estimates.

Information on natural bitumen deposits indicates that very large resources are present in Eastern Siberia in the Lena-Tunguska basin. This is harsh terrain and only the Olenek deposit has been studied in sufficient detail to permit an estimation of discovered bitumen in place. The Siligir deposit has been frequently cited in reports of world bitumen deposits, but the origin of the primary source for these citations is unknown. It may be assumed that the estimate of more than 51 billion barrels for the basin is conservative. This area is so remote, and Russia's conventional oil and gas resources so great, that it is unlikely that attempts will be made in the near future to exploit this natural bitumen. Most of the other Russian bitumen deposits are located in the Timan-Pechora and Volga-Urals Basins, which are geologically analogous to the Western Canada Sedimentary Basin. However, these deposits are scattered and the recoverable portions are not quantitatively large. The deposits in the Tatar Republic have been studied extensively and efforts to exploit them may be conducted in the future.

Switzerland

The Val de Travers natural bitumen deposit in Switzerland is small but representative of many such occurrences in Western European countries. Most of these have been known for centuries and a few have been mined, mainly for road material. Asphalt was mined at Travers from 1712 until 1986.

Syria (Arab Republic)

The Babenna natural bitumen in Syria has been mined for many years. It is one of many such deposits throughout the Middle East, those in Syria and Iraq being especially prominent since antiquity. They are of little interest as synthetic oil sources. Syria's annual output of natural bitumen has been 110–120 thousand tonnes in recent years.

Trinidad & Tobago

Trinidad and Tobago is rich in heavy oil but only a small amount of it, perhaps 300 million barrels of oil in place, is extra-heavy. The country has more than 1.1 billion barrels of oil in place in natural bitumen deposits, including Asphalt (Pitch) Lake. All these deposits are located in the Southern Basin, which is small, strongly faulted, but highly productive.

Asphalt (Pitch) Lake, at La Brea, contains a semi-solid emulsion of soluble bitumen, mineral matter, and other minor constituents (mainly water), and has been mined, mostly for use as road surfacing material, since at least 1815. The Lake contains 60 million barrels of bitumen, a sufficient supply for the foreseeable future.

Lake Asphalt of Trinidad and Tobago (1978) Ltd (TLA), a state-owned company, produces between 10 000 and 15 000 metric tons per year, most of which is exported. In combination with asphalt from refined crude oil, the product is used for road construction. In addition, it can be used in a range of paints and coatings and for making cationic bitumen emulsions. Production of these emulsions of bitumen, water, and soap began in late 1996 and the emulsions are now used widely throughout the industrialised world in place of solvent-based bitumen emulsions.

United Kingdom

The United Kingdom has two offshore extra-heavy oil deposits. One is a discovery in the West of Shetlands Basin, for which few data are available. The other is the producing Piper field in the North Sea Graben, which contains oil between 8.7 and 37° API gravity.

United States of America

The United States was endowed with very large petroleum resources, which are to be found in nearly all the various types of geological basins. The resources of extra-heavy oil and natural bitumen are likewise distributed in numerous geological settings, but the amounts in each are strikingly different.

More than 98% of the extra-heavy oil is found in basins which evolved along the rift-faulted, convergent continental margin of California. The island arcs which originally trapped the sediments against the land mass to the east have been destroyed.

About 73% of the undiscovered and poorly known natural bitumen in place and 80% of that in discovered deposits has accumulated in deposits of the Western Canada type, reflecting the fact that such basins possess ideal conditions for occurrences of degraded oil.

Distillation of oil from Casmalia tar sands in California was attempted in 1923. Many tar sands deposits in the United States have served as sources of road asphalt, but this industry disappeared with the advent of manufactured asphalt tailor-made from refinery stills.

A 1985 review (Marchant, 1988), with tar sands defined as reservoirs containing oil with a minimum viscosity of 10 000 cP, listed 43 projects, down from 52 in 1984. The projects included 34 in-situ and 9 mining and plant extraction. Marchant's review used as a mining criterion a ratio of overburden to reservoir thickness of less than one. On this basis only about 15% of the US natural bitumen resource would qualify.

In October 2003 the US House of Representatives approved a measure to remove certain leasing uncertainties relative to tar sands. In effect the bill would separate tar sands leases from oil and gas leases covering the same areas (Oil & Gas Journal Online, 2003, November 19). The objective is to revive tar sands exploitation, which has essentially ceased since the mid-1980s.

Uzbekistan

A single occurrence of extra-heavy oil is reported from the Khaudag deposit in the Amu Daria Basin. Its size is unknown.

Venezuela

A certain amount of extra-heavy oil is found in the Maracaibo Basin but the resources of worldwide significance lie in the Orinoco Oil Belt along the southern, up-dip edge of the Eastern Venezuela Basin.

One natural bitumen deposit, Guanoco Lake, is found near the Caribbean coast, on the north side of the Eastern Venezuela Basin. Wells drilled in its vicinity failed to discover recoverable oil resources. The deposit has been estimated to contain 62 million barrels of oil in place (Walters, 1974).

Four joint ventures for the exploitation of extra-heavy crude have been approved and two have been proposed. As of 2003, these projects suggest future production of about 732 000 barrels per day but this does not indicate the proposed production of all at maturity. All the projects, in one way or another, involve production, transportation, and upgrading facilities. Petróleos de Venezuela (PDVSA), the state oil company, has a minority interest in all four.

- Initial production from the *Petrozuata* project, a joint venture between ConocoPhillips and PDVSA, began during the fourth quarter of 1998. The project is designed to transport extra-heavy oil (EHO) from the Zuata region to the north coast where it is upgraded into a 19–26° API SCO.
  Commercial operations began in April 2001 following the completion of a 120 000 b/d upgrader. The resultant gasoil, LPG, sulphur and petroleum coke are marketed whilst the SCO is used as a feedstock for the Lake Charles and Cardón refineries in the USA and Venezuela respectively.
  In the second quarter of 2003 it was reported that production of EHO was averaging 130 000 b/d, yielding approximately 112 000 b/d of SCO.
- The *Cerro Negro* project, a joint venture between ExxonMobil, PDVSA and Veba Oel, has taken nearly 40 years to come to fruition from the time the resource was first identified. Initial production began in November 1999 and by end-June 2003 was averaging more than 121 000 b/d of EHO.
  The 8.5° API crude is first diluted with naphtha, pumped to the upgrader (located at the port of José on the north coast, completed August 2001) where it is converted to a 16° API crude oil. The installed capacity of 120 000 b/d is

capable of producing 108 000 b/d of upgraded heavy crude, sulphur and coke. The heavy-sour crude, especially well-suited for coking/high conversion refineries which possess cat feed hydrotreating, is refined at the ExxonMobil Chalmette refinery in the USA.

In mid-2003 ExxonMobil announced that in the short term it was planned to expand the project by 10–20%, and in the long term to double capacity.

- Following construction of the *Sincor* project, a joint venture between TotalFinaElf, PDVSA and Statoil, commercial production of SCO began in early 2002. The objective of the project is to produce approximately 200 000 b/d of 8.5° API extra heavy oil, creating a lighter crude (17° API) by means of dilution, transporting it by pipeline to the upgrader and converting it into 180 000 b/d of 32° API low-sulphur crude for international marketing. In addition, about 860 tonnes per day of sulphur and 6000 tonnes of coke are produced as by-products—the former for pharmaceutical use and the latter destined for the electric power industry.
- At the present time the *Hamaca* project is undergoing development. It is a joint venture between ConocoPhillips, ChevronTexaco and PDVSA and is designed to upgrade up to 190 000 b/d EHO that will produce approximately 180 000 b/d SCO for export.

Drilling of development wells began in January 2001 with production of EHO beginning in the fourth quarter of that year. By end-2001 production of the 8.7° API crude was running at 35 000 b/d, being blended with 30 000 b/d of lighter crudes for the international market. During 2002 the project exported in excess of 9 million barrels of EHO and by year-end, the wells were capable of producing 85 000 b/d EHO. Construction of the heavy-oil upgrader, pipelines and production facilities began in June 2000. Completion is scheduled for mid-2004 and thereafter it is expected that the 26° API syncrude, intended for the US third-party market, will be produced in commercial quantities.

In the early 1980s Intevep, the research affiliate of the state oil company PDVSA, developed a method of utilising some of the hitherto untouched potential of Venezuela's extra-heavy oil/natural bitumen resource. Natural bitumen (7.5–8.5° API) extracted from the reservoir is emulsified with water (70% natural bitumen, 30% water, <1% surfactants), the resulting product being called Orimulsion®. Orimulsion® can be pumped, stored, transported and burnt under boilers using conventional equipment with only minor modifications. Initial tests were conducted in Japan, Canada and the UK and exports began in 1988.

Orimulsion® is processed, shipped and marketed by Bitúmenes del Orinoco S.A. (Bitor), a PDVSA subsidiary, but with the fuel's relatively high sulphur content and its emission of particulates, Intevep continues to seek improvements in its characteristics in order to match increasingly strict international environmental regulations. Bitor operates an Orimulsion® plant at Morichal in Cerro Negro with a capacity of 5.2 million tonnes per year. The company hopes to produce 20 million tonnes per year by 2006.

Following manufacture at the plant, the Orimulsion® is transported by pipeline about 320 km to the José export terminal for shipment. During the 1990s other markets were developed and currently Barbados, Brazil, Canada, China, Costa Rica, Denmark, Finland, Germany, Guatemala, Italy, Japan, Lithuania, Northern Ireland, Philippines, Singapore, South Korea, Taiwan, Thailand and Turkey either consume or are considering consuming the product.

In addition to being used in conventional power plants using steam turbines, Orimulsion® can be used in diesel engines for power generation, in cement plants, as a feedstock for Integrated Gasification Combined Cycle and as a 'reburning' fuel (a method of reducing NO_x by staging combustion in the boiler).

During third quarter 2003, the Venezuelan government announced that it intended to absorb

Bitor's operations into PDVSA East (one half of the decentralised PDVSA). The uncertain climate generated by the Venezuelan economic upheaval in 2002/2003 has resulted in a general lack of information regarding PDVSA's activities.

Orimulsion® is a registered trademark belonging to Bitúmenes Orinoco S.A.

Chapter 5

Natural Gas

World Energy Council
CONSEIL MONDIAL DE L'ENERGIE

COMMENTARY

Looking at the energy market over the next 10 years, the important question will be how to secure new sources of energy to meet the needs of a growing population. Hydrocarbons will continue to be a major source of energy, and natural gas in particular has a vital role to play in the energy mix and in providing for long-term energy security.

Several factors will support the role of natural gas:

- the progress of liberalisation on the world's gas markets;
- the development of technology that increases the economic benefits;
- changing views of LNG investments due to the integration and growth of leading energy companies;
- the economic and environmental superiority of natural gas in answering growing environmental concerns.

As Lord Browne of Madingley stated in his Keynote Address to the 22nd World Gas Conference in Tokyo, 'Flexibility—in physical and commercial terms—lies at the heart of what will be a new order in gas'. (see Proceedings of the 22nd World Gas Conference, 2003).

During the coming 30 years gas will become the world's most important energy source. It has therefore now become an important political issue worldwide. One of the consequences is that strong political forces seek to make gas as cheap as possible for consumers, through active regulation of large areas of the gas business and through the unbundling of gas companies. On the other hand, this does not stop politicians, especially in Europe, from heavily taxing energy, not least gas, on the grounds of resource management and environmental protection.

But the gas business is also being politicised by those committed to climate and environmental protection. Likewise with the aid of regulatory market intervention, they want to enable gas to play a prominent part in transforming the energy industry into one based on low-carbon energy supplies. Both these political strategies will have to be reconciled with the requirements of long-term supply security, which can in the final analysis only be achieved on the basis of long-term supply contracts. The most prominent project in the gas industry until now, namely the development of the Troll gas field in the Norwegian North Sea, would certainly not have materialised on the basis of spot markets. It is thus evident that the gas industry is exposed worldwide, though with regional variations, to contradictory demands and developments:

- the political demand that gas be made as cheap as possible for customers;

2004 Survey of Energy Resources

- the rising cost of implementing major new gas projects and developing infrastructure;
- gas is assigned by politicians—especially in Europe—the role of acting as a bridge on the road to a less CO_2-intensive energy industry.

It is of paramount importance that the opportunity that gas offers to decrease the growth of CO_2 and other emissions be disseminated to policy makers all over the world. Since natural gas produces the smallest amount of CO_2 per kWh of all the fossil fuels, replacing coal and oil by natural gas where economically possible, notably in generating electricity, is a prime necessity. In the long term, a decrease in CO_2 emissions can be realised if CO_2 sequestration is further developed.

There is little disagreement about the desirability of gas taking on the role of fuel of choice, notably during the ongoing century. This creates a formidable challenge: it means that the gas business will have to more than double in size during the next 30–50 years and that virtually all markets will need substantial volumes of new gas supplies at competitive prices.

Exploration and Production

To fulfil this increasing demand for natural gas, it will be necessary for exploration to move to deeper formations, deeper waters and more remote areas, which are situated far from the main gas-consuming regions. Development of these fields will depend on new advanced technologies and the market situation. Research and development on novel exploration and production technologies will certainly contribute to making the development of such fields economically feasible in the future.

For the reason mentioned above, unconventional gas resources will acquire greater importance. These attract more and more attention in developed countries, due to the exhaustion of conventional gas fields and the proximity of large unconventional gas resources. Gradual growth of natural gas production costs has already led to the development of some unconventional gas resources (for example, gas from deep formations or low-permeability reservoirs). Sometimes resources previously considered as irrecoverable are now being classed as proved reserves. Development of unconventional gas sources such as natural gas hydrates, dense gas-bearing reservoirs and coal bed methane could provide humanity with vast reserves of clean and environmentally friendly energy for centuries.

Future exploration and production trends will be strongly dependent on national economic and political situations. Production cost trends could be the most important indicator for assessing the development of the future gas upstream sector. The general overview of the upstream gas activities presented in this commentary shows the broad range and complex nature of the technologies currently used in exploration and production. A common feature shared by these technologies is the fact that they are in constant progress, with new developments and innovations continually being reported. This trend is expected to continue at a faster pace in the future, stimulated by the need for better and more efficient ways to overcome the challenges waiting ahead and by the spectacular and rapid progress taking place in related sciences and technologies (see Proceedings of the 22nd World Gas Conference, 2003).

In drilling, a major breakthrough took place with the advent of horizontal wells, geo-steering and real-time data acquisition, making it possible to navigate within a reservoir for accurate placement of the drain and to achieve complex multilateral architectures. New breakthroughs will follow with innovations like expandable tubular, allowing the drilling of mono diameter wells, and casing drilling, allowing the use of casing instead of drill pipe in drilling operations.

In seismic technology, 3D and powerful work stations have opened up a new chapter in exploration, development and production of hydrocarbon fields and are expanding rapidly all over the world. Refinements and new

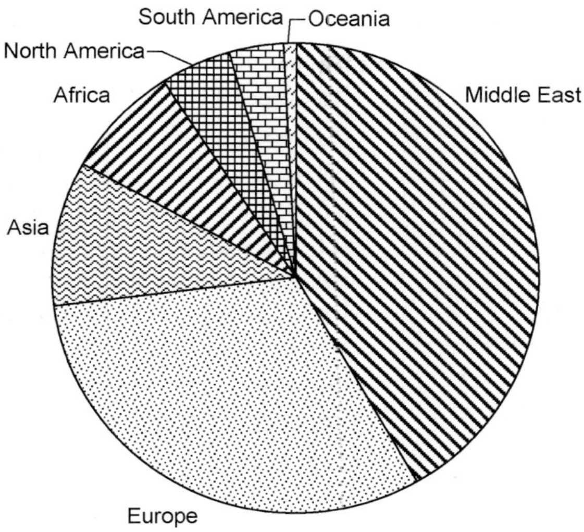

FIGURE 5.1 Proved gas reserves at end-2002—regional distribution.

developments such as 4D, 3D amplitude versus offset techniques, high resolution, well seismic, etc. will make it one of the major technologies in use far into the future.

Future trends in well and plant monitoring are already apparent, with the emergence of intelligent wells and plants, opening up new perspectives in the real-time monitoring and automation of field operations. In reservoir engineering, advances in numerical simulation coupled with the development of massive and cheaper computer power have opened the way for a new generation of numerical models with millions of cells, whereas a decade ago models with just a few thousand cells were considered a great achievement. It is not unlikely that at this pace models with hundreds of millions of cells will be a reality in the not too distant future. Although improvements in natural gas processing will be concerned mainly with costs and energy efficiency, advances in some processes (such as membrane technology) are expected. Likewise continued progress in gas-to-liquids (GTL) technology could bring new solutions for the development of stranded gas reserves. Real-time field data acquisition, processing and dispatching to remote control centres through sophisticated data transmission systems and in connections to the Web Network are promising a new dimension in daily field operations management and decision making.

A major trend that will keep expanding in the future is the multidisciplinary approach, integrating several interrelated technologies that can be used in synergy to achieve a problem-solving capacity and speed far beyond that of the individual technologies used separately. An example of such an approach could be the integration of intelligent wells, reservoir and surface facilities simulation models, 4D seismic, real-time data acquisition and data transmission systems. (Ter-Sarkisov, Proceedings of the 22nd World Gas Conference, 2003).

Transmission from Source to Market

LNG

There is a growing recognition that the world's gas industry does not have any problems as regards resources and reserves (see Tables 5.1 and 5.3, Nakicenovic, 2000; Ananenkov, 2003). Reserves versus production ratios from different sources show similar levels. Present proven gas reserves versus the present production levels result in a ratio of 50–60 years. If one considers potential (unconventional) gas reserves versus the current record production levels, gas reserves are adequate to support the current level of production for the next 200 years (Thorn, 2003; Urano, 2003). The lowest ratios are found in North America, illustrating the need for large investments in import schemes.

It is expected that LNG will be a major means of meeting the need for gas imports into North America. LNG will become more and more important in bringing distant markets and reserves together.

The LNG industry shows very satisfactory results concerning production and consumption. The increase of 110 times in worldwide liquefaction capacity and the increase of 100 times in the volume of LNG transported between 1964 and 2002 illustrate this. All LNG plants

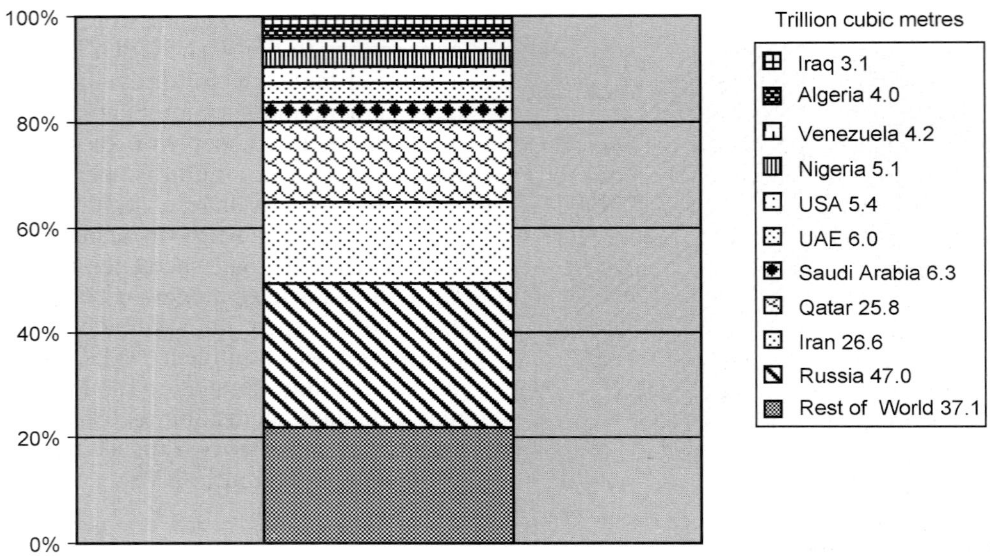

FIGURE 5.2 Proved gas reserves at end-2002—the leading countries.

and receiving terminals are operated in accordance with environmental and safety rules, making the LNG industry one of the safer and less polluting industries. The forecast for the LNG industry for 2014 (50 years after its inception) indicates that it will experience substantial growth, promising technological development and a non-negligible reduction in the costs of building and operating LNG installations. The Asian market is expected to remain the main LNG consumer, reinforced by new developments in China and India.

Currently not all of the LNG potentially available can be absorbed by the market, resulting in a very competitive gas price. But there is a possibility that in the future, in certain circumstances, LNG could be in an under-supply condition. If this situation happened, it would jeopardise the security of national energy supply on the buyer's side. The balance between supply and demand should be maintained by market mechanisms, to enable both producers and consumers to obtain continuity and stability in LNG supply and prices.

LNG price competition will eventually result in cost competition of feed gas. Therefore, if LNG price competition continues for long, the winners would be the projects based on huge gas fields, which can acquire feed gas cheaply.

In selecting LNG suppliers, buyers should take into consideration factors other than just prices, such as the diversity and reliability of LNG supply sources. On the other hand, suppliers should make efforts to cut costs throughout the LNG supply chain.

Forums to bring together producers, buyers and technology suppliers will be necessary in the future, to exchange information and to discuss the balancing of LNG demand and supply and the implementation of technical innovations.

Discoveries of huge natural gas fields are usually concentrated in a few countries. This condition makes LNG supply diversification difficult to achieve. There are small silent gas fields scattered throughout some countries, which contain what is called stranded gas. Currently these gas fields are not economic to be commercialised because of their small size. New technology is expected to bring this stranded gas into the LNG trade sometime in the future.

The pressure to reduce LNG production costs will continue. Technical innovations in liquefaction processes and upstream gas production are needed to achieve this. The development of a floating LNG plant means a potential saving in transport and onshore development costs. Application of electric motors as compressor drivers allows larger train sizes, with standardised equipment supplied from a wider vendor base. Electrical drivers also have the advantage of high availability and high efficiency, when supplied by a combined-cycle power station. Developments in liquids extraction will increase flexibility for producing LNG with different grades of heating value to suit different markets. As there will not be one concept that meets all requirements, train sizes between 4 and 7 mtpa are to be expected in the future.

LNG technology has matured over the past decades, mostly by capitalising on economies of scale. But in the future, LNG plants will be more diverse, both in size and technology. LNG technology providers and contractors will increasingly have to rely on a flexible portfolio of processes, drivers and plant sizes to achieve fit-for-purpose solutions.

In the long term, high-quality management is also about being cost efficient. Short cuts may be profitable in the short term, but with the prevailing industry standards, cutting corners will be unacceptable and will ultimately lead to greater costs and perhaps loss of business in some cases.

Technical innovations can bring the costs down throughout the entire chain and hence increase LNG's competitiveness. This can then increase the overall growth in the business and subsequently make the existing LNG business a safer place to operate in and create attractive new business opportunities in the years to come. (Travesset, Proceedings of the 22nd World Gas Conference, 2003)

GTL

Pipelines and LNG are by far the dominant means of transporting gas, generally via pipelines for up to 3 000 km and LNG for greater distances. There is also the possibility of transporting liquids from a GTL plant, also for distances above 3 000 km. Known already from its successful application in South Africa, this increasingly becomes an option for other markets.

In Qatar USS 10 billion is being invested in two GTL plants by Shell and ConocoPhillips. Their expected capacity will be in the order of 10 million m^3/day or 34 000 barrels liquid per day. Since Qatar is in a position to deliver equally as well to the Atlantic Basin as to the Asia Pacific region and has a huge reserve of 26 000 bcm, with a present production level of 30 bcm/yr, this development may lead to the rise of a world gas-based market by the year 2010 (see Tractebel, 2003).

Utilisation

The further development of technology is not just necessary to bring the gas to market. The utilisation of gas also needs further development in order to improve the efficiency in use of gas.

Although gas-based electricity production has the advantage with respect to capital, operation and maintenance cost, the fuel costs are the highest. Therefore, further improvements in the efficiency of combined-cycle gas turbines and combined heat and power (CHP) systems are paramount.

An interesting new development in this respect is the virtual power plant. This comprises a multitude of decentralised, grid-connected, micro and mini-CHP units using fuel cell, gas turbine, gas internal combustion engine or Stirling engine technology, installed in single family or multi-family houses, small enterprises, public facilities, etc. for combined heating and cooling and flexible electricity production. The multitude of decentralised units can be centrally controlled and managed as part of an interconnected network, resulting in a virtual power plant—with great flexibility in terms of fuel choice (natural gas, biogas, LPG, hydrogen)—contributing to meeting peak energy demand in the public electricity sector with potentially very high overall efficiencies. This could result in

From-To	Project
Argentina - Uruguay	Argentinian natural gas reached Montevideo through the Cruz del Sur pipeline crossing the River Plate from the southern end of the Buenos Aires high pressure ring
Nigeria - Algeria	Contract has been signed for studies on the Nigal pipeline
Algeria - Spain	Studies are in progress on the Medgas pipeline (1 200 km)
Algeria - Italy	GALSI project crossing Sardinia
Libya - Italy (southern)	Green Stream
Russia - Turkey	The Blue Stream pipeline was brought to stream in 2002 (16 bcm per year)
Norway (western) - UK (Easington)	Ormen Lange (1 200 km)
Netherlands - UK	BBL (235 km)
Russia - Germany	NTG from St Petersburg 2 000 km offshore (by 2007)
	Shtokman field to LNG plant North West Russia 2007
Iran/Turkey border and Georgia/Turkey border - Central Europe	Nabucco project through Turkey to be realised 2004-2009
Iran	950 km, 56 inch IGAT-4 from South Pars reservoir to southern Tehran (10 compressor stations) to be realised 2004-2005
Iran	470 km, 48 inch IGAT-6 from South Pars reservoir to Bid Boland treating plant in Khoozestan Province (2 compressor stations) to be realised 2004-2005
Iran	IGAT-8 from South Pars reservoir to southern Tehran, 1 050 km, 56 inch (10 compressor stations) to be realised 2007
Iran	360 km, 56 inch IGAT-9 from Kangan treating plant to Pataveh to be realised 2004-2005
Iran (Saveh) - Iran (northwestern)	Saveh/Hamadan/Mian-Doab 490 km, 48/42/40 inch extension from IGAT-3 (6 compressor stations) to be realised 2006
Iran	Parchin/Mashhad main pipeline, 1 000 km, 42 inch from Tehran to Mashhad (5 compressor stations) to be realised 2006
Iran	Arsanjan/Sar Cheshmeh 273 km, 42 inch extension from IGAT-4 (1 compressor station) to be realised 2004-2005
Europe (south)	South European Gas Ring, from Turkey, Greece, FYR of Macedonia, Serbia and Montenegro, Bosnia-Herzegovina, Croatia, Slovenia, Austria. For the time being, the Protocol (signed 8 April 2003) between the interested parties* expressed their intention to participate in a performance of a pre-feasibility study, based on mutually agreed pipeline routes, with the aim of reaching a common point of view about the current technical, economic and market conditions.
	* BOTAS (Turkey), DEPA (Greece), Ministry of Industry and Energy of Albania, Makpetrol (FYROM), NIS-GAS and NIS (Serbia and Montenegro), BH-Gas (Bosnia-Herzegovina), Plinacro (Croatia) and Geoplin (Slovenia)
Russia (northeast and Siberia)	Shaklin 1 and 2
USA	A number of provisions to increase supplies are in the current energy legislation now before the US Congress, including provisions to build a pipeline from Alaska to the lower 48 states.
Thailand	Transmission system has been expanded. The future infrastructure expansion includes the Gas Pipeline Master Plan III project and the Trans Thai-Malaysia gas pipeline project (2005)

FIGURE 5.3 Recent and future pipeline projects. (Source: IGU Members.)

primary energy savings of 10–30% and a greenhouse gas (CO_2) emission reduction of 20–50% (see Virtual Power Plant, 2003). If installed worldwide, this could lead to total world CO_2 emissions well below the limits aimed at by the Kyoto treaty.

In automotive use, natural gas can also play an important role, especially in contributing to the improvement of air quality in major cities. Natural gas vehicles (NGVs) are becoming more and more popular, with over 3 million vehicles to date (see International Association for Natural Gas Vehicles). Moreover, GTL-based fuel for cars may contribute to a cleaner environment, since this fuel is more homogenous and it is therefore easier for engine manufacturers to decrease harmful exhaust emissions.

Investment

There is increasing concern that regulatory concepts and their practical effects are impeding the development of infrastructure and the expansion of gas use. This jeopardises new mega-projects in production and transmission that are planned or have already been started.

Another difficulty is the dominance of a shareholder-value strategy in global capital markets. This clearly places the emphasis on high, short-term returns at the expense of returns on investments that are moderate but more reliable in the long run. But if investments in the international gas business are of limited attractiveness to investors, this must be compensated for by higher, sustained equity financing on the part of gas companies. It is envisaged that the expected two- to three-fold increase in world gas trade by 2030 will require US$ 1–3 trillion for additional investments and maintenance of existing infrastructure (Thorn, 2003; Urano, 2003). A stable framework of legal, fiscal and regulatory rules will be a prerequisite for such investments.

Consumption

The upward trend in gas consumption continued worldwide in 2002. Available statistics indicate that a continued, though moderate, increase in world gas production and consumption can be assumed. The worldwide share of gas in primary energy consumption has gone up slightly and is now just under a quarter. Gas is thus No. 3 in world energy supplies, coming after oil and lying just behind coal.

Trade

International gas trade in 2002 was 5% higher than in the preceding year. It amounted to a good quarter of world gas production and mainly took the form of intra-regional trade, primarily due to deliveries from Canada to the United States (up 3%) and flows within Western Europe and in the Asia/Pacific region. Inter-regional deliveries included principally Russian gas supplies to western, central and eastern Europe and Algerian exports to Europe and North America, as well as deliveries from the Persian Gulf to the Far East. A good three-quarters of the volume traded across borders took the form of pipeline gas and just under a quarter was LNG.

Prices

The development of gas prices during 2002 mainly followed the price trend of the international oil markets. Over the year as a whole, this was an upward trend, with brief downward dips. In the course of the first 10 months of 2003, a sustained high price level emerged, despite a brief significant price decrease in the period March/April.

Relationships differ between the development of gas prices and that of rival fuels. Steep price rises have been reported by some countries

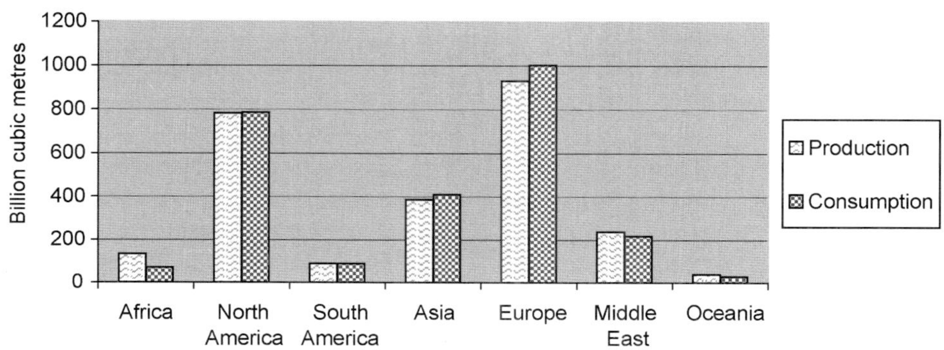

FIGURE 5.4 Gas production and consumption, 2002: regional distribution.

in central and eastern Europe which have been successively adjusting their prices to market conditions. For instance, Slovakia has registered an almost 44% rise in the regulated gas price for residential customers since the beginning of 2003, pointing out that even this price increase is not sufficient to make a profit. In Russia, two significant price rises in 2002 were likewise not enough to offset the losses sustained in gas supply operations.

Regulation

Although nowadays natural gas, as the fuel of choice, has taken a large share of world (commercial) energy supplies there is not (yet?) a world gas market like the global oil market. Furthermore, there are important geographic discrepancies between gas demand regions and gas supply regions. So regulatory harmonisation of a regional (supra-national) gas market (in respect of its gas transport and supply flexibility activities) in general does not encompass in an equitable way both the supply and the demand side.

Regulatory harmonisation can be achieved when a region is adhering to a common political concept and policy, which at least takes time: the more so within a set of demand and supply regions. Regulation in one (demand) region and no similarly aimed regulation in the supply region could lead to misunderstandings and tensions, which would not improve the climate for investment and business. Thus, special efforts are needed to maintain a mutual understanding, in which the gas industry and the financial market are willing and capable of realising in a timely manner the huge investments needed.

Should regulatory harmonisation in the (regional) gas market play as important a role as in the (regional) electricity market? It should be noted that in electricity markets supply and demand are much more in structural balance within the same region (or nation), and thus within the same political realm; one can decide to establish a power plant wherever desired, which is not possible with gas fields.

It is thereby realistic to appreciate that gas supply regions do have choices of where to direct new sales. Russia is not bound to Western Europe (and moreover its own market could grow hugely); the Middle East region has a choice between the Pacific Rim, Europe and the Indian subcontinent; the South-East Asian suppliers can send their LNG-carriers in quite different directions, as can the African suppliers. For some time to come it will be a greater challenge to acquire gas supplies in time at competitive prices for the demand regions than it will be to convince the energy consumers to choose gas (Verberg, 2003).

Many countries report on progress in market liberalisation, restructuring of gas industries and development of national regulatory authorities (Kabelitz, 2003).

In Europe, the development of the gas market has for a good 5 years now been largely shaped by the goal of market opening and the actual development of the legal framework. The EU Commission seeks to document the progress made in market opening in its annual benchmarking reports. It is acknowledged there that Germany, Austria and the United Kingdom are playing leading roles in opening up gas markets in Europe.

Rob Aptroot
International Gas Union

References

Ananenkov, A.G., 2003. Trends in the world of gas industry development, presentation at the IGU Council meeting, Cape Town, 29 October 2003.
IEA World Energy Outlook 2002.
International Association for Natural Gas Vehicles (www.iangv.org).
Kabelitz, K.R., 2003. Overview of recent developments in the world gas industry, presentation at the IGU Council meeting, Cape Town, 29 October 2003.
Nakicenovic, N., 2000. Global Natural Gas Perspectives, IIASA/IGU.
Proceedings of the 22nd World Gas Conference, Tokyo, June 2003.
Thorn, T.H., 2003. International Oil and Gas Finance Review, October 2003.
Tractebel, Gas Summit, Paris, October 2003.
Urano, H., 2003. President's Summary, 22nd World Gas Conference Tokyo, June 2003.
Tractebel, Gas Summit, Paris, October 2003.
Urano, H., 2003. President's Summary, 22nd World Gas Conference Tokyo, June 2003.
Verberg, G.H.B., 2003. Presentation in plenary session 3 of the World Forum on Energy Regulation, Rome, 7 October 2003.
Virtual Power Plant, Brochure of IGU, December 2003.

DEFINITIONS

Natural gas is a mixture of hydrocarbon and small quantities of non-hydrocarbons that exists either in the gaseous phase or is in solution in crude oil in natural underground reservoirs, and which is gaseous at atmospheric conditions of pressure and temperature.

Natural gas liquids (hydrocarbons that exist in the reservoir as constituents of natural gas but which are recovered as liquids in separators, field facilities or gas-processing plants) are discussed in Chapter 2—Crude Oil and Natural Gas Liquids.

Proved amount in place is the resource remaining in known natural reservoirs that has been carefully measured and assessed as exploitable under present and expected local economic conditions with existing available technology.

Proved recoverable reserves are the volume *within* the proved amount in place that can be recovered in the future under present and expected local economic conditions with existing available technology.

Estimated additional amount in place is the volume *additional to* the proved amount in place that is of foreseeable economic interest. Speculative amounts are not included.

Estimated additional reserves recoverable is the volume *within* the estimated additional amount in place that geological and engineering information indicates with reasonable certainty might be recovered in the future.

Production—where available, gross and net (marketed) volumes are given, together with the quantities re-injected, flared and lost in shrinkage (due to the extraction of natural gas liquids, etc.).

Consumption—natural gas consumed within the country, including imports but excluding amounts re-injected, flared and lost in shrinkage.

R/P (reserves/production) ratio is calculated by dividing proved recoverable reserves at the end of 2002 by production (gross less re-injected) in that year. The resulting figure is the time in years that the proved recoverable reserves would last if production were to continue at the 2002 level.

As far as possible, natural gas volumes are expressed in standard cubic metres, measured dry at 15 °C and 1 013 mb, and the corresponding cubic feet (at 35.315 cubic feet per cubic metre).

TABLE 5.1

Natural gas: proved recoverable reserves at end-2002

	Billion cubic metres	Billion cubic feet
Algeria	4 000	141 260
Angola	113	4 000
Benin	1	43
Cameroon	110	3 900
Congo (Brazzaville)	91	3 200
Congo (Dem. Rep.)	1	35
Côte d'Ivoire	24	845
Egypt (Arab Republic)	1 657	58 500
Equatorial Guinea	70	2 472
Ethiopia	25	883
Gabon	33	1 165
Ghana	24	848
Libya/GSPLAJ	1 314	46 404
Morocco	1	43
Mozambique	127	4 500
Namibia	62	2 200
Nigeria	5 055	178 517
Rwanda	57	2 000
Senegal	11	388
Somalia	6	200
South Africa	37	1 300
Sudan	85	3 002
Tanzania	28	989
Tunisia	78	2 755
Total Africa	**13 010**	**459 449**
Barbados	N	5
Canada	1 664	58 771
Cuba	71	2 500
Guatemala	3	109
Mexico	424	14 985
Trinidad & Tobago	589	20 801
United States of America	5 366	189 494
Total North America	**8 117**	**286 665**
Argentina	664	23 449
Bolivia	813	28 711
Brazil	237	8 370
Chile	98	3 460
Colombia	120	4 238
Ecuador	10	345
Peru	245	8 655
Venezuela	4 179	147 585
Total South America	**6 366**	**224 813**

TABLE 5.1 (*Continued*)

	Billion cubic metres	Billion cubic feet
Afghanistan	100	3 530
Armenia	176	6 215
Azerbaijan	1 370	48 382
Bangladesh	301	10 615
Brunei	391	13 800
China	1 676	59 188
Georgia	8	300
India	751	26 522
Indonesia	2 183	77 093
Japan	51	1 801
Kazakhstan	1 841	65 000
Kyrgyzstan	6	200
Malaysia	2 124	75 000
Myanmar (Burma)	445	15 715
Nepal	N	N
Pakistan	801	28 288
Philippines	70	2 465
Taiwan, China	76	2 700
Tajikistan	6	200
Thailand	441	15 578
Turkey	10	353
Turkmenistan	2 010	71 000
Uzbekistan	1 875	66 200
Vietnam	193	6 800
Total Asia	**16 905**	**596 945**
Albania	10	353
Austria	24	844
Belarus	3	100
Bulgaria	6	210
Croatia	31	1 095
Czech Republic	4	141
Denmark	87	3 072
France	15	513
Germany	326	11 513
Greece	1	18
Hungary	61	2 154
Ireland	20	700
Italy	227	8 000
Netherlands	1 756	62 000
Norway	2 120	74 868
Poland	77	2 719
Romania	163	5 756
Russian Federation	47 000	1 659 805
Serbia & Montenegro	24	863

(*continued on next page*)

TABLE 5.1 (*Continued*)

	Billion cubic metres	Billion cubic feet
Slovakia	15	530
Slovenia	N	N
Spain	N	N
Ukraine	1 110	39 200
United Kingdom	604	21 330
Total Europe	**53 684**	**1 895 784**
Bahrain	92	3 249
Iran (Islamic Republic)	26 570	938 320
Iraq	3 100	109 477
Israel	39	1 377
Jordan	6	212
Kuwait	1 557	54 985
Oman	859	30 336

TABLE 5.1 (*Continued*)

	Billion cubic metres	Billion cubic feet
Qatar	25 783	910 527
Saudi Arabia	6 343	224 003
Syria (Arab Republic)	371	13 102
United Arab Emirates	6 003	211 996
Yemen	396	13 985
Total Middle East	**71 119**	**2 511 569**
Australia	860	30 371
New Zealand	42	1 475
Papua New Guinea	428	15 115
Total Oceania	**1 330**	**46 961**
Total World	**170 531**	**6 022 186**

Notes:
(1) The relationship between cubic metres and cubic feet is on the basis of 1 cubic metre = 35.315 cubic feet throughout.
(2) Sources: WEC Member Committees, 2003; *Oil & Gas Journal*, 22 December 2003; *Natural Gas in the World 2001/2002*, Cedigaz; *Annual Statistical Report 2003*, OAPEC; *World Oil*, August 2003; various national sources.

TABLE 5.2

Natural gas: resources at end-2002

	Proved amount in place (bcm)	Estimated additional amount in place (bcm)	Estimated additional reserves recoverable (bcm)	Proved amount in place (tcf)	Estimated additional amount in place (tcf)	Estimated additional reserves recoverable (tcf)
Africa						
Algeria	6 000	2 000	960	211.9	70.6	33.9
Côte d'Ivoire	40			1.4		
Egypt (Arab Republic)			1 741			61.5
North America						
Canada		12 800	8 880		452.0	313.6
Mexico	892	475		31.5	16.8	
South America						
Argentina			306			10.8
Brazil			95			3.4
Asia						
China	3 002	11 440		106.0	404.0	
Indonesia	3 865			136.5		
Philippines	93	2	2	3.3	0.1	0.1
Thailand			258			9.1
Turkey	16			0.6		

TABLE 5.2 (*Continued*)

	Proved amount in place (bcm)	Estimated additional amount in place (bcm)	Estimated additional reserves recoverable (bcm)	Proved amount in place (tcf)	Estimated additional amount in place (tcf)	Estimated additional reserves recoverable (tcf)
Europe						
Albania	14	10		0.5	0.4	
Austria	24			0.8		
Croatia	31			1.1		
Czech Republic	10		2	0.4		0.1
Denmark	493	92	50	17.4	3.2	1.8
France	375			13.2		
Germany			200			7.1
Hungary	159	41–132	29–93	5.6	1.4–4.7	1.0–3.3
Italy	208	174		7.3	6.1	
Norway	3 670	2 510		129.6	88.6	
Poland	139	400–600		4.9	14.1–21.2	
Romania	630	101	79	22.2	3.6	2.8
Serbia & Montenegro	60			2.1		
Spain	N	N	N	N	N	N
Middle East						
Israel	53	70	50	1.9	2.5	1.8
Jordan	15	N	6	0.5		0.2
Oceania						
Australia			3 136			110.7
New Zealand	158			5.6		

Notes:

(1) The data on resources are those reported by WEC Member Committees in 2003. They thus constitute a sample, reflecting the information available in particular countries: they should not be considered as complete, or necessarily representative of the situation in each region. For this reason, regional and global aggregates have not been computed.

(2) Sources: WEC Member Committees, 2003.

TABLE 5.3

Natural gas: 2002 production

	Gross (bcm)	Re-injected (bcm)	Flared (bcm)	Shrinkage (bcm)	Net (bcm)	Net (bcf)	R/P ratio
Algeria	160.6	71.7	4.2	5.4	79.3	2 799	45.0
Angola	8.4	3.3	4.3	0.2	0.6	22	22.2
Cameroon	1.6		1.6		0.0	0	67.9
Congo (Brazzaville)	3.4	2.0	1.3	0.1	0.0	0	67.4
Côte d'Ivoire	1.3				1.3	47	17.9
Egypt (Arab Republic)	31.3	0.7	3.2	0.8	26.7	941	54.2
Equatorial Guinea	3.4	1.0	0.7	0.4	1.3	45	29.5
Gabon	2.3	0.6	1.6	0.1	0.1	3	18.9

(*continued on next page*)

TABLE 5.3 (*Continued*)

	Gross (bcm)	Re-injected (bcm)	Flared (bcm)	Shrinkage (bcm)	Net (bcm)	Net (bcf)	R/P ratio
Libya/GSPLAJ	10.4	2.5	1.4	0.4	6.2	219	>100
Morocco	0.1				0.1	2	20.0
Mozambique	0.1				0.1	2	>100
Nigeria	38.1	3.3	18.9	1.7	14.2	501	>100
Senegal	0.1				0.1	2	>100
South Africa	2.5		0.2		2.3	81	15.0
Tunisia	2.5		0.3	0.1	2.2	76	31.2
Total Africa	**265.9**	**85.1**	**37.5**	**9.1**	**134.2**	**4 741**	**70.4**
Barbados	N				N	1	4.7
Canada	220.6	12.5	2.3	18.2	187.6	6 625	8.0
Cuba	0.6		0.1	0.1	0.4	12	>100
Mexico	45.7	4.6	2.8	0.4	38.0	1 342	10.3
Trinidad & Tobago	19.2		1.9		17.3	611	30.7
United States of America	683.3	104.7	2.4	36.8	539.4	19 047	9.3
Total North America	**969.4**	**121.8**	**9.4**	**55.5**	**782.6**	**27 639**	**9.6**
Argentina	45.8	3.2	0.9	5.6	36.1	1 275	15.6
Bolivia	8.9	2.2	0.3	0.6	5.8	205	>100
Brazil	15.5	3.4	2.1	0.6	9.4	332	19.5
Chile	3.0	1.6	0.1	0.1	1.2	42	72.6
Colombia	15.1	8.0	0.5	0.5	6.2	218	16.9
Ecuador	1.1	0.2	0.8		0.1	4	11.1
Peru	1.0	0.4	0.2		0.4	16	>100
Venezuela	59.4	20.5	3.1	6.0	29.8	1 052	>100
Total South America	**149.7**	**39.5**	**7.9**	**13.3**	**89.0**	**3 143**	**57.8**
Afghanistan	0.1				0.1	2	>100
Azerbaijan	13.1	0.8	6.8	0.3	5.2	182	>100
Bangladesh	11.5				11.5	404	26.3
Brunei	12.9	1.8		0.3	10.9	383	35.2
China	32.6				32.6	1 152	51.4
Georgia	N				N	1	>100
India	27.2		0.8	1.4	25.0	883	27.7
Indonesia	86.9	6.8	4.6	5.1	70.4	2 484	27.3
Japan	2.8				2.8	99	18.1
Kazakhstan	12.6		1.4		11.2	394	>100
Kyrgyzstan	N				N	1	>100
Malaysia	52.8			4.3	48.5	1 713	40.2
Myanmar (Burma)	8.8		0.1	0.3	8.4	297	50.6
Pakistan	26.6	0.8		2.9	22.9	809	31.1
Philippines	2.4		0.3	0.3	1.8	62	29.2
Taiwan, China	0.9				0.9	31	85.4
Tajikistan	N				N	1	>100
Thailand	21.0			1.1	20.0	705	21.0
Turkey	0.4				0.4	14	24.4
Turkmenistan	53.5				53.5	1 889	37.6

TABLE 5.3 (*Continued*)

	Gross (bcm)	Re-injected (bcm)	Flared (bcm)	Shrinkage (bcm)	Net (bcm)	Net (bcf)	R/P ratio
Uzbekistan	58.4				58.4	2 062	32.1
Vietnam	2.6		0.4		2.3	80	73.9
Total Asia	**426.8**	**10.2**	**14.3**	**15.9**	**386.4**	**13 649**	**40.1**
Albania	N	N			N	N	>100
Austria	1.7				1.7	61	13.9
Belarus	0.3				0.3	9	11.5
Bulgaria	N				N	1	>100
Croatia	2.1				2.1	75	14.6
Czech Republic	0.5	0.3			0.2	5	26.7
Denmark	11.5	2.9	0.2		8.4	297	10.1
France	2.7		N	1.0	1.8	62	5.5
Germany	22.2		N	0.6	21.6	763	14.7
Greece	0.1			N	N	1	14.3
Hungary	3.3	0.1	N	0.1	3.1	110	19.1
Ireland	0.8				0.8	28	25.0
Italy	14.9				14.9	528	15.2
Moldova	0.1				0.1	2	
Netherlands	71.2				71.2	2 516	24.6
Norway	105.8	32.9	0.6	3.4	68.9	2 433	29.1
Poland	5.2				5.2	184	14.8
Romania	12.6	N	0.2	0.2	12.1	428	13.0
Russian Federation	607.8			12.5	595.3	21 023	77.3
Serbia & Montenegro	0.5				0.5	16	53.3
Slovakia	0.2				0.2	7	75.0
Slovenia	N				N	N	5.0
Spain	0.6		N		0.6	20	5.2
Ukraine	18.8				18.8	664	59.0
United Kingdom	112.7	2.1	1.6	6.9	102.1	3 607	5.5
Total Europe	**995.5**	**38.3**	**2.7**	**24.6**	**929.8**	**32 839**	**56.1**
Bahrain	12.2	2.7		0.3	9.3	327	9.7
Iran (Islamic Republic)	121.0	27.5	8.2	18.0	67.3	2 378	>100
Iraq	3.6		1.0	0.2	2.4	83	>100
Israel	N				N	N	>100
Jordan	0.3				0.3	9	23.1
Kuwait	9.7		0.4	1.0	8.3	293	>100
Oman	21.7	3.1	0.9	2.7	15.0	530	46.2
Qatar	38.1	3.8	0.3	4.5	29.5	1 042	>100
Saudi Arabia	60.0	0.1	0.2	3.0	56.7	2 002	>100
Syria (Arab Republic)	8.2	1.8	0.3	0.4	5.8	205	58.0
United Arab Emirates	63.8	15.2	0.5	4.8	43.4	1 532	>100
Yemen	20.1	19.7		0.4	0.0	0	>100
Total Middle East	**358.6**	**73.9**	**11.7**	**35.2**	**237.9**	**8 401**	**> 100**
Australia	38.8			4.2	34.6	1 222	22.2

(*continued on next page*)

TABLE 5.3 (Continued)

	Gross (bcm)	Re-injected (bcm)	Flared (bcm)	Shrinkage (bcm)	Net (bcm)	Net (bcf)	R/P ratio
New Zealand	6.4	0.3	0.1	0.1	5.9	210	6.9
Papua New Guinea	0.1	N			0.1	4	>100
Total Oceania	**45.3**	**0.3**	**0.1**	**4.3**	**40.7**	**1 436**	**29.5**
Total World	**3 211.1**	**369.0**	**83.5**	**157.9**	**2 600.6**	**91 847**	**59.8**

Notes:
(1) Sources: WEC Member Committees, 2003; *Natural Gas in the World 2001/2002*, Cedigaz; various national sources.

TABLE 5.4

Natural gas: 2002 consumption

	Billion cubic metres	Billion cubic feet
Algeria	23.3	824
Angola	0.6	22
Côte d'Ivoire	1.4	50
Egypt (Arab Republic)	25.2	891
Equatorial Guinea	1.3	45
Gabon	0.1	3
Libya/GSPLAJ	5.6	197
Morocco	0.7	23
Mozambique	0.1	2
Nigeria	6.4	225
Senegal	0.1	2
South Africa	2.3	81
Tunisia	3.8	136
Total Africa	**70.9**	**2 501**
Barbados	N	1
Canada	91.0	3 214
Cuba	0.4	12
Mexico	43.0	1 519
Puerto Rico	0.6	22
Trinidad & Tobago	12.0	423
United States of America	638.1	22 534
Total North America	**785.1**	**27 725**
Argentina	30.3	1 069
Bolivia	1.0	36
Brazil	14.4	509
Chile	6.5	230
Colombia	6.2	218
Ecuador	0.1	4
Peru	0.4	16
Uruguay	N	1
Venezuela	29.8	1 052
Total South America	**88.7**	**3 135**
Afghanistan	0.1	2
Armenia	1.4	49
Azerbaijan	10.0	351
Bangladesh	10.7	379
Brunei	1.7	59
China	30.2	1 069
Georgia	1.1	40
Hong Kong, China	2.4	83
India	25.0	883
Indonesia	33.9	1 197
Japan	75.3	2 657
Kazakhstan	13.0	457
Korea (Republic)	24.0	848
Kyrgyzstan	0.8	30
Malaysia	28.0	990
Myanmar (Burma)	2.2	78
Pakistan	22.9	809
Philippines	2.0	71
Singapore	1.8	62
Taiwan, China	7.9	278

TABLE 5.4 (Continued)

	Billion cubic metres	Billion cubic feet
Tajikistan	0.5	18
Thailand	25.6	904
Turkey	17.7	625
Turkmenistan	14.2	501
Uzbekistan	56.1	1 982
Vietnam	2.3	80
Total Asia	**410.8**	**14 502**
Albania	N	1
Austria	8.2	290
Belarus	15.5	546
Belgium	16.1	569
Bosnia-Herzogovina	0.2	6
Bulgaria	2.7	95
Croatia	2.9	102
Czech Republic	9.5	334
Denmark	4.9	173
Estonia	0.9	32
Finland	4.5	159
FYR Macedonia	0.1	4
France	42.2	1 489
Germany	98.3	3 471
Greece	2.1	75
Hungary	13.4	473
Ireland	4.3	152
Italy	70.4	2 486
Latvia	1.3	46
Lithuania	2.7	96
Luxembourg	0.9	32
Moldova	2.2	76
Netherlands	37.9	1 338
Norway	6.7	237
Poland	11.6	410
Portugal	2.9	102
Romania	15.6	551
Russian Federation	422.1	14 905
Serbia & Montenegro	2.1	73
Slovakia	7.1	251
Slovenia	1.0	34
Spain	22.7	800
Sweden	1.3	44
Switzerland	3.0	107
Ukraine	69.7	2 461
United Kingdom	95.0	3 355
Total Europe	**1 002.0**	**35 375**
Bahrain	9.3	327
Iran (Islamic Republic)	79.2	2 798
Iraq	2.4	83
Israel	0.5	18
Jordan	0.3	9
Kuwait	8.3	293
Oman	6.7	238
Qatar	11.2	396
Saudi Arabia	56.7	2 002
Syria (Arab Republic)	5.8	205
United Arab Emirates	36.2	1 280
Total Middle East	**216.6**	**7 649**
Australia	23.8	841
New Zealand	5.9	210
Papua New Guinea	0.1	4
Total Oceania	**29.8**	**1 055**
Total World	**2 603.9**	**91 942**

Notes:

(1) Sources: WEC Member Committees, 2003; *Natural Gas in the World 2001/2002*, Cedigaz; other international and national sources; estimates by the editors.

COUNTRY NOTES

The following Country Notes on Natural Gas provide a brief account of countries with significant gas resources. They have been compiled by the editors, drawing upon a wide variety of material including information received from WEC Member Committees, national and international publications.

The principal published sources consulted were:

- *Natural Gas in the World, 2001/2002 Survey*; Cedigaz;
- *BP Statistical Review of World Energy, June* 2003;
- *Energy Balances of OECD Countries 2000–2001;* 2003; International Energy Agency;
- *Energy Balances of Non-OECD Countries 2000–2001;* 2003; International Energy Agency;
- *Energy Statistics of OECD Countries 2000–2001;* 2003; International Energy Agency;
- *Energy Statistics of Non-OECD Countries 2000–2001;* 2003; International Energy Agency;
- *Oil & Gas Journal, 22 December, 2003*, PennWell Corporation;
- *OPEC Annual Statistical Bulletin 2002;* 2003, OPEC
- *Secretary-General's Twenty-Ninth Annual Report, A.H. 1422–1423/AD 2002;* 2003, OAPEC
- *World Oil, August 2003*, Gulf Publishing Company

Brief salient data are shown for each country, including the year of first commercial production of natural gas (where ascertained).

Algeria

Proved recoverable reserves (bcm)	4 000
Production (net bcm, 2002)	79.3
R/P ratio (years)	45.0
Year of first commercial production	1961

For the purposes of the present *Survey*, the Algerian WEC Member Committee has reported a proved amount in place of 6 000 bcm, of which 4 000 bcm is classified as proved recoverable reserves. Gas reserves non-associated with crude oil account for 80% of proved recoverable reserves. An additional amount in place of 2 000 bcm, of which 960 bcm is deemed to be recoverable, has also been reported by the Algerian MC.

Net production of natural gas in 2002 was the fifth highest in the world, after Russia, the USA, Canada and the UK. About 45% of gross production was re-injected, while much smaller proportions (in the order of 3% of production in each case) were flared or abstracted as NGLs. About 74% of net production was exported: 46% of gas exports were in the form of LNG, consigned to France, Spain, Turkey, Belgium, Italy, the USA and Greece. Exports by pipeline in 2002 went to Italy, Spain, Portugal, Tunisia, Morocco and Slovenia. Apart from oil and gas industry use, the main internal markets for Algerian gas are power stations, industrial fuel/feedstock and households.

Argentina

Proved recoverable reserves (bcm)	664
Production (net bcm, 2002)	36.1
R/P ratio (years)	15.6

Until 1993 responsibility for the transportation, distribution and commercialisation of natural gas in Argentina rested with the state company Gas del Estado. This responsibility was then transferred to the private sector; the role of the State changing from one of entrepreneur to one of Regulator, Supervisor and Judge of gas industry activity, through the creation of the regulation agency 'ENARGAS'. Therefore, under this new industrial organisation, transportation, distribution and commercialisation are in the hands of the private sector, operating within a competitive market structure. Free access to transport and distribution gas pipelines is granted and there are no restrictions related to the import and export of natural gas.

The proved recoverable reserves reported by the Argentinean WEC Member Committee are some 11% lower than those contributed to the 2001 *Survey*. Additional reserves, not yet proven but considered to be eventually recoverable, now stand at 306 billion m^3.

Gas extraction takes place in five sedimentary basins. The greatest production corresponds to the Neuquina basin which provides 51.9% of the total, followed by the Austral basin with 22.4% and the Northwest basin with 19.5%; minor contributions are made by the Golfo San Jorge and Cuyana basins. About 7% of current gross production is re-injected. Marketed production (after relatively small amounts are deducted through flaring and shrinkage) is the highest in South America.

For many years, gas supplies have been augmented by imports from Bolivia, but this flow ceased in October 1999, as the focus of Bolivia's gas exports shifted to Brazil. In a further re-orientation of the South American gas supply structure, Argentina has become a significant exporter in its own right, with a number of pipelines supplying Chile and others to Uruguay and Brazil.

Consumption of indigenous and imported gas in 2001 was almost equally divided between the power generation market (25%), industrial fuel/feedstock (22%), residential/commercial uses (24%) and gas industry own use/loss (24%); about 5% was consumed in road transport.

Australia

Proved recoverable reserves (bcm)	860
Production (net bcm, 2002)	34.6
R/P ratio (years)	22.2
Year of first commercial production	1969

The level of proved recoverable reserves quoted above is that reported by the Australian WEC Member Committee; it corresponds to 'Remaining commercial reserves at 1 January 2001' as given in *Oil and Gas Resources of Australia 2001*, published by Geoscience Australia in 2002. Doubtless due to the adoption of differing definitions of 'proved reserves', other published sources quote substantially higher levels, ranging (in terms of bcm) from *World Oil's* 2 407 and *Oil & Gas Journal's* 2 548 to OPEC at 3 550 and Cedigaz at 3 930.

The Australian MC also reports 3 136 bcm as 'estimated additional reserves recoverable': this corresponds with 'non-commercial reserves' in the Geoscience Australia publication cited above. Australia's principal gas reserves are located in the Carnarvon, Gippsland, Browse, Bonaparte and Cooper Basins.

Gross production grew by over 60% between 1990 and 1996, reflecting in part higher domestic demand but more especially a substantial increase in exports of LNG (almost all to Japan) from the North West Shelf fields. Production growth has continued in recent years, but at a slower pace.

The main gas-consuming sectors in Australia are public electricity generation, the non-ferrous metals industry and the residential sector.

Azerbaijan

Proved recoverable reserves (bcm)	1 370
Production (net bcm, 2002)	5.2
R/P ratio (years)	>100

Azerbaijan is one of the world's oldest producers of natural gas. After years of falling production the outlook has been transformed by recent developments. Proved reserves of gas, as quoted by Cedigaz, remain at the end-1999 level of 1 370 bcm.

Marketed production in 2002 was 5.2 bcm, of which much the greater part came from offshore fields. Over half of current gross production is reported to be flared or vented.

Bangladesh

Proved recoverable reserves (bcm)	301
Production (net bcm, 2002)	11.5
R/P ratio (years)	26.3
Year of first commercial production	pre-1971

Whilst the published volumes of proved gas reserves (e.g. *Oil & Gas Journal* 301 bcm, Cedigaz 340 bcm) are not particularly large, much of Bangladesh is poorly explored and the potential for further discoveries is thought to be substantial.

Gas production has followed a rising trend for many years and now exceeds 10 bcm per annum. Natural gas contributes about two-thirds of Bangladesh's commercial energy supplies; its principal outlets are power stations and fertiliser plants. Consumption by other industrial users and the residential/commercial sector is growing rapidly.

Bolivia

Proved recoverable reserves (bcm)	813
Production (net bcm, 2002)	5.8
R/P ratio (years)	>100
Year of first commercial production	1955

Recent years have seen annual increases in Bolivia's proved reserves of natural gas, as reported by Cedigaz, rising some 57% from 518 bcm at end-1999 to 813 bcm at end-2002. In addition, official assessments of 'probable reserves', published by the Instituto Nacional de Estadística, have been raised from 394 to 742 bcm over the same period.

Exports to Argentina used to be the major outlet for Bolivia's natural gas, but this flow ceased during 1999. The focus of Bolivia's gas export trade has now shifted towards Brazil, with the inauguration of two major export lines, one from Santa Cruz de la Sierra to south-east Brazil in 1999 and another in 2000 from San Miguel to Cuiaba. Exports in 2002 amounted to 4.8 bcm.

Internal consumption of gas is still on a small scale (only about 1 bcm/yr), and confined almost entirely to electricity generation and industrial fuel markets, residential use being minimal at present.

Brazil

Proved recoverable reserves (bcm)	237
Production (net bcm, 2002)	9.4
R/P ratio (years)	19.5
Year of first commercial production	1954

Brazil's natural gas industry is relatively small at present compared with its oil sector. Proved reserves, as reported by the Brazilian WEC Member Committee, are the fifth largest in South America, having increased only marginally over the past 3 years. Of the latest assessment of proved recoverable reserves, approximately 27% is non-associated with crude oil. Additional recoverable reserves, not classified as proved, are put at just over 95 bcm.

Over one-third of current gross production of natural gas is either re-injected or flared. Marketed production is mostly used as industrial fuel or as feedstock for the production of petrochemicals and fertilisers. As a consequence of Brazil's huge hydroelectric resources, use of natural gas as a power station fuel has been minimal until fairly recently. The consumption picture is now changing as imported gas (from Bolivia and Argentina) fuels the increasing number of gas-fired power plants that are being built in Brazil.

Brunei

Proved recoverable reserves (bcm)	391
Production (net bcm, 2002)	10.9
R/P ratio (years)	35.2

Natural gas was found in association with oil at Seria and other fields in Brunei. For many years this resource was virtually unexploited, but in the 1960s a realisation of the resource potential, coupled with the availability of new technology for producing and transporting liquefied natural gas, enabled a major gas export scheme to be devised. Since 1972 Brunei has been exporting LNG to Japan, and more recently to the Korean Republic.

Occasional spot cargoes are also made—in 2002 to Spain and the USA.

Despite annual exports approaching 10 bcm, proved reserves (as published by *Oil & Gas Journal*) have remained virtually steady at just under 400 bcm since 1992.

Nearly 90% of Brunei's marketed production is exported, the balance being mostly used in the liquefaction plant, local power stations and offshore oil and gas installations. Small quantities are used for residential purposes in Seria and Kuala Belait.

Canada

Proved recoverable reserves (bcm)	1 664
Production (net bcm, 2002)	187.6
R/P ratio (years)	8.0

Canada's gas reserves are the third largest in the Western Hemisphere. The proved recoverable reserves reported by the Canadian WEC Member Committee for the present *Survey* corresponded with 'remaining established reserves' of marketable natural gas at 31 December, 2001, as assessed by the Canadian Association of Petroleum Producers (CAPP). The level adopted for the *Survey* reflects the end-2002 data subsequently published by CAPP.

'Initial established reserves' of marketable natural gas (which include cumulative production to date) are quoted by CAPP as 5 450 bcm at end-2002. The Canadian WEC Member Committee reports that (based on a 1999 assessment) Canada is estimated to have 12 800 bcm of 'undiscovered in-place resources' of gas, of which 8 880 bcm is considered to be recoverable.

At end-2002, 85% of proved recoverable reserves consisted of non-associated deposits: the provinces with the largest gas resources were Alberta (with 75% of remaining established reserves), British Columbia (15%) and Saskatchewan (5%).

Gross production of Canadian natural gas was the third highest in the world in 2002. Of the net output remaining after allowance for re-injection, flaring and shrinkage, approximately 58% was exported to the United States. The largest users of gas within Canada are the industrial, residential and commercial sectors. A relatively small proportion is consumed in electricity generation, reflecting Canada's wealth of hydroelectric power.

China

Proved recoverable reserves (bcm)	1 676
Production (net bcm, 2002)	32.6
R/P ratio (years)	51.4
Year of first commercial production	1955

Past gas discoveries have been fewer than those of crude oil, which is reflected in the fairly moderate level of proved reserves. Gas reservoirs have been identified in many parts of China, including in particular the Sichuan Basin in the central region, the Tarim Basin in the north-west and the Yinggehai (South China Sea). China's gas resource base is thought to be enormous: estimates by the Research Institute of Petroleum Exploration and Development, quoted by Cedigaz, put total resources at some 38 000 bcm, of which 21% are located offshore. Most of the onshore gas-bearing basins are in the central and western parts of China.

The level of proved reserves adopted for the present *Survey* has been taken from a personal communication from an expert professional source in China, as no data were available from the WEC Member Committee. The reported level of 1 676 bcm is 11% higher than the end-2002 currently quoted by *Oil & Gas Journal* (1 510), whilst other published assessments fall within a broad band, ranging from *World Oil's* 1 321 bcm to Cedigaz's 2 600 bcm; OAPEC (1 510) and OPEC (1 560) are close to *Oil & Gas Journal*.

The major outlets for natural gas within China are as industrial fuel/feedstock (40%), oil/gas industry own use/loss (27%) and the residential sector (21%). Natural gas has relatively small shares in the generation of electricity and bulk heat. In January 1996, China began delivering natural gas to the Castle Peak power station in Hong Kong via a pipeline from the offshore Yacheng field; deliveries in 2002 were about 2.4 bcm.

Colombia

Proved recoverable reserves (bcm)	120
Production (net bcm, 2002)	6.2
R/P ratio (years)	16.9

The early gas discoveries were made in the north-west of the country and in the Middle and Upper Magdalena Basins; in more recent times, major gas finds have been made in the Llanos Basin to the east of the Andes. Proved reserves at end-2002 are quoted by Cedigaz as 120 bcm, a reduction of 38% from the figure quoted in the 2001 *Survey*, which emanated from the same source. Other published sources range from *World Oil's* 120 bcm to OPEC's 198. The Colombian state oil and gas company Ecopetrol quotes 2001 'remaining reserves' as 7 490 bcf (equivalent to 212 bcm).

At present a high proportion of Colombia's gas output (53% in 2002) is re-injected in order to maintain or enhance reservoir pressures. The major outlets for natural gas are own use by the gas industry and the power generation market, each of which represented about 32% of total gas consumption in 2001. Chemicals, cement and other industrial users had a 19% share, while residential/commercial consumers accounted for 16%. CNG use in road transport is small but growing.

Denmark

Proved recoverable reserves (bcm)	87
Production (net bcm, 2002)	8.4
R/P ratio (years)	10.1
Year of first commercial production	1984

The Danish WEC Member Committee quotes the Danish Energy Authority which does not use the terms proved and additional reserves, but uses the categories ongoing, approved, planned and possible recovery.

The figure for proved reserves (87 bcm) has been calculated as the sum of 'ongoing' and 'approved' reserves, while the figure for additional reserves (50 bcm) has been calculated as the sum of 'planned' and 'possible' reserves. Of the reported proved recoverable reserves, 56% is non-associated with crude oil.

Egypt (Arab Republic)

Proved recoverable reserves (bcm)	1 657
Production (net bcm, 2002)	26.7
R/P ratio (years)	54.2
Year of first commercial production	1964

Proved reserves are the third largest in Africa, having been increased by 35% since the 2001 *Survey*, according to the latest data reported by the Egyptian Member Committee. Since the end of 2000, Egypt's reserves have exceeded those of its neighbour Libya. About 97% of the reported reserves is non-associated with crude oil. An estimated 61.5 tcf (1 741 bcm) is reported as additional reserves recoverable, over and above the proved reserves.

The major producing area is the Mediterranean Sea region (mostly from offshore fields), although output of associated gas from a number of fields in the Western Desert and the Red Sea region is also important.

Marketed production has grown steadily in recent years and is now the second largest in Africa. The main outlets at present are power stations, fertiliser plants and industrial users such as the iron and steel sector and cement works.

Germany

Proved recoverable reserves (bcm)	326
Production (net bcm, 2002)	21.6
R/P ratio (years)	14.7

Although it is one of Europe's oldest gas producers, Germany's remaining proved reserves are still sizeable, and (apart from the Netherlands) they now rank as the largest onshore reserves in Western Europe. The principal producing area is in north Germany, between the rivers Weser and Elbe; westward from the Weser to the Netherlands border lies the other main producing zone, with more mature fields.

The proved recoverable reserves reported by the German WEC Member Committee are some 15% higher than the corresponding level in the 2001 *Survey*. An additional 200 bcm is considered to be eventually recoverable.

Indigenous production provides less than a quarter of Germany's gas supplies; the greater part of demand is met by imports from the Russian Federation, the Netherlands, Norway, the UK and Denmark.

India

Proved recoverable reserves (bcm)	751
Production (net bcm, 2002)	25.0
R/P ratio (years)	27.7
Year of first commercial production	1961

A sizeable natural gas industry has been developed on the basis of the offshore Bombay gas and oil/gas fields. Proved reserves at 1 April, 2002 have been reported by the Indian WEC Member Committee as 751 bcm. This level appears to be consistent with the series of 'proved and indicated balance recoverable reserves' published by the Ministry of Petroleum and Natural Gas. The Western Offshore (including the Bombay High) accounted for 58% of India's gas reserves in 2001, with the eastern state of Assam possessing 20% and the western states of Gujarat and Rajasthan 11%.

According to Cedigaz, out of a gross production of 27 bcm in 2002, 3% was flared or vented and 5% lost as a result of shrinkage, etc.

Marketed production is principally used as feedstock for fertiliser and petrochemical manufacture, for electricity generation and as industrial fuel. The recorded use in the residential and agricultural sectors is exceedingly small.

Indonesia

Proved recoverable reserves (bcm)	2 183
Production (net bcm, 2002)	70.4
R/P ratio (years)	27.3

The Indonesian WEC Member Committee reports proved recoverable gas reserves as 77 100 bscf (2 183 bcm), within a proved amount in place of 136 500 bscf (3 865 bcm). Other published assessments of Indonesia's proved reserves vary widely—they range from *World Oil's* 2 081 bcm to Cedigaz and OAPEC with around 3 800 bcm, with *Oil & Gas Journal* and OPEC quoting intermediate levels in the region of 2 600 bcm. Differences in definitions and coverage may account for some of the discrepancies.

The latest level reported by the Member Committee for proved reserves is 1.3% lower than that advised for the 2001 *Survey of Energy Resources*, whilst the latest assessment of the proved amount in place is nearly 16% higher than the previous one.

Indonesia's gas production is the highest in Asia. The main producing areas are in northern Sumatra, Java and eastern Kalimantan.

Exports of LNG from Arun (Sumatra) and Bontang (Kalimantan) to Japan began in 1977–1978. Indonesia has for many years been the world's leading exporter of LNG. Shipments in 2002 were chiefly to Japan (67%) but also to the Republic of Korea (20%) and Taiwan, China (13%). Indonesia exports more than half of its marketed production, including (from early 2001) supplies by pipeline to Singapore.

The principal domestic consumers of natural gas (apart from the oil and gas industry) are power stations and fertiliser plants: the residential and commercial sectors have relatively small shares.

Iran (Islamic Republic)

Proved recoverable reserves (bcm)	26 570
Production (net bcm, 2002)	67.3
R/P ratio (years)	>100
Year of first commercial production	1955

Iran's proved reserves are second only to those of the Russian Federation, (although now closely approached by those of Qatar). They account for 16% of the world total, and exceed the combined proved reserves of North America, South America and Europe (excluding the Russian Federation). The Iranian WEC Member Committee reports that at the end of 2002 proved reserves of natural gas were 26 570 billion m^3, approximately 9% higher than the end-1999 level reported for the 2001 *Survey of Energy Resources*.

Of the end-1999 reserves, 63% were non-associated with crude oil. For many years only minute quantities of associated gas output were utilised as fuel in the oil fields or at Abadan refinery: by far the greater part was flared. Utilisation of gas in the industrial, residential and commercial sectors began in 1962 after the construction of a pipeline from Gach Saran to Shiraz.

In 2002, 56% of Iran's gross production of 121 bcm of gas was marketed; about 23% was re-injected into formations in order to maintain or enhance pressure; about 7% was flared or vented and 14% lost through shrinkage and other factors. The marketed production volume of about 67 bcm was augmented by 5 bcm of gas imported from Turkmenistan, whilst a small quantity (0.7 bcm in 2002) was exported to Turkey. Iran's principal gas-consuming sectors are electricity generation (36% of total consumption in 2001), residential users (29%) and industry (23%).

Iraq

Proved recoverable reserves (bcm)	3 100
Production (net bcm, 2002)	2.4
R/P ratio (years)	>100
Year of first commercial production	1955

Gas resources are not particularly large, by Middle East standards: proved reserves (as reported by OAPEC) account for less than 5% of the regional total. Other published sources all quote proved reserves in the range of 3 100–3 200 bcm.

According to data reported by Cedigaz, 70% of Iraq's proved reserves consist of associated gas, whilst cap gas and non-associated gas account for 15% each. A high proportion of gas output is thus associated with oil production: some of the associated gas is flared.

Between 1986 and 1990 Iraq exported gas to Kuwait. Currently all gas usage is internal, as fuel for electricity generation, as a feedstock and fuel for the production of fertilisers and petrochemicals, and as a fuel in oil and gas industry operations.

Kazakhstan

Proved recoverable reserves (bcm)	1 841
Production (net bcm, 2002)	11.2
R/P ratio (years)	>100

The estimated proved reserves of 1.8–1.9 trillion m^3 quoted by *Oil & Gas Journal* and Cedigaz include 1.3 trillion for the giant Karachaganak field, located in the north of Kazakhstan, near the border with the Russian Federation. Another major field is Tengiz, close to the north-east coast of the Caspian Sea.

Kazakhstan exports natural gas to Russia from its western producing areas and imports gas into its south-eastern region from Turkmenistan and Uzbekistan.

Electricity generation is estimated to have accounted for about one-third of total internal gas consumption in 2001.

Kuwait

Proved recoverable reserves (bcm)	1 557
Production (net bcm, 2002)	8.3
R/P ratio (years)	>100
Year of first commercial production	1960

Note: Kuwait data include its share of Neutral Zone.

Gas reserves (as quoted by OAPEC) are relatively low in regional terms and represent only about 2% of the Middle East total. All of Kuwait's natural gas is associated with crude oil, so that its availability is basically dependent on the level of oil output.

After allowing for a limited amount of flaring and for shrinkage due to the extraction of NGLs, Kuwait's gas consumption is currently about 8 bcm/yr, one-third of which is used for electricity generation and desalination of seawater.

Libya/GSPLAJ

Proved recoverable reserves (bcm)	1 314
Production (net bcm, 2002)	6.2
R/P ratio (years)	>100
Year of first commercial production	1970

Proved reserves—the fourth largest in Africa—have been essentially unchanged since 1991, according to OAPEC. Utilisation of the resource is on a comparatively small scale: net production in 2002 was less than a quarter that of Egypt.

Since 1970 Libya has operated a liquefaction plant at Marsa el Brega, but LNG exports (in recent years, only to Spain) have fallen away to only 0.6 billion m^3 per annum.

Local consumption of gas is largely attributable to petrochemical/fertiliser plants and oil and gas industry use.

Malaysia

Proved recoverable reserves (bcm)	2 124
Production (net bcm, 2002)	48.5
R/P ratio (years)	40.2
Year of first commercial production	1983

Exploration of Malaysia's offshore waters has located numerous fields yielding natural gas or gas/condensates, mainly in the areas east of the peninsula and north of the Sarawak coast. Proved reserves (as quoted by *Oil & Gas Journal*) now stand at 75 tcf (2 124 bcm) and rank as the second highest in Asia, after Indonesia. Other published reserve assessments are somewhat higher, ranging from OPEC's 2 390 bcm to Cedigaz at 2 478 bcm and *World Oil* at 2 492 bcm.

Malaysia became a major gas producer in 1983, when it commenced exporting LNG to Japan. This trade has continued ever since, supplemented in recent years by LNG sales to the Republic of Korea and Taiwan, China and by gas supplies via pipeline to Singapore. In 2002, spot sales of LNG were made to Spain and the USA.

Domestic consumption of gas has become significant in recent years, the major market being power generation. The other principal outlet for natural gas, apart from own use within the oil/gas industry, is as feedstock/fuel for industrial users. Small amounts of CNG are used in transport, reflecting an official programme to promote its use.

Mexico

Proved recoverable reserves (bcm)	424
Production (net bcm, 2002)	38.0
R/P ratio (years)	10.3

The Mexican WEC Member Committee reports that proved recoverable reserves of dry natural gas at the end of 2002 were 424 billion m^3. The proved amount in place is reported as 902 bcm, equivalent to the sum of 'proved' plus 'probable' reserves as stated by Petróleos Mexicanos (Pemex) in their *Informe Estadístico de Labores 2002*. An additional amount in place of 480 bcm reflects the Pemex level of 'possible' reserves. Within the total amount of proved reserves, just over 50% are located in the southern region, 22% in the northern region, 18% in the marine north-east region and 10% in the marine south-west region.

The level and distribution of Mexico's reserves quoted above reflect a radical restatement carried out by Pemex in order to conform with the definitions laid down by the Securities and Exchange Commission of the United States.

In this context, proved reserves have been drastically reduced, while 'probable' and 'possible' reserves have been revised upwards.

Production of natural gas has been on a slowly declining trend in recent years. The greater part of Mexico's gas production (73% in 2002) is associated with crude oil output, mostly in the southern producing areas, both onshore and offshore.

The largest outlet for gas is as power station fuel (37% of total inland disposals in 2001); the energy industry consumed about 36%, industrial fuel/feedstock 24%; and households about 2%. Mexico habitually exports relatively small amounts of gas to the USA and imports somewhat larger quantities.

Myanmar

Proved recoverable reserves (bcm)	445
Production (net bcm, 2002)	8.4
R/P ratio (years)	50.6

Myanmar has long been a small-scale producer of natural gas, as of crude oil, but its resource base would support a substantially higher output of gas. Proved reserves are put at 10 tcf (283 bcm) by *Oil & Gas Journal*, but for the purposes of the present *Survey* the higher level (445 bcm) published more recently by *World Oil* and Cedigaz has been utilised.

Until 2000, gas production tended to oscillate around a slowly rising trend. With the commencement of exports of natural gas to Thailand from two offshore fields, first Yadana and subsequently Yetagun, Myanmar's gas industry has entered a new phase. As offtake by Thailand's 3 200 MW Ratchaburi Power Plant has built up, gas production in Myanmar has moved onto a significantly higher level than in the past. Domestic consumption of gas is mainly for power generation.

Netherlands

Proved recoverable reserves (bcm)	1 756
Production (net bcm, 2002)	71.2
R/P ratio (years)	24.6

In the absence of a report from the Netherlands WEC Member Committee, the level of proved reserves reflects that quoted by *Oil & Gas Journal*, which is marginally higher than that reported for end-1999. They still represent one of the largest gas resources in Western Europe. The giant Groningen field in the north-west of the Netherlands accounts for almost two-thirds of the country's proved reserves.

Gas production has tended to fluctuate in recent years, depending on weather conditions in Europe, thus demonstrating the flexibility that enables the Netherlands to play the role of a swing producer. Nearly 60% of 1999 output came from onshore fields, with Groningen contributing about 40%.

Over half of Netherlands gas output is exported, principally to Germany but also to Italy, Belgium, France, the UK and Switzerland. The principal domestic markets are electricity and heat generation, the residential sector and industrial fuel and feedstock.

New Zealand

Proved recoverable reserves (bcm)	42
Production (net bcm, 2002)	5.9
R/P ratio (years)	6.9
Year of first commercial production	1970

The Maui offshore gas/condensate field (discovered in 1969) is the largest hydrocarbon deposit so far discovered in New Zealand: it presently accounts for 36% of the country's economically recoverable gas reserves. Effective utilisation of its gas resources has been a key factor in New Zealand's energy policy since the early 1980s.

The proved recoverable reserves adopted for the present *Survey* correspond with the estimates of 'proven and probable' reserves (or P50 values) that have been compiled by the Ministry of Economic Development, on the basis of information provided by field operators. These reserves have been assessed within the context of 'ultimate recoverable reserves' of about

158 billion m^3. The latest assessment of proved reserves is substantially lower than that for end-1999, largely due to a major reduction in Maui's reserves.

The Maui field came into commercial production in 1979 when a pipeline to the mainland was completed. Three plants were commissioned in the 1980s to use indigenous gas, producing (respectively) methanol, ammonia/urea and synthetic gasoline. Nine gas fields were in production in 2002, with Maui accounting for over 70% of total output.

An extensive transmission and distribution network serves industrial, commercial and residential consumers in the North Island. Small (and declining) amounts of CNG are used in motor vehicles.

Nigeria

Proved recoverable reserves (bcm)	5 055
Production (net bcm, 2002)	14.2
R/P ratio (years)	>100
Year of first commercial production	1963

Published assessments of Nigeria's proved reserves of natural gas at the end of 1999 all fell within a narrow band (3 511–3 568 bcm). Since then disparities have emerged in respect of the level of reserves at end-2002: although *Oil & Gas Journal* has remained at the bottom of the range, the other published sources now quote substantially higher figures: OAPEC and OPEC give 4 503 bcm, whilst Cedigaz and *World Oil* say 5 055 bcm. The level adopted for the present *Survey* is that quoted by Cedigaz in their 2001–2002 Survey.

Nigeria's proved reserves on this basis are now the largest in Africa, ahead of those of Algeria, but historically its degree of gas utilisation has been very low. Much of the associated gas produced has had to be flared, in the absence of sufficient market outlets. Efforts are being made to develop gas markets, both locally and internationally, and to reduce flaring to a minimum. There are projects to replace non-associated gas by associated gas in supplies to power stations and industrial users. Virtually 50% of Nigeria's gross gas production of 38.1 bcm in 2002 was flared or vented.

The Bonny LNG plant (commissioned in the second half of 1999) exported 7.8 bcm of natural gas as LNG during 2002, chiefly to Italy, Spain and Turkey, with smaller quantities going to France, Portugal and the USA. An expansion of the plant was completed towards the end of 2002 and a further enlargement of Bonny is scheduled to come on stream by end-2005. A project is underway for the construction of a pipeline to supply Nigerian associated gas to power plants in Benin, Togo and Ghana.

Norway

Proved recoverable reserves (bcm)	2 120
Production (net bcm, 2002)	68.9
R/P ratio (years)	29.1
Year of first commercial production	1977

Norway's proved reserves are the highest in Europe (excluding the Russian Federation). The bulk of reserves is located in the North Sea, the rest having been discovered in the Norwegian Sea and the Barents Sea. The level of proved recoverable reserves reported by the Norwegian WEC Member Committee has risen sharply from 1 245 bcm at end-1999 to 2 120 bcm at end-2002. The latter figure is set within the context of a slightly reduced 'proved amount in place' of 3 670 bcm. In addition, some 2 510 bcm of non-proved gas is believed to be in situ, but no estimate is available of the quantity of gas likely to be recoverable therefrom.

Norway's gas production continues to follow a rising trend. A high proportion (31% in 2002) of output is re-injected; nearly 90% of marketed production is exported. In 2002 supplies went to 10 European countries, principally Germany, France, Belgium, Italy, the UK and the Netherlands. Apart from gas industry own use, Norway's internal consumption of gas is still at a very low level, being largely confined to minor feedstock use.

Oman

Proved recoverable reserves (bcm)	859
Production (net bcm, 2002)	15.0
R/P ratio (years)	46.2
Year of first commercial production	1978

Oman is one of the smaller gas producers in the Middle East, with moderate proved reserves which have risen slowly since 1995, according to OAPEC. The levels of reserves quoted in other published sources fall within a fairly narrow range, from *Oil & Gas Journal's* 829 bcm to Cedigaz at 946, with *World Oil* roughly midway at 878.

Oman has developed its utilisation of gas to such an extent that oil has long been displaced as the Sultanate's leading energy supplier. Currently, the principal outlet for marketed gas is the power generation/desalination complex at Ghubrah. Other gas consumers include mining and cement companies.

The Oman LNG project began operating in early 2000, with the first shipment (to the Republic of Korea) taking place in April. Regular shipments of LNG are also being made to Japan, whilst during 2002 additional supplies (including spot cargoes) were delivered to Spain, France, Belgium and the USA.

Pakistan

Proved recoverable reserves (bcm)	801
Production (net bcm, 2002)	22.9
R/P ratio (years)	31.1
Year of first commercial production	1955

The level of proved reserves reported by the Pakistan WEC Member Committee for end-2002 (28 288 bcf, equivalent to 801 bcm) is substantially higher than that reported for end-1999 as input to the 2001 *Survey of Energy Resources* (581 bcm). The increase reflects the inclusion of several new fields and the re-appraisal of many others. Currently, the major gas-producing fields are Sui in Balochistan and Qadirpur and Mari in Sindh. Only 4% of natural gas output is associated with oil production.

Production of natural gas increased by 32% over the 5 years up to 2001-2002. The major markets for gas (excluding own use) in that year were power generation (38%), fertiliser plants (22%), households and commercial consumers (20%) and industrial users (19%). Small (but growing) quantities of CNG are consumed as a transport fuel.

Papua New Guinea

Proved recoverable reserves (bcm)	428
Production (net bcm, 2002)	0.1
R/P ratio (years)	>100
Year of first commercial production	1991

The Hides gas field was discovered in 1987 and brought into production in December 1991. Other resources of non-associated gas have been located in PNG, both on land and offshore. Published assessments of proved reserves range between *Oil & Gas Journal's* 346 and the 428 bcm quoted by Cedigaz, with *World Oil* positioned roughly midway at 381 bcm: for the present *Survey*, the level given by Cedigaz has been retained.

Up to the present, the only marketing outlet for Hides gas has been a 42 MW gas-turbine power plant serving the Porgera gold mine; offtake averages 14–15 million cubic feet/day. Associated gas produced in the Kutubu area is mostly re-injected into the formation.

The Highlands Gas Project for a gas export pipeline to Australia recorded some progress in 2003, with the signing of conditional agreements with two Australian prospective customers. The proposed pipeline includes a 500 km undersea section across the Torres Strait and 2 100 km of line following a route southwards close to the coastline of Queensland.

Peru

Proved recoverable reserves (bcm)	245
Production (net bcm, 2002)	0.4
R/P ratio (years)	>100

There is virtual consensus among the main published sources with regard to Peru's gas reserves: for the present Survey, *Oil & Gas Journal's* assessment of 8 655 bcf (only marginally lower than the end-1999 figure of 9 000 bcf) has been adopted. Gas output is mostly associated with oil production; an appreciable proportion of production (36% in 2002) is re-injected, whilst about 18% is flared or vented.

Marketed production of gas has averaged around 0.4 bcm/yr in recent times. Consumption is divided approximately equally between power stations and the upstream operations of the oil and gas industry.

Qatar

Proved recoverable reserves (bcm)	25 783
Production (net bcm, 2002)	29.5
R/P ratio (years)	>100
Year of first commercial production	1963

Qatar's gas resources far outweigh its oil endowment: its proved reserves of gas of almost 26 trillion m^3 (as quoted by Cedigaz, OAPEC, OPEC and *World Oil*) are only exceeded within the Middle East by those reported by Iran, and account for about 15% of global gas reserves. Although associated gas has been discovered in oil fields both on land and offshore, the key factor in Qatar's gas situation is non-associated gas, particularly that in the offshore North Field, one of the largest gas reservoirs in the world.

Production of North Field gas began in 1991 and by 2002 Qatar's total annual gross production had risen to about 38 bcm; 10% was re-injected and nearly 12% lost through shrinkage. The gas consumed locally is principally for power generation/desalination, fertiliser and petrochemical production and gas industry own use.

Since the end of 1996, Qatar has become a substantial exporter of LNG; in 2002, shipments exceeded 18 billion m^3 of gas, of which 46% was consigned to Japan, 38% to the Republic of Korea, 9% to Spain and smaller percentages to the USA, Belgium, Puerto Rico and Italy.

Romania

Proved recoverable reserves (bcm)	163
Production (net bcm, 2002)	12.1
R/P ratio (years)	13.0

The Romanian WEC Member Committee reports proved recoverable reserves of 163.3 billion m^3, a substantial reduction on the 405.6 bcm reported for the 2001 *Survey*. The reported additional amount of 'unproved' gas in place has fallen from 370.6 to 100.6 bcm. The proportion of proved recoverable reserves that is non-associated with crude oil is reported to be 76.5%.

After peaking in the mid-1980s, Romania's natural gas output has been in secular decline, falling to around 12 bcm in 2002, only about one-third of its peak level. Indigenous production currently supplies about 75% of Romania's gas demand; the principal users are power stations, CHP and district heating plants, the steel and chemical industries and the residential/commercial sector.

Russian Federation

Proved recoverable reserves (bcm)	47 000
Production (net bcm, 2002)	595.3
R/P ratio (years)	77.3

The gas resource base is by far the largest in the world: current estimates of Russia's proved reserves (quoted as 47 000 bcm by Cedigaz) are about 75% greater than those of Iran and nine times those of the USA. Other major published sources quote figures similar to the level given by Cedigaz. The greater part (77%) of the Federation's reserves are located in West Siberia, where the existence of many giant, and a number of super-giant, gas fields has been proved.

The Federation's net natural gas production of 595.3 bcm in 2002 accounted for almost 23% of the world total.

Russia is easily the largest exporter of natural gas in the world: in 2002, according to Cedigaz, its exports totalled 207 bcm, of which 86 bcm went to Western Europe, 41 bcm to Central

Europe and the balance to former republics of the Soviet Union.

Saudi Arabia

Proved recoverable reserves (bcm)	6 343
Production (net bcm, 2002)	56.7
R/P ratio (years)	>100
Year of first commercial production	1961

Note: Saudi Arabia data include its share of Neutral Zone.

Most of Saudi Arabia's proved reserves and production of natural gas are in the form of associated gas derived from oil fields, although a number of sources of non-associated gas have been discovered. In total, proved reserves of gas (as given by OAPEC) rank as the third largest in the Middle East. Other published assessments of Saudi Arabian reserves range fairly narrowly from *Oil & Gas Journal's* 6 363 bcm to OPEC's 6 646, with Cedigaz (6 540) and *World Oil* (6 643) at intermediate levels.

Output of natural gas has advanced fairly steadily during the past 10 years. A significant factor in increasing the utilisation of Saudi Arabia's gas resources has been the operation of the gas-processing plants set up under the Master Gas System, which was inaugurated in the mid-1980s. These plants produce large quantities of ethane and LPG, which are used within the country as petrochemical feedstock; a high proportion of LPGs is exported. The main consumers of dry natural gas (apart from the gas industry itself) are power stations, desalination plants and petrochemical complexes.

Thailand

Proved recoverable reserves (bcm)	441
Production (net bcm, 2002)	20.0
R/P ratio (years)	21.0
Year of first commercial production	1981

Thailand's WEC Member Committee reports proved recoverable reserves at end-2002 as 15 578 bcf (equivalent to 441 billion m^3). Non-associated gas accounts for 98% of the reported level of proved reserves. In addition to the proven quantities, the Committee reports an additional 9 096 bcf (258 bcm) as considered to be recoverable in due course. Other published assessments of Thailand's proved gas reserves fall short of the level reported for the present *Survey*: they range from *World Oil's* 365 bcm to Cedigaz at 385 bcm, with *Oil & Gas Journal* in between at 378.

Since its inception 20 years ago, Thailand's natural gas output has grown almost unremittingly year after year. Much the greater part of Thailand's gas output is used for electricity generation; industrial use for fuel or chemical feedstock is relatively small, whilst transport use (CNG) is at present minimal.

Thailand began to import natural gas from Myanmar in 1999; in 2002 the volume involved was 6.2 bcm.

Trinidad & Tobago

Proved recoverable reserves (bcm)	589
Production (net bcm, 2002)	17.3
R/P ratio (years)	30.7

Trinidad's proved reserves of natural gas, as assessed by Cedigaz, fell by about 2% between end-1999 and end-2002. Other published assessments are reasonably close: OPEC at 558 bcm, *World Oil* at 576 bcm and *Oil & Gas Journal* somewhat higher at 664.

Marketed production of gas has increased rapidly during recent years, as exports from the Atlantic LNG plant (inaugurated in 1999) have built up. Local consumption is also on the increase, reflecting a government policy of promoting the utilisation of indigenous gas through the establishment of major gas-based industries: fertilisers, methanol, urea and steel. In 2001 the chemical and petrochemical industries accounted for about 39% of Trinidad's gas consumption, power stations for 19% and other industry (including iron and steel) for 11%;

the balance of consumption is accounted for by use/loss within the gas supply industry.

Turkmenistan

Proved recoverable reserves (bcm)	2 010
Production (net bcm, 2002)	53.5
R/P ratio (years)	37.6

Apart from the Russian Federation, Turkmenistan has the largest proved reserves of any of the former Soviet republics: for the present *Survey*, the level quoted by *Oil & Gas Journal* (and also by OAPEC) has been adopted. Cedigaz gives Turkmenistan's proved reserves as 2 900 bcm and states that its total gas resources have been evaluated at 22.9 trillion m^3. Many gas fields have been discovered in the west of the republic, near the Caspian Sea, but the most significant resources have been located in the Amu-Daria Basin, in the east.

Gas deposits were first discovered in 1951 and by 1980 production had reached 70 bcm/yr. It continued to rise throughout the 1980s, but by 1992 a serious contraction of the republic's export markets had set in and output fell sharply. Natural gas output recovered in 1999, and has since advanced to 53.5 bcm in 2002. Exports to Iran amounted to 4.9 bcm in that year.

Ukraine

Proved recoverable reserves (bcm)	1 110
Production (net bcm, 2002)	18.8
R/P ratio (years)	59.0

In the absence of a report from the Ukrainian WEC Member Committee, the adopted level of proved reserves at end-2002 reflects that quoted by Cedigaz (*Oil & Gas Journal* gives a similar figure). For the purpose of the 2001 *Survey of Energy Resources*, the Ukrainian WEC Member Committee reported proved recoverable reserves as 825 bcm at end-1999, within a proved amount in place of 1 118 bcm. Gas associated with crude oil was stated to account for only about 4% of the proved reserves. Over and above the proved quantities, there were estimated to be about 368 bcm of gas in place, of which only some 10 bcm was likely to be recoverable.

Ukraine's output of natural gas has been virtually flat for the past 10 years. The republic is one of the world's largest consumers of natural gas: demand reached 137 bcm in 1990. Although consumption had fallen back to about 70 bcm by 2002, indigenous production met only 27% of local needs; the balance was imported from Russia and Turkmenistan. The principal areas of consumption are households, industry and the generation of electricity and bulk heat.

United Arab Emirates

Proved recoverable reserves (bcm)	6 003
Production (net bcm, 2002)	43.4
R/P ratio (years)	>100
Year of first commercial production	1967

Four of the seven emirates possess proved reserves of natural gas, with Abu Dhabi accounting for by far the largest share. Dubai, Ras-al-Khaimah and Sharjah are relatively insignificant in regional or global terms. Overall, the UAE accounts for about 8% of Middle East proved gas reserves.

OAPEC's published level of UAE gas reserves (6 003 bcm) is unchanged from that quoted in the 2001 *Survey of Energy Resources*. Apart from *World Oil*, which gives a figure of 5 778 bcm, the other main published sources (Cedigaz, *Oil & Gas Journal* and OPEC), all quote UAE reserves within a narrow band (6 000–6 100 bcm).

Two major facilities—a gas liquefaction plant on Das Island (brought on-stream in 1977) and a gas-processing plant at Ruwais (in operation from 1981)—transformed the utilisation of Abu Dhabi's gas resources. Most of the plants' output (LNG and NGLs, respectively) is shipped to Japan. In 2002, other LNG customers comprised Spain and the Republic of Korea.

Within the UAE, gas is used mainly for electricity generation/desalination, and in plants producing aluminium, cement, fertilisers and chemicals.

United Kingdom

Proved recoverable reserves (bcm)	604
Production (net bcm, 2002)	102.1
R/P ratio (years)	5.5
Year of first commercial production	1955

The UK is Europe's leading offshore gas producer, but its proved reserves are much smaller than those of Norway. The data on gas resources and reserves adopted for the present *Survey* are based on those reported in *The Digest of United Kingdom Energy Statistics 2003*, published by the Department of Trade and Industry.

Proved recoverable reserves are deduced as 604 bcm, being the remainder obtained after subtracting 'cumulative production to end of 2002' (1 726 bcm) from 'total initial reserves in present discoveries' (2 330 bcm). In this context the DTI defines proven reserves as those which on the available evidence are virtually certain to be technically and economically producible (i.e. those reserves with a better than 90% chance of being produced). 'Probable' reserves (with a better than 50% chance of being technically and economically producible) are put at 375 bcm, whilst 'possible' reserves (with a significant, but less than 50%, chance) are estimated as 330 bcm.

Natural gas production rose year by year during the 1990s, reflecting burgeoning consumption in the power generation sector and higher exports at the end of the decade, following the commissioning of the Interconnector pipeline between Bacton in the UK and Zeebrugge in Belgium, in October 1998. Total output peaked in 2000, with both 2001 and 2002 registering annual decreases of around 2.5%.

United States of America

Proved recoverable reserves (bcm)	5 366
Production (net bcm, 2002)	539.4
R/P ratio (years)	9.3

The USA possesses the world's sixth largest proved reserves of natural gas, and accounts for just over 3% of the global total. Apart from the Russian Federation and the United States, all other countries in the top 10 for gas reserves are members of OPEC.

During the 3 years since the last edition of the *Survey of Energy Resources*, US gas reserves have registered an increase of 19 540 bcf, or about 553 bcm. Net additions to reserves in 2000-2002 totalled 77.9 tcf, an amount 33.5% greater than the 58.4 tcf of gas produced during the same period. The figure of 5 366 bcm tabulated above is derived from total proved reserves of dry natural gas at end-2002, as given by the Energy Information Administration in its *U.S. Crude Oil, Natural Gas and Natural Gas Liquids Reserves 2002 Annual Report*. For the purposes of the present *Survey*, the original data in billion cubic feet at 14.73 psia and 60 °F have been transformed into standard *SER* terms (1 013 mb and 15 °C) by means of separate adjustments for pressure and temperature.

The increase in reserves was due partly to discoveries (field extensions, new field discoveries and new reservoir discoveries in old fields, totalling 59.7 tcf in 2000-2002), partly to revisions and adjustments to estimates for old fields (11.2 tcf) and partly to the net balance of sales and acquisitions (7.0 tcf). Total discoveries during 2002 amounted to nearly 18 tcf; the largest component comprised field extensions, notably in Texas, Wyoming, Colorado, Oklahoma and New Mexico.

About 84% of proved reserves consist of non-associated gas. The states with the largest gas reserves at end-2002 were Texas (23.7% of the USA total), New Mexico (9.3%), Wyoming (11.0%) and Oklahoma (8.0%). Reserves in the Federal Offshore areas in the Gulf of Mexico accounted for 13.5% of the total.

Uzbekistan

Proved recoverable reserves (bcm)	1 875
Production (net bcm, 2002)	58.4
R/P ratio (years)	32.1

The republic's first major gas discovery (the Gazlinskoye field) was made in 1956 in the Amu-Daria Basin in western Uzbekistan.

Subsequently, other large fields were found in the same area, as well as smaller deposits in the Fergana Valley in the east.

For the present *Survey*, proved recoverable reserves have been retained at the level quoted by *Oil & Gas Journal*.

Uzbekistan is a major producer of natural gas: its 2002 net output was, for example, greater than that of Saudi Arabia or Malaysia. It exports gas to its neighbouring republics of Kazakhstan, Kyrgyzstan and Tajikistan.

The principal internal markets for natural gas are the residential/commercial sector, power stations, CHP and district heating plants, and fuel/feedstock for industrial users. Some use is made of CNG in road transport.

Venezuela

Proved recoverable reserves (bcm)	4 179
Production (net bcm, 2002)	29.8
R/P ratio (years)	>100

Venezuela has by far the biggest natural gas industry in South America, possessing two-thirds of its proved reserves and accounting for 33% of regional marketed production in 2002. In its 2000 Annual Report, the Venezuelan state oil and gas company Petróleos de Venezuela, S.A. (PDVSA) states that its proved reserves of natural gas at the end of 2000 were 147 585 bcf (4 179 bcm). As the latest figure released by the state oil and gas company, this figure has been adopted for the present *Survey*. Published sources quote very similar levels for end-2002 (all in bcm): OAPEC 4 174; *Oil & Gas Journal* 4 191; OPEC 4 191; Cedigaz 4 195; *World Oil* 4 225.

Substantial quantities of Venezuela's natural gas (amounting to nearly 35% of gross output in 2002) are re-injected in order to boost or maintain reservoir pressures, while smaller amounts (5%) are vented or flared; about 10% of production volumes are subject to shrinkage as a result of the extraction of NGLs.

The principal outlets for Venezuelan gas are power stations, petrochemical plants and industrial users, notably the iron and steel and cement industries. Residential use is on a relatively small scale.

Yemen

Proved recoverable reserves (bcm)	396
Production (net bcm, 2002)	–
R/P ratio (years)	>100

Yemen has appreciable reserves of natural gas—currently quoted by OAPEC as 396 bcm—but no commercial utilisation has so far been established. Cedigaz, *Oil & Gas Journal* and *World Oil* give a somewhat higher level of proved reserves (c. 480 bcm).

A project for the construction of an LNG plant at Bal Haf, with a capacity of 6.2 million tonnes of LNG per annum, has been stalled since mid-2002, when two of the partners in the Yemen LNG consortium withdrew.

Chapter 6

Part I: Uranium

World Energy Council
CONSEIL MONDIAL DE L'ENERGIE

COMMENTARY

Overview

Since the turn of the millennium, the international uranium industry has witnessed quite a turnaround in world market prices. Uranium spot prices rebounded from an historical low of US\$ 7.10/lb U_3O_8 in late 2000, when prices approached production costs, or perhaps even dropped below for some uranium producers, to US\$ 14.50/lb U_3O_8 by the end of 2003 (see Fig. 6.1).

The natural uranium market remained characterised by a large disparity between global reactor requirements and mine production. The demand–supply disparity emerged in the early 1990s (see Fig. 6.2) when secondary supplies from uranium stocks that had accumulated from surplus production prior to 1990 began to supplement freshly mined uranium. Nuclear fuel produced by reprocessing spent reactor fuels, from surplus military plutonium, from down-blending highly enriched military uranium (HEU) and from re-enrichment of depleted uranium tails also began to curtail demand for freshly mined uranium. The disparity that began in the early 1990s has widened rather steadily through the end of 2003. Currently, freshly mined uranium accounts for just over half of the global annual reactor demand with the balance provided by secondary sources.

With short-run growth in global reactor requirements expanding at only 2.6% annually and existing and committed global uranium production capabilities more than 30% greater than current production levels, what then caused the price increase in 2003? Are secondary sources, including inventories, about to dry up? Or has the recent upswing in price anything to do with a perceived future supply gap on the part of market participants and the belief that bid-up prices are needed to stimulate production and avoid a supply shortfall in the future? The answer is not really. What the market has witnessed is a 'production gap' or thinness of spot supplies (Ux Weekly, 2003), largely caused by the preceding low price period and exacerbated by events such as the temporary shutdown of the McArthur River mine in Canada due to flooding in April 2003, the contract dispute between Globe Nuclear Services & Supply and Russia's Techsnabexport, uncertainty about the long-term prospects for the Rössing uranium mine in Namibia, and the regulatory problems at Honeywell's conversion facility in Illinois. In short, prices have been driven up by low inventories and the relative scarcity of spot supplies and not by a perceived future supply scarcity.

The 'production gap' is the result of the roller-coaster price decay that started in mid-1996 due

2004 Survey of Energy Resources

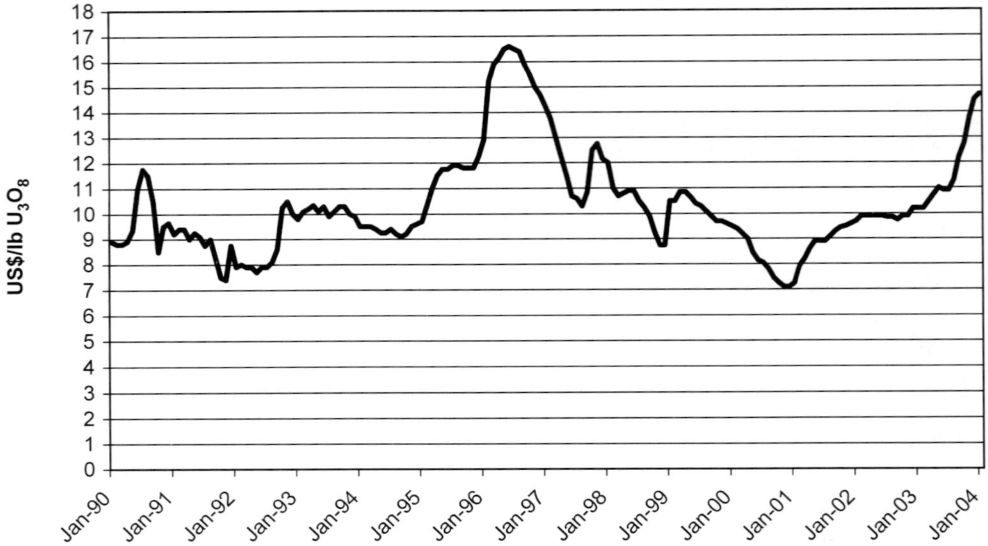

FIGURE 6.1 Uranium spot prices, 1990–2003 (Source: The Ux Consulting Company, LLC).

to market surpluses which caused the closure of mines, reduction of production and postponement of new mine projects. Because production has not yet responded to the recent price signals, the market, in the short run, will essentially depend on secondary supplies. There have been calls for the additional release of government-controlled HEU to mitigate further spot price increases. However, if higher prices are needed to stimulate production, pitching more HEU at the market might dampen the industry's propensity to invest in the additional production capacity required for adequate longer term supplies.

In the long term the supply of uranium will have to come from primary production. Present

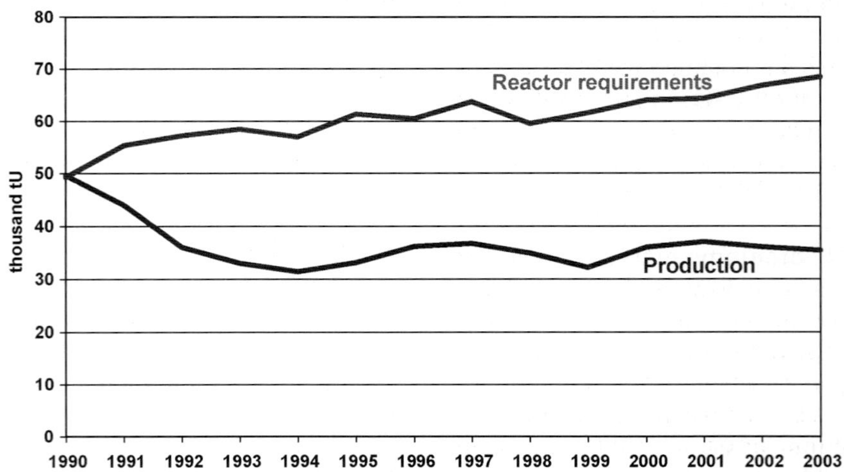

FIGURE 6.2 Annual global reactor requirements and world uranium mine production (Source: Uranium 2003: Resources, Production and Demand).

levels will not be increased, however, until producers see an adequate price and return on their investment. Known uranium reserves are more than enough to cover the requirements of existing reactors during their lifetimes and beyond and are likely to continue to increase as exploration progresses, but new production takes many years to be brought online and, until recently, has not been encouraged by the low prices of uranium. In these circumstances, periods of imbalance between supply and demand causing future shortages and considerable price volatility cannot be excluded. By the same token, any future surplus capacity or increase in secondary supplies will exert downward pressures on uranium market prices.

Whilst the price rise in 2003 was not due to the perception of a supply gap, it may have provided the market with a glimpse of what may happen in the future if supplies fall short of meeting demand.

Production

Since 2000, annual world uranium mine production has fluctuated around 36 000 tU, with a high of 37 020 tU in 2001 and a low of approximately 35 400 tU expected for 2003 (see Fig. 6.2). In 2002, uranium was produced in 20 countries, one less than in 2001 as production ceased in Portugal. The number declined further to 19 producing countries in 2003 as production ceased also in Spain that year. Canada and Australia account for around 50% of world production (51% in 2002 and 48% in 2003). Australia and Canada together with five other countries (Kazakhstan, Namibia, Niger, the Russian Federation and Uzbekistan) account for almost 90% of globally mined uranium.

Open-pit and underground mining and conventional milling continue to be the dominant uranium production technologies, accounting for 70.3% of total production in 2001, and 69.9% in 2002. These values are slightly lower than the 2000 value of 71% due to the closure of open-pit and underground operations in France, Portugal, Spain and Kazakhstan, as well as decreased production in Namibia. The increased in situ leaching (ISL) component in 2002 is the result of increased production in Kazakhstan and Australia (Beverley) offsetting decreased production in the United States and Uzbekistan. One further mining technique is uranium production as a co-product or by-product from copper and gold operations. The volumes of by-product uranium depend on the market situations of the respective main products. Small amounts of uranium are also recovered from water treatment and environmental restoration activities.

Canada remained the world's leading producer (11 607 tU or 32% of world production), as increased McArthur River and Cluff Lake production exceeded the decline in Rabbit Lake output in 2002. Production in 2003 is expected to decline to 9 700 tU, as Cluff Lake was definitively closed at the end of 2002 and operations were suspended at the McArthur River mine in April 2003 for 3 months due to flooding. Australia, the only producing country in the Pacific region, reported a decline in output from 7 579 tU in 2000 to 6 854 tU in 2002 (19% of world production), due to constrained production at Olympic Dam following a fire in the solvent extraction circuits. In 2001, ISL production started at the Beverley mine, and production in Australia is expected to increase to 7 070 tU in 2003.

Production in the Russian Federation increased from 2 760 tU in 2000 to 2 850 tU in 2002. Most of this production came from the Priargunski mine, although 100 tU were produced in 2002 at the new Dalur ISL facility at the Dalmatovskoe deposit in the Transural district. Production in Russia is expected to increase to 3 070 tU in 2003.

Three countries in Africa, Namibia, Niger and South Africa, contributed about 17% to world production in 2002. Although production in Namibia decreased due to unfavourable exchange rates from 2 715 tU in 2000 to 2 333 tU in 2002, the opening of a new mine at Langer Heinrich could increase production by 1 000 tU in 2006. Despite the closure of the uranium recovery facility at the Palabora copper mine, production in South Africa remained essentially unchanged. Niger's production increased from 2 914 tU in 2000 to 3 080 tU in 2002 through stepped up output from the Arlit and Akouta mines.

Production in the United States declined to 902 tU in 2002, down from 1 522 tU in 2000. Almost all the production came from three ISL operations, with a small amount recovered from ISL restoration and mine water treatment activities. The last conventional uranium mills closed in 2002. In late 2003, Uranium Resources Inc., a company that had been on the verge of going out of business, announced that it was planning to resume uranium production in Texas and had signed several new contracts.

Brazil was the only producing country in Central and South America in 2001 and 2002. Estimated production increased to 272 tU in 2002, as the Lagoa Real production centre reached full capacity. In Argentina, the Sierra Pintada mine, which was placed on standby in 1999, is expected to restart production in 2005.

In Germany, 27 tU were recovered from mine rehabilitation efforts in 2001 which increased to 221 tU in 2002. It is expected that 150 tU will be recovered in 2003, as mine flooding has reached levels containing sulphuric acid and dissolved uranium from previous operations. The remaining Western European output (from France and Spain) was also derived from clean-up operations and water treatment.

Production in Kazakhstan rose from 1 870 tU in 2000 to 2 822 tU in 2002, and is expected to further increase to 3 315 tU in 2003 as market conditions and export relations continue to improve. Mine tests have started at the Katco and Inkai ISL mines. During the same period, production in Uzbekistan decreased from 2 028 tU in 2000 to 1 859 tU in 2002, but is expected to increase to 2 300 tU in 2003. India and Pakistan do not report production information, but their 2002 output is estimated to have increased slightly from 2000 by 230 and 38 tU, respectively. China, the only producing country in East Asia, does not report official production figures. Production is estimated to have been 700 tU in 2001 and 730 tU in 2002.

Exploration

Worldwide uranium exploration continues to be unevenly distributed geographically, with the majority of exploration expenditures being concentrated in areas considered to have the best likelihood for the discovery of economically attractive deposits, mainly unconformity related and sandstone-type deposits.

Annual expenditures in uranium exploration increased from 1999 to 2000 where a total of 21 countries reported domestic expenditures of about US$ 115.2 million (see Fig. 6.3), 16% higher than in the previous year. In 2001, the declining trend in exploration expenditures resumed, with only 18 countries reporting exploration activities amounting to about US$ 89 million. In 2002, domestic exploration expenditures totalled about US$ 95.1 million, an increase of about 7% compared to the 2001 total. 17 countries reported exploration expenditures in 2002, though only 9 countries, i.e. Australia, Canada, China, Egypt, India, Kazakhstan, Niger, Russia and Uzbekistan accounted for about 96% of total domestic exploration expenditures.

Resources

Recent and detailed information on uranium resources is reported in the publication *Uranium 2003: Resources, Production and Demand* (Red Book), a joint report of the OECD Nuclear Energy Agency and the International Atomic Energy Agency. The Red Book contains information on 47 countries with reported uranium resources. The resources are classified by the level of confidence in the estimates, and by production cost categories. The known conventional resources are classified as Reasonably Assured Resources (RAR) and Estimated Additional Resources I (EAR-I) and are reported as recoverable quantities, while undiscovered

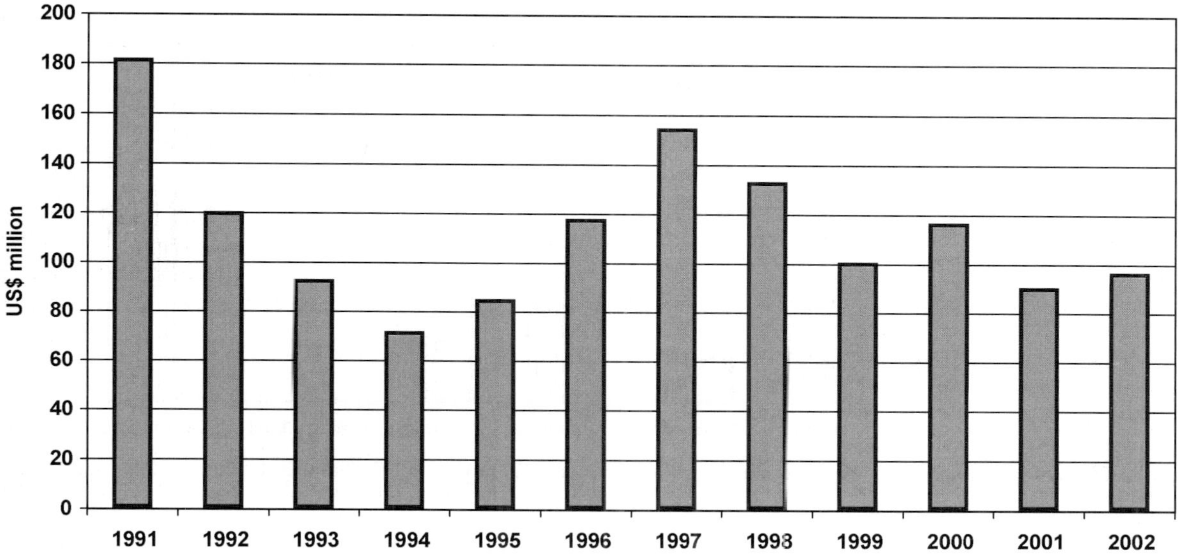

FIGURE 6.3 Trend in annual exploration expenditures (Source: Uranium 2003: Resources, Production and Demand.).

resources classified as Estimated Additional Resources II (EAR-II) and Speculative Resources (SR) are reported as in situ.

As of 1 January, 2003, total known conventional resources (RAR and EAR-I) with recovery costs of less than US$ 80/kgU (or US$ 36/lbU) amounted to about 3 537 000 tU, and to about 4 589 000 tU for recovery costs of less than US$ 130/kgU (or US$ 59/lbU). Compared to 1999 levels, recoverable quantities increased significantly for both cost categories by 188 000 and 210 000 tU, respectively. Since 1999 known resources in the less than US$ 40/kgU cost category have grown by about 66% or 687 000 to 1 730 500 tU, mainly due to increases in this category reported by Australia, Canada, China and Niger. In fact, Australia, which holds 40% of these low-cost resources, reported for the first time in this category in 2001. The resources in the less than US$ 40/kgU cost category (which represent about 25 years of 2002 global reactor uranium requirements) provide the potential for maintaining the economic competitiveness of nuclear electric programmes by helping to assure that low-cost fuel supplies are available for a sustained period.

Total undiscovered conventional resources (EAR-II and SR) in 2003 amount to about 9 794 000 tU, a decline of some 2 477 000 tU from the total reported in 2001, mainly due to reductions reported by China and the Russian Federation. However, reporting of SR is incomplete, as only 28 countries reported, compared to 43 countries that reported known resources.

Outlook

Market conditions are the primary driver of decisions to develop new, or expand existing, primary production centres. As market prices increase, or expectations of a sustained price increase develop, new production could be developed in order to meet increased demand. Yet a key element influencing market price is the availability of secondary sources of uranium, particularly the level of stocks available and the length of time these stocks will last.

In the short run, the outlook for uranium producers remains fraught with uncertainty. The large natural uranium, HEU and plutonium stocks held in various forms by the military in

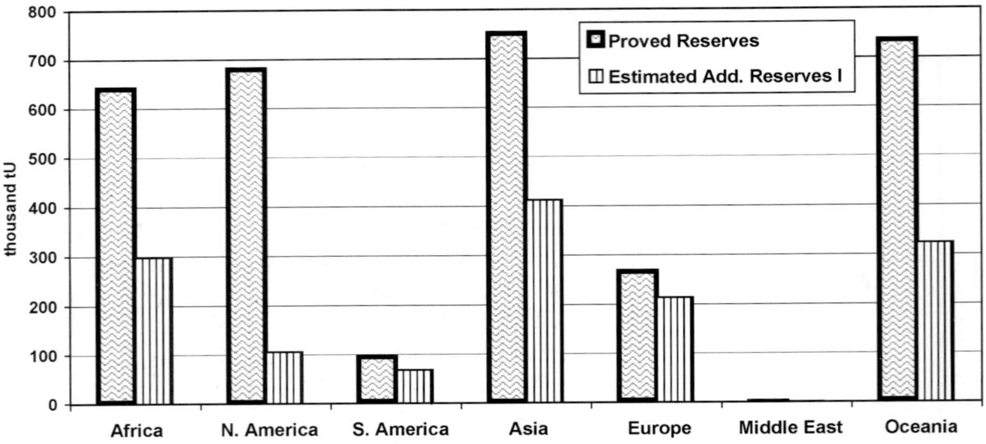

FIGURE 6.4 Known conventional uranium resources as of January 1, 2003 (Source: Uranium 2003: Resources, Production and Demand.).

the Russian Federation and the United States are equivalent to several years of global civilian reactor requirements. Accurate information on secondary sources of uranium, especially uranium inventory levels, is not readily available. How much, how fast and how soon these secondary sources will be commercially available are the crucial questions affecting the propensity of the industry to invest in capacity expansion. From known quantities of secondary sources, the current insufficiency of world primary uranium production capacity to meet global requirements is expected to continue, possibly to 2020. Moreover, in the absence of sufficiently high price expectations of at least US$ 15/lb, which should suffice to bring online a portion of currently idle production capabilities, the uranium market may witness periods of volatile spot prices. Current production (around 36 000 tU) amounts to some 75% of the stated production capability (47 260 tU) of the uranium industry.

The addition of planned and prospective production centres by 2005 could increase total capability to 51 155 tU. Further subsequent planned and prospective centres are projected to more than compensate for expected closures of existing mines due to resource depletion, thereby bringing the total to 63 939 tU/yr in 2010. This level of uranium production capability is projected to remain relatively stable throughout the following decade.

World uranium requirements were about 68 400 tU in 2003 and are projected to lie within a range of 72 000–86 000 tU/yr by 2020. In short, secondary sources will have to be used to balance demand and supply. Provided non-production supplies are forthcoming, primary and secondary sources should be able to meet all the requirements. Nevertheless, unexpected delays or interruptions of secondary supplies could create market instability and price volatility.

It should be unequivocally noted that conventional world uranium resources are more than adequate to meet nuclear fuel requirements for decades to come and pose no limitation on the future growth of nuclear energy. Exploitation of undiscovered conventional resources, especially as the geographical coverage of uranium exploration is far from complete, could increase this to several hundreds of years, although significant exploration and development efforts would be required to move these resources to more definitive categories.

However, one crucial consideration in evaluating future supply and demand balances is the time needed to discover uranium and develop new production capability. The lead time for discovery and the development of new uranium production facilities has historically been in the order of one to two decades, owing to a number of factors, e.g. business decisions, environmental challenges and technical difficulties.

Notwithstanding the variety of causes, these long lead times underscore the importance of making decisions to pursue new production capabilities well in advance of any supply shortfall. Without accurate information on secondary sources, there exists the potential for supply–demand imbalances as secondary sources become exhausted. This could result in significant upward pressure on uranium prices.

Unconventional resources, including phosphate deposits and seawater, contain vast amounts of uranium and their use could fuel nuclear energy for millennia if advanced reactor and fuel cycle technologies are deployed. Thus, sufficient nuclear fuel resources exist to meet energy demands at current and increased levels well into the future. However, to reach this potential considerable research, development and demonstration efforts and investment are required, both to develop new extraction technologies in a timely manner and also to allow promising technologies to reach their potential.

H.-Holger Rogner
International Atomic Energy Agency

DEFINITIONS

Uranium does not occur in a free metallic state in nature. It is a highly reactive metal that interacts readily with non-metals, and is an element in many intermetallic compounds.

This *Survey* uses the system of ore classification developed by the Nuclear Energy Agency (NEA) of the Organisation for Economic Cooperation and Development (OECD) and the International Atomic Energy Agency (IAEA). However, the names given to the classes as defined below are different because the WEC tries to use similar terms to define comparable classes of reserve for each of the energy sources covered in the *Survey*.

Estimates are divided into separate categories according to different levels of confidence in the quantities reported.

The estimates are further separated into categories based on the cost of uranium recovered at ore-processing plants. The cost categories are: less than US$ 40/kgU; US$ 40–80/kgU and US$ 80–130/kgU. Costs include the direct costs of mining, transporting and processing uranium ore, the associated costs of environmental and waste management, and the general costs associated with running the operation (as defined by the NEA). The resource data quoted in the present *Survey* reflect those published in the 2003 'Red Book'. Cost categories are expressed in terms of the US dollar as at 1 January, 2003.

The WEC follows the practice of the NEA/IAEA and defines estimates of discovered reserves in terms of uranium recoverable from mineable ore and not uranium contained in the ore (i.e. to allow for mining and processing losses). Although some countries continue to report in situ quantities, the major producers generally conform to these definitions.

All resource estimates are expressed in terms of tonnes of recoverable uranium (U), not uranium oxide (U_3O_8).

Note: 1 tonne of uranium = approximately 1.3 short tons of uranium oxide;

US$ 1 per pound of uranium oxide = US$ 2.60 per kilogram of uranium;

1 short ton U_3O_8 = 0.769 tU.

Proved reserves correspond to the NEA category 'Reasonably Assured Resources' (RAR), and refer to recoverable uranium that occurs in known mineral deposits of such size, grade and configuration that it could be recovered within the stated production cost ranges with currently proven mining and processing technology. Estimates of tonnage and grade are based on specific sample data and measurements of the deposits, together with knowledge of deposit characteristics. Proved reserves have a high assurance of existence.

Estimated additional amounts recoverable corresponds to the NEA category 'Estimated Additional Resources—Category I' (EAR-I), and refers to recoverable uranium (in addition to proved reserves) that is expected to occur (mostly on the basis of direct geological evidence) in extensions of well-explored deposits and in deposits for which geological continuity has been established, but where specific data and measurements of the deposits and knowledge of their characteristics are considered to be inadequate to classify the resource as a proved reserve.

Such deposits can be delineated and the uranium subsequently recovered, all within the stated production cost ranges. Estimates of tonnage and grade are based primarily on knowledge of the deposit characteristics as determined in its best-known parts or in similar deposits. Less reliance can be placed on the estimates in this category than on those for proved reserves.

Other amounts expected to be recoverable at up to US$ 130/kgU refers to uranium in addition to proved reserves and estimated additional amounts recoverable, and corresponds to the sum of the two NEA categories, 'Estimated Additional Resources—Category II' (EAR-II) and 'Speculative Resources'

(SR). This category includes estimates of undiscovered uranium resources. These may refer to deposits believed to exist in well-defined geological trends or areas of mineralisation with known deposits. Estimates of such deposits are on the basis that they can be discovered, delineated and the uranium subsequently recovered at up to US$ 130/kgU. Estimates of tonnage and grade are based primarily on the knowledge of the deposit characteristics in known deposits within the respective trends or areas and on such sampling, geological, geophysical or geochemical evidence as may be available. They include deposits that are thought to exist mostly on the basis of indirect evidence and geological extrapolations relating to deposits discoverable with existing exploration techniques.

Annual production is the production output of uranium ore concentrate from indigenous deposits, expressed as tonnes of uranium.

Cumulative production is the total cumulative production output of uranium ore concentrate from indigenous deposits, expressed as tonnes of uranium, produced in the period from the initiation of production until the end of the year stated.

TABLE 6.1

Uranium: proved reserves (RAR) as at 1 January, 2003 (conventional resources recoverable at up to US$ 130/kgU)

	Recoverable at (thousand tonnes of uranium)				Total recoverable at up to US$ 130/kgU (thousand tonnes of uranium)
	<US$ 40/kgU	US$ 40–80/kgU	<US$ 80/kgU	US$ 80–130/kgU	
Algeria			19.5		19.5
CAR			6.0	6.0	12.0
Congo (DR)			1.4		1.4
Gabon				4.8	4.8
Malawi			8.8		8.8
Namibia	57.3	82.0	139.3	31.2	170.5
Niger	89.8	12.4	102.2		102.2
Somalia				5.0	5.0
South Africa	119.2	112.5	231.7	83.7	315.3
Zimbabwe			1.4		1.4
Total Africa			**510.2**		**640.8**
Canada	297.3	36.6	333.8		333.8
Greenland				20.3	20.3
Mexico				1.3	1.3
USA			102.0	243.0	345.0
Total N. America			**435.8**		**700.4**
Argentina	4.8	0.1	4.9	2.2	7.1
Brazil	26.2	60.0	86.2		86.2
Chile					0.6
Peru			1.2		1.2
Total S. America			**92.3**		**95.0**
China	26.2	8.8	35.1		35.1
India					41.0
Indonesia		0.3	0.3	4.3	4.6
Japan					6.6
Kazakhstan	280.6	104.0	384.6	145.8	530.5
Mongolia	8.0	38.3	46.2		46.2
Thailand				N	N
Turkey		6.8	6.8		6.8
Uzbekistan	61.5		61.5	18.1	79.6
Vietnam					1.0
Total Asia			**534.6**		**751.4**
Bulgaria	1.7	4.2	5.9		5.9
Czech Republic		0.8	0.8		0.8
Finland				1.1	1.1
Germany				3.0	3.0

TABLE 6.1 (*Continued*)

	Recoverable at (thousand tonnes of uranium)				Total recoverable at up to US$ 130/kgU (thousand tonnes of uranium)
	<US$ 40/kgU	US$ 40–80/kgU	<US$ 80/kgU	US$ 80–130/kgU	
Greece	1.0		1.0		1.0
Italy			4.8		4.8
Portugal			7.5		7.5
Romania				3.3	3.3
Russian Fed.	52.6	71.4	124.1	19.0	143.0
Slovenia		2.2	2.2		2.2
Spain		2.5	2.5	2.5	4.9
Sweden				4.0	4.0
Ukraine	15.4	19.3	34.6	30.0	64.7
Total Europe			**183.3**		**246.2**
Iran (Islamic Rep.)				0.4	0.4
Total M. East					**0.4**
Australia	689.0	13.0	702.0	33.0	735.0
Total Oceania			**702.0**		**735.0**
Total World			**2 458.2**		**3 169.3**

Notes:

(1) Data for the <US$ 40, US$ 40–80 and US$ 80–130 categories are not available for all countries; thus regional and global aggregates have not been computed for these categories.

(2) Source: Uranium 2003: Resources, Production and Demand, 2004, OECD Nuclear Energy Agency and International Atomic Energy Agency.

TABLE 6.2

Uranium: estimated additional resources (EAR-I) and undiscovered resources (EAR-II and SR) as at 1 January, 2003 (conventional resources recoverable at up to US$ 130/kgU)

	Estimated additional resources recoverable at (thousand tonnes of uranium)					Undiscovered resources recoverable at up to US$ 130/kgU (thousand tonnes of uranium)
	<US$ 40/kgU	US$ 40–80/kgU	<US$ 80/kgU	US$ 80–130/kgU	up to US$ 130/kgU	
Congo (DR)			1.3		1.3	
Egypt Arab Rep.						0.1
Gabon				1.0	1.0	
Namibia	57.1	16.4	73.6	13.5	87.1	
Niger	125.4		125.4		125.4	9.5
Somalia				2.6	2.6	
South Africa	49.3	17.6	66.9	13.4	80.3	1 223.2
Zambia						22.0
Zimbabwe						25.0
Total Africa			**267.2**		**297.6**	**1 279.8**
Canada	86.6	18.2	104.7		104.7	850.0
Greenland				12.0	12.0	60.0
Mexico				0.5	0.5	13.0
USA	NA	NA	NA	NA	NA	2 613.0
Total N. America			**104.7**		**117.2**	**3 536.0**
Argentina	2.9		2.9	5.7	8.6	1.4
Brazil		57.1	57.1		57.1	620.0
Chile					0.9	4.7
Colombia						228.0
Peru			1.3		1.3	26.3
Venezuela						163.0
Total S. America			**61.3**		**67.9**	**1 043.4**
China	5.9	8.8	14.7		14.7	7.7
India					18.9	32.5
Indonesia				1.2	1.2	4.1
Kazakhstan	131.2	106.6	237.8	79.4	317.2	810.0
Mongolia	8.3	7.5	15.8		15.8	1 390.0
Thailand				N	N	
Uzbekistan	31.8		31.8	7.1	38.8	231.6
Vietnam			0.8	4.6	5.4	237.9
Total Asia			**300.8**		**412.0**	**2 713.8**
Bulgaria	1.7	4.7	6.3		6.3	18.2
Czech Republic		0.1	0.1		0.1	179.2
France				9.5	9.5	

TABLE 6.2 (Continued)

	Estimated additional resources recoverable at (thousand tonnes of uranium)					Undiscovered resources recoverable at up to US$ 130/kgU (thousand tonnes of uranium)
	<US$ 40/kgU	US$ 40–80/kgU	<US$ 80/kgU	US$ 80–130/kgU	up to US$ 130/kgU	
Germany				4.0	4.0	74.0
Greece			6.0		6.0	6.0
Hungary				13.8	13.8	
Italy				1.3	1.3	10.0
Portugal			1.5		1.5	6.5
Romania				3.6	3.6	6.0
Russian Fed.	15.9	18.4	34.3	87.0	121.2	649.5
Slovenia		5.0	5.0	5.0	10.0	1.1
Spain				6.4	6.4	
Sweden				6.0	6.0	
Ukraine	0.9	3.8	4.7	6.7	11.4	256.6
Total Europe			**57.8**		**201.1**	**1 207.1**
Iran (Islamic R.)				0.7	0.7	13.9
Total M. East					**0.7**	**13.9**
Australia	276.0	11.0	287.0	36.0	323.0	
Total Oceania			**287.0**		**323.0**	
Total World			**1 078.8**		**1 419.4**	**9 794.0**

Notes:

(1) Data for the <US$ 40, US$ 40–80 and US$ 80–130 categories are not available for all countries; thus regional and global aggregates have not been computed for these categories.

(2) 'Other amounts recoverable at up to US$ 130/kgU' are reported as in situ resources and include some speculative resources with their cost range unassigned.

(3) Source: Uranium 2003: Resources, Production and Demand, 2004, OECD Nuclear Energy Agency and International Atomic Energy Agency.

TABLE 6.3

Uranium: annual and cumulative production at end-2002

	2002 production (tonnes of uranium)	Cumulative production to end-2002 (tonnes of uranium)
Congo (DR)		25 600
Gabon		26 612
Namibia	2 333	76 699
Niger	3 080	87 859
South Africa	824	152 543
Zambia		102
Total Africa	**6 237**	**369 415**
Canada	11 607	364 652
Mexico		49
USA	902	355 713
Total N. America	**12 509**	**720 414**
Argentina		2 509
Brazil	272	1 369
Total S. America	**272**	**3 878**
China	730	8 865
India	230	7 733
Japan		84
Kazakhstan	2 822	95 078
Mongolia		535
Pakistan	38	921
Uzbekistan	1 859	99 562
Total Asia	**5 679**	**212 778**
Belgium		686
Bulgaria		16 720
Czech Republic	465	108 197
Estonia		65
Finland		30
France	18	73 866
Germany	221	219 090
Hungary	10	21 050
Poland		660
Portugal		3 721
Romania	90	17 904
Russian Fed.	2 850	119 963
Slovenia		382
Spain	37	5 028
Sweden		200
Ukraine	800	11 647
Total Europe	**4 491**	**599 209**
Australia	6 854	105 731
Total Oceania	**6 854**	**105 731**
Total World	**36 042**	**2 011 425**

Notes:

(1) Data for China, India, Pakistan, Romania and Ukraine are estimated.

(2) The cumulative production shown for China covers only the period 1990–2002 inclusive, as data for earlier years are not available.

(3) The cumulative production shown for Ukraine covers only the period 1992–2002 inclusive, as data for earlier years are not available.

(4) Source: Uranium 2003: Resources, Production and Demand, 2004, OECD Nuclear Energy Agency and International Atomic Energy Agency.

COUNTRY NOTES

The Country Notes on Uranium have been compiled by the editors, drawing principally upon the following publication: *Uranium 2003: Resources, Production and Demand* (known as the Red Book); 2004; OECD Nuclear Energy Agency and International Atomic Energy Agency.

Information provided by WEC Member Committees and from other sources has been incorporated when available.

Algeria

In situ RAR have been assessed at 26 000 tonnes U, of which an estimated 75% is recoverable at less than US$ 80/kgU, but no production has ensued.

Argentina

Exploration for uranium started in the early 1950s, since when deposits have been discovered in a number of locations, mostly in the western part of the country and in the southern province of Chubut in Patagonia. During the 1990s, a countrywide programme of exploration directed at the evaluation of areas with uranium potential was undertaken. Regional assessment of uranium potential continues, with selected areas of interest being studied in greater depth.

Uranium has been produced on a small scale since the mid-1950s, with cumulative production reaching 2 509 tonnes by the end of 1999. Since then, output has been in abeyance. The production centre at San Rafael in the province of Mendoza, which processed ore from the Sierra Pintada deposit, has been placed on a standby basis. The facility has been under reassessment by the state agency CNEA, which since 1996 has owned and operated Argentina's uranium industry.

Proved reserves of uranium, recoverable at less than US$ 80/kgU, were 4 880 tonnes at end-2002, a slight decrease on the end-2000 estimate, which is attributable to the devaluation of the Argentinian currency and a revised assessment of the Sierra Pintada deposit. Further known conventional resources consist of 2 200 tonnes of RAR, recoverable at US$ 80–130/kgU and 8 560 tonnes of EAR-I recoverable at less than US$ 130/kgU. Undiscovered resources (at the latter cost level) remain at 1 440 tonnes.

Australia

Exploration activities between 1947 and 1961 led to a number of uranium discoveries, including the deposits at Mary Kathleen (Queensland), Rum Jungle (Northern Territory) and Radium Hill (South Australia). A decrease in uranium requirements for defence purposes induced a virtual cessation in exploration between 1961 and 1966. Activity picked up again during the late 1960s, as civilian export demand accelerated, and numerous major deposits were located.

In 1983 the Government introduced the so-called 'three mines' policy, which permitted uranium exports only from the Nabarlek, Ranger and Olympic Dam mines. This restrictive measure, with its dampening effect on uranium exploration, lasted until 1996. Exploration expenditure and drilling activity rose in the latter half of the 1990s, but declined to historic lows in 2001 and 2002. Exploration in recent years has been concentrated in parts of the Northern Territory and South Australia.

Australia produced nearly 6 900 tonnes of uranium in 2002, bringing cumulative output to more than 105 000 tonnes since 1954. Three uranium production centres were in operation in 2002: Ranger (capacity 4 660 tU/yr), Olympic Dam (capacity 3 930 tU/yr) and Beverley (848 tU/yr). A new centre has been constructed at Jabiluka, but the site was placed on a standby and environmental maintenance basis in 2000. An ISL production centre is planned for the Honeymoon deposit.

Proved reserves of uranium (RAR) are reported in the Red Book as 702 000 tonnes at less than US$ 80/kgU and 33 000 tonnes at US$ 80–130/kgU. Estimated additional amounts recoverable at these cost levels are 287 000 and 36 000 tonnes, respectively.

Brazil

Exploration activity over a period of some 40 years, ending in 1991, resulted in the discovery of occurrences and deposits of uranium in eight different states of Brazil. Known conventional resources are substantial, consisting of proved reserves (= RAR) of 162 000 tonnes (recoverable at less than US$ 80/kgU) plus EAR-I of about 100 000 tonnes. Undiscovered conventional resources (EAR-II and SR) are put at 120 000 tonnes U, recoverable at under US$ 80/kgU and 500 000 tonnes of SR with no cost range assigned. All these tonnages are quoted on an in situ basis and are unchanged since the 2001 *Survey*.

Although Brazil's RAR are very substantial, and backed up by massive additional resources, its uranium output has never been on a commensurately large scale: cumulative production at end-2002 was less than 1 400 tonnes.

After 2 years on standby, the 360 tU/yr Poços de Caldas production centre in Minas Gerais state was definitively shut down in 1997 and is now being decommissioned. It has been replaced by a new plant (now called Caetité) at Lagoa Real in the eastern state of Bahia. The Caetité plant has a current nominal production capacity of 340 tU/yr.

Another production centre, planned for construction at Itataia in north-eastern Brazil, is at the feasibility stage: its annual uranium production capacity, as a by-product of phosphate output, would be 325 tonnes. Implementation of this project will depend on the way the markets for both products are seen as developing.

Brazil's conventional resources are supplemented by unconventional resources, for which there are at present no plans for recovery:

- carbonatite (containing 13 000 tonnes U);
- marine phosphates (28 000 tonnes U);
- quartz-pebble conglomerates (2 000 tonnes U).

Canada

From 1942, uranium was obtained from the Port Radium deposit of pitchblende in the Northwest Territories, which had previously been mined for radium. Exploration directed specifically towards finding uranium led to the discovery of many deposits, the most important being in the Blind River/Elliot Lake area of southern Ontario and the Athabasca Basin in northern Saskatchewan.

Uranium production peaked at 12 200 tonnes in 1959, when the last defence contracts were signed, and output fell rapidly to less than 3 000 tonnes in 1966. Increases in uranium demand and rising prices led to renewed growth from the mid-1970s, with the focus of production moving westwards. Three out of four production centres in Ontario were phased out in the early 1990s, and the last closed in mid-1996, leaving Saskatchewan the sole producing province.

Canadian primary uranium output totalled 11 607 tonnes in 2002, by far the largest in the world and equivalent to 32% of global production. Its RAR (at up to US$ 80/kgU) amount to some 334 000 tonnes, or 13.6% of the world total. EAR-I in the same cost-band are some 105 000 tU, while undiscovered resources consist of EAR-II at 150 000 tU and 700 000 tonnes of SR.

All Canadian uranium mining takes place in northern Saskatchewan. The Cluff Lake uranium production centre was closed down in 2002, after producing some 24 000 tonnes of uranium in its 22 years of operation. Three centres are currently in production: McArthur River/Key Lake (owned by Cameco and CRI), McClean Lake (CRI and others) and Rabbit Lake (Cameco).

New production centres are being developed in Saskatchewan: Cigar Lake (start-up possibly in 2006) and Midwest (estimated start-up in 2010).

Both have cleared the environmental review process.

China

Nearly 50 years of exploration for uranium has resulted in the discovery of deposits in various parts of the country. The major resources are in Jiangxi and Guangdong provinces in the south-east, in Liaoning province to the north-east of Beijing and in the Xinjiang Autonomous Region of north-western China.

Total known resources in nine locations are stated to be 77 000 tonnes (in situ), and are now reported with a classification by production cost. RAR are given as 49 200 tonnes, of which some 35 000 tU would be recoverable at up to US$ 80/kgU, with EAR-I as 20 100 tonnes (14 700 recoverable) in the same cost bracket. Undiscovered resources are reported as 3 600 tonnes of EAR-II plus 4 100 tonnes of speculative, a long way from the estimate of 1.77 million tonnes for SR quoted in previous editions of the Red Book.

There are five operational production centres, with an aggregate capacity of 840 tU/yr. Construction of a new production centre at Fuzhou is under way. Output in 2002 was 730 tonnes.

Colombia

Although no resource data were reported to the IAEA for their 2003 Red Book, Colombia is still quoted as possessing 11 000 tonnes of uranium in category EAR-II and 217 000 tU of SR, both amounts on an in situ basis, at less than US$ 130/kgU. No production of uranium has so far been recorded.

Czech Republic

After an early start in 1946, uranium exploration in the republic was systematic and intensive during a period of more than 40 years. From 1990, however, expenditure decreased sharply, with field exploration coming to an end early in 1994.

There are 23 uranium deposits, of which 20 have been mined-out or closed. The Rozná deposit is being mined and two others may be exploited in the future. The Straz production centre has been closed but some ISL extraction is continuing under a remediation regime. Output from Czechoslovakian mines began in 1946 and until 1990 was all exported to the Soviet Union. Production in 2002 amounted to 465 tonnes, giving a cumulative output of about 108 000 tonnes.

As a result of the Straz deposit being deemed uneconomic, and of the depletion of resources at the Rozná production centre, RAR declined to 830 tU at the end of 2002 and EAR-I to only 90 tU, both recoverable at up to US$ 80/kgU.

Undiscovered resources (on an in situ basis) comprise 180 tonnes of EAR-II recoverable at up to US$ 80/kgU and 179 000 tonnes of SR, unassigned to a cost category.

Finland

Exploration for uranium took place during the period 1955–1989, resulting in the identification of four uranium provinces. Proved reserves (RAR at US$ 80–130/kgU) amount to 1 500 tonnes, of which 75% is regarded as recoverable. Unconventional resources are represented by possible by-product production of 3 000–9 000 tU from Talvivaara black schists and 2 500 tU from Sokli carbonatite.

Finland's past production of uranium has been limited to the minor quantity (c. 30 tU) produced by a pilot plant at the Paukkajanvaara mine in eastern Finland, which was operated from 1958 to 1961.

France

Exploration for uranium commenced in 1946 and during the next 40 years a number of deposits were located. Since 1987, exploration activities have been on the decline, as has the level of production. Total output in 2002 was only 18 tonnes, bringing the cumulative tonnage to 73 866 tonnes. Since the closure of France's last uranium mine (Jouac) in 2001, RAR have been put at zero; estimated additional resources are 11 740 tonnes, of which 9 510 tU would be recoverable at US$ 80–130/kg.

The last French ore-processing plant, at Le Bernardan in the north-western part of the Massif Central, ceased operations in 2001.

Gabon

Exploration by the French Commissariat à l'Energie Atomique (CEA) led to the discovery in 1956 of a substantial deposit of uranium ore near Mounana in south-eastern Gabon. Further deposits in the Franceville Basin were located during 1965–1982. Exploratory activity continued until the late 1990s.

Uranium production from the Mounana production centre began in 1961 and built up to a peak of around 1 250 tpa by the end of the 1970s. Subsequently output followed a declining trend, ceasing altogether in early 1999. The last underground mine, exploiting the Okelobondo deposit (discovered in 1974), closed down in November 1997. An open-pit operation at the Mikouloungou deposit (discovered in 1965) was in production from June 1997 to March 1999, since when Gabon has ceased to be a uranium producer.

Gabon's cumulative production of nearly 27 000 tonnes of uranium indicates its historic significance as one of the leading minor producers.

Known conventional resources of uranium in Gabon amount to just under 6 000 tonnes, comprising 4 830 tonnes of RAR recoverable at less than US$ 130/kgU, and 1 000 tonnes of EAR-I in the same price category.

Germany

Prior to Germany's reunification in 1990, the GDR had been a major producer of uranium, with a cumulative output of some 213 000 tonnes. All uranium mines have now been closed and the only production relates to uranium recovered in clean-up operations in the former mining/milling areas: 2002 output from this source was 221 tonnes, obtained during the decommissioning of the Königstein mine in Saxony.

Greenland

Exploration for uranium was carried out for more than 30 years (1955–1986), with moderate success. Fairly sizeable quantities of in situ uranium resources have been reported for Greenland in the 2003 Red Book: 27 000 tU of RAR and 16 000 tU of EAR-I, 75% of both being recoverable at US$ 80–130/kgU, together with an in situ 60 000 tU in the speculative category, most of which is deemed to be recoverable at less than US$ 130/kgU. No production of uranium has yet taken place.

Hungary

Uranium exploration commenced in the early 1950s, with the Mecsek deposit in southern Hungary being discovered in 1954. An underground mine came into production at Mecsek in 1956. Initially the raw ore produced was shipped to the USSR, but from 1963 onwards it passed through a processing plant at Mecsek before being shipped as uranium concentrates.

Mining and milling operations at the Mecsek site were shut down at the end of 1997. Cumulative production of uranium, including a relatively small amount derived from heap leaching, was about 21 000 tonnes.

Hungary's remaining known conventional resources of uranium, as reported to the IAEA/NEA, are 18 399 tonnes of EAR-II, 75%

of which would be recoverable at less than US$ 130/kgU.

India

Exploration for uranium began in 1949, since when deposits have been located in many parts of the country. Exploratory activity is continuing, with expenditure of around US$ 13 million per annum. Uranium has been produced at the Jaduguda mine in the eastern state of Bihar since 1967. In 2002, output from this and two other mines in the same area was some 230 tonnes. The recovery of uranium as a by-product of copper refining has been temporarily suspended.

RAR (with their cost range unassigned) are approximately 54 600 tonnes. Other known conventional resources consist of just over 25 200 tonnes classified as EAR-I, also without an assigned cost range. Both these amounts are expressed on an in situ basis, thus recoverable tonnages would be substantially lower. Undiscovered conventional resources (in situ) consist of about 15 500 tonnes of EAR-II and 17 000 tonnes of SR. Unconventional resources have been estimated to amount to about 6 600 tonnes, recoverable from copper mine tailings in the Singhbhum district of the state of Jharkhand.

At the beginning of 2003 there were three production centres in operation, fed by underground mines in the Jaduguda area. A new uranium production centre was under construction at Turamdih in the Singhbhum (East) district of Jharkhand, based on underground mining, and two others—Banduhurang, Jharkhand (based on open-pit mining), and Lambapur-Peddagattu, Andhra Pradesh (to use both open-pit and underground mining)—were at the planning stage.

Indonesia

The Nuclear Minerals Development Centre of the Indonesian National Atomic Energy Agency (BATAN) began exploring for uranium in the 1960s. Since 1988, exploratory work has been concentrated in the vicinity of Kalan in West Kalimantan, with a significant drilling programme being completed in 1992. Exploration work has continued, but since 1997 budgetary constraints have severely limited operations.

At the beginning of 2003, RAR, on an in situ basis and recoverable at less than US$ 130/kgU, amounted to 6 797 tonnes, of which about 7% fell within the less than US$ 80 bracket; estimated additional resources (at up to US$ 130) were 1 699 tonnes. Over and above these amounts, SR were put at 4 090 tonnes.

Iran (Islamic Republic)

Exploratory work has been undertaken for more than 20 years and a number of prospects have been defined, mostly in the central province.

RAR (in situ) amount to 491 tonnes, with EAR-I assessed as 936 tonnes, with an estimated 75% of both being recoverable at US$ 80–130/kgU. Undiscovered conventional resources (in situ) consist of 3 350 tonnes in category EAR-II, plus 4 500 tonnes of SR, both recoverable at less than US$ 130/kgU. An additional 6 000 tonnes of SR, with cost range unassigned, is also reported.

Japan

Between 1956 and 1988, the Power Reactor and Nuclear Fuel Development Corporation (PNC) and its predecessor undertook domestic exploration for uranium, resulting in the discovery of deposits at two locations on the island of Honshu. Total discovered reserves, reported as RAR recoverable at up to US$ 130/kgU, are some 6 600 tonnes.

Cumulative production of uranium in Japan amounts to only 84 tonnes, produced by a test pilot plant operated by PNC at the Ningyo-toge mine between 1969 and 1982.

Kazakhstan

Uranium exploration commenced in 1948 and since then a large number of ore deposits have been located, initially in the districts of Pribalkhash (in south-eastern Kazakhstan), Kokchetau in the north of the republic, and Pricaspian near the Caspian Sea. Since 1970 extensive low-cost resources have been discovered in the Chu-Sarysu and Syr-Darya basins in south-central Kazakhstan.

Production started in 1953, initial output being processed in Kyrgyzstan. Production centres in Kazakhstan were started up by the Tselinny Mining and Processing Company in 1958 (based on underground-mined ore) and by the Kaskor Company in 1959 (based on open-pit mining). Economic pressures forced the closure of the Kaskor plant in 1993 and of the Tselinny plant in 1995. All subsequent uranium production has utilised ISL methods.

At the beginning of 2003 there were six ISL production centres in operation in Kazakhstan, with an aggregate production capacity of 4 700 tU/yr, together with one linked with an underground mine and operated by KazSubton (the foreign-owned successor to the Tselinny company), with a capacity of 1 250 tU/yr. Total output of uranium in 2002 was 2 822 tonnes, and cumulative national production now exceeds 95 000 tonnes.

Kazakhstan was the fifth largest producer in 2002, but its RAR (427 360 tonnes (in situ), of which about 90% would be recoverable at up to US$ 80/kg) put it in a much higher ranking—second only to Australia—and give it a 16% share in global resources at that cost level. In addition, there are well over 500 000 tonnes of other known resources: 162 000 tonnes of RAR and 352 000 tonnes of EAR-I, 90% of which is deemed to be recoverable at costs of less than US$ 130/kgU.

Undiscovered resources (in situ) recoverable at the same cost level are also massive: 310 000 tonnes of EAR-II and 500 000 tonnes of SR.

Malawi

Exploration during the 1980s led to the discovery of a uranium deposit at Kayelekera in northern Malawi. No exploratory activity has been reported in recent years.

The uranium resources (in situ) in the Kayelekera deposit amount to 11 700 tonnes, classified as RAR, 75% of which the IAEA estimates to be recoverable at less than US$ 80/kgU. No other uranium resources, either known or undiscovered, have been reported.

Mexico

Exploration for uranium came to an end in 1983: at that point, known in situ resources totalled 2 400 tonnes recoverable at US$ 80–130/kgU, comprising 1 700 tonnes of RAR and 700 tonnes of EAR-I: the IAEA estimates that 75% of these tonnages would be recoverable. Additional undiscovered resources (in situ) amounted to 13 000 tonnes, the bulk of which (10 000 tonnes) were speculative.

Unconventional resources contained in marine phosphates in Baja California amount to about 150 000 tU, as assessed in the early 1980s.

For a short period (1969–1971), molybdenum and by-product uranium were recovered from a variety of ores at a plant in Aldama, Chihuahua state. Uranium output totalled 49 tonnes: there are presently no plans for resuming production.

Mongolia

In situ resources have been assessed as 61 600 tonnes of RAR and 21 000 tonnes of EAR-1, all at up to US$ 80/kgU, plus 1.39 million tonnes of SR at less than US$ 130/kgU. In assessing recoverable resources, the IAEA applies a recovery factor of 75% to Mongolia's in situ RAR and EAR-I tonnages. Despite the extent of the established resources, recorded cumulative production of uranium amounts to only 535 tonnes.

Namibia

Although uranium mineralisation had been detected in the Rössing Mountains in the Namib Desert in 1928, extensive exploration for uranium did not get under way until the late 1960s. The major discovery was the Rössing deposit, located to the north-east of Walvis Bay; other discoveries were made in the same area of west-central Namibia, notably the Trekkopje and Langer Heinrich deposits, but Rössing is the only one that has so far been developed. The Langer Heinrich deposit was acquired by an Australian company in August 2002.

A large open-pit mine operated by Rössing Uranium Ltd (56.3% owned by Rio Tinto Zinc, 3.5% by the Namibian Government and 40.2% by other interests) has been in production since 1976; output in 2002 was 2 333 tonnes, with cumulative production amounting to 76 699 tonnes. The 2002 output level represented 60.7% of the 3 845 tU/yr design capacity of Rössing's processing plant.

In December 2003 Rössing reported that its plan was to mine out the present pit (expected by end-2007), after which operations would cease. The company announced that an earlier closure could occur if the mine's financial situation did not improve by mid-2004. Comprehensive studies for a Phase 2 extension of the open pit have been completed although this too would be subject to an improvement in economic circumstances.

Namibia is currently the sixth largest uranium producer in the world. Its reasonably assured reserves of 139 297 tonnes (at up to US$ 80/kgU) are equivalent to nearly 6% of the global total. RAR recoverable at US$ 80–130/kgU are over 31 000 tonnes; estimated additional resources are also substantial, exceeding 107 000 tonnes (in situ), of which about 87 000 tU would be recoverable at up to US$ 130/kgU.

Niger

Exploration for uranium began in 1956, resulting in the discovery of a number of deposits in the Aïr region of north-central Niger. There are currently two uranium production centres, one near Arlit processing ore from the Ariege, Arlette, Tamou and Taza deposits and operated by Société des Mines de l'Aïr (Somaïr), and the other at Akouta processing ore from the Akouta and Akola deposits and operated by Compagnie Minière d'Akouta (Cominak). Niger's participation in the producing companies is 36.6% in Somaïr, and 31% in Cominak.

Somaïr has been producing uranium from open-pit operations since 1970, while Cominak has carried out underground mining since 1978. In 2002, Somaïr produced 1 074 tonnes, recording a cumulative output of around 39 000 tonnes; Cominak's output was 2 006 tonnes in 2002, with a cumulative total of more than 48 000 tonnes. The two companies have current production capabilities of 1 500 and 2 300 tU/yr, respectively. Niger is the world's third largest producer of uranium, accounting for about 9% of global output.

The 2003 Red Book quotes radically revised tonnages for Niger's uranium resources, compared with the previous edition. RAR recoverable at up to US$ 80/kgU now stand at 102 227 tU, compared with an in situ figure of 29 603 tU in the 2001 book, whilst EAR-I, in the same cost bracket, now show as 125 377 tU recoverable as against 25 529 in situ. EAR-II, also in the same cost bracket, are now put at less than 10 000 tU compared with over 16 000 tU, whilst SR are no longer quoted.

Pakistan

Extensive exploration for uranium has been carried out. Recent discoveries reported in the 1999 Red Book related to the Kamlial Formation in the Salt Range and the Maraghzar area in the Swat district, but no uranium resources have been reported to the IAEA. A number of previously discovered deposits have been mined out. Cumulative output of uranium, all recovered using ISL technology, now exceeds 900 tonnes.

Peru

During the course of exploration carried out up to 1992, the Peruvian Nuclear Energy Institute (IPEN) discovered over 40 occurrences of uranium in the Department of Puno, in the south-east of the republic.

Known conventional resources (in situ) in the Macusani area in northern Puno are estimated to amount to 3 650 tonnes, of which 1 790 are classified as RAR and 1 860 as EAR-I: approximately two-thirds of each category is reckoned to be recoverable. Undiscovered resources (in situ) consist of 6 610 tonnes in the EAR-II category (recoverable at less than US$ 80/kgU), plus 19 740 tonnes of SR (recoverable at less than US$ 130/kgU).

Portugal

Uranium had been mined since 1951 from a large number of small deposits in two areas of central Portugal. Uranium production came to an end in 2001, after cumulative production of about 3 700 tonnes. RAR (at up to US$ 80/kgU) are put at almost 7 500 tonnes. Other known conventional resources consist of EAR-I of 1 450 tonnes, recoverable at less than US$ 80/kgU; undiscovered conventional resources recoverable at below US$ 130/kgU comprise 1 500 tonnes of EAR-II and 5 000 tonnes of SR.

Romania

Since 1952, when Romania started to produce uranium, cumulative output has reached nearly 18 000 tonnes. There are deposits in three principal areas: the Apuseni Mountains in the west, the Banat Mountains in the south-west and the Eastern Carpathians. Since 1978, all of Romania's production of uranium ore has been processed at the Feldioara mill in the centre of the country.

Uranium output in 2002 was 90 tonnes, with remaining RAR (at up to US$ 130/kgU) estimated as 3 325 tonnes (recoverable). Further known conventional resources recoverable at the same cost level are 3 608 tonnes of EAR-I; in situ undiscovered resources comprise 3 000 tonnes of EAR-II plus an equal tonnage of SR.

Russian Federation

Uranium exploration has been undertaken since 1944; 11 ore-bearing districts have been identified east of the Urals and four in the European part of Russia. Exploration and development activity in recent years has been largely concentrated on three east-of-Urals uranium districts (Transural, West Siberia and Vitim) in which there are deposits suitable for the application of ISL.

Mining and processing of uranium ore started in 1951 in the Stavropolsky region of European Russia, a source which had been exhausted by the late 1980s, after producing 5 685 tonnes. Between 1968 and 1980, the Sanarskoye deposit in the Transural district produced 440 tonnes of uranium, using ISL technology.

For more than a decade, the most important uranium producing area has been the Streltsovsky region near Krasnokamensk in the Chitinskaya Oblast of eastern Siberia. The state concern responsible for production in the Krasnokamensk area is the Priargunsky Mining-Chemical Production Association; its production centre has a nominal production capacity of 3 500 tU per annum. In 2002, the Dalur production centre in the Kurgan region started commercial ISL extraction from the Dalmatovskoe deposit. By 2010, it is planned that additional ISL sites at this deposit and at Khokhlovskoe will increase Dalur's annual production capacity to 700 tU. Another production centre is planned for the Khiagda deposit in the Vitim district.

Total national output in 2002 was 2 850 tU, nearly all of which was derived from ore obtained by underground mining, the balance being obtained from low-grade ore by heap- or in-place leaching. The Russian Federation was the world's fourth largest producer of uranium in 2002, accounting for 7.9% of global output. Its RAR (estimated to be recoverable at up to US$ 80/kgU) of 124 000 tonnes represented 5.0% of the global total at end 2002.

The balance of known conventional resources recoverable at less than US$ 80/kgU consisted of 34 260 tonnes of EAR-I. Undiscovered resources (in situ) are estimated to be exceedingly large: nearly 105 000 tonnes of EAR-II at up to US$ 130/kgU plus 545 000 tonnes of SR in the same price bracket.

Slovenia

Exploration of the Zirovski Vrh area began in 1961, followed some 20 years later by the commencement of mining and eventually by the production of yellow cake in 1985. Exploration expenditure ceased in 1990 and uranium production came to an end 2 years later, with cumulative output of 382 tU.

The reported uranium resources are fairly modest: RAR of 2 200 tU and EAR-I of 5 000 tU, both recoverable at under US$ 80/kgU, plus 5 000 tU of EAR-I and about 1 000 tU of EAR-II, both of which are deemed recoverable at US$ 80–130/kgU.

South Africa

Between the late 1940s and the early 1970s uranium exploration was pursued as an adjunct to exploration for gold, centred on the quartz-pebble conglomerates in the Witwatersrand Basin in the Transvaal. The 1973–1974 oil crisis triggered intensified exploration for uranium, leading to the country's first primary uranium mine (Beisa) being commissioned in 1981. Output as a by-product of gold mining had begun 30 years previously, and by 1959 26 mines in the Witwatersrand Basin were supplying 17 processing plants, resulting in an annual output of nearly 5 000 tonnes.

Between the late 1980s and the early 1990s, a substantial reduction in production capacity took place; subsequent closures brought the total of operational production centres at the beginning of 2002 down to two, each served by a single mine. The companies in production were Vaal River Operations at Klerksdorp, and Palabora Mining Company in the Northern Province; uranium production by the latter company, as a by-product of copper mining, ceased during the year. Total uranium output in 2002 was 824 tonnes, the ninth largest national level in the world. The cumulative output of uranium in South Africa up to the end of 2002 exceeded 152 000 tonnes.

The country's RAR (at up to US$ 80/kgU), consisting to a considerable extent of quartz-pebble conglomerates, came to nearly 232 000 tonnes by the end of 2002, equivalent to 9.4% of the world total. Further resources are on a commensurately large scale: almost 84 000 tU of RAR recoverable at US$ 80–130/kgU, over 80 000 tU of recoverable EAR-I, 110 000 tU of EAR-II and more than 1.1 million tU in the speculative category.

Spain

The first uranium discoveries were made in the western province of Salamanca in 1957–1958. Subsequently other finds were made further to the south and, in one instance, in central Spain. Production began in 1959 and by the end of 2002, a cumulative total of over 5 000 tonnes had been produced. Ore mining ceased in December 2000 and the production of uranium concentrates was terminated 2 years later.

At end-2002, remaining RAR (at less than US$ 80/kgU) were put at 2 460 tonnes. Further known conventional resources recoverable at US$ 80–130/kgU comprised 2 465 tonnes of RAR and 6 380 tonnes of EAR-I.

Sweden

Exploration for uranium was carried out from 1950 until 1985, when low world prices for the metal brought domestic prospecting to a halt. Four principal uranium provinces were identified, two in south/central Sweden and two in the north.

Proved reserves are reported as 4 000 tonnes recoverable at less than US$ 130/kgU, with additional amounts recoverable comprising 6 000 tonnes in the same cost bracket.

There are substantial unconventional resources of uranium in alum shale, but the deposits are very low grade and recovery costs would exceed US$ 130/kgU. During the 1960s, a total of 200 tonnes of uranium was recovered from alum shale deposits at Ranstad, in the Billingen district of Västergötland, southern Sweden. This mining complex has now been rehabilitated, the open pit being transformed into a lake and the tailings area treated to prevent the formation of acid.

Ukraine

Since the start of exploration for commercial resources of uranium in 1944, a total of 21 deposits have been discovered, mostly located in south-central Ukraine, between the rivers Bug and Dnepr. The most important ore bodies are Vatutinskoye, Severinskoye and Michurinskoye, all in central Ukraine. Uranium has been produced since 1947, initially by the Prednieprovskiy Chemical Plant and since 1959 also by the Zheltiye Vody production centre. The first plant ceased producing uranium in 1990; the 2002 output of the other facility was 800 tonnes, 80% of its nominal production capacity. All currently processed ore comes from underground operations at the Ingul'skii mine on the Michurinskoye deposit and from the Vatutinskii mine on the Vatutinskoye deposit.

In 2002 Ukraine was the 10th largest producer of uranium, accounting for just over 2% of the world total. Its uranium resources have been re-assessed for the 2003 edition of the Red Book, particularly in respect of EAR-I. RAR (at up to US$ 80/kgU) are now put at 45 150 tonnes in situ, of which 34 630 is deemed to be recoverable. Further known conventional resources are represented by 39 000 tonnes of in situ RAR (30 030 recoverable) at US$ 80–130/kgU and a much reduced 14 850 tonnes of in situ EAR-I (11 410 recoverable) at up to US$ 130/kgU.

Undiscovered resources (in situ) comprise 1 600 tonnes of EAR-II recoverable at up to US$ 130/kgU plus 255 000 tonnes of SR (with cost range unassigned).

United States of America

Between 1947 and 1970 the US Atomic Energy Commission (AEC) promoted the development of a private-sector uranium exploration and production industry; in late 1957 the AEC concluded its own exploration and development activities. Private-sector efforts accelerated in the 1970s in a context of rising prices and anticipated growth in the demand for the metal to fuel civilian power plants.

This exploration activity revealed the existence of extensive ore deposits in the western half of the United States, particularly in the states of Wyoming, Nebraska, Utah, Colorado, Arizona and New Mexico and in the Texas Gulf Coastal Plain. Numerous production centres were erected over the years, but many have now been closed down and either dismantled or put on standby.

Current production is mainly reliant on ISL; some uranium is obtained by other methods, such as mine water treatment and environmental restoration. At the beginning of 2003, two ISL plants (with an aggregate capacity of 1 150 tU/yr) were operational; four ISL plants were undergoing restoration and four conventional mills were on standby. US uranium output in 2002 amounted to an estimated 902 tonnes, the eighth highest in the world.

RAR (at up to US$ 80/kgU) were estimated to be 102 000 tonnes at end-2002, equivalent to 4.1% of the global total; RAR recoverable at US$ 80–130/kgU were 243 000 tonnes.

Estimated additional resources (not specified separately for EAR-I and EAR-II) were 839 000 tonnes at up to US$ 80/kgU and 434 000 at US$ 80–130/kgU. SR at up to US$ 130/kgU were 858 000 tonnes, with additional SR (with a cost range of US $130–260/kgU) amounting to 482 000 tonnes.

Uzbekistan

Deposits of uranium ores have been found in at least 25 locations since the early 1950s, mostly lying in the central Kyzylkum area running from Uchkuduk in the north-west to Nurabad in the south-east. Although there was some production in the Fergana valley area, starting in 1946, commercial mining began in 1958 at Uchkuduk from open-pit and underground operations. ISL recovery methods were brought into use from 1965 and gradually came to dominate the production scene. The last of the open-pit and underground mines were closed in 1994, after conventional mining had produced a cumulative total of nearly 56 000 tonnes, 65% of which had come from open-pit operations.

Uranium output in 2002 by the state-owned Navoi Mining and Metallurgical Complex (NMMC), the sole producer, totalled 1 859 tonnes—equivalent to just over 5% of global output. In operation during 2002 were three ISL production centres, which sent their output by rail to the NMMC processing plant at Navoi (nominal production capacity 3 000 tU/yr).

The republic's RAR (at up to US$ 80/kgU) amounted to just under 84 000 tonnes at the end of 2002, of which about 73% is considered to be recoverable. The balance of known conventional resources consists of some 25 000 tonnes of in situ RAR (of which about 18 000 is considered recoverable at US$ 80–130/kgU) and over 55 000 tonnes of in situ EAR-I (with nearly 39 000 tU recoverable at up to US$ 130/kgU). Undiscovered conventional resources (on an in situ basis) total about 230 000 tonnes, of which EAR-II recoverable at up to US$ 130/kgU account for 85 000 tonnes, the balance being SR without a cost range assigned.

Vietnam

Exploration for uranium in selected parts of the republic began in 1955, and since 1978 a systematic regional programme has been undertaken. Virtually the entire country has now been explored, with a number of occurrences and anomalies subjected to more intensive investigation. Since 1997, exploration activity has been concentrated on the Nong Son basin in the Quang Nam province of central Vietnam.

RAR recoverable at up to US$ 130/kgU (on an in situ basis) are 1 337 tonnes; EAR-I (on the same basis) are 7 244 tonnes: in both categories, approximately 75% is estimated to be recoverable. Undiscovered in situ conventional resources in the same cost bracket consist of 7 860 tonnes in the EAR-II category, plus 100 000 tonnes of SR. Further SR (without a cost range assigned) amount to 130 000 tonnes.

Unquantified amounts of unconventional resources have been reported to be present in deposits of coal, rare earths, phosphates and graphite.

No production of uranium has so far been achieved.

Chapter 6

Part II: Nuclear

World Energy Council
CONSEIL MONDIAL DE L'ENERGIE

COMMENTARY

Nuclear Power Today

2004 marks the 50th anniversary of the first delivery of nuclear-generated electricity to an electrical grid, at Obninsk in the USSR in June 1954. Commercial nuclear power capacity grew rapidly in industrialised countries in the next few decades, but after 1986, the year of the Chernobyl accident, expansion slowed to essentially the rate of global electricity growth, and nuclear power's share of worldwide electricity supplies has since held steady at 16–17%. The blue bars in Fig. 6.5 show annual capacity additions since 1966. The red line shows total annual nuclear electricity generation. The steepest increase in actual generation coincides with the mid-1980s peak in capacity additions, but particularly in the 1990s generation rose faster than capacity additions owing to increases in availability. The average energy availability factor for the global nuclear fleet has increased from 74.2% in 1991 to 83.7% in 2002.

As of 31 December, 2003 there were 440 nuclear power plants (NPPs) operating in 31 countries. With now over 11 000 reactor years of commercial experience (see Fig. 6.6), the nuclear industry has shown that well designed, constructed and operated NPPs are reliable, safe, economical and environmentally benign.

Current expansion and growth prospects are centred in Asia. Nineteen of the 32 reactors under construction are located in China; Taiwan, China; India; Japan; the Republic of Korea; and the Democratic People's Republic of Korea (DPRK). Nineteen of the last 28 reactors to be connected to the grid are in the Far East and South Asia. The one connection of a new NPP to the grid in 2003 was in China, and the one new construction start was in Japan.

In Western Europe, capacity is likely to remain relatively constant despite nuclear phase-outs in Belgium, Germany and Sweden. The most advanced planning for new nuclear capacity is in Finland. The utility Teollisuuden Voima Oy signed a contract in 2003 with Areva and Siemens for a 1 600 MW_e European pressurised water reactor and formally applied for a construction licence in January 2004.

The emphasis in the USA is on licence extensions and upratings. By the end of 2003, the US Nuclear Regulatory Commission (NRC) had approved 19 licence extensions of 20 years each (for a total licensed life of 60 years for each NPP). It has also streamlined procedures to encourage new construction, and three companies have now applied for the NRC's new early site permits, which can be reserved for future use. In Canada, near-term expansion is taking

2004 Survey of Energy Resources

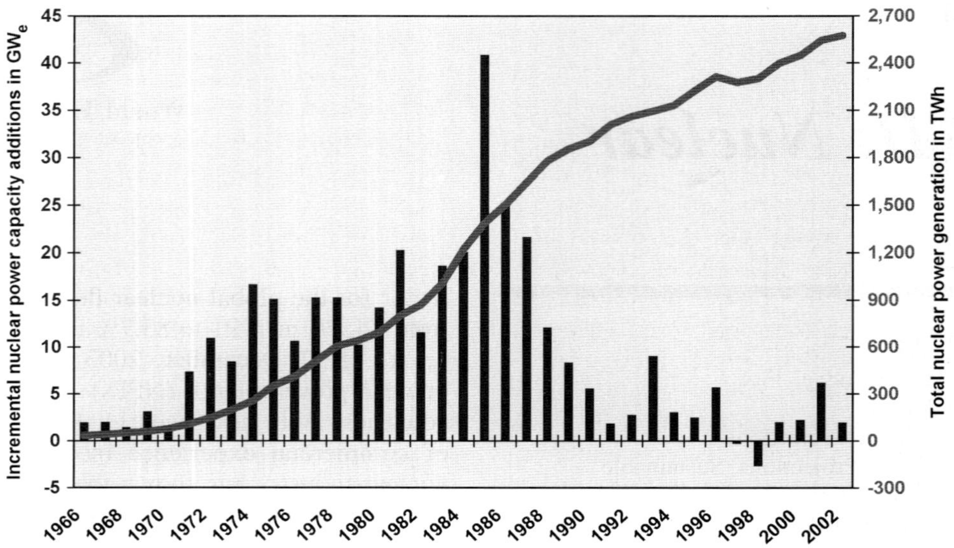

FIGURE 6.5 Annual changes in global nuclear capacity (bars, left axis) and annual nuclear electricity generation (line, right axis), 1966–2002.

the form of restarting some or all of the eight nuclear units (out of a Canadian total of 22) that have been shut in recent years. The first two such restarts took place in 2003.

In the Russian Federation, which has 30 NPPs in operation and three under construction, Rosenergoatom has begun a programme to extend licences at 11 NPPs. Extensions for Novovoronezh-3 and Kola-1 had been approved by the end of 2003.

Nuclear expansion, like the expansion of electricity generation in general, is motivated by growth in electricity demand. All independent analyses and forecasts of global energy

Country	Years	Months	Country	Years	Months
Argentina	50	7	Mexico	23	11
Armenia	36	3	Netherlands	59	0
Belgium	191	7	Pakistan	35	10
Brazil	25	3	Romania	7	6
Bulgaria	129	2	Russian Federation	761	4
Canada	486	11	Slovakia	103	0
China	39	1	Slovenia	22	3
Czech Republic	74	10	South Africa	38	3
Finland	99	4	Spain	219	2
France	1 346	2	Sweden	311	1
Germany	648	1	Switzerland	143	10
Hungary	74	2	Taiwan, China	134	1
India	223	5	Ukraine	279	10
Japan	1 124	4	United Kingdom	1 329	8
Korea (Republic)	220	7	United States of America	2 871	8
Lithuania	36	6	**Total**	**11 146**	**8**

FIGURE 6.6 Nuclear power reactors total operating experience to 31 December, 2003.

needs project large increases in the century ahead. The principal drivers are global population growth and economic development in today's developing countries. Electricity demand will grow even faster because in all sectors users prefer its cleanliness at the point of end-use, its convenience and its flexibility. But the rate of growth, and the options for meeting that growth, differ from country to country. For nuclear power, therefore, the motivation for expansion tends to be strongest in countries projecting rapid increases in electricity demand (e.g. China and India) or having few indigenous alternatives (e.g. Japan and the Republic of Korea), and weakest in countries experiencing slower growth and more able to afford alternatives (as in North America and the European Union). Such variations across countries are reflected in national differences in the public acceptance of nuclear power and in differences in national economic assessments of generation alternatives. They show up in all four of the key issues affecting near-term nuclear expansion: economics, safety, waste and proliferation resistance.

Economics: the front-loaded cost structure of NPPs means that existing amortised well-run plants can be quite profitable while new NPPs are often more expensive than alternatives. The economic attractiveness of today's well-run amortised plants is driven partly by the increasing energy availability factors cited at the beginning of this commentary, and it is reflected in the increasing emphasis, noted above, on licence extensions, particularly in the USA. But economic attractiveness differs for different countries (as noted above), investors and markets. New NPPs are most attractive where energy demand growth is rapid, alternative resources are scarce, energy supply security is a priority or nuclear power is important for reducing air pollution and greenhouse gas (GHG) emissions. NPPs are also more attractive to government investors responsible for energy security, GHG emissions and long-term development than for private investors who need rapid returns and receive no financial benefit from nuclear power's low GHG emissions or contribution to energy security. Thus in deregulated, slower growth markets in the West, new NPPs are generally less attractive. Anticipating entry-into-force of the Kyoto Protocol, Europe is creating a GHG emissions market, and future investors may thus realise a tangible benefit from nuclear power's low GHG emissions. The USA is exploring alternative ways to adjust near-term market incentives to encourage nuclear expansion in line with the longer term US National Energy Policy. Furthermore, reactor designers are continually working on reducing capital costs through improved designs, as discussed in the 'Advanced Reactors and Fuel Cycles' section.

Safety: Although the Chernobyl accident still hangs over nuclear power, the industry's current safety record is very good. Statistics for 2002 from the World Association of Nuclear Operators continue to show the trend seen throughout the 1990s towards lower industrial accident rates and fewer unplanned automatic scrams. The safety debate today is largely in the context of the European Union's efforts to accelerate closure of first-generation water cooled WWER and graphite-moderated reactors in Eastern Europe. Most recently, Bulgaria shut down Kozloduy-1 and -2 at the end of 2002 and Lithuania agreed to shut Ignalina-2 in addition to Ignalina-1.

Spent fuel and waste: As shown in Fig. 6.7, inventories of spent fuel are growing, owing to limited reprocessing and delays in disposal. For high-level waste, the most progress on disposal facilities has been made in Finland, Sweden and the USA. Finland's Government and Parliament have approved a decision 'in principle' to build a final repository for spent fuel near Olkiluoto. Separate construction and operating licences will also be required. Construction should start in 2011 and operation about 10 years later. Sweden has begun detailed geological investigations at three candidate sites. These should run for 5 or 6 years, and the Swedish nuclear fuel and waste management company, SKB, hopes to make a final site proposal by about 2007. The Waste Isolation Pilot Plant in the USA began

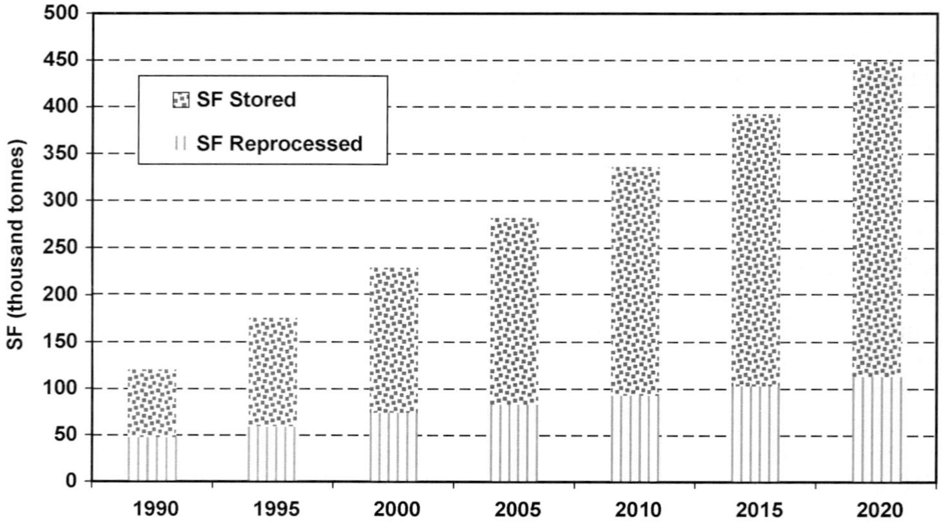

FIGURE 6.7 Cumulative worldwide spent fuel (SF) reprocessing and storage, historical and projected, 1990–2020.

accepting military transuranic waste in 1999 for permanent disposal in bedded salt. In 2002, the US President and Congress decided to proceed with the Yucca Mountain disposal site, operations at which are planned to begin in 2010.

Although the present focus remains on establishing national repositories, there is renewed interest in the possibility of regional or international repositories. One reason is the interest noted below in increasing international control of nuclear material as one effort to strengthen the global non-proliferation regime. The other is the reality that for countries with no good waste sites, or with small research and power programmes, individual national disposal sites make no economic sense.

Proliferation resistance: The international non-proliferation regime consists of the Treaty on the Non-Proliferation of Nuclear Weapons (NPT) and comprehensive International Atomic Energy Agency (IAEA) safeguards agreements, including additional protocols now in force in 38 countries[1], international verification measures (the safeguards system of the IAEA plus regional agreements and bilateral agreements) and export controls. The system is robust, nearly universal and in the main scrupulously applied. The cornerstone is the NPT, to which all of the UN's 191 countries (plus the Holy See) are parties except India, Israel and Pakistan—although the status of the DPRK is confusing. In February 2003 the DPRK notified the UN Security Council of its withdrawal from the NPT, but the UN continues to list the DPRK as a party to the agreement.

While the system has been largely effective for more than 30 years, violations by the DPRK, Libya and Iraq and the questions raised in 2003 about Iran suggest that additional measures are needed. New emphasis is being given to proposals to limit the use of and access to fissile material. One possibility is to limit facilities for legitimate uses to multinational consortia that are carefully and transparently constrained. Another complementary possibility is to develop new nuclear reactor designs and fuel cycles with engineered features to increase proliferation resistance. Current development programmes are summarised in the 'Advanced Reactors and Fuel Cycles' section.

Cutting across all these factors is the issue of preserving, transferring and expanding nuclear

[1] As of 18 December, 2003.

knowledge and skills. All are critical to ensuring safety and security, encouraging innovation, and ensuring that the nuclear option remains available for countries that wish to use it. However, in some industrialised countries, the nuclear workforce is ageing, fewer young people are studying nuclear science, nuclear engineering and related fields, and a number of universities have given up nuclear education programmes altogether. Programmes to reverse such trends have recently begun in quite a few countries, and the IAEA and others have become active at the international level.

Looking Ahead

Medium-Term Projections

Updated projections for nuclear power, produced by the IAEA in 2003, are shown in Fig. 6.8. The low projection essentially assumes no new NPPs beyond those that are already being built or are firmly planned today, plus the retirement of old NPPs. This projection shows a 20% increase in global nuclear generation up until the end of 2020, followed by a decrease, resulting in global nuclear generation in 2030 only 12% higher than in 2002. Increases are most substantial in the Far East, and decreases are greatest in Western Europe.

In the high projection, global nuclear generation steadily increases by 70% through 2030. There are increases in all regions, led by the Far East. However, overall electricity generation increases even faster than nuclear power, causing nuclear power's share of overall electricity to decline from 16% in 2002 to 11% in 2030.

The latest update, in 2002, of the Reference Scenario of the OECD International Energy Agency (IEA) follows essentially the same pattern as the IAEA's low projection, since both are based on current official government policies and plans for energy development. Nuclear generation in the IEA Reference Scenario, however, declines somewhat more after 2010 and has a final 2030 value 6% lower than in the IAEA's low projection.

Longer-Term Projections, Sustainable Development and Climate Change

Long-term scenarios, however, paint a more optimistic picture for nuclear expansion. The 40 reference scenarios in the Intergovernmental Panel on Climate Change's (IPCC's) *Special Report on Emissions Scenarios* (SRES), published in 2000, projected global nuclear capacities in 2050 ranging from the current value of 350 GW_e up to more than 5 000 GW_e, with a median of more than 1 500 GW_e. This would require adding 50–150 GW_e per year from 2020–2050. But none of the SRES reference scenarios includes policies to limit GHG emissions. A scenario that does include this objective, as well as objectives concerning energy security, diversification and access, is the 'SD Vision Scenario' published by the IEA in 2003. In it, nuclear energy expands 14-fold between 2000 and 2050, near the top end of the SRES range.

This 'projection gap' between the medium-term IAEA projections and the long-term scenarios of SRES and the IEA is largely due to differing assumptions about political constraints, cost improvements and innovation. The medium-term scenarios assume relatively hostile or indifferent political environments, no strong sustainable development constraints, no innovation and little or no progress on new NPP costs. The long-term SRES scenarios generally assume, first, that nuclear technologies, like other technologies, are not static and, second, that in the long term investments are made ultimately on the basis of economics. The IEA scenario also assumes strong sustainable development constraints.

The 'Advanced Reactors and Fuel Cycles' section summarises current R&D on innovative, lower cost NPPs. With respect to political constraints, recent important international debates have taken place within the UN Commission on Sustainable Development (CSD) and at the 2002 World Summit on Sustainable Development (WSSD).

The CSD was established in 1992 to follow up implementation of *Agenda 21* as negotiated at the UN Conference on Environment and

Country Group	2002			2010			2020			2030		
	Total Electricity TWh	Nuclear		Total Electricity TWh	Nuclear		Total Electricity TWh	Nuclear		Total Electricity TWh	Nuclear	
		TWh	%		TWh	%		TWh	%		TWh	%
North America	4 779	851.1	17.8	5 034	874	17	5 784	870	15	6 451	844	13
				5 444	894	16	6 709	939	14	8 146	944	12
Latin America	1 078	28.6	2.7	1 178	29	2.5	1 628	47	2.9	2 227	30	1.3
				1 427	38	2.7	2 291	50	2.2	3 758	92	2.4
Western Europe	3 084	880.2	28.5	3 352	858	26	3 634	823	23	3 942	564	14
				3 609	893	25	4 687	961	20	6 061	1 090	18
Eastern Europe	1 758	298.5	17.0	1 884	319	17	2 174	423	19	2 463	378	15
				2 074	399	19	2 867	552	19	4 133	611	15
Africa	459	12.0	2.6	538	13	2.5	699	14	2.0	876	14	1.6
				612	14	2.3	973	24	2.4	1 530	60	3.9
Middle East & South Asia	1 176	19.6	1.7	1 342	41	3.1	1 805	53	3.0	2 327	70	3.0
				1 626	47	2.9	2 596	100	3.9	3 946	194	4.9
South East Asia & the Pacific	600			736			934			1 162		
				786			1 119	5.5	0.5	1 584	18	1.2
Far East	3 157	484.3	15.3	3 399	695	20	4 199	855	20	5 073	981	19
				4 296	702	16	6 605	1 125	17	9 830	1 361	14
World total Low Estimate	**16 090**	**2574.2**	**16.0**	**17 463**	**2 830**	**16**	**20 857**	**3 085**	**15**	**24 520**	**2 881**	**12**
High Estimate				**19 873**	**2 987**	**15**	**27 848**	**3 756**	**13**	**38 989**	**4 369**	**11**

FIGURE 6.8 Estimates of total electricity generation and contribution by nuclear power.

Development (UNCED). In April 2001, at its ninth session, the CSD focussed on energy issues for the first time. It conducted a comprehensive debate on nuclear energy and came to two principal conclusions. The first is that the parties agreed to disagree on nuclear energy's role in sustainable development. The final text notes that some countries see nuclear energy as a substantial contributor to sustainable development while others consider the two to be fundamentally inconsistent. Second, the parties agreed that, 'the choice of nuclear energy rests with countries'. These conclusions on nuclear power were reinforced a year and a half later at the WSSD in Johannesburg, South Africa.

In connection with rising international concerns about global warming, implementation of the Kyoto Protocol would be an important step towards attaching a tangible economic value to nuclear power's avoidance of GHG emissions. Currently the fact that nuclear power produces virtually no greenhouse gases is an advantage that is invisible to investors. Except for a very few instances, there have been no restrictions or taxes on GHG emissions and thus no economic value to their avoidance. Particularly in liberalised energy markets, binding restrictions on GHG emissions are needed if nuclear power's advantage of very low emissions is ever to matter to investors. And at the moment, the Kyoto Protocol is the world's only operative route toward widespread, coordinated restrictions.

Given the current status of ratifications, the Kyoto Protocol will enter into force if and only if either Russia or the USA ratifies. The US Government has emphatically said it will not ratify. Deliberations are under way in Russia, but both the timing and outcome are uncertain.

Advanced Reactors and Fuel Cycles

Many countries are working to improve the economics, safety and proliferation resistance of advanced reactor–fuel cycle systems. Efforts are focussed on making plants simpler to operate, inspect, maintain and repair. In the near term, most new NPPs are likely to be evolutionary designs building on proven systems while incorporating technological advances and often economies of scale. For the longer term, the focus is on innovative designs, several of which are in the small-to-medium range (up to 700 MW_e). These envision construction with factory-built components, including complete modular units for fast on-site installation, creating possible economies of series production instead of economies of scale. Other advantages foreseen for smaller units are easier financing, greater suitability for small electricity grids or remote locations, and their potential for district heating, seawater desalination and other non-electric applications. All should increase their attractiveness for developing countries.

Important efforts on large evolutionary LWR designs are under way in China, France, Germany, Japan, the Republic of Korea, the Russian Federation and the USA. The main efforts on small and medium-size evolutionary LWR designs are in China, France, Japan, the Russian Federation and the USA. Innovative LWR designs (i.e. those that incorporate radical conceptual changes in design approaches or system configuration) are being developed in Argentina, Japan, the Republic of Korea, the Russian Federation and the USA.

Both Canada and India are working on advanced heavy water reactor designs, and a number of advanced gas cooled reactor designs are being developed with participation from China, France, Germany, Japan, the Russian Federation, South Africa, the UK and the USA. For liquid metal cooled fast reactors, development activities are under way in China, France, India, Japan, the Republic of Korea and the Russian Federation. Development activities for lead alloy and sodium liquid metal cooled fast reactor systems and for gas (helium) cooled fast reactors are being conducted within the Generation IV International Forum (GIF) and in the Russian Federation. Research on fast neutron spectrum hybrid systems (e.g. accelerator driven systems) is under way in the Republic of Korea, the Russian Federation, the USA and eight EU countries.

Complementing the many initiatives above are two major international efforts to promote innovation—the US-initiated GIF and the IAEA's International Project on Innovative Nuclear Reactors and Fuel Cycles (INPRO). The former is currently geared towards specific technology development for reactors intended to come online around 2030. The latter is focussed more on the establishment of user requirements to guide fuel cycle and reactor R&D programmes targeted on prospective global mid-century energy markets.

H.-Holger Rogner
International Atomic Energy Agency

Table Notes

The majority of the data shown in Table 6.4 were provided by WEC Member Committees in 2003. If information was not available from this source, data have been derived from the following published sources:

- *Nuclear Power Reactors in the World*; April 2003; International Atomic Energy Agency, Vienna;
- *Elecnuc: Les Centrales Nucléaires dans le Monde 2003*; Commissariat à l'énergie atomique, Paris.

TABLE 6.4

Nuclear energy: capacity and generation

	In operation in 2002		Under construction at end-2002		Net generation in 2002 (TWh)	Nuclear share of electricity generation in 2002 (%)
	Units (number)	Capacity (MW_e)	Units (number)	Capacity (MW_e)		
South Africa	2	1 800			12.0	5.8
Total Africa	**2**	**1 800**			**12.0**	
Canada	14	10 018			70.9	12.5
Mexico	2	1 365			9.4	4.1
USA	104	98 564			780.1	20.3
Total N. America	**120**	**109 947**			**860.4**	
Argentina	2	935	1	692	5.8	7.6
Brazil	2	1 901			13.8	4.0
Total S. America	**4**	**2 836**	**1**	**692**	**19.6**	
Armenia	1	376			2.1	40.5
China	7	5 318	4	3 275	23.5	1.4
India	14	2 503	7	3 420	17.8	3.7
Japan	52	43 893	3	3 696	314.3	33.9
Korea (DPR)			1	1 040		
Korea (Republic)	18	14 890	2	1 920	113.1	38.6
Pakistan	2	425			1.8	2.5
Taiwan, China	6	4 884	2	2 700	33.9	20.5
Total Asia	**100**	**72 289**	**19**	**16 051**	**506.5**	
Belgium	7	5 760			44.7	57.3
Bulgaria	6	3 538			18.8	47.3
Czech Republic	5	2 556	1	912	17.6	25.0
Finland	4	2 656			21.4	25.6
France	59	63 273			416.5	77.9
Germany	19	21 283			162.3	29.9
Hungary	4	1 770			13.1	39.7
Lithuania	2	2 370			12.9	80.1

(continued on next page)

TABLE 6.4 (*Continued*)

	In operation in 2002		Under construction at end-2002		Net generation in 2002 (TWh)	Nuclear share of electricity generation in 2002 (%)
	Units (number)	Capacity (MW_e)	Units (number)	Capacity (MW_e)		
Netherlands	1	450			3.7	4.0
Romania	1	708	1	708	5.4	9.8
Russian Federation	30	20 793	3	2 825	130.0	16.0
Slovakia	6	2 460	2	820	16.5	54.6
Slovenia	1	656			5.5	38.0
Spain	9	7 579			60.3	25.7
Sweden	11	9 424			65.6	44.0
Switzerland	5	3 127			25.7	39.5
Ukraine	13	11 207	4	3 800	73.4	45.7
United Kingdom	31	12 486			81.1	22.1
Total Europe	**214**	**172 096**	**11**	**9 065**	**1 174.5**	
Iran			1	1 000		
Total Middle East			**1**	**1 000**		
Total World	**440**	**358 968**	**32**	**26 808**	**2 573.0**	

Note:

(1) The capacity and output of the Krsko nuclear power plant, shown against Slovenia in the table, is shared 50/50 between Slovenia and Croatia.

COUNTRY NOTES

The Country Notes on Nuclear have been compiled by the editors, largely on the basis of material published in:

- *Nuclear Power Reactors in the World*, April 2003, International Atomic Energy Agency, Vienna;
- *Elecnuc: Les Centrales Nucléaires dans le Monde 2003*, Commissariat à l'énergie atomique, Paris;
- *Daily Press Review*, IAEA, Vienna;
- *WNA News Briefing*, World Nuclear Association, London.

Information provided by WEC Member Committees has been incorporated when available.

Argentina

There are two NPPs: Atucha-I, a 335 MW_e PHWR supplied by Germany, and Embalse, a Canadian-designed 600 MW_e PHWR: Atucha-I came online in 1974, Embalse in 1983. In 2002 the two nuclear stations provided 7.6% of Argentina's electricity output.

The construction of a third unit (Atucha-II), a 692 MW_e PHWR, has been interrupted since 1995. The project had advanced 80%, and the estimated time to complete the work from the date it is restarted is 4 years. At present the extension of the useful life of the existing nuclear power stations and the conditions that should be observed to complete Atucha II Nuclear Power Station are being studied.

Armenia

A NPP came into operation at Medzamor, 64 km from the capital Yerevan, in 1976 but it was closed down in 1989 following an earthquake the previous year. Concern over the station's safety from a seismic point of view was exacerbated by the repercussions of the Chernobyl incident.

One of the two original WWER units (Medzamor-2) has been upgraded and refurbished, coming back into commercial operation in 1996 with a capacity of 376 MW_e. It provided about 40% of Armenia's electricity output in 2002. In February 2003, Russia assumed financial management of Medzamor-2 in settlement of fuel debts owed by Armenia.

In December 2003, it was reported that the IAEA and local companies were examining the possibility of constructing a new reactor at Medzamor NPP to meet growing energy demand in the period to 2020.

Belgium

A total of seven reactors were constructed between 1975 and 1985, four units at Doel and three at Tihange; they are all of the PWR type, with an aggregate net generating capacity of 5 760 MW_e. In 2002, nuclear power provided about 57% of Belgium's electricity generation.

In January 2003, Belgium's Senate voted for a nuclear phase-out law which stipulates that all seven units shall be closed after completing 40 years of operation. The first reactors are thus due to be shut down in 2015, the last in 2025.

Brazil

At the end of 2002, Brazil had two NPPs in operation: Angra-1, a 626 MW_e net PWR, and Angra-2 (1 275 MW_e net). In an electricity market dominated by hydropower, nuclear's share of generation in 2002 was only 4%.

A project for a third unit at Angra, of similar size to Angra-2, is under discussion in the Ministry of Mines and Energy, which is responsible for the definition of energy policies. Many aspects are being analysed, including the project's environmental and economic viability.

Hence the construction of the third NPP has not yet been authorised by the Brazilian Electricity Regulatory Agency.

Bulgaria

Six WWER units have been constructed at Kozloduy, in the north-west of the country, close to the border with Romania. Four units (each of 408 MW_e net capacity) were brought into operation between 1974 and 1982, and two others (each of 953 MW_e capacity) were commissioned in 1987 and 1989, respectively. The combined output of the Kozloduy reactors provided 47% of Bulgaria's electricity generation in 2002. Kozloduy-1 and -2 were shut down in December 2002 and the Government has agreed to the EU's demand that units 3 and 4 be closed down by 2006. The Bulgarian WEC Member Committee foresees a total nuclear capacity of 3 000 MW_e in 2015, with four units in operation.

A legal, regulatory and pricing framework will be set for the implementation of every single new project on a fair and genuinely competitive basis. Bulgaria will continue to rely on nuclear energy and to develop it further in compliance with the up-to-date requirements on safety, cost efficiency and reliability, nuclear safety and radiation protection.

Construction of unit 1 on a site at Belene started in 1987, but at present this project is on hold. Any decision on the decommissioning of nuclear facilities will be based on a comprehensive analysis of the country's capacity to maintain and upgrade safety levels and to operate nuclear facilities in compliance with national legislation and commitments arising from the Convention on Nuclear Safety.

Russia has expressed interest in participating in the construction of a seventh unit at Kozloduy or one at Belene, about 110 km to the east. The European Commission has granted a loan of €212.5 million to modernise and upgrade safety at Kozloduy-5 and -6.

Canada

There are currently 22 nuclear power reactors in Canada, which are operated by public utilities and private companies in Ontario (20), Quebec (1) and New Brunswick (1). Of the 22 reactors installed, 14 reactors are currently in full commercial operation; all plants are of the PHWR type, providing total nuclear generating capacity of almost exactly 10 GW. They generate, on average, around 12.5% of Canada's electricity needs.

Of the eight nuclear reactors that were laid-up at the end of 2002, four were at Pickering A station and four at Bruce A station. It is anticipated that all four units at Pickering A and two units at Bruce A will be brought back into service by 2007, subject to final regulatory approvals. (In fact, Bruce A-4 returned to service in October 2003, and the A-3 unit reached criticality in December of the same year.) Bruce Power, the Canadian consortium currently leasing the Bruce units, has indicated that the other two units will be restarted if a proper business case can be made for resurrecting them.

Canada will continue to rely on nuclear energy to help satisfy its electricity requirements in the future. While the units presently laid-up are expected to be back in service by 2007, no new nuclear reactors are currently planned to be built over this period.

The Federal Government regulates the nuclear industry through the Canadian Nuclear Safety Commission (CNSC) and provides financial support for the research and development programme of Atomic Energy of Canada Limited (AECL).

China

China's first NPP, a 279 MW_e PWR, came online at Qinshan, near Shanghai, in December 1991. Two larger PWRs (each 944 MW_e net) were brought into operation at Daya Bay (Guangdong province) in 1993–1994. Four more plants came into operation during 2002: Qinshan 3-1, a 665 MW_e PHWR and three PWRs, Lingao 1 and 2, each 938 MW_e, and Qinshan 2-1 (610 MW_e). At end-2002, China's nuclear generating capacity stood at 5 318 MW_e; with output from the seven units providing 1.4% of its electricity generation during the year.

Four more nuclear units (a 665 MW_e PHWR and three PWRs, one of 610 MW_e and two 1 000 MW_e) were under construction at the end of 2002, with an aggregate net capacity of 3 275 MW_e.

Croatia

There are no nuclear plants on Croatian soil at present but the republic has a 50% share in the 656 MW_e Krsko PWR located across the border in Slovenia.

Czech Republic

There are four 411 MW_e (net) reactors at Dukovany, which came into operation between 1985 and 1987. Two 912 MW_e (net) units have been constructed at Temelín: the first unit came online in December 2000, the second during 2003. In 2002, nuclear power provided 25% of the republic's net electricity generation.

The construction and commissioning of the Russian-designed WWER units at Temelín have aroused considerable anxiety in neighbouring countries, particularly Austria, owing to fears over their operational safety. The reactors are based on a Soviet design, subsequently updated with Western control systems.

The commissioning of Temelín-2 marks the completion of the Czech Republic's current nuclear programme. Any further developments will depend upon political decisions at government level. Continued use of nuclear energy for electricity generation is foreseen during the next 20 or more years in the document entitled 'Energy Policy of the Czech Republic'.

Egypt (Arab Republic)

In June 2002, the Egyptian Minister of Electricity and Energy announced plans for the country's first nuclear station, to be constructed on the Mediterranean coast, 150 km west of Alexandria. The WEC Member Committee has reported that, in cooperation with the IAEA, Egypt is currently carrying out studies to determine the type and size of the installation, which will provide both electricity generation and desalination of seawater. The plant is planned to be in commercial operation before 2020.

Finland

Four nuclear reactors were brought into operation between 1977 and 1980: two 488 MW_e WWERs at Loviisa, east of Helsinki, and two 840 MW_e BWRs at Olkiluoto. In 2002 the four units accounted for 25.6% of Finland's electricity output.

The Finnish Parliament ratified in May 2002 the Government's earlier favourable Decision-in-Principle (DiP) on a fifth NPP unit. The application for the DiP had been filed in November 2000 by the nuclear power company Teollisuuden Voima (TVO). Following the ratification of the DiP, TVO was authorised to continue preparations for the construction of a new NPP. A new nuclear power unit of 1 000–1 600 MW_e could be in commercial use by 2010.

In October 2003, TVO announced that its preferred site for the fifth nuclear station was at Olkiluoto, 250 km north-west of Helsinki. A construction contract was signed with a Framatome and Siemens consortium in December of the same year.

France

France has pursued a vigorous policy of nuclear power development since the mid-1970s and now has by far the largest nuclear generating capacity of any country in Europe, and is second only to the USA in the world. At end-2002 there were 59 reactors in operation, with an aggregate net capacity of over 63 000 MW_e. NPPs provide some 78% of France's electricity output. Apart from a single fast reactor (Phenix), PWRs account for the whole of current nuclear capacity.

There are no nuclear reactors presently under construction: the completion of Civaux-2 in December 1999 marked the end of the current French nuclear programme; no more units are likely to be built before 2015.

Germany

A total of 19 reactor units, with an aggregate net generating capacity of 21 283 MW_e, were operational at the end of 2002. Nuclear power provided 30% of Germany's net electricity generation in that year.

In June 2000, the Federal Government concluded an agreement with the German utility companies that provides for an eventual phasing-out of nuclear generation. The agreement specifies a maximum of 2 623 TWh for the lifetime production of all existing nuclear reactors, which implies an average plant lifetime of 32 years. As the newest German reactor (Neckarwestheim-2) was connected to the grid in January 1989, it could be expected to survive until 2021; however, utilities will be allowed to switch productive capacity between stations, so that the life of the newer, more economic plants could be extended by prematurely shutting down other units. Moreover, the calculated 32-year average lifespan is predicated on a capacity factor of over 90%; using a somewhat lower (and more realistic) level of, say, 85% the average plant lifetime would approach 35 years.

Hungary

Four WWER reactors, with a current aggregate capacity of 1 770 MW_e, came into commercial operation at Paks in central Hungary, between 1983 and 1987. Their combined output in 2002 accounted for nearly 40% of total net electrical generation.

The Hungarian WEC Member Committee reports that there are two plans in connection with the present nuclear power installation:

- extension of its lifetime by 20 years (from 2017 to 2037);
- increasing its capacity (up to 500 MW_e rated capacity for each block), over the period to 2008.

India

At the end of 2002, India had 14 reactor units in operation, with an aggregate net generating capacity of 2 503 MW_e. Twelve were PHWRs, the other two being of the BWR type: all were relatively small units, with individual capacities up to 202 MW_e. Output from India's nuclear plants represented 3.7% of total electricity generation in 2002.

Three 202 MW_e PHWRs are currently under construction: Kaiga-3 and -4, and Rajasthan-5; also two larger PHWRs—Tarapur-3 and -4 (each of 490 MW_e net capacity) and two 917 MW_e WWERs (Kudankulam 1 and 2). In all, these seven units will add 3 420 MW_e to India's nuclear generating capacity over the next 5 years.

Construction of a 300 MW_e advanced heavy water reactor (AHWR) is scheduled to begin in 2004, with a view to exploring the use as fuel of thorium, of which India has substantial reserves, much larger than its known uranium resources.

Indonesia

The Minister of Research and Technology announced plans in January 2003 for the construction of Indonesia's first NPP. Construction is aimed to start in 2010, with completion by 2015. A year later, the plan was publicised again, this time by the Director-General of Electricity at the Ministry of Mines and Energy, who outlined a project for a 6 000 MW_e NPP to be built at Gunung Muria, Central Java, between 2011 and 2016.

In April 2003, during a visit to Moscow, the Indonesian President was offered a 40 MW_e floating nuclear plant, for completion in around 2015–2017. Russia plans to install a seaborne NPP off its Barents Sea port of Severodvinsk during the next 5 years.

Iran (Islamic Republic)

Construction of two 1 200 MW_e PWRs started at Bushehr in the mid-1970s, but work was suspended following the 1979 revolution. France's Commissariat à l'énergie atomique reports (end-2003) that Iran currently has two units under construction: Bushehr-1 (915 MW_e net) and Bushehr-2 (1 196 MW_e net). In October 2003, Russia, the supplier of the reactor equipment, announced a year's delay to the projected start-up date of the first Iranian reactor, putting it back until 2006. The Iranian WEC Member Committee expects both units to be in operation in 2015.

Japan

At the end of 2002 there were 52 operable nuclear reactors, with an aggregate gross generating capacity of 45 742 MW_e (43 893 MW_e net). Within this total there were 29 BWRs (26 376 MW_e gross, 25 468 MW_e net), and 23 PWRs (19 366 MW_e gross, 18 425 MW_e net). The Monju prototype fast-breeder reactor (260 MW_e net) has not yet been put back into operation, 8 years after a serious leak of sodium caused it to be shut down. In January 2004, the Nuclear and Industrial Safety Agency approved modifications to Monju designed to enable the reactor to be restarted.

In 2002, the output from Japan's NPPs provided about 34% of its net generation of electricity. At the end of the year, there were three units under construction (excluding the restoration of Monju), with an aggregate generating capacity of 3 838 MW_e gross (3 696 MW_e net). The Japanese WEC Member Committee expects that by the end of 2015 there will be 67 nuclear reactors in operation, with a total gross capacity of 65 415 MW_e (approximately 63 000 MW_e net).

It was reported in July 2003 that Japan's new long-term energy plan would continue to promote the use of nuclear power as a key player in the drive to alleviate global warming.

Construction of Shimane-3 (a 1 375 MW_e ABWR) has been delayed and is now expected to start in March 2005, with completion planned for 6 years later.

Kazakhstan

The only NPP to have operated in Kazakhstan was BN-350, a 70 MW_e fast breeder reactor located at Aktau on the Mangyshlak Peninsula in the Caspian Sea. It came into service in 1973 and was eventually shut down in June 1999. Reflecting its small generating capacity, and its additional use for desalination and the provision of process heat, BN-350's contribution to the republic's electricity supply was minimal: over its lifetime of operation, its average annual output was only about 70 GWh.

In June 2003, the Minister of Energy and Mines announced plans for the construction of a NPP within the next 15 years. The two–three unit NPP is to be established on the shores of Lake Balkhash in the Karaganda region of central Kazakhstan.

Korea (Democratic People's Republic)

A 1 040 MW_e PWR—known as LWR Project Unit 1—has been under construction since August 2002.

Korea (Republic)

At end-2002, there were 18 nuclear reactors (14 PWRs and 4 PHWRs) in operation, with an aggregate net capacity of 14 890 MW_e. Two PWRs (Ulchin-5 and -6) were under construction, with a total capacity of 1 920 MW_e; it is anticipated that by the end of 2015, 28 units with a total generating capacity of about 25 GW_e will be in service: this would represent a 70% increase in Korea's nuclear capacity.

Lithuania

Two LWGRs (each of 1 500 MW_e gross capacity) were built at Ignalina, north-east of Vilnius, in the mid-1980s: one was commissioned in December 1983 and the other in August 1987. After the accident at Chernobyl, the capacity of the Ignalina NPP was derated to 2 600 MW_e gross (2 370 MW_e net) for safety reasons. The two units accounted for 80% of Lithuania's electricity generation in 2002.

The National Energy Strategy approved by Parliament (the Seimas) in 1999 provided that Unit 1 of Ignalina NPP would be closed before 2005, taking into account conditions of long-term financial assistance from the European Union and the G-7 countries, as well as from international financial institutions. On the basis of the same assumptions, and taking into account that the Member States of the European Union are ready to provide adequate additional assistance for decommissioning, the National Energy Strategy approved by the Seimas in 2002 provides for the closure of Unit 2 of Ignalina NPP in 2009. However, the Strategy declares that, seeking to use the existing infrastructure at Ignalina and to remain a nuclear energy state, Lithuania will legally, financially and politically support investments into the construction of a new unit or reactor (complying with modern safety requirements).

Mexico

There is a single nuclear power station with two BWR units of total net capacity 1 365 MW_e, located at Laguna Verde in the eastern state of Veracruz. The first unit was brought into operation in April 1989 and the second in November 1994. Laguna Verde's electricity output accounted for just over 4% of Mexico's total net generation in 2002.

As present there is no officially declared nuclear policy. Nevertheless, nuclear power is recognised as part of the country's energy mix. The continued operation of the two reactors at the Laguna Verde NPP is contemplated until the end of their operational lifetime. There are currently no plans for expansion of nuclear energy use during the next 20 years.

Netherlands

Two NPPs have been constructed in the Netherlands: a 55 MW_e BWR at Dodewaard (connected to the grid in 1968) and a 450 MW_e PWR at Borssele (online from 1973). The BWR was taken out of service in 1997 but the PWR will not be shut down until 2013. Borssele's output accounted for 4% of Dutch electricity generation in 2002.

Pakistan

A small (125 MW_e) PHWR plant was commissioned in 1971. Known as Kanupp (Karachi Nuclear Power Plant), this facility has made a minor contribution (less than 1%) to the national annual electricity supply. In June 2000, it was joined by a second plant (Chasnupp 1), a 300 MW_e PWR constructed at Chasma. Together, the two plants contributed 2.5% to Pakistan's power supplies in 2002.

In view of the limited availability of indigenous fossil fuels and seasonal variations in hydro resources, Pakistan is keen to expand the use of nuclear power to meet its future electricity requirements. Nuclear power development has remained slow, owing to restrictions on its international trade and financial difficulties. The successful functioning of the two existing NPPs has given the Pakistan Atomic

Energy Commission (PAEC) confidence to plan for more NPPs, in a manner that would progressively lead to a higher degree of self-reliance. Kanupp has completed its design life of 30 years and is in the process of obtaining a licence for a life extension of 10–15 years. A third NPP (Chasnupp-2, or C-2), with a net capacity of 300 MW_e, is in the planning stage. PAEC is planning to construct additional nuclear units after C-2.

Romania

Romania's first nuclear plant—a 708 MW_e PHWR supplied by AECL of Canada—came online in 1996 at Cernavoda in the east of the republic. In 2002, it supplied about 10% of Romania's electricity generation.

The Romanian WEC Member Committee reports that the first nuclear group (unit) in Romania, using the Canadian CANDU technology, has produced very good results. A second group is under construction, with its completion expected in 2006. A national development programme for nuclear energy has been prepared which foresees the construction and exploitation of another two groups. At national level, there is no opposition towards nuclear energy development from the public and organisations in the field of environmental protection.

Russian Federation

There were 30 nuclear units installed at nine different sites at the end of 2002, with an aggregate net generating capacity of 20 793 MW_e. The reactor types represented consisted of eleven 925 MW_e LWGRs, eight 950 MW_e WWERs, four 411 MW_e WWERs, four 11 MW_e LWGRs, two 385 MW_e WWERs and one 560 MW_e FBR. In all, NPPs provided 16% of the Russian Federation's electricity output in 2002.

Three reactor units, with an aggregate capacity of 2 825 MW_e, were under construction at the end of 2002.

It was reported in April 2003 that Russia was aiming to double nuclear generation by 2020: the increase in capacity would be achieved through 'the construction of new production units, modernisation of existing units and the use of new techniques'.

Russia plans to build a floating NPP, with a reactor of the type installed in nuclear submarines. The plant, designed to be mounted on a barge, is said to have a capacity of 70 MW of electricity and 140 Gcal of thermal energy.

Slovakia

Four 408 MW_e WWERs were brought into service at Bohunice between 1978 and 1985; a slightly smaller (388 MW_e net) WWER came into operation at Mochovce in 1998. Mochovce-2 (also 388 MW_e) was connected to the grid just before the end of 1999 and went commercial in April 2000. Together, these six reactors are reported to have a current net capacity of 2 460 MW_e and to have provided 54.6% of the republic's electricity output in 2002. Two more blocks (total net capacity 820 MW_e) are under construction, but completion is probably a long way off.

Slovenia

A bi-national PWR (current capacity 656 MW_e net) has been in operation at Krsko, near the border with Croatia, since 1981. Its output is shared 50/50 with Croatia. Slovenia's share provided 38% of its electricity generation in 2002.

South Africa

There is a single nuclear power station at Koeberg, about 40 km north of Cape Town. The plant has two 900 MW_e PWR units which were commissioned in 1984–1985. The plant, which is owned and operated by Eskom, the national utility, provided nearly 6% of South Africa's electricity in 2002.

No expansion of the PWR is envisaged. A feasibility study on the development and construction of a pebble bed modular reactor (PBMR) has been conducted. The Government has recently accepted the environmental impact study. A nuclear licence needs to be issued and the final decision to construct the reactor to be taken. The PBMR concept envisages a number of small (~ 100 MW) reactors operating in tandem. If all approvals are given, a demonstration unit could be operational by 2012. This is expected to be erected at the Koeberg site.

Spain

Nine nuclear reactors were brought into commission between 1968 and 1988: at the end of 2002, they had an aggregate net capacity of 7 579 MW_e and in 2002 provided nearly 26% of Spain's electricity generation. Two of the units are BWRs (total capacity 1 489 MW_e), the rest being PWRs. Planned uprating of capacity brought the total up to 7 620 MW_e by end 2003. Upgrading programmes to increase the capacity of the Spanish nuclear power plants have resulted in a total increase of 464 MW_e since 1990.

The Spanish WEC Member Committee foresees an aggregate nuclear capacity of 7 581 MW_e by end-2015 (taking into account the shutdown of Spain's oldest NPP, José Cabrera (153 MW_e) in 2006).

At present, the construction of new NPPs is not foreseen. The present policy concerning the existing NPPs is to continue their operation as long as they are safe, economic and reliable. The current life management programme will allow them to exceed the usual 40-year mark by a substantial number of years, in line with the trends prevailing in various leading countries.

Sweden

Between 1971 and 1985 a total of 12 nuclear reactors (nine BWRs and three PWRs) commenced operation. The 11 units remaining in service at end-2002 had an aggregate net capacity of 9 424 MW_e. Nuclear power provided 44% of Sweden's net output of electricity in 2002.

In June 1997, the Swedish Parliament took a decision to start the phasing-out of nuclear power. The decision specified that the two units, 600 MW_e each, at the Barsebäck nuclear station were to be closed by end-June 1998 and end-June 2001, respectively; an earlier decision with regard to a final date for total nuclear phase-out by 2010 was explicitly removed, without specifying an alternative final date.

The execution of the first closure was delayed by legal conflicts between the owner, Sydkraft, and the Government. During November 1999 an agreement was reached concerning the level of compensation, and Barsebäck-1 was permanently taken out of operation at the end of the month, without the closure being enforced by law.

The phasing-out of Barsebäck-2 is, however, conditional upon its replacement by sufficient capacity from renewable sources and/or proven results from electricity conservation. Sweden's nuclear capacity in 2015 is forecast by the WEC Member Committee to total 9 400 MW_e from 10 units, implying that the station is indeed taken out of service in the short/medium term.

Switzerland

There are three PWRs and two BWRs in operation, with a total net generating capacity of 3 127 MW_e. All five reactors were commissioned between 1969 and 1984. Their output in 2002 accounted for almost 40% of Switzerland's total power generation.

In a referendum in May 2003, Swiss voters rejected a proposal by environmental groups to phase-out nuclear power in Switzerland. No new nuclear power stations are planned, while the existing plants are expected to continue in operation through 2015.

Taiwan, China

There are six reactors in service at three locations (Chinshan, Kuosheng and Maanshan), with an aggregate net generating capacity of 4 884 MW_e; the four BWRs and two PWRs were all brought online between 1977 and 1985. In 2002 nuclear plants provided just over 20% of Taiwan's electricity generation.

Two more BWRs, with a total net capacity of 2 700 MW_e, are under construction at a fourth location (Lungmen). Owing to the intense political controversy generated by this project, its progress and eventual completion are subject to considerable uncertainty.

Ukraine

At end-2002 there were 13 nuclear reactors (with a total net generating capacity of 11 207 MW_e) in service at four sites: they had come into operation between 1980 and 1995. Nuclear plants accounted for nearly 46% of Ukraine's power output in 2002.

Four 925 MW_e RBMK reactors were installed at Chernobyl between 1977 and 1983. In April 1986 the last unit to be completed, Chernobyl-4, was destroyed in the world's worst nuclear accident. Chernobyl-2 was closed down in 1991, Chernobyl-1 in 1996 and Chernobyl-3 in December 2000.

The European Bank for Reconstruction and Development has granted a loan to Ukraine to finance the completion of two 950 MW_e nuclear reactors (Khmelnitski-2 and Rovno-4 (also known as K2R4)) to replace the electricity output lost as a result of the shutdown of Chernobyl-3. Two further WWERs (Khmelnitski-3 and -4) are also currently under construction.

United Kingdom

The UK had 31 nuclear reactor units in service at the end of 2002, with an aggregate net generating capacity of 12 486 MW_e. In 2002, nuclear power accounted for 22% of net electricity generation. No new plants are under construction, on order or planned.

The Government's main energy policy objective is to ensure secure, diverse and sustainable supplies of energy at competitive prices. Nuclear power is playing an important role in meeting that objective. The Government believes that existing nuclear power stations should continue to contribute both to electricity supply and to the reduction of emissions, as long as they can do so to the high safety and environmental standards which are currently observed.

The 'Energy White Paper—Our Energy Future—creating a low carbon economy' published by the Department of Trade and Industry in February 2003 states that 'Nuclear power is currently an important source of carbon-free electricity. However, its current economics make it an unattractive option for new, carbon-free generating capacity and there are also important issues of nuclear waste to be resolved. These issues include our legacy waste and continued waste arising from other sources. This white paper does not contain specific proposals for building new nuclear power stations. However, we do not rule out the possibility that at some time in the future new nuclear build might be necessary if we are to meet our carbon targets. Before any decision to proceed with the building of new nuclear power stations, there will need to be the fullest public consultation and the publication of a further white paper setting out our proposals.' (page 12, paragraph 1.24)

United States of America

At the end of 2002, there were 104 nuclear reactor units connected to the grid, with an aggregate net generating capacity of 98 564 MW_e (equivalent to about 27% of total world nuclear capacity). The totals include Brown's Ferry-1 (1 065 MW_e), which has been shutdown since March 1985 but is still fully licensed to operate. Nuclear plants accounted for about 20% of US electricity output in 2002.

Only two reactors have come online since 1990: Comanche Peak 2 (1 110 MW_e) in 1993, and Watts Bar-1 (1 177 MW_e) in 1996. No commercial reactors are under construction in the United States. Although construction permits have been issued for three units (total capacity 3 360 MW_e), these are not expected to come online. The U.S. WEC Member Committee foresees that by the end of 2015 there will be 101 reactors connected to the grid, with an overall net generating capacity of 99 500 MW_e.

NPPs in the United States are largely owned and operated by private sector entities, although there are several plants owned by a Federal Government agency, the Tennessee Valley Authority. If the owners of nuclear plants are regulated electric utilities (which is usually the case), they are generally subject to economic regulation by state or local public utilities commissions. Some utilities have divested themselves of their nuclear assets in order to take advantage of incentives offered by State governments. As a result some nuclear plants are being operated by 'non-utilities' (i.e. private companies). Under deregulation, countries other than the United States can become involved in plant ownership.

The construction, operation and decommissioning of NPPs is closely regulated to ensure public health and safety by a Federal Government agency, the Nuclear Regulatory Commission (NRC). The NRC also regulates the handling and transportation of nuclear materials, including nuclear fuels. Whether a NPP is operated by a utility or non-utility, the power plant itself continues to be regulated by the NRC.

The NRC will generally grant a completed plant an operating licence for a period of years consistent with the expected operating life of the plant. It is now undertaking research to determine the conditions under which existing nuclear plants may be modified to safely extend their operating lives and permit the NRC to relicence these plants.

On September 10, 2003, Secretary of Energy Abraham re-affirmed the Administration's continued support of commercial nuclear power expansion. In a letter to the Chairman of the House, Senate Conference on HR 6, the Secretary emphasised: 'the Administration supports the expansion of nuclear energy in the United States as a major component of our national energy policy. Nuclear energy emits virtually no air pollutants or greenhouse gases, and advanced nuclear technologies offer the potential of efficient, safe, and proliferation resistant reactor designs.' He also recommended that the Congress permanently reauthorise the Price-Anderson Act, which includes provisions limiting liability in a commercial nuclear incident. The Administration has actively encouraged utilities to begin construction of at least one new reactor by 2010. As of September 25, 2003, two companies have submitted applications for an Early Site Permit (the first step in the process) to the Nuclear Regulatory Commission.

Chapter 7

Hydropower

World Energy Council
CONSEIL MONDIAL DE L'ENERGIE

COMMENTARY

Present Situation and Future Potential

The 2003 *International Energy Outlook* of the US Energy Information Administration indicates that generation from hydropower and other renewable energy sources is projected to grow by 56% over the next 24 years.

Hydropower currently provides 17% of the world's electricity supply. The contribution of other renewables is currently very small; according to the European Commission's *World Energy, Technology and Climate Policy Outlook 2030*, in the year 2000 wind and solar power contributed 0.16 and 0.01%, respectively. The report projects these values to have increased by 2010 to 0.6% for wind and 0.12% for solar.

It is clear, therefore, that hydropower will remain the major contributor in the renewable power sector for the foreseeable future. The current installed hydropower capacity is some 730 GW and annual generation is around 2 700 TWh. In addition, some 100 GW of new capacity is currently under construction.

In world terms, only 33% of the technically and economically feasible potential has been developed; there is great variation, however, in the extent of development in each of the continental regions. Europe and North America have developed the majority of their potential; the main priorities in these regions are related to extending the life of existing schemes, and adding capacity where possible. Considerable potential for new development remains in Africa, Asia and South America. This is in direct correlation to the regions with the most pressing needs for water and energy; therefore, hydropower can significantly contribute to the Millennium Development Targets of the United Nations (Fig. 7.1).

Water, Energy, Health, Agriculture and Biodiversity

With good planning and good management, hydropower is a catalyst for the sustainable improvement of people's lives. At the World Summit on Sustainable Development (WSSD) in September 2002, Kofi Annan, UN Secretary-General, defined water, energy, health, agriculture and biodiversity (WEHAB) as five key areas in which progress is possible with the resources

2004 Survey of Energy Resources

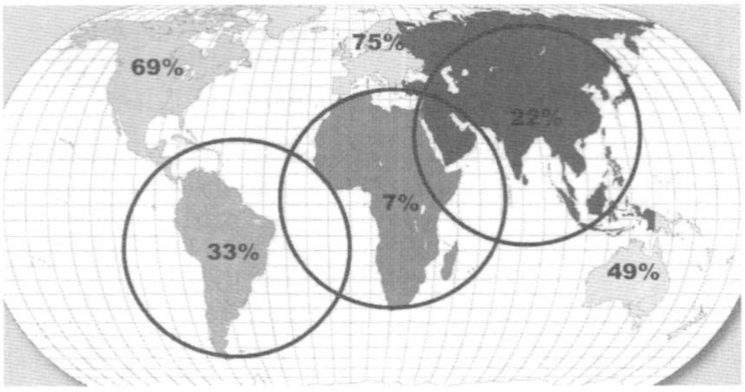

FIGURE 7.1 Percentage of technical and economic hydropower potential exploited in 2002, by region (Source: World Atlas of Hydropower & Dams, 2003).

and technologies at our disposal today. Hydropower stands at the crossroads of two of these key areas: water and energy. The integrated use of water and energy is an important component of sustainable development. Wherever suitable sites are available, hydropower offers the possibility to assist in meeting both of these basic human needs. In addition to water and energy security, well-conceived schemes can foster developments relating to better health services, increased agricultural productivity and environmental management. Many schemes have been designated sites of heritage and some have been chosen as sites of special scientific interest (SSSI) because of the ecosystems that form in their reservoir areas. The Achanalt reservoir in Scotland is SSSI designated because it hosts outstanding nesting grounds for many species. The recently developed Miyagase reservoir in Japan has several managed 'biotope' wetland areas that have been intensively studied and shown to host a broad range of flora and fauna.

Some groups have moved to exclude hydropower schemes greater than 10 MW from renewables incentives and reporting systems. The WSSD settled this issue by clarifying that there was no scientific or technical justification for this. 'WSSD identified all hydro as a renewable source of energy to be supported by the international community' (John Briscoe, World Bank, reporting on the WSSD Implementation Plan, Item 19e).

The WSSD also called for better integration and optimisation in the use of water. Reservoirs built to manage water can have far-reaching and long-term multiple benefits. For example, the Hoover dam and Lake Mead scheme in USA have been responsible for fostering the economic prosperity of what is today one of the richest regions of the world. For more than six decades, the scheme has reliably supplied a significant amount of power (some 4 TWh/yr). In addition, Lake Mead is responsible for

- 18 million people in the region being supplied with fresh water;
- commerce in Arizona, Nevada and California benefiting from secure water supplies;
- 6 000 km^2 of agricultural land being irrigated in USA and Mexico;
- 12 million people visiting the scheme each year for recreational purposes.

The use of an indigenous and renewable resource requires long-term planning. Often financial challenges are encountered because of the high initial capital cost; however, with operating costs of hydropower typically only 1% of the investment cost, the long-term sense is clear. In 2003, the Hoover dam was depicted as a tombstone in advertising funded by the World Wide Fund for Nature. It is not clear whose tombstone the advertising was intended to depict, but perhaps it is that of the people in poor nations denied development opportunities by short-term decision making?

The World Energy Council's *Living in One World* estimates that there are 1.7 billion people without access to electricity living in developing countries. It states that 'reliable and affordable access to modern energy services is an indicator of sustainable development, for without it basic needs cannot be satisfied'.

Hydropower Typology

Hydropower can be developed within a wide range of scale and type, according to site constraints, prevailing needs and market conditions.

Small-scale, decentralised development continues to be responsible for bringing light and power to communities throughout the world. In China 43 000 small schemes with an aggregate installed capacity of 26 GW are currently in operation. This has contributed substantially to the provision of electricity to the nation's rural areas. Larger hydropower schemes in China feed the regional grid systems, substantially reducing the burden on coal-fired generation, which is the predominant power supply source. According to the 2003 *International Energy Outlook*, electricity consumption in China is projected to grow by an average of 4.3% per annum, almost tripling over the next 24 years. Because of this pace of development, there continues to be a significant challenge for hydropower and other renewables to maintain their relative contributions in such countries.

In Europe and North America, large schemes (run-of-river and reservoir types) play an important role. Nine countries in Europe rely on hydropower for more than half of their electricity supply, while the USA and Canada currently generate one quarter of the world's hydropower. In these regions, hydropower makes up the majority of the national renewable power generation portfolios. Additional small schemes and the expansion of existing larger ones are the main areas of activity. In addition, reservoir and pumped-storage schemes play an essential role in the optimisation and stability of the power system networks.

The argument that only small schemes are useful is now defunct. At the UN Environment Programme's Dams and Development Forum in September 2003, Klaus Toepfer, UNEP's Executive Director announced that he was 'no longer concerned by the small or the large, but the well planned and well managed'.

Future development of decentralised small schemes will remain essential for rural electrification programmes throughout the world. Efforts through technology transfer to increase the domestic component of such schemes must continue to be encouraged. However, the vast majority (some 95%) of population growth in the coming decades is likely to be in and around cities. Currently, there are some 400 million people living in urban areas without water and power services. The World Energy Council's *Living in One World* predicts that by the middle of the century there will be some 2 billion 'urban dwellers' in developing countries. Future sustainable solutions will need to include medium- and large-scale, interconnected power plants (Fig. 7.2).

The *World Energy Outlook* of the International Energy Agency predicts that some 1 000 GW of existing thermal electricity generating plants will need to be decommissioned before 2030. One of the great strengths of hydropower is its longevity. The civil structures relating to hydro schemes have proved to last for a century or more. Modern, technically advanced electro-mechanical equipment can be used to refurbish existing schemes to utilise the same flows at greater efficiency. For example the 43 MW Rannoch scheme in the UK, which was commissioned in the 1930s, was recently upgraded with new generating equipment. The 7% efficiency improvement now provides some 14 GWh/yr of additional generation. The scheme is expected to operate for another 30 years without the need for further investment.

The *World Energy Outlook* also indicates that about 3 000 GW of new electricity generating capacity will be required before 2020. It is clear that no single generation technology will meet this demand. It is essential that all technologies are integrated and optimised to suit mixed-energy, national and regional power systems (Fig. 7.3).

Improving the Performance of Thermal Plants

Hydro schemes store potential energy in reservoirs and release water for power generation when it is required. Within a typical power system, the power demand can peak for short periods at twice the mean value. The capacity of hydro storage facilities to respond at short notice is a great advantage. In commercial electricity-supply markets, the price paid to generators

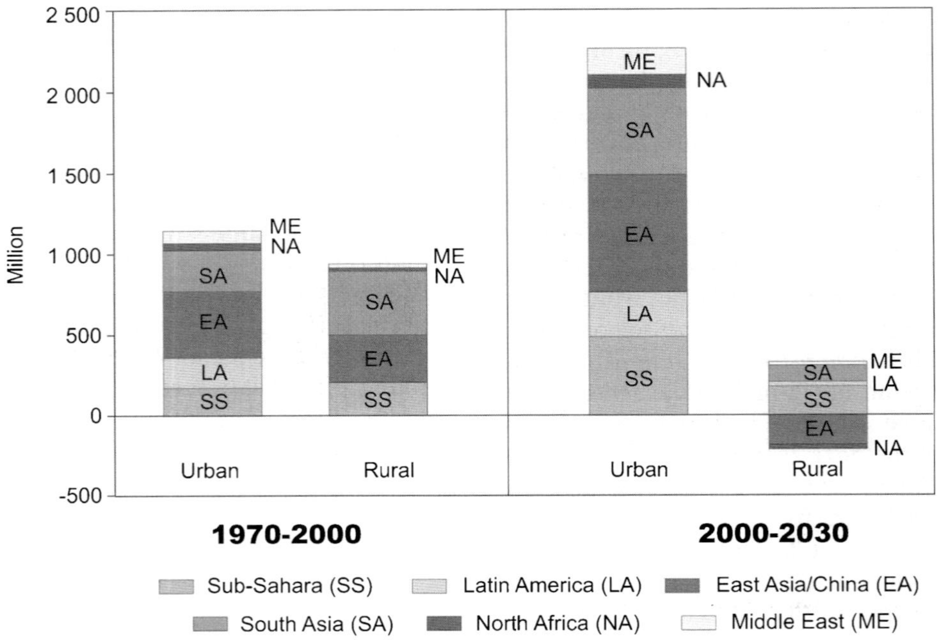

FIGURE 7.2 Evolution of urban and rural population, by region: 1970–2000 and 2000–2030 (Source: World Energy Outlook, IEA).

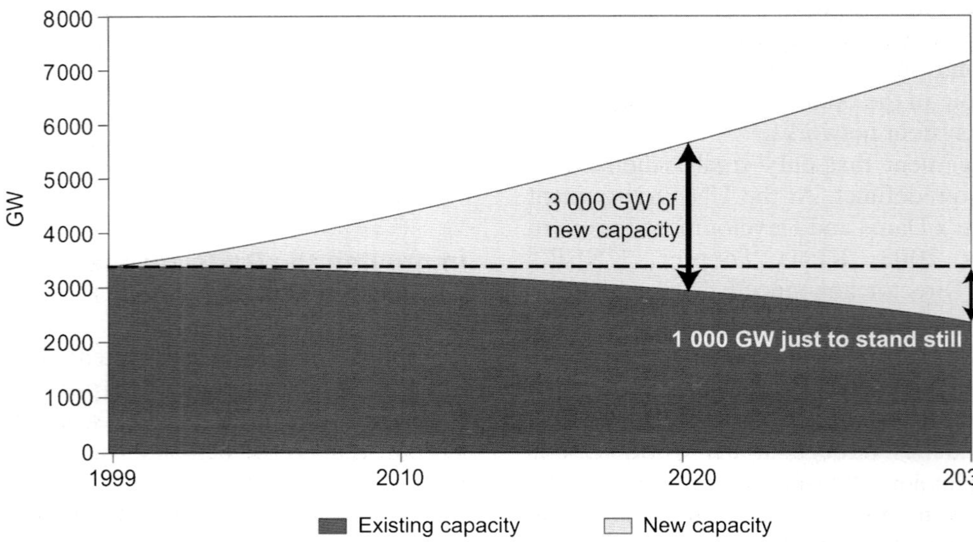

FIGURE 7.3 Generating capacity requirements 2030 (Source: World Energy Outlook, IEA).

changes significantly during the various demand periods of the day, and it is not surprising that hydro peaking plants are highly valued. The flexibility of hydropower to follow rapid changes in demand enables thermal power plants in the network to continue to operate at their point of best efficiency, thereby minimising harmful emissions and fuel consumption. When sites for traditional hydropower storage schemes are not sufficient, pumped-storage schemes can provide the flexibility to ensure quality and security of supply. Storage schemes can

- optimise the efficient use of thermal plants;
- backup intermittent renewable sources;
- ensure system reliability and provide independent start-up capacity;
- maintain quality of supply (for example, frequency control and voltage regulation);
- exploit price differentials in the power market.

The EU has 134 pumped-storage schemes in operation in 12 of the 15 member states, the most recently commissioned being the 1 060 MW Goldisthal plant in Germany. It is equipped with four units, two of which utilise adjustable-speed technology. Conventionally, pumped-storage units use synchronous generator-motors and their operating speed is constant in both generating and pumping modes. Adjustable-speed technology, pioneered in the Japanese power market, has been developed to operate synchronous machines at an arbitrary speed within a range of plus/minus several per cent of synchronous speed (Fig. 7.4). Adjustable-speed pumped-storage units have the following advantages:

- pump input power can be changed by adjusting the rotational speed. This enables frequency control in pumping as well as in generating mode;
- efficiency of pump-turbines in turbine operation is improved by 2–3%, and the operable load range can be extended;
- adjustable-speed units can follow very rapid load fluctuation in the network. When system disturbances arise, they change their speed rapidly, discharging the energy from inertia of the rotor like a flywheel generator (and vice versa).

FIGURE 7.4 An adjustable-speed reversible power set (Source: VATech Hydro).

In water-scarce South Africa, the 400 MW Palmiet pumped-storage scheme, one of two currently in operation, with a third about to begin construction, operates on a weekly cycle. During the weekend low-demand period, water is pumped to Palmiet's upper reservoir. In the weekdays, during peak demand and at times of system instability, the water is released to generate in accordance with the needs. The operator, Eskom, is paid not only for power generation but also ancillary services such as spinning reserve, voltage- and frequency control. In addition, the scheme is used to deliver water from its upper reservoir to the Cape Town municipality. Therefore, the scheme also ensures water security in the area. In October 2003, Palmiet was awarded the Blue Planet Prize by the International Hydropower Association (IHA) in recognition of its significant role in providing power and water security. In addition, the Prize recognises the scheme's synergy with the natural environment (the scheme is located within an area designated as a UNESCO biosphere).

Synergy with the Intermittent Renewables

One of the challenges of the increased use of wind and solar technologies is the intermittent nature of their supply. For the time being, the backup for solar power remains battery storage, however, the commercial expansion of photovoltaic technology may lead to tests involving energy storage through hydropower.

Wind power is constrained if winds are too gentle or too strong and, therefore, the technology cannot follow electricity demand. The development of new wind capacity is increasing rapidly. However, wind will always need a backup technology, therefore, it should be seen as an important supplementary technology, rather than a panacea. The increase in wind power capacity is quite complementary to hydropower, because storage schemes are the renewable partner for supporting an increasing commitment to intermittent sources connected to a power network. In short, hydro reservoirs provide the firming capability to cover the fluctuating nature of the wind source (Fig. 7.5).

Wind/hydro schemes are being developed to produce entirely renewable systems for power trading; the largest of which is the programme currently under way in Tasmania. This development is made possible because it is supported by existing hydropower storage schemes. The developer, Hydro Tasmania, will participate in the national power market through a submarine transmission line that will link the island state to the mainland of Australia.

Optimising Existing Infrastructure

Only about 25% of the reservoirs in the world have any associated hydropower facilities. The majority of reservoirs have been developed for water supply, primarily for irrigation. In Asia and Africa only 6–7% of the existing dams have hydropower as their main purpose. Unfortunately, water and energy policies have not been well coordinated in the past and authorities have been reluctant to engage in cross-sector activities. In the USA, a recent resource assessment concluded that a capacity of about 20 GW could be gained by adding hydropower facilities at the country's existing dams.

With increasing incentives for renewable energy, many owners are revisiting the possibilities of recovering energy from the transfer of water. The hydropower industry now offers an array of equipment that is suitable for these types of schemes. One example is the Matrix turbine,

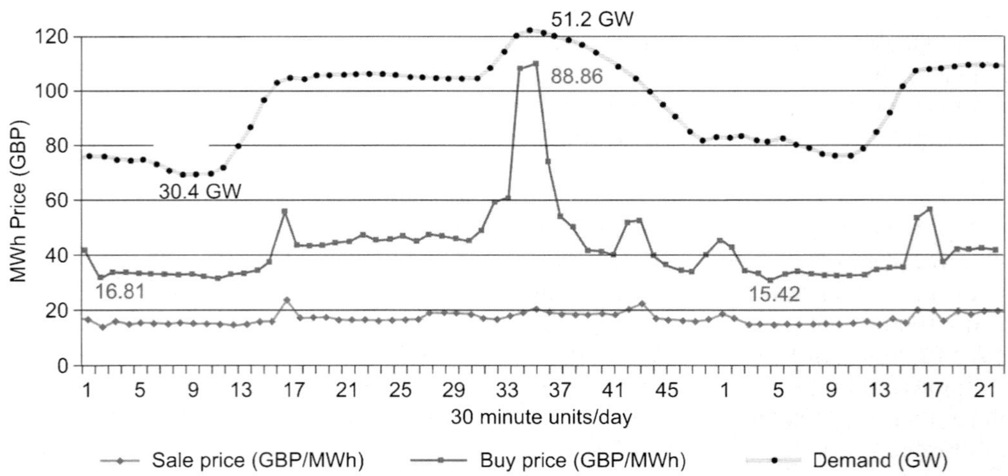

FIGURE 7.5 UK NETA prices, 15/16 December 2003 (Source: NETA, Office of Gas and Electricity Markets, UK).

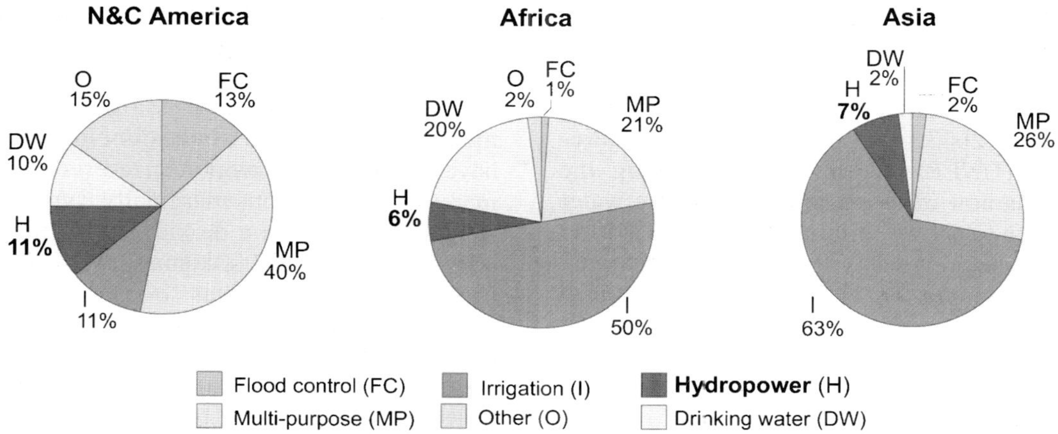

FIGURE 7.6 Main uses of dams, by region (Source: International Commission on Large Dams).

which comprises sets of small reaction turbines to capture the energy from water that would normally be discharged through gates or valves. At the Freudenau scheme on the Danube river, Austria, a 25-unit (5 MW) 'Matrix' system has been installed in the navigation lock.

As water-use optimisation becomes increasingly important, the industry will need to continue its innovation in the introduction of power generation equipment to recover energy at weirs, locks and other water management structures (Figs. 7.6 and 7.7).

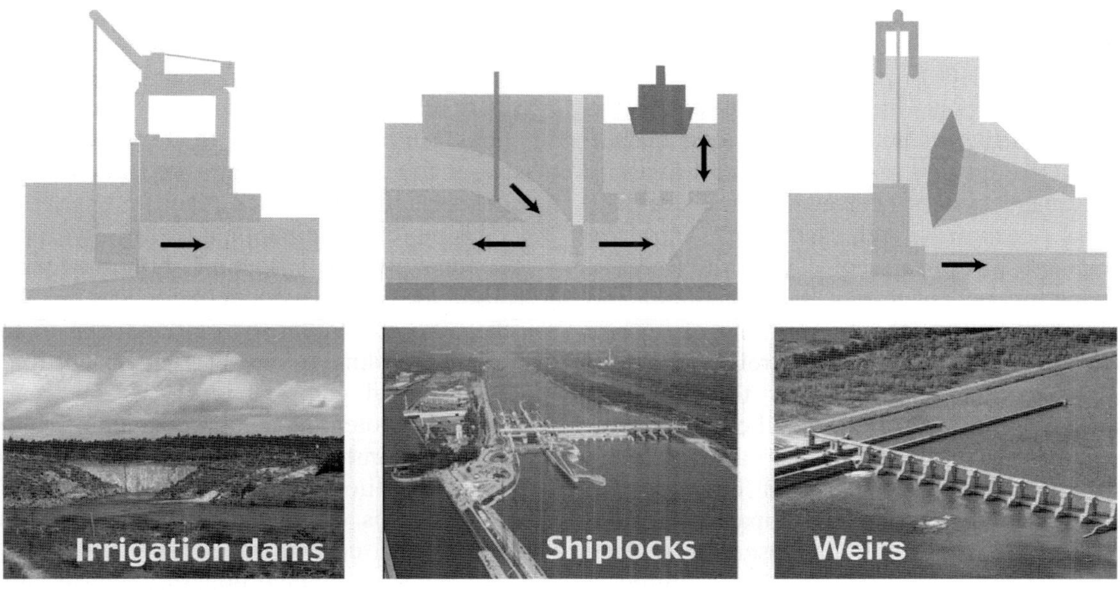

FIGURE 7.7 Typical applications of dams (Source: International Commission on Large Dams).

Climate Change and Reservoir Emissions

At the UN Climate Change negotiations in December 2003, the UN Environment Programme (UNEP) issued a report putting the cost of natural disasters in 2003, mostly water related, at US$ 60 billion, up from US$ 55 billion in the previous year. The high losses were described as part of 'a worrying trend linked with climate change'. At the same meeting, the International Energy Agency reported that 'climate change mitigation will require profound modification in energy production and use worldwide'. It is clear that adaptation will require the increased management of water systems for security against both extreme floods and drought. The use of reservoirs will play a fundamental role in this.

In recent years greenhouse gas (GHG) released from reservoirs has been investigated as a source of anthropogenic emissions. Initial studies indicated emissions of both carbon dioxide and methane from the surface of reservoirs. Some commentators were quick to allege that carbon was in some way being manufactured within reservoirs. Studies have demonstrated that all ecosystems (especially wetlands and seasonally flooded forests and plains) emit GHG as part of the natural carbon cycle. The carbon is temporarily fixed by the flora within the watershed and is then flushed and transported in streams, rivers and lakes in the drainage system. Throughout the drainage process, carbon is released back into the atmosphere in the form of carbon dioxide and methane, or it is encapsulated in lake, delta and marine sediments. The proportions of carbon dioxide and methane that are released to the atmosphere are determined by the site conditions, especially ecosystem and climate type.

If the watershed contains a man-made reservoir, the pre-impoundment emissions of the area would need to be compared with the emissions recorded after the reservoir scheme has been completed. This would be required to determine if there is a difference in the emissions of carbon dioxide and methane. Methane is important because its global-warming potential is 23 times that of carbon dioxide. In slow-moving anoxic water bodies, the proportion of methane is increased. Studies in North America have shown that hydropower reservoirs tend to increase the emissions marginally (typical emissions are 99% carbon dioxide to 1% methane), and a value of 10 000 tonnes/TWh of carbon dioxide equivalent has been allocated to schemes in this region. Because of a lack of data confirming the situation in warmer, especially, tropical climates, and the variation in reservoir area per unit of power, a value of 40 000 tonnes/TWh has been proposed as an international average value for hydropower.

These values do not take into account the sequestration of carbon in the reservoir sediments; studies have indicated that carbon sequestration in reservoirs is more substantial than that which occurred before impoundment. This research indicates that the 40 000 tonnes value for hydro may be a substantial overestimate; however, further studies are required to confirm this.

Even excluding the role of reservoir carbon sequestration, hydropower still compares very favourably with fossil-fuel generation. Using the international value presented above, hydropower emits 10 times less GHG than combined-cycle gas-fired plants and 25 times less than coal-fired plants. For hydro in northern climates, the values are 40 times less than gas-fired plant and 100 times less than coal-fired plant.

Using the international value, hydropower currently avoids some 2.1 billion tonnes of carbon dioxide equivalent every year that would otherwise be emitted by stations run on fossil fuels (according to the current proportions of coal, oil and gas generation). If the realistic remaining potential of hydropower were to be developed instead of fossil-fuelled plants, a further 7 billion tonnes of emissions could be avoided. This would be equivalent to saving a third of the current man-made emissions, or three times the annual emissions of all cars on the planet.

It should be emphasised that hydropower generation also avoids the emission of sulphur and nitrogen oxides, which are important contributors to acid rain and photochemical air pollution, together with particulate emissions, associated with fossil-fuel generation.

Hydropower Sustainability Guidelines

The Ministerial Declaration of 170 Countries at the World Water Forum in Kyoto 2003 states: 'We recognise the role of hydropower as one of the renewable and clean energy sources, and that its potential should be realised in an environmentally sustainable and socially equitable manner.'

In accordance with the above, and in response to the recommendations of the World Commission on Dams and the WSSD, the IHA has developed a set of Sustainability Guidelines and a Compliance Checklist. The Guidelines, covering both the development of new schemes and the management of existing ones, have been through several consultative cycles, involving industry, financiers and stakeholders involved with the UNEP's Dams and Development Forum. In November 2003, the Hydropower Sustainability Guidelines were formally adopted by the IHA membership, which spans 82 countries. Subsequently, these Guidelines have been submitted to international funding agencies and UN organisations, with the proposal that they are used in the evaluation of future projects and in the screening of applications for credit relating to existing schemes.

Richard Taylor
International Hydropower Association

DEFINITIONS

This chapter is restricted to that form of hydraulic energy that results in the production of electrical energy as a result of the natural accumulation of water in streams or reservoirs being channelled through water turbines. Energy from tides and waves is discussed in Chapters 14 and 15.

Annual generation and capacity attributable to pumped storage is excluded. Where such installations produce significant energy from natural run-off, the amount is included in the total for annual generation.

It must be recognised that for some countries it is not possible to obtain comprehensive data corresponding exactly to the definitions. This particularly applies to small hydro schemes, many of which are owned by small private generators. Also, not all countries use the same criteria for the distinction between small and large hydro. In this Survey, small hydro mainly applies to schemes of less than 10 MW. However, some countries and other sources of data make the distinction between small and large schemes at other levels.

In the tables, the following definitions apply:

Gross theoretical capability is the annual energy potentially available in the country if all natural flows were turbined down to sea level or to the water level of the border of the country (if the watercourse extends into another country) with 100% efficiency from the machinery and driving waterworks. Unless otherwise stated in the notes, the figures have been estimated on the basis of atmospheric precipitation and water run-off.

Gross theoretical capability is often difficult to obtain strictly in accordance with the definition, especially where the data are obtained from sources outside the WEC. Considerable caution should therefore be exercised when using these data.

Where the gross theoretical capability has not been reported, it has been estimated on the basis of the technically exploitable capability, assuming a capacity factor of 0.40. Where the technically exploitable capability is not reported, the value for economically exploitable capability has been adopted, preceded by a ' > ' sign.

Technically exploitable capability is the amount of the gross theoretical capability that can be exploited within the limits of current technology.

Economically exploitable capability is the amount of the gross theoretical capability that can be exploited within the limits of current technology under present and expected local economic conditions. The figures may or may not exclude economic potential that would be unacceptable for social or environmental reasons.

Capacity in operation is the total of the rated capacities of the electric generating units that are installed at all sites which are generating, or are capable of generating, hydroelectricity.

Actual generation is the net output (excluding pumped-storage output) in the specified year.

Probable annual generation is the total probable net output of electricity at the project sites, based on the historical average flows reaching them (modified flows), net heads, and the plant capacities reported, making allowance for plant and system availability.

Capacity planned refers to all sites for which projects have been proposed and plans have been drawn up for eventual development, usually within the next 10 years.

Capacity under construction and planned relates to all units not operational but which were under construction, ordered or about to be ordered at the end of 2002.

TABLE 7.1

Hydropower: capability at end-2002

	Gross theoretical capability (TWh/yr)	Technically exploitable capability (TWh/yr)	Economically exploitable capability (TWh/yr)
Algeria	12	5	
Angola	> 150	108	65
Benin	2	1	
Burkina Faso	1	N	N
Burundi	6	2	1
Cameroon	294	115	103
Central African Republic	7	3	
Chad	N	N	
Congo (Brazzaville)	> 125	> 50	
Congo (Democratic Republic)	1 397	774	419
Côte d'Ivoire	46	12	2
Egypt (Arab Republic)	> 125	> 50	50
Ethiopia	650	> 260	260
Gabon	200	80	33
Ghana	17	11	7
Guinea	26	19	15
Guinea-Bissau	1	N	N
Kenya	> 30	9	
Lesotho	5	2	
Liberia	28	11	
Madagascar	321	180	49
Malawi	15	6	
Mali	> 12	> 5	
Mauritius	N	N	
Morocco	12	5	4
Mozambique	50	38	32
Namibia	25	10	2
Niger	> 3	> 1	1
Nigeria	43	32	30
Rwanda	1	N	
Senegal	11	4	2
Sierra Leone	17	7	
Somalia	2	1	
South Africa	73	14	5
Sudan	48	19	2
Swaziland	4	1	N
Tanzania	39	20	3
Togo	4	2	
Tunisia	1	N	N
Uganda	> 18	> 13	
Zambia	52	29	11
Zimbabwe	19	18	
Total Africa	**> 3 892**	**> 1 917**	

(continued on next page)

TABLE 7.1 (Continued)

	Gross theoretical capability (TWh/yr)	Technically exploitable capability (TWh/yr)	Economically exploitable capability (TWh/yr)
Belize	1	N	N
Canada	1 284	948	522
Costa Rica	223	43	20
Cuba	3	1	
Dominica	N	N	N
Dominican Republic	50	9	6
El Salvador	7	5	2
Greenland	800	14	
Grenada	N	N	N
Guatemala	54	22	
Haiti	4	1	N
Honduras	16	7	
Jamaica	1	N	
Mexico	135	49	32
Nicaragua	33	10	7
Panama	26	> 12	12
United States of America	4 485	1 752	501
Total North America	**7 122**	**> 2 873**	
Argentina	172	130	
Bolivia	178	126	50
Brazil	3 040	1 488	811
Chile	227	162	
Colombia	1 000	200	140
Ecuador	167	134	106
Guyana	64	> 26	26
Paraguay	111	85	68
Peru	1 577	> 260	260
Surinam	32	13	
Uruguay	32	10	
Venezuela	320	246	130
Total South America	**6 920**	**> 2 880**	
Armenia	22	8	6
Azerbaijan	44	16	7
Bangladesh	5	2	
Bhutan	263	70	56
Cambodia	88	11	5
China	5 920	1 920	1 270
Cyprus	59	24	
Georgia	139	68	32
India	2 638	660	
Indonesia	2 147	402	40
Japan	718	136	114
Kazakhstan	163	62	27
Korea (Republic)	52	26	19

TABLE 7.1 (*Continued*)

	Gross theoretical capability (TWh/yr)	Technically exploitable capability (TWh/yr)	Economically exploitable capability (TWh/yr)
Kyrgyzstan	163	99	55
Laos	233	63	
Malaysia	230	123	
Mongolia	56	22	
Myanmar (Burma)	877	130	
Nepal	727	394	221
Pakistan	307	263	
Philippines	47	20	18
Sri Lanka	9	7	5
Taiwan, China	103	20	8
Tajikistan	527	> 264	264
Thailand	18	16	15
Turkey	433	216	126
Turkmenistan	24	5	2
Uzbekistan	88	27	15
Vietnam	300	100	90
Total Asia	**16 400**	**> 5 174**	
Albania	40	15	6
Austria	75	> 56	56
Belarus	7	3	1
Belgium	1	N	N
Bosnia–Herzegovina	60	24	19
Bulgaria	27	15	12
Croatia	10	9	8
Czech Republic	12	4	
Denmark	N	N	N
Estonia	2	N	
Faroe Islands	1	N	N
Finland	48	25	20
FYR Macedonia	9	6	
France	270	100	70
Germany	120	25	20
Greece	80	15	12
Hungary	7	5	
Iceland	184	64	40
Ireland	1	1	1
Italy	340	105	65
Latvia	7	6	5
Lithuania	6	2	1
Luxembourg	N	N	N
Moldova	2	1	1
Netherlands	1	N	N
Norway	600	200	187
Poland	23	14	7
Portugal	32	25	20

(*continued on next page*)

TABLE 7.1 (*Continued*)

	Gross theoretical capability (TWh/yr)	Technically exploitable capability (TWh/yr)	Economically exploitable capability (TWh/yr)
Romania	70	40	30
Russian Federation	2 295	1 670	852
Serbia & Montenegro	37	27	24
Slovakia	10	7	7
Slovenia	13	9	6
Spain	138	70	41
Sweden	176	130	90
Switzerland	144	41	35
Ukraine	45	24	17
United Kingdom	40	3	1
Total Europe	**4 933**	**> 2 741**	
Iran (Islamic Republic)	176	> 50	50
Iraq	225	90	67
Israel	125	50	
Jordan	N	N	N
Lebanon	2	1	N
Syria (Arab Republic)	5	4	4
Total Middle East	**533**	**> 195**	
Australia	265	> 30	30
Fiji	3	1	
French Polynesia	N	N	N
New Caledonia	2	1	N
New Zealand	46	37	24
Papua New Guinea	175	49	15
Solomon Islands	2	> 1	
Western Samoa	N	N	
Total Oceania	**493**	**> 119**	
Total World	**> 40 293**	**> 15 899**	

Notes:

(1) A quantification of hydropower capability is not available for Comoros, Equatorial Guinea, Mauritania, Réunion, São Tomé & Príncipe, Guadeloupe, Puerto Rico, St Vincent & the Grenadines, French Guiana, Afghanistan, Korea (Democratic People's Republic) and Palau.

(2) As the data available on economically exploitable capability do not cover all countries, regional and global totals are not shown for this category.

(3) Sources: WEC Member Committees, 2003; *Hydropower & Dams World Atlas 2003*, supplement to The International Journal on Hydropower & Dams, Aqua~Media International; estimates by the editors.

TABLE 7.2

Hydropower: status of development at end-2002 (all schemes)

	In operation		Under construction		Planned	
	Capacity (MW)	Actual generation in 2002 (GWh)	Capacity (MW)	Probable annual generation (GWh)	Capacity (MW)	Probable annual generation (GWh)
Algeria	275	58				
Angola	430	1 000	520			
Benin	1	9				
Burkina Faso	32	79				
Burundi	32	134				
Cameroon	719	3 474				
Central African Republic	19	130				
Comoros	1	2				
Congo (Brazzaville)	89	352				
Congo (Democratic Republic)	2 515	6 000	5			
Côte d'Ivoire	604	1 800			328	1 400
Egypt (Arab Republic)	2 745	15 130	65	473	65	429
Equatorial Guinea	1	2				
Ethiopia	451	2 151	559			
Gabon	171	878				
Ghana	1 072	6 100				
Guinea	139	497				
Kenya	677	2 833	60			
Lesotho	76	200				
Madagascar	104	540	42			
Malawi	283	1 100				
Mali	114	500				
Mauritania	18	50				
Mauritius	71	100				
Morocco	1 300	2 500				
Mozambique	2 184	11 548				
Namibia	249	1 250				
Nigeria	1 938	6 986	64			
Réunion	125	500				
Rwanda	27	89				
São Tomé & Príncipe	6	10				
Senegal	38	100				
Sierra Leone	4	15	50			
Somalia	5	10				
South Africa	687	2 357				
Sudan	323	520	1 250			
Swaziland	41	192	19			
Tanzania	561	2 602				
Togo	66	170				
Tunisia	54	60				
Uganda	318	1 650	330			

(continued on next page)

TABLE 7.2 (Continued)

	In operation		Under construction		Planned	
	Capacity (MW)	Actual generation in 2002 (GWh)	Capacity (MW)	Probable annual generation (GWh)	Capacity (MW)	Probable annual generation (GWh)
Zambia	1 670	8 196	60			
Zimbabwe	754	3 000	1			
Total Africa	**20 989**	**84 874**	**3 025**			
Belize	25	80				
Canada	67 406	345 597	2 160	11 033	2 853	14 573
Costa Rica	1 263	5 660	318			
Cuba	57	75	11			
Dominica	8	32				
Dominican Republic	412	745	3			
El Salvador	407	1 218				
Greenland	30	192	1	6		
Guadeloupe	5	15				
Guatemala	600	2 500	154			
Haiti	70	280				
Honduras	435	1 914	150			
Jamaica	24	78				
Mexico	9 223	24 931	1 680	1 228	1 710	4 940
Nicaragua	111	437				
Panama	553	2 499	120			
Puerto Rico	85	260				
St Vincent & the Grenadines	6	23				
United States of America	79 842	263 642			21	
Total North America	**160 562**	**650 178**	**4 597**			
Argentina	9 734	36 022	191	820	1 013	5 900
Bolivia	330	1 688	98			
Brazil	65 311	284 944	7 153	34 400	6 807	32 800
Chile	4 000	22 605	580			
Colombia	9 323	37 000				
Ecuador	1 712	7 070	18			
French Guiana	116	560				
Guyana	1	1	105			
Paraguay	7 500	52 000	1 350			
Peru	3 000	15 500	426			
Surinam	189	600				
Uruguay	1 538	9 535				
Venezuela	13 760	54 797	4 530	24 850	7 429	34 601
Total South America	**116 514**	**522 322**	**14 451**			
Afghanistan	200	530				
Armenia	960	1 500				
Azerbaijan	992	1 301	20			

TABLE 7.2 (*Continued*)

	In operation		Under construction		Planned	
	Capacity (MW)	Actual generation in 2002 (GWh)	Capacity (MW)	Probable annual generation (GWh)	Capacity (MW)	Probable annual generation (GWh)
Bangladesh	230	750				
Bhutan	430	2 000	1 117			
Cambodia	1	4	12			
China	82 700	257 500	35 000			
Cyprus	1	2				
Georgia	2 800	6 000	700			
India	25 751	73 954	5 274			
Indonesia	4 300	9 370	363			
Japan	27 348	94 250	988	1 743	19 026	47 451
Kazakhstan	2 200	7 000	305			
Korea (DPR)	4 750	17 000				
Korea (Republic)	1 576	2 228	9			
Kyrgyzstan	2 910	10 644	240			
Laos	622	1 475	150			
Malaysia	2 079	5 720	2 700			
Mongolia	3	1	1			
Myanmar (Burma)	390	1 528	1 862			
Nepal	563	1 864	72	312	39	171
Pakistan	5 010	18 940	1 786	8 045	16 032	81 966
Philippines	2 518	7 104	695		701	
Sri Lanka	1 128	2 577	70	303	150	530
Taiwan, China	1 719	5 068	83			
Tajikistan	4 054	15 000	670			
Thailand	2 936	7 471	1 011	143	38	
Turkey	12 241	33 684	3 338	10 845	19 951	70 786
Uzbekistan	1 710	6 835	249			
Vietnam	4 156	18 197	1 064			
Total Asia	**196 278**	**609 497**	**57 779**			
Albania	1 640	3 651			400	2 200
Austria	11 668	41 837				
Belarus	10	30	1			
Belgium	95	435				
Bosnia–Herzegovina	2 380	5 900	126			
Bulgaria	2 729	2 778	80	200		
Croatia	2 076	5 400			83	309
Czech Republic	992	2 482				
Denmark	11	32	N	N	N	N
Estonia	3	20	2			
Faroe Islands	31	76				
Finland	2 980	10 660	28	23	56	176

(*continued on next page*)

TABLE 7.2 (Continued)

	In operation		Under construction		Planned	
	Capacity (MW)	Actual generation in 2002 (GWh)	Capacity (MW)	Probable annual generation (GWh)	Capacity (MW)	Probable annual generation (GWh)
FYR Macedonia	436	975	80			
France	25 282	65 300				
Germany	4 831	23 900				
Greece	3 061	3 438				
Hungary	48	194				
Iceland	1 151	6 972	690	4 540	100	630
Ireland	249	725				
Italy	16 727	48 796				
Latvia	1 536	2 430				
Lithuania	116	354	3	12	115	400
Luxembourg	39	107				
Moldova	60	318				
Netherlands	38	60				
Norway	27 628	129 728	422	993	2 685	5 374
Poland	848	2 355				
Portugal	4 330	14 375	236			
Romania	5 911	15 640	900	2 900		
Russian Federation	44 700	164 300	8 400			
Serbia & Montenegro	3 452	11 456				
Slovakia	2 514	5 680			100	370
Slovenia	844	3 403	32			
Spain	15 677	28 437			1 000	2 750
Sweden	16 100	66 000	N	N	N	N
Switzerland	13 300	33 870	6	17		
Ukraine	4 731	10 868				
United Kingdom	1 577	4 788				
Total Europe	**219 801**	**717 770**	**11 006**			
Iran (Islamic Republic)	2 006	5 077	7 896	16 296	8 334	21 995
Iraq	910	586				
Israel	7	10			2	2
Jordan	7	53				
Lebanon	283	332				
Syria (Arab Republic)	1 520	7 000				
Total Middle East	**4 733**	**13 058**	**7 896**			
Australia	6 936	16 185	29		546	
Fiji	79	421				
French Polynesia	47	188				
New Caledonia	78	336				
New Zealand	5 245	24 361			570	3 000
Palau	10	30				

TABLE 7.2 (*Continued*)

	In operation		Under construction		Planned	
	Capacity (MW)	Actual generation in 2002 (GWh)	Capacity (MW)	Probable annual generation (GWh)	Capacity (MW)	Probable annual generation (GWh)
Papua New Guinea	222	547				
Solomon Islands	N	N				
Vanuatu	1	5				
Western Samoa	12	54				
Total Oceania	**12 630**	**42 127**	**29**			
Total World	**731 507**	**2 639 826**	**98 783**			

Notes:

(1) A quantification of hydropower development is not available for Chad, Guinea-Bissau, Liberia, Niger, Grenada and Turkmenistan.

(2) As the data available on the probable annual generation of capacity under construction, and on planned capacity and generation, do not cover all countries, regional and global totals are not shown for these categories.

(3) Data on planned capacity and generation are as reported by WEC Member Committees.

(4) Sources: WEC Member Committees, 2003; *Hydropower & Dams World Atlas 2003*, supplement to The International Journal on Hydropower & Dams, Aqua~Media International; *Energy Statistics Yearbook 2000*, United Nations; national and international published sources; estimates by the editors.

TABLE 7.3

Hydropower: status of development at end-2002 for small-scale schemes (<10 MW)

	Economically exploitable capability (GWh/yr)	In operation		Under construction and planned	
		Capacity (MW)	Actual generation in 2002 (GWh)	Capacity (MW)	Probable annual generation (GWh)
Africa					
Algeria		69	5		
Côte d'Ivoire		5	17	6	20
Egypt (Arab Republic)	49	6	49	12	83
South Africa	350	34	83		
North America					
Canada	41 157	1 136		200	1 140
Greenland				7	27
Mexico	121	385	1 488		

(*continued on next page*)

TABLE 7.3 (*Continued*)

	Economically exploitable capability (GWh/yr)	In operation		Under construction and planned	
		Capacity (MW)	Actual generation in 2002 (GWh)	Capacity (MW)	Probable annual generation (GWh)
United States of America	31 943	2 749	10 624		
Venezuela		1 607	7 036	5 130	22 468
South America					
Argentina		91	334		
Brazil	17 169	975	3 800	2 280	1 250
Asia					
Japan		3 478	18 809	98	442
Korea (Republic)	2 900	42	111		
Nepal		47	124	N	1
Pakistan		6	20	649	
Philippines		90	243		
Sri Lanka		44	115	35	105
Thailand	80	55	170	72	343
Turkey	3 623	175	654	674	2 969
Europe					
Albania		14	6		
Austria	10 000	838	4 233	200	1 000
Bulgaria	755	65	355		
Croatia	250	31	88		
Czech Republic		276	1 010		
Denmark		11	32		
Finland	700	400	900		
France		2 000	7 700		
Germany		1 421	7 126		
Greece			56		
Hungary		9	28		
Iceland		44	249		
Italy		2 233	8 657		
Latvia	70	21	30	5	10
Lithuania	185	15	37	3	12
Luxembourg		28	84		
Norway	14 000	1 025	4 810	73	324
Poland		468	1 300		
Portugal		281	770	40	
Romania		278	416		
Serbia & Montenegro		13	21		
Slovakia	1 000	66	160	40	150
Spain	7 000	1 493	3 811	720	2 230

TABLE 7.3 (*Continued*)

	Economically exploitable capability (GWh/yr)	In operation		Under construction and planned	
		Capacity (MW)	Actual generation in 2002 (GWh)	Capacity (MW)	Probable annual generation (GWh)
Sweden		6 700	3 500		
Switzerland		603			
Middle East					
Iran (Islamic Republic)		8	20	631	3 302
Israel		7	10	2	2
Jordan	90	7	53		
Lebanon		38			
Oceania					
Australia		299		29	
New Zealand	450	118	430	12	8

Notes:

(1) The data on small-scale schemes are those reported by WEC Member Committees. They thus constitute a sample, reflecting the information available in particular countries: they should not be considered as complete, or necessarily representative of the situation in each region. For this reason, regional and global aggregates have not been computed.

(2) Sources: WEC Member Committees, 2003.

COUNTRY NOTES

The Country Notes on Hydropower have been compiled by the editors, drawing principally upon the 2003 edition of the *Hydropower & Dams World Atlas*, supplement to *The International Journal on Hydropower & Dams*, published by Aqua ~ Media International, together with information provided by WEC Member Committees in 2003 and various national published sources.

Albania

It has been reported that the exploitation of hydro energy through small hydropower plant schemes is of interest. Up to 1988, 83 small HPPs were built in Albania, with capacities varying from 5 to 1 200 kW and a total capacity of 14 MW. The purpose of construction of these HPPs was, at the beginning, to supply isolated mountainous areas with electricity. The average life of such HPPs is 25 years. A development programme for these HPPs has been part of Government energy policy and the law on privatisation of small HPPs has created some possibilities to bring them back into effective operation. Taking into account the average life of the existing HPPs, the damages they suffered during the transition period post-1990, the qualifications of the employees who have worked in these HPPs, and their maintenance cost, considerable investments, particularly on their mechanical and electrical equipment, are required.

Argentina

Hydropower & Dams World Atlas quotes Argentina's gross theoretical hydropower potential as 172 000 GWh/yr; its technically feasible potential is put at 130 000 GWh/yr, of which about 24% has so far been exploited.

Hydro output in 2002 was 36.0 TWh, but this was an exceptionally high level, reflecting unusually favourable hydrological conditions. With an installed capacity of 9 734 MW at end-2002, a 'normal' year's hydro output would be around 30.8 TWh.

A substantial portion of Argentina's hydro capacity is accounted for by its 50% share in two bi-national schemes: Salto Grande (installed capacity 1 890 MW), shared with Uruguay and Yacyretá (3 100 MW), shared with Paraguay. The latter plant is currently operating at a reduced head, with its capacity restricted to 1 800 MW.

The total amount of hydro capacity reported to be under construction at the end of 2002 was 191 MW, with a further 1 013 MW at the planning stage. The Secretariat of Energy has accorded priority to the drawing up of a Catalogue of Hydroelectric Projects. This task will involve organising a Projects Library, updating and improving cost-estimation procedures, reviewing existing projects and evaluating newly identified resources.

Australia

It should be noted that Australia is the driest inhabited continent on earth, with over 80% of its landmass receiving an annual average rainfall of less than 600 mm/yr and 50% less than 300 mm/yr. A high variability in rainfall, evaporation rates and temperatures also occurs between years, resulting in Australia having very limited and variable surface and groundwater resources.

The economically exploitable capability is estimated by *Hydropower & Dams World Atlas* as 30 TWh/yr, of which more than half has already been harnessed. The Australian WEC Member Committee reports that a mere 29 MW of hydro capacity is presently under construction, although as much as 546 MW is planned for future installation.

The prospects for large-scale hydroelectric projects in Australia are limited, principally because most available sites have already been developed and, in some cases, have required a compromise to be reached on wilderness preservation or other environmental factors. There are, however, development opportunities associated with hydropower projects. These opportunities are in the refurbishment of existing plant and equipment. Many turbines are now on average 45 years old and to remain serviceable need maintenance and/or refurbishment. Indeed, one of the intentions of the federal Government's *Renewable Energy (Electricity) Act 2000* has been to give recognition to large-scale hydro as a renewable energy resource, and to the possibility that hydro could play an important role in meeting the target of 9 500 GWh of new renewable energy by 2010.

In addition to refurbishment opportunities, there are also prospects for increased contributions from mini-hydro projects. Private development of small-scale grid-connected hydro has been taking place in Australia since the mid-1980s, with the first significant project undertaken by Melbourne Water on the Thompson Dam. The installed capacity of mini hydro across the country is about 300 MW, producing around 700–750 GWh per annum.

The synergy from combining wind and hydropower as renewable energy resources is being fully explored in Tasmania, where up to 1 000 MW of wind energy potential has been identified. The key to unlocking this opportunity has been securing development approval for the AUD 500 million Basslink interconnector, currently being constructed by the UK company National Grid Transco. This undersea cable across Bass Strait will connect Tasmania to the national electricity market via Victoria, and allow Tasmania to supply peaking power at premium prices to Victoria at times of high demand in summer and winter. It will also enable the export of new renewable energy from wind farm developments to the national electricity market.

Bolivia

Bolivia has a considerable hydro potential, its technically feasible potential being assessed at 126 TWh/yr, of which 50 TWh/yr is considered to be economically exploitable. Only a minute proportion of the potential has been harnessed so far—2002 hydro capacity was 330 MW, with an output of about 1.7 TWh.

At the end of 2002, 98 MW of additional hydro capacity was under construction. The approximately 700 MW of hydro capacity planned includes plants at San José (126 MW), Misicuni (120 MW) and Palillada (80 MW).

Brazil

Hydroelectric power is one of Brazil's principal energy assets: the republic has by far the largest hydropower resources on the continent. The Brazilian WEC Member Committee reports that gross theoretical capability exceeds 3 000 TWh/yr, with an economically exploitable capability of over 800 TWh/yr, of which nearly 40% has been harnessed so far. Hydro output in 2002 was 285 TWh, which accounted for 89% of Brazil's electricity generation.

Hydroelectricity plants (above 30 MW capacity) represent 78% of the overall Brazilian installed capacity. Because of that, electricity generation in Brazil is strongly influenced by the natural flows of rivers and other watercourses. In 2001, due to unexpectedly low natural flows, a power outage occurred in Brazil. However, in 2002 reservoir levels rose, new power plants came into operation and consumers made more efficient use of energy, all of which served to considerably reduce the risk of a new power outage in the short term.

Hydro generating capacity became more than doubled between 1980 and 1999, partly through gradual commissioning of the huge Itaipú scheme (total capacity 12 600 MW), which came into operation between 1984 and 1991. Brazil shares Itaipú's output with its neighbour Paraguay, which sells back to Brazil the surplus power remaining after its own electricity needs have been satisfied.

At the end of 2002, Brazil had over 7 GW of hydro capacity under construction, including a major (4 125 MW) extension of capacity at Tucuruí and two additional 700 MW units at

Itaipú. Nearly 7 GW of further hydro capacity is planned for future development.

Within the overall picture outlined above, small-scale hydro (since 1998, defined in Brazil as plants with a capacity of 1–30 MW) has an economically exploitable capability of about 17 TWh/yr, some 27% of which had been exploited by capacity installed as at end-2002. The 975 MW of small-scale hydro currently in place will be augmented by 2 280 MW additional capacity which is under construction or planned. Under current legislation, certain incentives are given to owners/developers of small-scale hydro schemes, in order to improve competition in the electricity market.

Bulgaria

The Bulgarian WEC Member Committee reports that there are 102 hydropower plants in Bulgaria, 77 belonging to the National Electric Company (NEK), 24 to independent power producers and one to the Shumen District Heating Company. Total installed hydro capacity is about 2 700 MW, which is more than 15% of total Bulgarian generation capacity. The contribution of the hydropower plants to electricity output varies between 2 and 7%, depending on the weather conditions.

The reasons for the low capacity factors of the hydropower plants are various:

- Overestimation of the design water quantities.
- The plants are used predominantly for regulation purposes. The largest cascades like Belmeken-Sestrimo and Batak-Aleko are frequency regulators included in the Unit Control System (UCS) of NEK. An enhancement of the UCS is under way in order to incorporate two other large cascades—Vacha and Arda.
- The water is being used to supply households, agriculture and industry. Great quantities of water are thus diverted and not processed by the plants. This applies to the Belmeken-Sestrimo cascade, the Iskar cascade and some others. The Kokalyane plant is practically out of operation because the water is needed for Sofia's population. In addition, the water levels in the reservoirs are often lower than required and the water discharge higher.
- The bad condition of the equipment—a highly important reason. As a rule, hydropower plants, dams and reservoirs are the last priority of NEK. The equipment of these facilities is the most robust and in a situation of shortage the management of the company uses its financial resources for other purposes. As a result, the efficiency of the hydro-turbines is going down, the losses of water are high and safety problems arise.

Because of the above reasons, the hydropower plants use more water than if they had been in normal condition. It is worth mentioning that the capacity available at some of the plants exceeds their design capacity, but these are isolated cases such as Sestrimo HPP and Anton HPP. There are three hydropower plants with pumped-storage facilities: Belmeken, Anton and Chaira.

At the moment the contribution of small HPPs to the country's energy system is less than 2%, both with respect to capacity and to power generation, which could be regarded as insignificant (at the moment the installed capacity of micro HPPs is 65 MW). Moreover many existing micro HPPs are not in operation, for a variety of reasons, but they could be modernised at relatively low cost, especially in the case of small, typically rural and relatively remote installations.

Taking into account projects already planned, as well as some future ones which are combined with schemes for improving water supply and quality and are acceptable from an environmental point of view, additional installed capacity from small HPPs can realistically be expected in the short term (year 2005). There are possibilities for the construction of small HPPs on running water and in association with dams, as well as on irrigation channels in water improvement schemes.

Cameroon

The technically exploitable hydro capability (115 000 GWh) is the fourth largest in Africa but the current level of utilisation of this potential is, like that in other hydro-rich countries in the continent, very low. Within a total hydro capacity of 719 MW, Cameroon's major stations are Song Loulou (384 MW) and Edéa (263 MW), both of which are undergoing refurbishment. New hydro plants are being investigated for a number of other sites but no schemes are presently under construction.

Canada

Canada possesses enormous hydropower potential, with an economically exploitable capability second only to that of Brazil in the whole of the western hemisphere. About 60% of Canada's electricity generation is furnished by hydro plants, which in 2002 produced some 346 TWh. Hydroelectricity generation in 2002 represented 66% of the assessed economic potential of 522 TWh/yr.

The Canadian WEC Member Committee has reported that at the end of 2002, 2 160 MW of hydroelectric generating capacity was under construction, with the major projects (all located in the province of Quebec) comprising: Sainte Marguerite 3 (880 MW), Toulnustouc (526 MW), Eastmain 1 (480 MW) and Grand-Mère (220 MW). A total of 2 853 MW additional capacity was planned for future development, including Gull Island, Quebec (1 700 MW), Gull Rapids, Manitoba (600 MW) and Wuskwatim, Manitoba (200 MW).

Studies to establish the exploitable capability of small-scale hydropower were undertaken prior to 2000, mostly at a provincial level or as individual site studies. The reported technically exploitable capability for small hydro is some 103 TWh/yr, with the economically exploitable potential put at about 41 TWh/yr, both assuming a 60% capacity factor.

Installed capacity of hydro plants of less than 10 MW capacity totalled 1 136 MW at end-2002; 200 MW is planned for future installation.

Chile

There is substantial hydropower potential, with the technically exploitable capability estimated at about 162 TWh/yr, of which about 15% has so far been exploited. Hydro output in 2002 was 22.6 TWh, equivalent to about 53% of Chile's total electricity generation.

The largest hydro scheme currently in hand is the 570 MW Ralco project, which is expected to become operational in 2004. A number of projects, including long-term schemes, are planned: La Higuera/Tinguririca (260 MW), Baker (1 000 MW), Pascua (1 200 MW), Neltume (400 MW), Choshuenco (150 MW) and Punilla (100 MW).

China

China's hydroelectric resources are vast, however measured: its gross theoretical potential approaches 6 000 TWh/yr, while its economically feasible potential has been assessed as some 290 000 MW (1 270 TWh/yr)—in both instances, far larger than that of any other country in the world. Current hydro output exceeds 250 TWh/yr, contributing about 17% to the republic's electricity generation.

The total amount of hydro capacity under construction is about 35 000 MW: as large as the combined current building programme of the next four largest hydro developers (Brazil, the Russian Federation, Iran and India). By far the largest hydro scheme under way is the Three Gorges Project (18 200 MW), scheduled for commissioning between 2003 and 2009; other plants in course of construction include Longtan (4 200 MW, eventually 5 400 MW), Xiaowan (4 200 MW), Pubugou (3 300 MW), Shuibuya (1 840 MW) and Gongboxia (1 500 MW).

More than 50 GW of pure hydroelectric capacity is planned for construction, including five very large schemes: Xiluodu (12 800 MW) and Xiangjiaba (6 000 MW) in the Yangtze river basin, Nuozhadu (5 550 MW) in the Lancang basin, Laxiva (3 720 MW) and Jinping I(3 300 MW).

China has about 5 700 MW of pumped-storage capacity, with 6 120 MW under construction and about 58 GW planned.

Colombia

The theoretical potential for hydropower is very large, being estimated to be 1 000 TWh/yr, of which 20% is classed as technically feasible. The economically exploitable capability has been evaluated as 140 TWh/yr: hydro output in 2002 represented about a quarter of this potential, and accounted for around two-thirds of Colombia's electricity generation.

Two large hydro schemes have recently come into service—Porce II (392 MW) and La Miel I (400 MW). Around 10 000 MW of new capacity is at the planning stage, for medium- to long-term implementation, including Sogamoso (1 035 MW), Nechí (750 MW), Porce III (660 MW), La Miel II (411 MW) and Quimbo (400 MW).

Congo (Democratic Republic)

The assessed potential for hydropower is by far the highest in Africa, and one of the highest in the world. The gross theoretical potential is almost 1 400 TWh/yr, of which about 55% is regarded as technically feasible. The current level of hydroelectric output is equivalent to less than 1% of this latter potential. Hydro provides virtually the whole of the country's electricity.

The national power authority SNEL has 16 hydro plants, with a total rated capacity of 2 426 MW; its largest stations are Inga I (351 MW) and Inga II (1 424 MW). The effective capacity at SNEL's hydro plants has recently been only about half their rated level, owing to problems in maintenance and refurbishment.

A significant increase in capacity would be provided by the projected Inga III (3 500 MW), which is planned for 2010. There is also a huge scheme (Grand Inga) for the installation of up to 52 generators of 750 MW each, with as a second stage (Inga IV) the supply of electricity to Egypt and South Africa via new long-distance transmission lines. Both generating plant and transmission lines are the subject of feasibility studies.

Costa Rica

For a country with a surface area of only 51 100 km^2, Costa Rica has a surprisingly large hydroelectric potential. Its gross theoretical potential is estimated at 223 TWh/yr, within which 43 TWh/yr has been assessed as technically feasible.

Aggregate hydro capacity was 1 263 MW at end-2002, equivalent to about 77% of Costa Rica's generating capacity. Several new hydro plants are under construction or planned: the largest being Cariblanco (70 MW), due to be commissioned during 2006, and Pirris (128 MW), scheduled to come on line in 2007.

Czech Republic

The overall potential for all sizes of hydropower is quite modest (technically exploitable capability: 3 978 GWh/yr). Total hydroelectricity output in 2002 was 2 482 GWh, representing 62% of the technical potential. Hydropower furnishes about 3% of the republic's electricity generation.

A relatively high proportion (nearly 40%) of the technically exploitable capability is classified as suitable for small-scale schemes; installed capacity in this category at the end of 2002 was 276 MW, equivalent to about 28% of the Czech Republic's hydro capacity. Actual generation from small-scale schemes in 2002 accounted for 41% of hydro output.

Ethiopia

There are enormous resources for hydro generation, the gross theoretical potential (650 TWh/yr) being second only to Congo (Democratic Republic) in Africa. The economically feasible potential is stated to be 260 TWh/yr, of which 10% represents the potential for small-scale hydro installations. Hydro output in 2002 was about 2 TWh, a minute fraction of the assessed potential. Currently, hydroelectricity provides around 98% of Ethiopia's electricity.

At the end of 2002 approximately 450 MW of hydro capacity was in place, while the plants under construction comprised Gilgel Gibe (184 MW), Gojeb (150 MW) and Tekeze (225 MW).

France

France is one of Western Europe's major producers of hydroelectricity, but its technically feasible capacity has now been very largely exploited. No more hydro plants are under construction or planned.

The total installed capacity of small-scale (< 10 MW) plants is around 2 000 MW, which generated 7.7 TWh in 2001. There are, on the other hand, some 280 hydro plants of greater than 10 MW, with an aggregate installed capacity of 23 000 MW.

Ghana

There are 17 potential hydro sites, of which only Akosombo (912 MW) and Kpong (160 MW) have so far been developed. The most attractive hydro project is the 400 MW Bui dam on the Black Volta river, which is at a preparatory stage.

Electricity generation in Ghana is a responsibility of the Volta River Authority, established in 1961. The average annual output of its two existing hydro stations (6 100 GWh) is equivalent to about 58% of Ghana's technically exploitable hydro capability.

Iceland

Together with its geothermal resources, Iceland's hydropower potential represents virtually its only indigenous source of commercial primary energy. Gross theoretical potential of 184 TWh/yr includes 40 TWh of economically harnessable output. Hydroelectricity production in 2002 was 6 972 GWh, implying that about 17% of this economic potential has been developed. Hydro capacity at present under construction will add 690 MW to the existing installed capacity of 1 151 MW. A further 100 MW of hydro capacity is planned.

The technically exploitable capability of small-scale hydro plants is reported to be 12.3 TWh/yr, equivalent to about 19% of the level for total hydro. Installed capacity of small hydro at end-2002 was 44 MW, or 3.8% of total hydro capacity.

Hydropower provides 17% of Iceland's primary energy supply and 83% of its electricity generation.

India

India's gross theoretical hydropower potential (2 638 TWh/yr) and theoretically feasible potential (660 TWh/yr) are amongst the highest in the world. The public utilities' total installed hydroelectric capacity exceeded 26 500 MW at the end of 2002, with a corresponding generation of 68.5 TWh, equivalent to 12.9% of India's public sector electricity generation.

Hydropower & Dams World Atlas 2003 reports that about 5 GW of hydro capacity is under construction and a further 37 GW is planned. There are at least 17 plants of over 300 MW capacity being built, of which the largest are Nathpa Jhakri (1 500 MW), Sardar Sarovar (1 200 MW), Tehri Stage I (1 000 MW) and Narmada Sagar (1 000 MW).

The *Atlas* also reports that there are 267 small-scale hydro plants (up to 3 MW) in operation, with an aggregate installed capacity of about 210 MW; a further 140 MW of small-scale capacity is under construction. A total of 3 349 schemes, aggregating 2 852 MW, have been identified for possible future development.

Indonesia

At some 2 150 TWh/yr, Indonesia's gross theoretical hydro potential is the third largest in Asia. Its technically exploitable capability is just over 400 TWh/yr, of which about 10% is considered to be economically exploitable. Average annual hydro output is about 9 400 TWh, indicating the evident scope for further development within the feasible potential. Hydro presently provides approximately 12% of Indonesia's electricity supplies.

Hydropower & Dams World Atlas 2003 reports that at least 363 MW of hydroelectric generating capacity is under construction (including a 210 MW plant at Musi) and that approximately 2 000 MW of additional hydro capacity is at the planning stage.

Italy

Italy's theoretical resource base for hydropower is one of the largest in Western Europe, and its economically exploitable capability is nearly as much as that of France. Hydroelectric power has not, however, been developed to quite the same degree as in the case of its neighbour: about 72% of the assessed economic potential of 65 000 GWh/yr has so far been harnessed. Hydroelectricity accounts for about 18% of Italy's power generation.

The installed capacity of small-scale plants at end-2001 was 2 233 MW, representing about 13% of the overall hydro capacity of 16 727 MW. About 18% of the overall hydroelectric generation is attributable to small-scale plants.

Japan

A high proportion of Japan's large potential for hydro generation has already been harnessed. *Hydropower & Dams World Atlas 2003* quotes its gross theoretical capability as about 718 TWh/yr, of which 136 TWh is regarded as technically exploitable and 114 TWh as economically exploitable. Hydro output in 2001 is given as 94 250 GWh, representing nearly 9% of Japan's electricity.

The Japanese WEC Member Committee reports that just under 1 000 MW of hydro capacity was under construction at end-2002. Most of the sites suitable for the installation of large-scale conventional hydroelectric plants have now been developed. The great majority of the larger hydro projects presently under construction or planned in Japan are pumped-storage schemes.

The technically exploitable capability for small-scale hydro developments is reported by the Japanese Member Committee to be 47 TWh/yr, a relatively high proportion (34%) of the total hydro level. Developed small-hydro capacity at end-2002 was about 3.5 GW, equivalent to 12.7% of total hydro capacity. Capacity planned for construction totalled 98 MW, with a probable annual generation of 442 GWh.

Latvia

Although its hydro potential is quite modest—a gross theoretical capability of only about 7 TWh/yr—Latvia is of interest for its rapid development of small-scale hydro plants in recent years. In 1996 there were only 16 small hydro stations, which generated 4.5 GWh. By 1999, the number in service had grown to 53 and annual generation to 15 GWh. By 2002, the number in service was 138 and annual generation 70 GWh.

Madagascar

Madagascar has a considerable land area (greater than that of France, for example) and heavy annual rainfall (up to 3 600 mm). Consequently, the potential for hydropower is correspondingly large: gross theoretical potential is put at 321 TWh/yr, within which the technically feasible potential is 180 TWh/yr. With current installed capacity standing at 104 MW and annual hydro output about 540 GWh, the island's hydro capability has scarcely begun to be utilised. A small amount of hydro capacity (42 MW) is under construction, with more than 500 MW planned.

Malaysia

There is a substantial potential for hydro development, with a total technically feasible potential of about 123 TWh/yr, most of which is located in Sarawak (87 TWh/yr) and Sabah (20 TWh/yr); a considerable proportion of Peninsular Malaysia's technically feasible potential of 16 TWh/yr has already been developed.

After being halted in 1997 as an austerity measure, construction of the 2 400 MW Bakun hydro project in Sarawak is once again under way, with completion foreseen for 2007. There is also a 300 MW hydro plant under construction at Ulu Terengganu, Peninsular Malaysia. A 300 MW upgrading is planned for the Kenyir station, while a number of other schemes are the subject of feasibility studies.

Mexico

With a gross theoretical hydro capability of 135 TWh/yr and a technically exploitable capability of 49 TWh/yr (both as quoted by *Hydropower & Dams World Atlas 2003*), Mexico possesses a considerable hydroelectric potential. Its economically exploitable capability is given by the same source as 32.2 TWh/yr.

The Mexican WEC Member Committee reports that actual hydro generation in 2002 was 24.9 TWh, equivalent to 11.6% of total net generation. Nearly 1 700 MW of additional hydro capacity was reported to be under construction at end-2002, with approximately the same amount of capacity planned for future development. The principal plants involved are

- El Cajon (680 MW), scheduled for completion in 2007;
- La Parota (765 MW), planned for 2008;
- Copainalá (210 MW), also due in 2008.

A major extension of the Manuel Moreño Torres (Chicoasén) hydro plant is planned for completion in 2003; this will add three units, with a total incremental capacity of 900 MW.

At end-2002, installed capacity of small-scale hydropower is reported by the Mexican Member Committee to have been 385 MW, with output during the year 1 488 GWh.

Myanmar

The country is well endowed with hydro resources: its technically feasible potential is given by *Hydropower & Dams World Atlas 2003* as 37 000 MW. At an assumed annual capacity factor of 0.40, this level would imply an annual output capability of approximately 130 TWh; actual output in recent years has been extremely variable, reflecting drought conditions in certain years. Severe water shortages in 1999 brought about a drastic reduction in hydroelectric output, the year's total falling to less than half the normal level. Output recovered in 2000–2001 and is now probably in the vicinity of 2.0 TWh/yr. Given a sustained return to historical amounts of precipitation, there appears to be ample scope for substantial development of hydropower in the long term.

Current hydro capacity is about 390 MW; plants under construction will substantially increase this total within a few years. Under the Ministry of Electric Power's 5-year plan (2001/2002–2005/2006), 12 hydro plants with a total capacity of 1 862 MW are under construction. Longer term projects under consideration include several large export-orientated schemes, e.g. Tasang (7 110 MW), Tamanthi (1 200 MW) and Taninthayi (600 MW).

Nepal

There is a huge theoretical potential for hydropower, reported by the Nepalese WEC Member Committee to be some 727 TWh/yr, with a technically exploitable capability put at 394 TWh/yr and an economically exploitable capability of 221 TWh/yr.

Total hydro capacity at end-2002 was 563 MW; a further 72 MW of capacity was reported to be under construction at that time. Actual hydro generation in 2002 was 1 864 GWh, of which 124 GWh (6.7%) was produced by small-scale plants (schemes of less than 10 MW).

Nepal's topography gives it enormous scope for the development of hydroelectricity, which probably provides the only realistic basis for its further economic development. Small-scale hydro plants are the most viable option for rural electrification. Large projects, however, in view of Nepal's limited financial resources, would probably require power export contracts with India as a prerequisite.

Norway

Norway possesses Western Europe's largest hydro resources, both in terms of its current installed capacity and of its economically feasible potential. The WEC Member Committee for Norway reports a gross theoretical capability of 600 TWh/yr, of which 187 TWh is economically exploitable. The hydro generating capacity installed by the end of 2002 had an output capability equivalent to around 63% of the economic potential. Actual hydro output in 2002 was nearly 130 000 TWh, providing virtually all of Norway's electric power.

Hydro capacity under construction at the end of 2002 amounted to 422 MW, while a further 2 685 MW of capacity was approved for development (development licence granted), under licencing or under planning.

The economically exploitable capability applicable to small-scale hydro schemes (1–10 MW) is reported to be 14 TWh/yr, equivalent to 7% of the overall level. Installed capacity of small hydro plants totalled 1 025 MW at end-2002, with an average annual output capability of 4.8 TWh. A further 73 MW of small-scale capacity was under construction or approved for development (development licence granted).

Pakistan

The Pakistan WEC Member Committee reports a total installed hydro capacity of 5 010.2 MW, which is 28.2% of the national installed capacity. The country has an identified hydro potential of 40 000 MW out of which 30 000 MW is economically exploitable. The main potential sources of hydropower are on the rivers Indus and Jhelum, plus sites at Swat and Chitral. Both hydro and thermal power plants are being developed to meet the country's demand for electricity, as part of the state utility WAPDA's 'Vision 2025' programme.

Hydro capacity in operation at the end of 2002 included major plants at Tarbela (3 478 MW) and Mangla (1 000 MW); output during the year was 18.9 TWh, accounting for 26% of Pakistan's electricity generation. Capacity reported to be under construction at end-2002 amounted to 1 786 MW; the major project is the Ghazi Barotha hydro station, with five units totalling 1 450 MW. Many other sites have been identified for development in the medium and longer term, the total capacity reported as planned being some 16 000 MW.

As regards small hydro, the provincial governments of North-West Frontier Province and Azad-Kashmir, together with the Northern Area Authorities, are developing many potential hydropower sites, mainly of less than 10 MW.

A programme for feasibility studies of small hydropower projects on canal falls and barrages was approved by the Government of Pakistan in 2001. A total of 591 sites with aggregate potential of 649 MW at various canal falls and barrages throughout the country were identified. Data for canal falls and distributaries having a discharge of 200 cusecs or more are being collected. Three sites, namely Machai, Pakpattan and Rohri, have been taken as pilot projects, for which feasibility reports have been completed. The feasibility of six other sites is under study. A Memorandum of Understanding has been signed between the Governments of Pakistan and China for the implementation of the above nine projects on a turnkey basis with financial facilities.

Paraguay

In the context of energy supply, Paraguay's outstanding natural asset is its hydroelectric potential, which is mainly derived from the river Paraná and its tributaries. The country's gross theoretical capability for hydroelectricity is about 111 TWh/yr, of which 68 TWh is estimated to be economically exploitable. Two huge hydroelectric schemes currently utilise the flow of the Paraná: Itaipú, which Paraguay shares with Brazil, and Yacyretá, which it shares with Argentina.

Itaipú is the world's largest hydroelectric plant, with a total generating capacity of 12 600 MW, of which Paraguay's share is 6 300 MW. This share is far in excess of its present or foreseeable needs and consequently the greater part of the output accruing to Paraguay is sold back to Brazil. Two further 700 MW units are being installed at Itaipú, with completion expected in 2004.

The bi-national plant at Yacyretá, downstream from Itaipú, has an installed capacity of 3 100 MW. There are 20 generating units, each of 155 MW capacity, but all are still operating at only 90 MW per unit, owing to the level of the reservoir being held below that originally planned. The planned addition of 255 MW of hydro capacity on the Añacuá, a tributary of the Paraná, would raise the level of the water in the Yacyretá reservoir, leading to improved rates of utilisation for the bi-national scheme's turbines.

Paraguay has a wholly owned 256 MW hydro plant (Acaray), which will probably be uprated during the next few years. The state electric utility, ANDE, also plans to install two 100 MW units at Yguazu. An environmental impact study has been conducted for the projected bi-national Corpus Christi dam (2 880 MW, to be shared with Argentina), sited on the Paraná, downstream of Itaipú and upstream of Yacyretá.

Peru

Peru's topography, with the Andes running the length of the country, and many fast-flowing rivers, endows the republic with an enormous hydroelectric potential. Its hydro capability is assessed as one of the largest in the whole of South America: its economically exploitable capability is some 260 TWh/yr. Current utilisation of this capability is very low—at around 6%. Hydro provides about 75% of Peru's electric power.

Plants under construction at end-2002 were Chimay (142 MW), Ocona (150 MW) and Yúncan (134 MW). About 1 500 MW of hydro capacity is planned for development over the short/medium term, including Olmos (624 MW), Cheves (525 MW) and El Platanal (two plants totalling 270 MW).

Russian Federation

Russia's hydro resource base is enormous—the gross theoretical potential is some 2 295 TWh/yr, of which 852 TWh is regarded as economically feasible. The bulk of the Federation's potential is in its Asian regions (Siberia and the Far East). Hydro generation (including pumped-storage output) in 2002 (164 TWh) represented 19% of the economic potential and accounted for 18% of total electricity generation.

At the end of 2002 installed hydroelectric generating capacity was 44.7 GW; according to *Hydropower & Dams World Atlas 2003*, 8.4 GW of additional capacity was under construction and nearly 12 GW of further capacity was planned for installation.

The largest plants under construction are Boguchany (1 920 MW, eventually rising to 3 000 MW) on the Angara river in Siberia and Bureya (2 000 MW) on the river Bureya in the Far East.

Sweden

Sweden has one of the highest hydro potentials in Western Europe: the Swedish WEC Member Committee reports a gross theoretical capability of 176 TWh/yr, of which 90 TWh is economically exploitable. The average annual capability of the 16 100 MW hydro capacity installed at the end of 2002 was 65 TWh, about 72% of the economic potential. Actual hydro output in 2002 was 66 TWh, which provided nearly half of Sweden's electricity generation.

The construction of new hydro plants has virtually stopped, on account of environmental and political considerations. Future activity is likely to be very largely confined to the modernisation and refurbishment of existing capacity.

Tajikistan

The terrain and climate are highly favourable to the development of hydropower. Apart from the Russian Federation, Tajikistan has the highest potential hydro generation of any of the FSU republics. Its economically feasible potential is estimated to be 263.5 TWh/yr, of which only about 6% has been harnessed so far. Hydropower provides over 95% of Tajikistan's electricity generation.

There is just over 4 GW of hydro capacity installed, the principal site being Nurek (2 700 MW). The Sangtuda plant, currently under construction, will add another 670 MW to Tajikistan's capacity. *Hydropower & Dams World Atlas 2003* reports that plans exist for installing a further 4.5 GW. Work on the huge Rogun scheme (3 600 MW) on the river Vakhsh has been suspended, but is expected to go ahead in due course.

Turkey

The Turkish WEC Member Committee reports a gross theoretical hydropower potential of 433 TWh/yr, a technically feasible potential of 216 TWh/yr and an economically feasible potential of 126 TWh/yr. Hydro output of 33.7 TWh in 2002 points to a considerable degree of development potential.

At end-2002, operational hydro capacity amounted to just over 12.2 MW. A further 3.3 GW of capacity was under construction at that point in time. The largest plant involved was Deriner (670 MW), which is scheduled for completion in 2005. At least eight other hydro stations with individual capacities of more than 100 MW were in the course of being built. Some 20 000 MW of additional capacity is planned for development over the next 25 years.

United States of America

The hydro resource base is huge: the gross theoretical potential was assessed in 1979 as 512 GW, equivalent to 4 485 TWh/yr. The United States WEC Member Committee has reported for the present *Survey* that the annual technically exploitable capability is 1 752 TWh, of which 501 TWh is economically exploitable, both based on recent publications of the Idaho National Environmental and Engineering Laboratory. The end-2002 US hydro capacity of 79.8 GW had an average annual capability of about 295 TWh, equivalent to 59% of the assessed economic potential.

Actual hydroelectric output of 263.6 TWh in 2002 accounted for 6.9% of US electricity generation. The US Member Committee reports that there were no hydro plants under construction at end-2002, and that 21 MW of additional hydroelectric generating capacity was at the planning stage.

The reported economically exploitable capability of small-scale hydropower is almost 32 TWh. The installed generating capacity of small hydro plants totalled 2.75 GW at end-2002; actual generation in 2002 was about 10.6 TWh, equivalent to 4% of total US hydro output.

The US Department of Energy's small-scale hydropower research programme is focussed on reducing the barriers to development of an estimated 30 000 MW of undeveloped low head/low power hydropower resource. The Department completed a detailed resource assessment of the US hydrologic regions, and plans to do a technology assessment to match technologies against the resource, and identify technology gaps. To further facilitate the development of small-scale hydropower, in FY2004, the Department of Energy issued a small business solicitation for innovative low-head hydropower conceptual designs.

Uruguay

Hydropower is Uruguay's only indigenous source of commercial primary energy, but even this is on a relatively limited scale. According to *Hydropower & Dams World Atlas 2003*, the technically exploitable potential is 10 TWh/yr, within a gross theoretical potential of 32 TWh. On the other hand, the Uruguayan WEC Member Committee quotes a gross theoretical capability of only 16 TWh. As it also reports that 2002 output was 9 535 GWh, it would appear that there is only a moderate amount of incremental capacity available (in principle) for exploitation in the future.

During the 1980s almost all of Uruguay's incremental generating capacity was in the form of hydropower, with the commissioning of the bi-national Salto Grande (1 890 MW) plant on the river Uruguay; the republic shares its output with Argentina. No hydro plants are under construction or planned: future increases in generating capacity are likely to be fuelled by natural gas.

Venezuela

The Venezuelan WEC Member Committee reports a gross theoretical capability of 320 TWh/yr, of which 130 TWh/yr is considered as economically exploitable. Hydroelectric output in 2002 was 54.8 TWh, but this was a relatively low level, constrained by water availability. Hydro output in an average year would be 64–65 TWh, indicating that half the realistic potential has already been harnessed. About 70% of the republic's electricity requirements are met from hydropower.

A large increase in hydroelectric capacity occurred during the 1980s, the major new plant being Guri (Raúl Leoni), on the river Caroní in eastern Venezuela—its capacity of 10 300 MW makes it currently the world's second largest hydro station, after Itaipú (shared by Brazil and Paraguay). Its total capacity is currently being expanded to 10 700 MW.

At the end of 2002, total hydroelectric generating capacity is reported to have been 13.76 GW; 4.5 GW was under construction and a further 7.4 GW of hydro capacity was planned for future development.

The 2 160 MW Caruachi project, sited 59 km downstream from Guri, is scheduled for phased entry into operation between 2003 and 2006. Two other major projects planned for the river Caroní are Tocoma (2 160 MW) and Tayucay (2 450 MW).

Vietnam

Vietnam has abundant hydro resources, particularly in its central and northern regions. Its gross theoretical potential is put at 300 000 GWh/yr, with an economically feasible potential of some 90 000 GWh/yr. Total installed hydro capacity was nearly 4 200 MW in 2002 and output of about 18 TWh provided over half of Vietnam's power supplies.

The principal areas of hydro potential are the rivers Da in the north, Sesan in central Vietnam and Dongmai in the south. *Hydropower & Dams*

World Atlas 2003 reported that at end-2002, 1 064 MW of hydro capacity was under construction, the major sites being: Tuyen Quang (432 MW), Dai Ninh (300 MW) and Se San 3 (300 MW). Projects are at various stages of planning for some 20 further sites; the largest scheme is Son La (2 400 MW), with construction expected between 2005 and 2011.

Chapter 8

Peat

World Energy Council
CONSEIL MONDIAL DE L'ENERGIE

COMMENTARY

Production Capacity

Peat is used mainly for two different purposes: horticulture and energy. The peatland area under active production is approximately 230 000 ha. This is about 0.05% of the global peatland area of 400 million hectares. Assuming an average production figure of 150 tonnes/ha, the annual average production is of the magnitude of 35 million tonnes of peat. Of this amount some 50–60% is used for fuel and the remainder for horticulture.

Peat production is mostly concentrated in Europe: the total active production area is estimated to be about 200 000 ha, while in North America peat is produced on an area of about 25 000 ha. The other 5 000 ha includes production in Africa, Australia, New Zealand and South-East Asian countries. The bulk of European fuel peat production occurs in the following countries: Finland, Ireland, Russia, Belarus, Sweden, Estonia, Ukraine, Latvia and Lithuania. In North America peat is produced only for horticulture.

The Use of Fuel Peat

The role of peat in energy production has changed significantly in the countries concerned. A major trend in the East European countries has been the decreasing use of peat as an energy source, a development which accelerated during the 1990s. Nowadays, there again seems to be an increasing interest in fuel peat.

Currently, Belarus uses about 2 million tonnes of fuel peat yearly; the proportion of peat in the total energy balance has decreased from 14.5% in 1970 to less than 2% in the 1990s. After World War II peat was predominantly used for power generation, but a gradual changeover to fuel oil and natural gas in power plants led to a decrease in the production and use of fuel peat. At the present time the Belarussian peat industry concentrates on the production of peat briquettes, which are used mainly as fuel in municipal plants and in private houses. There are over 30 peat briquette plants in Belarus, which produce almost 2 million tonnes of peat briquettes each year. The share of peat briquettes in house heating in the 1990s was some 40–45%, the rest being firewood and coal.

In Ireland a significant development has occurred in the energy sector during the past few years. About 3 million tonnes of milled peat are used for power generation in four condensation power plants, one of which was built a couple of years ago by the Finnish company Fortum Oy. Three other power stations owned by the Irish Electricity Supply Board are being

2004 Survey of Energy Resources

phased out and replaced by two modern plants. The combined capacity of the three plants will be 370 MW_e. When all these plants are in operation, the use of milled peat will rise to 3.8 million tonnes per year. In addition to the electricity generation sector, 600 000 tonnes of sod peat and 260 000 tonnes of peat briquettes are used domestically for house heating.

In Finland, the use of fuel peat has increased rapidly since 2001, after a 4-year fall in 1997–2000. In 2003 the 12-month cumulative use of fuel peat has been at a record level of about 30 million cubic metres or some 27 TWh. The main reason for this development has been the cold winter of 2002–2003 and low rainfall in the Nordic countries. The market price for electricity has risen significantly and peat has been used more for condensation power generation. In Finland there is one condensation power plant and five larger CHP plants, in which condensation electricity can also be produced. The need for peat in heat and power generation has also increased owing to a shortage of wood fuel, the demand for which has greatly increased as a result of Finland's new energy policy. In Finland it is understood that increasing the proportion of local fuels in the total energy balance is possible only by promoting the joint combustion of peat and wood fuel and, to an increasing degree, also Canary grass cultivated on cutaway peatlands.

Energy Policy of the European Union

On 19 November 2003 the European Commission approved a proposal of the Swedish Government to include peat, used as fuel in CHP plants, in the official list of certified energy sources. The reasons for this important decision were promotion of the use of power production in CHP plants, increasing the use of wood fuel in combination with peat, lower emissions from peat compared with those from fossil fuels and technical (as well as economic) advantages of the joint combustion of wood and peat.

This decision of the European Commission was an indication of a positive attitude towards the use of peat in the energy sector. Partly, it can be understood from the background that the share of imported fuels is gradually increasing in the European Union and that the EU Commission supports efforts to increase the utilisation of local energy sources. In Sweden and Finland, as well as in the Baltic States, soon to become EU members, peat resources are very large and their use with other biomass is justified during future decades.

A real threat to the use of peat in the energy sector is constituted by carbon trading, which will start within the European Union at the beginning of 2005. If national solutions, such as taxation, are not used to balance the situation, calculated emissions from peat combustion based on the IPCC system will lead to a very high cost of power and heat production from peat. Depending on the price of CO_2 emissions in the carbon trade, it is estimated that the use of peat as fuel might, in the worst case, fall to one-third of the present level. Under Finnish circumstances, the only realistic alternative to compensate the use of peat in such a case would be coal, which would mean a return to the situation prevailing in Finland some 30 years ago.

Wise Use of Peatlands

Especially in Europe, strong objections have been voiced to the industrial utilisation of peatlands and many campaigns were organised in the 1990s to stop peat harvesting. In the United Kingdom, anti-peat campaigns have already led to the conservation of large areas of peat production sites. Now organisations for the protection of nature have focussed their attention on the Baltic States, claiming that they are destroying all their peat resources within a very short period of time, a claim which is far from the truth.

It is known that large areas of peatlands have been degraded in many countries around the world, but it is estimated that about 80% of the world's peatlands are still in their natural state. The worst situation prevails in Central Europe,

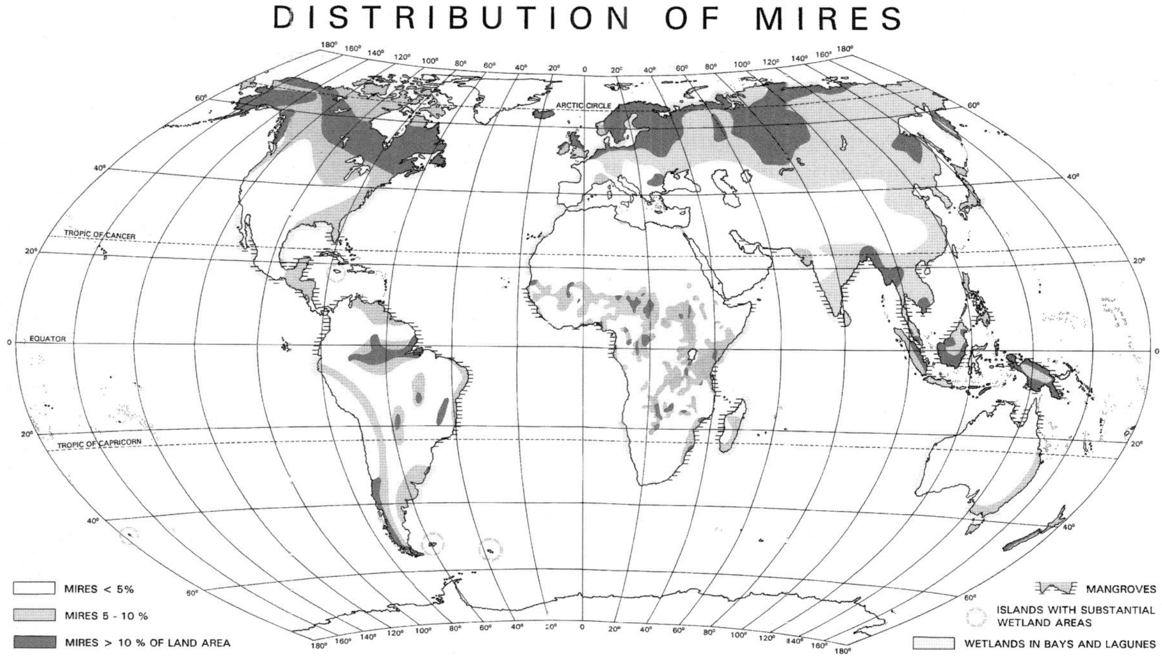

FIGURE 8.1 Distribution of mires (Source: International Peat Society).

where a large part of the natural peatlands has been utilised for various purposes during the past centuries. In the border states of the European Union, where large untouched mire areas still exist, the situation is, however, much better.

In order to avoid continuous conflicts between the interests of industry and conservation, the International Peat Society and the International Mire Conservation Group published in November 2002 a book called 'Wise Use of Mires and Peatlands'. In this publication, guidelines for the wise management of peatland ecosystems are introduced. Strict nature protection legislation in most of the countries nowadays forms a good 'shield' against unconsidered use of peatlands and secures the conservation of valuable nature monuments for future generations. There are good reasons to assume that, in countries with reasonably large peat resources, an acceptable balance between nature conservation and controlled use of peatlands for energy and horticulture will be achieved.

Raimo Sopo
International Peat Society

References

Brandel, M., 2003. Elcertifikaten I hamm, Press Release. Swedish Peat Producers' Association, 26 November 2003.
Hourican, J.D., 2003. The present situation and the outlook for the peat industry in Ireland. Future development of the production and use of peat. Proceedings of the 60th Anniversary Seminar of the Association of the Finnish Peat Industry, Jyväskylä, Finland.
International Peat Society. Peat Production Survey, Statistical Data for Various Years, Jyväskylä, Finland.
Lis, A.V. Use of Peat for Energy Purposes, Handout of Beltopgaz, Belarus.
Reedik, H., 2003. Peat production in the Baltic countries. Proceedings of the 60th Anniversary Seminar of the Association of the Finnish Peat Industry, Jyväskylä, Finland.
Sopo, R., 2003. Peat Campaign Continues, Peatlands International 2/2003. International Peat Society, Jyväskylä, Finland.

DEFINITIONS

Peat is a soft organic material consisting of partly decayed plant matter together with deposited minerals.

For the purposes of Table 8.1, **Peatland** is defined as follows: for land to be designated as peatland, the depth of the peat layer, excluding the thickness of the plant layer, must be at least 20 cm on drained and 30 cm on undrained land.

Peatland reserves are most frequently quoted on an area basis because initial quantification normally arises through soil survey programmes or via remotely sensed data. Even where deposit depths and total peat volumes are known, it is still not possible to quantify the reserves in energy terms because the energy content of in situ peat depends on its moisture and ash contents. However, the organic component of peat deposits has a fairly constant anhydrous, ash-free calorific value of 20–22 MJ/kg, and if the total quantity of organic material is known, together with the average moisture and ash contents, then the peat reserve may be equated with standard energy units.

The definitions applicable to peat resources and reserves (as discussed in some of the Country Notes) are as follows:

Proved amount in place is the tonnage that has been carefully measured and assessed as exploitable under present and expected local economic conditions, with existing available technology.

Proved recoverable reserves are the tonnage *within* the proved amount in place that is recoverable under present and expected local economic conditions, with existing available technology.

Estimated additional amount in place is the indicated and inferred tonnage *additional to* the proved amount in place that is thought likely to exist in unexplored extensions of known deposits or has been inferred from geological evidence. Speculative amounts are not included.

Estimated additional amount recoverable is the tonnage *within* the estimated additional amount in place that geological and engineering information indicates with reasonable certainty might be recovered in the future.

Types of Peat Fuel

There are three main forms in which peat is used as a fuel:

- **Sod peat**—slabs of peat, cut by hand or by machine, and dried in the air; mostly used as a household fuel;
- **Milled peat**—granulated peat, produced on a large scale by special machines; used either as a power station fuel or as raw material for briquettes;
- **Peat briquettes**—small blocks of dried, highly compressed peat; used mainly as a household fuel.

TABLE 8.1

Peat: areas of peatland

	Thousand hectares
Algeria	22
Angola	10
Burundi	14
Congo (Brazzaville)	290
Congo (Democratic Republic)	40
Côte d'Ivoire	32
Egypt (Arab Republic)	46
Guinea	525
Kenya	160
Liberia	40
Madagascar	197
Malawi	91
Mozambique	10
Nigeria	700
Rwanda	80
Senegal	7
South Africa	950
Sudan	100
Tunisia	1
Uganda	1 420
Zambia	1 106
Total Africa	**5 841**
Belize	90
Canada	111 328
Costa Rica	37
Cuba	658
El Salvador	9
Haiti	48
Honduras	453
Jamaica	12
Mexico	1 000
Nicaragua	371
Panama	5
Puerto Rico	10
Trinidad & Tobago	1
United States of America	21 400
Total North America	**135 422**
Argentina	50
Bolivia	1
Brazil	1 500
Chile	1 047
Colombia	339

TABLE 8.1 (Continued)

	Thousand hectares
Falkland Islands	1 151
French Guiana	162
Guyana	814
Paraguay	50
Peru	10
Surinam	113
Uruguay	3
Venezuela	1 000
Total South America	**6 240**
Afghanistan	12
Armenia	3
Bangladesh	60
Brunei	10
China	1 044
Georgia	25
India	100
Indonesia	27 000
Japan	200
Korea (Democratic People's Republic)	136
Korea (Republic)	630
Malaysia	2 536
Myanmar (Burma)	965
Pakistan	2
Philippines	240
Sri Lanka	5
Thailand	64
Turkey	56
Vietnam	100
Total Asia	**33 188**
Albania	10
Austria	22
Belarus	2 397
Belgium	20
Bulgaria	3
Czech Republic	27
Denmark	142
Estonia	902
Finland	8 900
France	100
Germany	1 420
Greece	10
Hungary	100

TABLE 8.1 (*Continued*)

	Thousand hectares
Iceland	1 000
Ireland	1 180
Italy	120
Latvia	640
Lithuania	483
Netherlands	280
Norway	2 370
Poland	1 200
Portugal	20
Romania	7
Russian Federation	56 800
Slovakia	4
Slovenia	100
Spain	38
Sweden	6 400
Switzerland	22
Ukraine	1 008
United Kingdom	1 926
Total Europe	**87 651**
Iran (Islamic Republic)	290
Iraq	1 790
Israel	5
Total Middle East	**2 085**
Australia	15
Fiji	4
New Zealand	260
Papua New Guinea	685
Total Oceania	**964**
Total World	**271 391**

Notes:

(1) Data for African countries are as given by *Global Peat Resources* and relate to total mire areas, which 'include coastal mangroves and other wetlands without any information about the thickness of peat or other organic soils'.

(2) The peatland area shown for Slovenia also includes those in Bosnia–Herzegovina, Croatia and Serbia & Montenegro.

(3) The peatland area shown for Australia is as reported by the Australian WEC Member Committee for the 1995 *Survey of Energy Resources*; mangrove swamps, tidal marshes and salt flats are excluded.

(4) Sources: WEC Member Committees; Lappalainen, E. (Ed.), 1996. *Global Peat Resources*. International Peat Society, Finland.

TABLE 8.2

Peat: 2002 production and consumption for fuel

	Production (thousand tonnes)	Consumption (thousand tonnes)
Burundi	13	13
Total Africa	**13**	**13**
Argentina	N	N
Falkland Islands	12	12
Total South America	**12**	**12**
China	600	600
Indonesia	536	520
Total Asia	**1 136**	**1 120**
Austria	1	1
Belarus	2 053	2 000
Denmark	N	N

TABLE 8.2 *(Continued)*

	Production (thousand tonnes)	Consumption (thousand tonnes)
Estonia	682	468
Finland	6 766	7 838
France	N	N
Germany	145	10
Ireland	2 709	3 605
Latvia	250	240
Lithuania	38	38
Norway	N	N
Poland	N	N
Romania	15	15
Russian Fed.	2 500	2 000
Sweden	1 071	1 330
Ukraine	454	450
UK	24	23
Total Europe	**16 708**	**18 018**
Total World	**17 869**	**19 163**

Notes:

(1) Data on production relate to peat produced for energy purposes; data on consumption (including imported peat) similarly relate only to fuel use.

(2) Annual production of peat in individual countries tends to vary considerably from year to year; the peat drying process is highly dependent on the weather, with below-average sunshine and/or wind, or above-average rainfall, depressing output (and vice versa). Demand for peat is generally much more stable than production: the resulting surpluses or deficits are borne by buffer stocks of dried peat.

(3) Data for China relate to 1990 and for Indonesia to 1996.

(4) Tonnages are generally expressed in terms of air-dried peat (35–55% moisture content).

(5) Sources: International Peat Society (Energy Peat Working Group), plus estimates based on pre-2002 data in *Energy Statistics Yearbook*, 2000, United Nations; *Survey of Energy Resources* 1992, 1998 and 2001; *Energy Statistics of OECD Countries 2000–2001*, *Energy Statistics of Non-OECD Countries 2000–2001*, International Energy Agency.

COUNTRY NOTES

The Country Notes on Peat have been compiled by the editors, drawing principally upon the following publications:

- Lappalainen, E. (Ed.), 1996. *Global Peat Resources*. International Peat Society, Finland.
- Couch, G.R., 1993. *Fuel Peat—World Resources and Utilisation*. IEA Coal Research, London.

Information from other sources (including WEC Member Committees) has been incorporated when available.

Argentina

There are some 500 km^2 of peat bogs on the Isla Grande de Tierra del Fuego at the southern tip of the republic. These deposits constitute some 95% of Argentina's peatlands: other peat bogs exist in the highland valleys of the Andean Cordillera and in other areas. However, economic exploitation of peat is almost entirely confined to Tierra del Fuego, where relatively small amounts (circa 3 000 m^3 per annum) are extracted, almost entirely for use as a soil-improvement agent. Consumption of peat for fuel is currently negligible.

Belarus

The peatlands of Belarus are by far the most extensive in Eastern Europe (excluding the Russian Federation), amounting to 24 000 km^2. The largest areas of peat formation are in the Pripyat Marshes in the south and in the central area around Minsk. Peat has been used as a fuel for many years, with the highest consumption during the 1970s and 1980s. The use of peat as a power station fuel ceased in 1986; fuel output in recent years has been largely confined to the production of peat briquettes, mainly for household use.

Out of a total fuel peat production of around 2 million tonnes per annum, deliveries to briquetting plants account for about 85%. Consumption of peat by heat plants amounts to about 200 000 tpa, with the balance of peat supply either being exported or consumed by a variety of small-scale consumers. Current annual output of peat briquettes is approximately 1.1 million tonnes, of which over 80% is directly consumed by residential users.

Brazil

The area of peatland has not been precisely established but it is believed to be at least 15 000 km^2, which makes it the largest in any South American country. There are extensive deposits in the Middle Amazon and in a large marshy plain (Pantanal) near the Bolivian border. Smaller areas of peatland exist in some coastal locations; those in the industrialised south-east of Brazil (in the states of Espírito Santo, Rio de Janeiro and São Paulo), and further north in Bahia state, have attracted interest as potential sites for the production of peat for energy purposes. The Irish peat authority Bord na Móna carried out preliminary surveys in Brazil in the early 1980s but no production of peat for fuel has yet been developed.

The total amount of peat in situ has been estimated as 25 billion tonnes. According to the Ministry of Mines and Energy, measured/indicated/inventoried resources of peat amounted to just over 129 million tonnes at end-1999, with an inferred/estimated additional amount of 358 million tonnes.

Burundi

There are appreciable areas of peatland, totalling about 140 km^2. The principal known deposits lie beneath the Akanyaru swamp complex in northern Burundi: these cover about 123 km^2 and are estimated to contain 1.42 billion cubic metres of peat in situ. The proved amount in place (expressed in terms of recoverable dry peat) was reported in 1992 to be 56 million tonnes.

Peat has been proposed as an alternative fuel to wood, in order to reduce deforestation, and a number of surveys have been conducted. Fuel peat is currently produced by semi-manual methods at four locations, but usage of the resource remains predominantly for agricultural purposes. The IEA estimates annual production and consumption of fuel peat as 13 000 tonnes.

Canada

Canada's peatlands are estimated to exceed 1.1 million km^2 in area and are the largest in the world.

There have been a number of assessments of the potential for using peat as a fuel (including for power generation) but at present there is virtually no use of peat for energy purposes and none is likely in the immediate future. Canada is, however, a major producer (and exporter) of peat for horticultural applications.

China

Peatlands totalling some 10 000 km^2 are quite widely distributed but do not have a high overall significance in China's topography, accounting for only about 0.1% of the country's land area. The principal peat areas are located in the region of the Qingzang Plateau in the south-west, in the north-east mountains and in the lower Yangtze plain in the east.

Peat has been harvested for a variety of purposes, including fuel use, since the 1970s. Some is used in industry (e.g. brick-making), but the major part of consumption is as a household fuel. Peat has been reported to be sometimes mixed with animal dung as input to biogas plants. No information is available on the current level of peat consumption for fuel. The Chinese WEC Member Committee reported production and consumption of 600 000 tonnes in 1990 for an earlier *Survey*.

Denmark

Human activities, chiefly cultivation and drainage operations, have reduced Denmark's originally extensive areas of peatland from some 20–25% of its land area to not much more than 3%. Out of a total existing mire area of some 1 420 km^2, freshwater peatland accounts for about 1 000 km^2, the remainder consisting of salt marsh and coastal meadow. Commercial exploitation of peat resources is at a low level: in 1995 the area utilised was some 1 200 ha, producing about 100 000 tonnes per annum. Almost all the peat produced is used in horticulture; fuel use is negligible.

Estonia

Peatlands are a major feature of the topography of Estonia, occupying about 22% of its territory. They are distributed throughout the country, with the largest mires being located on the plains.

Out of a total peatland area of over 9 000 km^2, commercial extraction of peat takes place on about 160 km^2. More than half of the output is used for horticultural purposes: the use of peat for fuel is currently in the order of 470 000 tonnes per annum, cut from about 60 km^2 of peat bogs. Most of the peat is consumed in the form of briquettes—there are three briquetting plants, each with an output capacity of 120 000 tonnes per year. In 2002 briquette production totalled 128 000 tonnes; 110 000 tonnes of briquettes were exported, the balance being very largely consumed in the residential sector. Most of the consumption of un-briquetted peat is accounted for by district heating and electricity generation. Some sod peat (about 80 000 tonnes in 2002) is exported.

Finland

With their total area of some 89 000 km^2, the Finnish peatlands are some of the most important in Europe and indeed globally—Finland has the highest proportion of wetlands of any nation in the world. Peat deposits are found throughout Finland, with a greater density to the west and north of the country.

The area of peat potentially suitable for commercial extraction is 6 220 km^2, of which about 22% contains high-grade peat suitable for horticulture and soil improvement. The remaining 78% (together with other deposits from which the surface layers have been harvested for horticultural use) is suitable for fuel peat production. In 1995, the total area used for peat production was only 530 km^2, from which 25.8 million m^3 were extracted for fuel use and 2.1 million m^3 for non-energy uses.

In 2001, CHP plants accounted for 49%, and power stations 32%, of the total national consumption of fuel peat; industrial users consumed 15%, the balance being used in heat plants (3%), and directly in the residential and agricultural sector (1%).

Germany

The majority of around 14 000 km^2 of peatlands is in the northern Länder of Lower Saxony, Mecklenburg–West Pomerania and Brandenburg. Most of Germany's fens have been drained, the land being used for agriculture, mainly grassland farming.

Out of the total area covered by raised bogs, approximately 60% is farmed, with only a small proportion (less than 10%) exploited for peat production. Energy use of peat is reported to be very limited at present, virtually all production being destined for agricultural/horticultural uses or for the manufacture of activated carbon. A small amount of energy-grade peat is exported.

Greece

Despite the drainage of large stretches of former fenland, and the loss of much peat through oxidation and self-ignition, peat resources in Greece are still quite considerable. The largest deposits are in the north of the country, at Philippi in eastern Macedonia and Nissi in western Macedonia. The Philippi peatland covers about 55 km^2 and is nearly 190 m deep—the thickest known peat deposit in the world.

Fuel Peat: World Resources and Utilisation quotes total reserves as 4 billion tonnes: the proportion of this amount that might be suitable for fuel use is indeterminate.

Peat resources in Greece have not so far been commercially exploited, either for use as fuel or for agricultural, horticultural or other purposes. Schemes for peat-fired electricity generation at Philippi and Nissi have been proposed in the past, but have subsequently been abandoned.

Iceland

Peatlands cover some 10 000 km^2 or about 10% of Iceland's surface area; the ash content of the peat is usually high (10–35%), owing to the frequent deposition of volcanic ash. Although peat has traditionally been used as a fuel in Iceland, present-day consumption is reported as zero. In the past, an important non-energy application of peat consisted of the use of 'peat bricks' in the construction of buildings.

Indonesia

The peatlands are by far the most extensive in the tropical zone (estimated as up to 270 000 km^2) and rank as the fourth largest in the world: they are located largely in the sub-coastal lowlands of Kalimantan and Sumatra. A feasibility study was carried out between 1985 and 1989 regarding the use of peat for electricity generation in central Kalimantan; no project resulted, but a small peat-fired power plant has operated in southern Sumatra.

Ireland

More than 17% of the republic's land surface is classified as peatland. Peat deposits totalling nearly 12 000 km^2 are widely distributed, being especially prominent along the western seaboard and across the Midland Plain in the centre of the island. Domestic consumption of peat for energy purposes in Ireland dates back to prehistoric times, with documentary evidence of its use existing from as early as the 8th century. After large stretches of the island's forests were cleared in the 17th century, peat became the only fuel available to the majority of households.

Mechanical methods of extraction were adopted on a large scale following World War II, both for the production of milled peat (used as a power-plant fuel and in the manufacture of peat briquettes) and to replace manual cutting of sod peat for household use. Out of current annual consumption of peat for energy purposes, nearly 70% is used in power stations and heat plants, 16% is briquetted and 13% consists of sod peat, used predominantly as a residential fuel. Peat briquettes are also almost all used as household fuel.

Since its foundation in 1946, the Irish Peat Development Authority (Bord na Móna) has promoted the economic development of Ireland's peat resources. A number of power stations and briquetting plants have been built near peat deposits. A programme is under way to replace several old peat-fired power plants with more efficient and more environmentally friendly peat-fired power plants. The first of the new stations, built by Edenderry Power Ltd near Clonbulloge, County Offaly, with a net output capacity of 118 MW, was commissioned in November 2000. It consumes approximately 1 million tonnes of milled peat per annum. The other new stations are being constructed at Lough Ree (100 MW), which is scheduled to replace the existing Lanesboro station in October 2004, and West Offaly (150 MW), which is due to replace Shannonbridge in January 2005.

Production of fuel peat in 2002 (as reported by the IPS Energy Peat Working Group) was about 2.7 million tonnes, with consumption of around 3.6 million tonnes, both at 40% moisture content.

Italy

There are significant resources of peat (circa 1 200 km^2) in Italy, mostly in Piedmont, Lombardia and Venezia in the north of the country. *Fuel Peat: World Resources and Utilisation* gives the estimated reserves as 2.5 billion tonnes: the proportion of this amount that might be suitable for fuel use is indeterminate.

Although peat has been used for fuel during the past, notably in the context of wartime shortages of other sources of energy, no present-day usage has been reported.

Latvia

Peatlands cover about 6 400 km^2, or almost 10% of Latvia's territory, with the major deposits being located in the eastern plains and in the vicinity of Riga.

Peat has been used in agriculture and as a fuel for several hundred years: output peaked in 1973, when fuel use amounted to 2 million tonnes. By 1990, the tonnage of peat extracted had fallen by 45% and fuel use was down to only about 300 000 tonnes. Consumption has tended to decline in recent years, with deliveries to CHP plants accounting for much the largest part. Small tonnages of peat are consumed by heat plants and in the production of peat briquettes for household use.

Lithuania

Peatlands (totalling nearly 5 000 km^2) are widespread, with the larger accumulations tending to be in the west and south-east of the country. Fuel use of peat fell from 1.5 million tonnes in 1960 to only about 0.1 million tonnes in 1985, since when consumption has declined further to around 40 000 tonnes per year. The principal peat consumers are heat plants, briquetting plants and households; the last-named also account for virtually all Lithuania's consumption of peat briquettes.

Norway

Although there are extensive areas of essentially undisturbed peatland, amounting to nearly 24 000 km^2, peat extraction (almost all for horticultural purposes) has been at a relatively low level in recent years.

Peat had traditionally been used as a fuel in coastal parts of the country; unrestrained cutting led to considerable damage to the peatland, which in 1949 resulted in legislation to control extraction.

Poland

The area of peatland is some 12 000 km^2, with most deposits in the northern and eastern parts of the country.

Much use was made of peat as a fuel in the years immediately after World War II, with some production of peat briquettes and peat coke; by the mid-1960s fuel use had, however, considerably diminished. Current consumption of peat is virtually all for agricultural or horticultural purposes.

Romania

There are just over 70 km^2 of peatlands. Peat production for energy purposes has been only a few thousand tonnes per annum in recent years, with consumption confined to the residential and agricultural sectors.

Russian Federation

According to *Global Peat Resources*, the total area of peatlands is some 568 000 km^2: the deposits are widely but unevenly distributed throughout the Federation. The principal peat areas are located in the north-western parts of European Russia, in West Siberia, near the western coast of Kamchatka and in several other far-eastern regions. The Siberian peatlands account for nearly 75% of the Federation total.

Total peat resources are quoted in *Global Peat Resources* as 186 billion tonnes, second only to Canada's in world terms. Of the total, 11.5 billion tonnes have been the subject of detailed surveys and a further 6.1 billion tonnes have been preliminarily surveyed.

The bulk of current peat production is used for agricultural/horticultural purposes. Peat deposits have been exploited in Russia as a source of industrial fuel for well over a hundred years. During the 1920s, the use of peat for power generation expanded rapidly, such that by 1928 over 40% of Soviet electric power was derived from peat. Peat's share of power generation has been in long-term decline, and since 1980 has amounted to less than 1%.

Approximately 5% of the exploitable peat deposits are used for fuel production, which currently amounts to around 2.5 million tonnes per annum.

Sweden

In Western Europe, the extent of Sweden's peatlands (64 000 km^2 with a peat layer thicker than 30 cm) is second only to Finland's: the deposits are distributed throughout the country, being particularly extensive in the far north.

According to data reported to the IEA, peat production in recent years has averaged about 0.9 million tonnes per annum. In 2001, CHP plants accounted for 62% of total consumption, heat plants for 35% and industrial users for the remaining 3%.

The largest peat-production unit is located at Sveg, central Sweden, at an altitude of over 400 m; it supplies a nearby briquetting plant, the only one in the country. This plant has an output capacity of about 300 000 tonnes per annum: production of briquettes (made from a mixture of peat, sawdust and wood chips) is currently about 220 000 tpa.

The use of peat as a household fuel has never been of much significance. Production of peat for industrial energy use began during the 19th century and, after reaching a peak level during World War II, declined to virtually zero by 1970. Use of peat as a fuel for power stations and district heating plants started in the mid-1980s and now constitutes by far the greater part of consumption.

Ukraine

There are over 10 000 km^2 of peatlands, more than half of which are located in Polesie, in the north of the country, where they account for 6.4% of the surface area. The other main area for peat deposits is the valley of the Dnieper, in particular on the east side of the river. Peat production rose during the period of the communist regime, reaching 7.5 million tonnes in 1970, when 73% was used in agriculture and 27% for fuel. In recent years consumption of peat for fuel purposes has fallen to less than 0.5 million tonnes per annum, all consumed by households.

United Kingdom

The peatlands of Great Britain cover an area of some 17 500 km^2, most deposits being in the northern and western regions; Scotland accounts for about 68% of the total area of peat, England for 23% and Wales 9%.

There are about 1 700 km^2 of peatland in Northern Ireland, mostly located in the western half of the province.

The total UK peatland area is nearly twice that of Ireland, but the extraction of peat is on a very much smaller scale: in Great Britain, commercialised peat extraction takes place on only some 5 400 ha (equivalent to about 0.3% of total peatland). Almost all peat industry output is for the horticultural market; there is, however, still quite extensive (but unquantified) use of peat as a domestic fuel in the rural parts of Scotland and Northern Ireland. About 20 000 tonnes per annum of air-dried sod peat is reported by the International Peat Society to be produced for energy purposes, some of which is exported to Sweden.

United States of America

In 1995 the total area covered by peat soils (known as histosols) was some 214 000 km^2, of which Alaska accounted for just over 50%. In the contiguous United States, the major areas of peat deposits are in the northern states of Minnesota, Michigan and Wisconsin, along the eastern seaboard from Maine to Florida and along the Gulf coastal region as far as Louisiana.

The potential uses of peat as fuel were evaluated during the 1970s; a Department of Energy study published in 1980 covered—in addition to direct combustion uses—the potential for producing liquid fuels from peat.

Interest in developing the use of peat for energy purposes has diminished since 1980. A small (23 MW) power plant was constructed in 1990 in Maine, to be fuelled by local peat. Initial problems associated with the use of inappropriate harvesting equipment were overcome but it was then difficult to obtain further permits to exploit the larger bog area required; the boilers were subsequently fuelled mainly by wood chips.

Chapter 9

Wood Fuels

COMMENTARY

A decade ago, the energy sector initiated a 'silent revolution' as a result of the deregulation and liberation policies implemented by many countries. The changes were followed more recently by reforms directed towards environmental protection and climate change mitigation (Kyoto Protocol, 11 December 1997) and poverty reduction (World Summit on Sustainable Development [WSSD], Johannesburg, 2–4 September, 2002). In fact, while the Kyoto Protocol opens up new opportunities for the development of renewable energy sources (including wood energy) as a major mechanism to displace fossil fuels use and to assist in the reduction of GHG emissions, the WSSD reaffirmed its commitment to sustainable development through 'The Johannesburg Declaration on Sustainable Development' and specifically addressed the bioenergy issue in several parts of the WSSD Plan of Implementation (PI), emphasising that 'access to energy facilitates the eradication of poverty'.

In this context, many changes can be observed in the energy sector. Large public utilities have now been fragmented and privatised. New smaller energy projects and initiatives have taken place, making use of local advantages in terms of expertise, manpower, facilities (and infrastructures), and resources (such as by-products) that were previously not possible to utilise. This 'revolution' opened up new opportunities for the development of renewable energy, in general, and wood energy, in particular. Thus, in addition to countries like Brazil, which pioneered the development of bioenergy as a local energy solution, countries such as Austria, Canada, Finland, Germany, the Netherlands, Sweden and the United States have also adopted new energy policies with an increased utilisation of wood energy within their 'energy mix'.

While wood energy is gradually penetrating the energy markets of industrialised countries as a clean, cost-effective and locally available source of energy, in developing countries wood-based fuels remain the dominant source of energy for over 2 billion people on low incomes.

The following sections provide an overview of the current wood energy situation and outline the major economic, environmental and social issues, which need to be addressed for the implementation of sustainable wood energy systems.

Overview of the Current Wood Energy Situation

Wood fuels consist of three main commodities: fuelwood, charcoal and black liquor. Fuelwood

2004 Survey of Energy Resources

and charcoal are traditional forest products derived from the forest, trees outside forests, wood-processing industries and recycled wooden products from society. Black liquors are by-products of the pulp and paper industry. See Fig. 9.1.

The following sections highlight the consumption of different wood fuel types. For the sake of simplification, wood fuel consumption is assumed to be more or less equal to wood fuel production.

How Much of the World's Energy Supply Comes from Wood?

Fig. 9.2 provides the amount of wood fuels production by region for 2002 derived from the FAO wood energy database (see Table 9.1 for detail by country).

In 2002, about 2 450 million m^3 (equivalent to 23 600 PJ) of wood fuels (fuelwood, charcoal and black liquor) were consumed worldwide for energy, which is about 560 million tonnes of oil

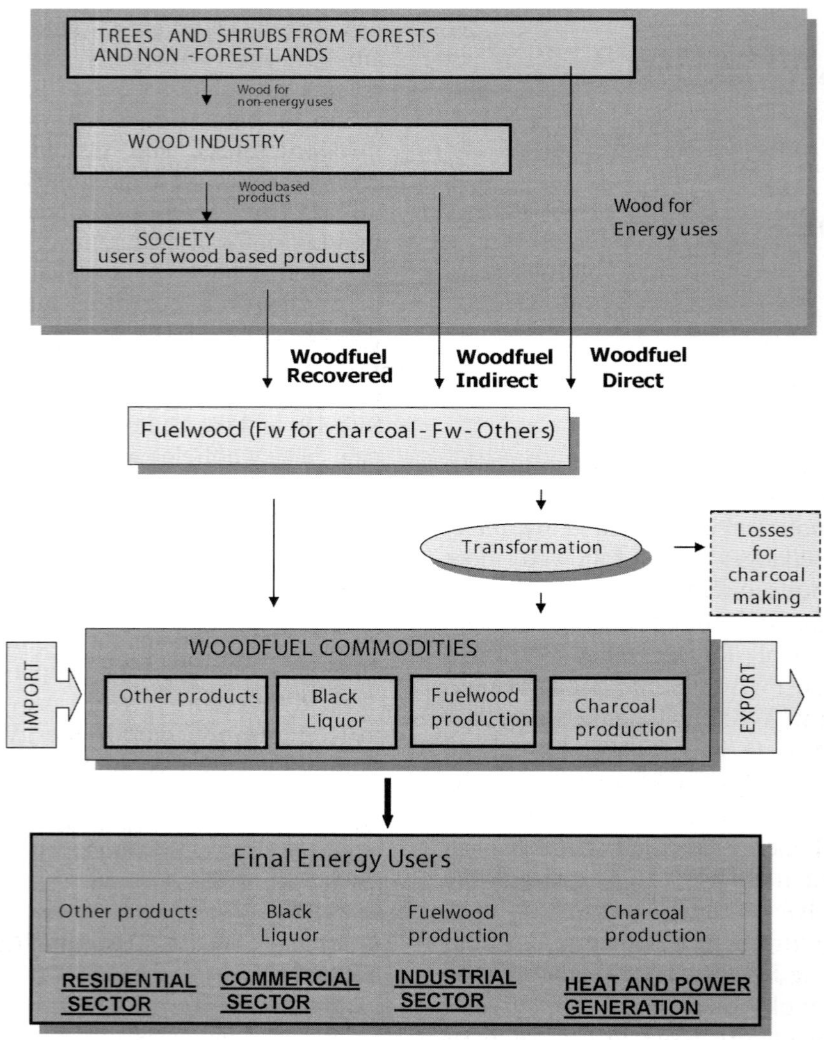

FIGURE 9.1 Wood fuel balance scheme from supply source to end-user (Source: FAO i-WEIS).

FUELWOOD	Million tonnes	Million m^3	PJ	Million toe	%
Africa	441	609	6 088	145.4	31.3
North and Central America	121	167	1 673	40.0	8.6
South America	111	153	1 528	36.5	7.9
Asia	671	925	9 254	221.0	47.6
Europe	58	81	806	19.2	4.1
Oceania	6	9	86	2.0	0.4
Total World	1 411	1 946	19 458	464.7	100.0

CHARCOAL	Million tonnes of charcoal	Million m^3 wood *	PJ	Million toe	%
Africa	15	89	453	10.8	50.5
North and Central America	2	12	64	1.5	7.1
South America	7	42	211	5.1	23.6
Asia	5	29	145	3.5	16.2
Europe	1	4	23	0.5	2.5
Oceania	0	0	1	0.0	0.1
Total World	29	176	897	21.4	100.0

* Estimated amount of wood used for charcoal production

BLACK LIQUOR	Million m^3 wood equivalent	PJ	Million toe	%
Africa	no data	no data	no data	no data
North and Central America	160	1 599	38.2	49.4
South America	60	601	14.4	18.6
Asia	41	414	9.9	12.8
Europe	59	592	14.1	18.3
Oceania	3	29	0.7	0.9
Total World	323	3 234	77.3	100.0

TOTAL WOODFUELS	Million tonnes	Million m^3	PJ	Million toe	%
Africa		698	6 541	156.2	27.7
North and Central America		340	3 335	79.7	14.1
South America		255	2 341	56.0	9.9
Asia		995	9 812	234.4	41.6
Europe		144	1 420	33.8	6.0
Oceania		12	115	2.7	0.5
Total World	1 771	2 443	23 589	563.4	100.0

Note : Total World figures include small amounts for Middle East

FIGURE 9.2 Estimated world total wood fuels consumption, 2002 (Source: FAO i-WEIS).

equivalent (mtoe) or about 5% of the world's total energy requirement.

Considering all the biofuels together, they represent 44.8% of world total renewables supply. The second largest renewable source is hydropower, providing 34.8% of renewable primary energy. With a 9.6% share, geothermal energy is the third largest renewable source[1].

The amount and types of wood fuel used vary considerably between regions, mainly due to different local situations and conditions. Asia is by far the largest consumer of fuelwood, with 48% of the world production. Africa has the next highest share at 31%, followed by North America (including Central America) and South America, at around 8.6 and 8%, respectively[2]. Europe consumes only around 4%, most of which is derived from the use of black liquor from the pulp and paper industries.

Africa is the most intensive user of charcoal, with around 50% of the total world charcoal production. Most of the charcoal is used by urban dwellers, although some quantities are used by commercial (restaurants) and industrial users (metallurgical applications). South America is also an important consumer, with approximately 24% of the total charcoal consumption. Brazil uses about 6 million tonnes of charcoal annually in its steel industry.

In Africa, almost all countries rely on wood to meet their basic energy needs. The share of wood fuels in African primary energy consumption is estimated at 60–86%, except in North African countries and South Africa[3]. On average, about 40% of the total energy requirement is met by fuelwood in Africa.

In Asia, about 7% of the total energy consumption is met by wood fuels. However, the wood energy situation varies from country to country. For example, many countries in South Asia and Southeast Asia, e.g. Nepal, Cambodia, Thailand and Indonesia, rely heavily on fuelwood: in Laos (PDR), Cambodia and Nepal, woodfuels supplied 97, 79 and 76%, respectively, of their total energy consumption[4]. In Latin America, about 10% of the total energy requirement is met by fuelwood.

In Europe, the share of fuelwood in total energy requirement is low, at 1.2%. However, in countries such as Finland and Sweden, per capita consumption is quite high if black liquor is included. In Austria, Finland and Sweden wood energy provides around 12–18% of their total primary energy supply (TPES)[5].

[1] IEA, 2003.
[2] FAO, 2001.
[3] FAO, 1999b, p. 1.

Household Wood Energy

Fuelwood, the most common form of wood fuel, continues to be widely used as a major source of energy for households, especially in developing countries. Charcoal is also increasingly used in many African countries by urban dwellers, as a result of a relentless process of migration of people from rural areas towards urban centres. The major energy end-use in households is cooking: about 86% of fuelwood consumed in urban households in India is for this purpose, while the rest is mostly used for water heating[6]. In Africa, more than 86% of total wood fuels consumption was attributed to the household sector in 1994[7]. Dependence on wood fuels to meet household energy needs is especially high in most of sub-Saharan Africa, where 90–98% of residential energy consumption is met from this source[8].

In short, the use of fuelwood and charcoal remains the dominant source of energy for most developing countries. It is estimated that over 2 billion poor people depend on fuelwood and/or charcoal for meeting their basic daily energy needs for cooking and heating. For them, wood fuels are not only vital to the nutritional stability of rural and urban households, but are also often essential in food-processing industries for baking, brewing, smoking, curing and electricity production.

[4] FAO, 2003, p. 23.
[5] FAO, 1997, p. 6.
[6] FAO, 1997a, pp. 25–26.
[7] FAO, 1999b, p. 12.
[8] FAO, 1999b, p. 1.

In places with high fuelwood and charcoal consumption (due to high population density with low income and/or severe climatic conditions) and weak supply sources, strong pressures are put on existing tree resources, and deforestation and devegetation problems remain a matter of great concern. In addition, urbanisation and economic development are bringing about changes in consumption patterns in developing countries, which in turn are leading to major changes in the household energy sector. A pronounced shift from fuelwood to charcoal, especially in Africa, is observed. This issue has raised concerns among environmentalists and those responsible for forest development and management because these charcoal-making activities are, in most cases, carried out illegally[9]. Moreover, they put a much higher pressure on natural forests than the extraction of fuelwood, which is often produced from trees outside forests or from other sources not involving the destruction of forests.

In developed countries, heat production by households also remains the major use of fuelwood. For instance, in the European Union wood fuels account for around 60% of the total wood energy consumed[10], although their utilisation as an industrial energy source for electricity and heat generation is increasing, as a result of new energy policies enacted in most countries to comply with climate change mitigation programmes.

Industrial and Commercial Wood Energy

Most of the non-household fuelwood consumption in developing countries is in commercial and industrial activities such as crop drying, tea processing and tobacco curing, as well as the brick and ceramic industries. Fuelwood consumption by these sectors is smaller than that in households; nevertheless, it cannot be overlooked, as it can constitute 10–20% of fuelwood use, as seen in some Asian countries[11]. In Africa, it is estimated that consumption of wood fuels in industry accounted for about 9.5% of the total in 1994[12].

In developed countries, fuelwood uses for electricity and heat generation at industrial sites or in municipal district heating facilities are rapidly rising as a substitute for fossil fuels. According to the information provided by the IEA, OECD countries' primary energy supply from renewable sources increased from 264.1 to 303.7 mtoe between 1990 and 2001, yielding an average annual growth of 1.3%. In 2001, renewable sources contributed 5.7% of TPES in the OECD.

Electricity generation from woody biomass grew from 59.5 to 79.6 TWh between 1990 and 2001, yielding a 2.7% average annual growth. As the second largest renewable electricity source after hydropower, solid biomass accounted for 5.6% of renewable electricity generation in 2001. This share is up from 4.6% in 1990. Of the electricity generated from solid biomass, 52.3% (41.6 TWh) is accounted for by the USA, where it makes up 14.6% of the country's renewable electricity production. The second largest producer of electricity from solid biomass is Finland (8.2 TWh), where it represents 37.8% of renewable electricity supply. Other big producers are Japan and Canada; in fact electricity is produced from solid biomass in most OECD member countries. This mainly involves the combustion of commercial woody fuels in modern devices—for example, wood-chip-fired co-generation plants for heat and power. Other applications are industrial heat supply and co-firing with coal for large-scale power generation[13].

Combustion is the most common way of converting biomass into energy. It is well understood, relatively straightforward and commercially available, and can be regarded as a proven technology. However, the desire to burn uncommon fuels, improve efficiencies, cut costs and decrease emission levels results in new technologies being continuously developed[14].

[9] FAO, 2002.

[10] FAO, 1997.

[11] FAO, 1997a, p. 28.

[12] FAO, 1999b, p. 12.

[13] IEA, 2003.

[14] IEA Bioenergy, Task 32: Biomass Combustion and Co-firing.

Many countries such as Slovenia, Croatia, Albania, the Czech Republic, Poland, Romania and Turkey are also large wood energy consumers. However, this issue is not properly recognised in their national energy and forestry statistics.

Charcoal continues to be used as an important industrial source of energy. For example, in Brazil, some 6 million tonnes of charcoal is produced every year for use in heavy industry, such as steel and alloy production[15]. The industrial demand for charcoal in the last few years has led to new, more efficient and larger-scale technologies, mainly aimed at improving charcoal yield and quality.

On the other hand, the production and consumption of black liquor (Fig. 9.2), which is a by-product of pulp and paper production, are concentrated in developed countries with large paper industries. In the pulp and paper industry, black liquors are widely used for heat and power production. Almost all this industry's energy needs are met by black liquors, with surplus electricity being sold to the public grid in some cases[16]. About 50% of black liquor consumption takes place in North America, followed by Europe and South America, both with about 19%, and Asia with 13%.

Some industry associations are also making special voluntary contributions to reduce the consumption of fossil fuels and increase the use of wood-based energy. For instance, the European paper industry aims to achieve an approximately 25% increase in the amount of wood fuel used for on-site heat and power production by 2010, and increase the share of wood in its on-site total primary energy consumption from 49 to 56%[17].

International Wood Fuel Trade

Although fuelwood is mainly a local source of energy, there are signs of an international trade in wood fuel developing between European and North American countries[18].

[15] FAO, 1995, p. 16.

[16] WEC/FAO, 1999, p. 120.

[17] CEPI. http://www.cepi.org/index.asp.

[18] FAO, 2003a.

How Much Wood is Used for Energy?

Approximately 61% (2 443 million m^3 for wood fuels versus 1 579 million m^3 of world industrial roundwood production in the year 2002[19]) of the world's total wood removals is used for energy purposes[20].

While only 24.7% of the wood produced in developed countries is used for energy (23% in Europe and 28.3% in North America), in developing countries that amount reaches 83%. In Africa, Asia and Latin America wood fuels account for 91, 82 and 69% of total wood consumption, respectively.

At the country level, the share of woodfuels within total wood consumption ranges from only 43% in Malaysia to 98% in Bangladesh, 95% in Nepal, 93% in Pakistan and in excess of 90% in most African countries.

In other words, energy is the main product of the forests, and energy and environmental policies which have been enacted present new opportunities for its further development. A combination of factors such as higher oil prices and technological developments in wood fuel production, transportation and combustion is also making wood fuels more attractive.

Issues for the Development of Sustainable Wood Energy Systems

The dynamics of wood fuel flows are complex and very site-specific. The development of sustainable wood energy systems remains one of the most critical issues to be addressed by policy makers and community planners in Energy Agencies and Forestry Services. With society giving increasing attention to sustainability issues, in the case of wood energy in both developing and developed countries,

[19] FAOSTAT, 2002.

[20] The percentage values reported in this section are based on total woodfuel consumption (i-WEIS—WETT estimates) as fraction of total wood removals (total woodfuel consumption + total industrial roundwood production (FAOSTAT, 2002)).

the following economic, environmental and social issues deserve particular attention[21].

Economic Sustainability

The development of an increased use of wood energy can have many positive economic benefits for the forestry sector including (among others):

- diversification of forestry activities, creating markets for forest wastes;
- improvement in the economic viability of silvicultural operations (e.g. thinning and harvesting);
- promotion of reforestation and afforestation activities, especially on marginal or unused land;
- redistribution of national income derived from the substitution of imported fossil fuels;
- creation of new incomes and jobs, specially in rural and forest areas.

As a result, a considerable amount of wood fuels (pellets and wood chips) is locally and internationally traded today. For instance, in Europe wood fuel trade had reached a level of almost 50 PJ/yr in 1999. This international trade has effects—both positive and negative—on the parties and actors involved, which need to be properly assessed and understood[22].

Environmental Sustainability

The production of fuelwood and charcoal may also cause positive and negative local environmental impacts, in particular the production of charcoal, which is still increasing in many developing countries. In addition, fuelwood derived from thinning and harvesting operations of forests is increasingly used in countries such as Sweden, Finland and Austria, using mechanised techniques. In such cases, forest productivity, biodiversity and climate change are also issues of major concern for the implementation of sustainable wood energy systems.

[21] WFC, 2003.
[22] FAO, 2003a.

Forest productivity relies on careful harvesting practices to reduce physical soil disturbance and compaction or the removal of organic matter layers on the soil surface, which can be seriously affected if certain rules are not properly applied. Science-based studies of site productivity and harvesting are now able to indicate which areas should *not* be harvested for wood fuel. Sites where nutrients are the primary limitation to tree growth should not be harvested, or harvesting should be limited to removal of stemwood. Avoiding harvesting on drought-stressed sites and limiting removals to once per rotation largely avoid the environmental impacts of harvesting.

Conservation of biodiversity is a central issue for forest management and a significant public policy issue. Experience has shown that a normal utilisation of residues as fuelwood after forest operations has little negative impact on biodiversity, while their energy use is environmentally beneficial because it displaces fossil fuels. However, intensive harvesting practices may lead to changes in wildlife habitats compared with conventional harvesting. Silvicultural techniques can overcome most of these effects through the connection of fragmented habitats by reforestation, alteration of the size of harvested areas, elimination of pesticides, encouragement of ground vegetation and the creation of a multi-aged, multi-species forest, which provides a diverse habitat for wildlife.

Climate change mitigation—greenhouse gas emissions can be considerably reduced using carbon substitution and sequestration options. Carbon substitution serves to combat climate change when wood fuels displace fossil fuels. In this case, the greenhouse gas balance of producing bioenergy is positive, so that the replacement of fossil fuels by wood fuels both reduces emissions and enhances carbon sequestration. On the other hand, carbon sequestration using short-rotation tree plantations or forests established on former agricultural land accumulates carbon in the soil (sinks).

Both carbon substitution and carbon sequestration can be easily implemented, together with many associated environmental benefits, as requested by the Kyoto Protocol. This is why wood energy and, more generally, bioenergy is so strongly promoted as a contributor to combat climate change.

Social Sustainability

The essence of social sustainability is how different societies can benefit from wood energy systems. Wood fuel production systems require people to operate them, thus creating jobs. Most residue-harvesting operations are conducted by contractors, who may supply fuelwood for a small district heating plant, or who collectively supply for larger plants. Although the number of jobs created is seriously affected by the level of mechanisation adopted, it is now clear that wood energy is the most labour-intensive technology and has the highest employment-creation potential among the renewable energy options. In addition, most of the jobs are primarily generated in rural areas, which is a key issue in line with 'The Johannesburg Declaration on Sustainable Development'[23].

On the other hand, long-standing problems of poor households using inefficient fuelwood and charcoal cooking stoves remain unsolved. This situation leads to poor indoor air conditions in kitchens and working places negatively affecting all the family members, especially women and children. Moreover, where these conditions prevail, the collection of fuelwood for household consumption is usually carried out by women and children, a task that is becoming more and more burdensome as the supply of fuelwood becomes scarce.

Institutional Issues

Nowadays the level of wood energy use is mainly a policy matter. Most of the technical, economic, environmental and social concerns can be easily overcome. Therefore, sound energy and forestry policies are vital for the development of sustainable wood energy systems and should make use of all the incentives available, especially those in climate change mitigation programmes. To be effective, these energy, environmental and forestry policies will need to be transparent, cost-effective in achieving their objectives and 'fair' as regards renewable versus non-renewable energy systems.

It is therefore vital that Governments and specialised governmental agencies create an environment in which realistic, coherent and flexible wood energy policies, oriented to poverty reduction and food security, can be established through improved wood energy services and an increased number of jobs, in line with the energy-related international negotiations of the Commission for Sustainable Development or the Millennium Development Goals.

The policies need to include adequate legislation and regulations to ensure sustainable production, marketing and use of wood as fuel. They should also abolish incentives and legislation that promote energy alternatives, which inhibit the competitiveness of locally available energy sources, such as wood fuels and other biomass. A sound institutional and regulatory infrastructure for all forest activities, not merely for wood energy harvesting, is therefore of paramount importance.

Governments and financial institutions have a key role to play in using fiscal levers to allocate investment funding to wood-based energy developments, not only on the demand side but also in support of those who supply wood for fuel. Forest owners (and farmers) need to be properly assisted and prepared to invest in new wood energy initiatives where new technologies and practices could complement conventional ones.

In addition, participation from all stakeholders, particularly from groups representing poor people's interests, needs to be encouraged. The change in attitude of people and NGOs to the utilisation of wood fuels (and to bioenergy in general) will also help to develop wood energy systems more capable of supplying the energy needed to keep economic activities running.

The multi-disciplinary character of wood energy systems requires that actions among the different agencies and ministries be coordinated, both at the national and the international level,

[23] FAO, 2003b.

and the forestry services and energy agencies have a key role to play.

Gaps in Wood Energy Statistics

The efforts undertaken during the last 5 years by FAO and other organisations such as the IEA and the WEC have considerably improved the information existing in the databases on wood energy[24].

In fact, wood energy terminology has been harmonised, facilitating data exchange between the major national and international organisations[25]. 'A guide for wood fuel surveys' is also available, which recommends simple methods to verify existing data rapidly, fill gaps in the information chain and conduct more reliable surveys.

In addition, assistance to countries is needed to improve their national wood energy information systems and enhance their capabilities for the development of wood energy policies, as well as to monitor the contribution of wood to energy needs and climate change mitigation[26].

Miguel Trossero
Rudi Drigo
Food and Agriculture Organization of the United Nations

[24] WEC, 1999.
[25] FAO, 2003c.
[26] WFC, 2003b.

References

CEPI, Confederation of European Paper Industries. Impacts of the increased use of renewable energy sources in Europe on the wood supply of the European pulp and paper industry (http://www.cepi.org/index.asp).

FAOSTAT 2002, 2002. FAO Forestry Department, Food and Agriculture Organization of the United Nations, Rome (http://apps.fao.org/page/collections?subset = forestry).

Forests, fuel and the future, Wood energy for sustainable development, FAO Forestry Department, Forestry Topics Report No. 5. Food and Agriculture Organization of the United Nations, Rome, 1995.

Girard, P., 2002. Unasylva 211, vol. 53. Food and Agriculture Organization of the United Nations, Rome.

Hall, P., 2003. IEA Bioenergy, Policy changes needed in Forestry Services, World Forestry Congress, Canada, October 2003.

Hillring, B., 2003a. International wood-fuel trade, FAO Forestry Department Working Paper. Food and Agriculture Organization of the United Nations, Rome, December 2003.

IEA Bioenergy, Task 32: Biomass Combustion and Co-firing.

Información y análisis para el manejo forestal sostenible: integrando esfuerzos nacionales e internacionales en 13 países tropicales en América Latina, FAO Forestry Department Working Paper. Food and Agriculture Organization of the United Nations, Rome, May 2001.

Renewables Information, IEA Statistics. International Energy Agency, Paris, 2003.

Socio-economic analysis of bioenergy systems: a focus on employment, FAO Forestry Department Working Paper. Food and Agriculture Organization of the United Nations, Rome, December 2003b.

The role of wood energy in Europe and OECD, FAO Forestry Department Working Paper FOPW/97/1. Wood Energy Today for Tomorrow (WETT) Regional Studies, Food and Agriculture Organization of the United Nations, Rome, 1997.

The role of wood energy in Asia, FAO Forestry Department Working Paper FOPW/97/2. Wood Energy Today for Tomorrow (WETT) Regional Studies, Food and Agriculture Organization of the United Nations, Rome, 1997a.

The role of wood energy in Africa, FAO Forestry Department Working Paper FOPW/99/3. Wood Energy Today for Tomorrow (WETT) Regional Studies, Food and Agriculture Organization of the United Nations, Rome, 1999b.

Trossero, M.A., 2003b. FAO Forestry Department, The Role of Forestry Services for the Development of Wood Energy, Barriers and Constraints, World Forestry Congress, Canada, October 2003.

Unified wood energy terminology, FAO Forestry Department Working Paper. Food and Agriculture Organization of the United Nations, Rome, December 2003c.

Wood Commentary, Paper prepared by Food and Agriculture Organization of the United Nations, Rome for the World Energy Council, 1999.

Wood Energy Information Analysis in Asia, FAO Forestry Department Working Paper. Food and Agriculture Organization of the United Nations, Rome, April 2003.

World Energy Outlook 2000. International Energy Agency, Paris, 2000.

DEFINITIONS

In Table 9.1, the following definition applies:

Fuelwood production is the tonnage of wood in the rough produced for direct use as a fuel or for conversion into charcoal. Wood residues recycled to energy use are excluded.

Table Notes

The data shown in Table 9.1 as estimated fuelwood production in 2002 are projections based on historical series in FAO's i-WEIS database. Apart from countries in the African region, for which fuelwood data were only available in terms of consumption, all countries shown in the table are based on production series. In each case, a number of trend-lines (generally exponential, polynomial and linear) were fitted to each series and one selected as a basis for extrapolation. The series available for analysis generally began in 1980 and extended to 1996/1997/1998, but in some instances obvious breaks in the series reduced the number of years available for trend-fitting.

Bearing in mind the uncertainty associated with virtually all wood energy statistics, the levels of fuelwood production shown in Table 9.1 should not be taken as definitive, precise assessments but as, in general, no more than indicative of the magnitude involved.

TABLE 9.1

Estimated fuelwood production in 2002

	Fuelwood production (million tonnes)
Algeria	1.7
Angola	4.9
Benin	2.8
Botswana	1.7
Burkina Faso	10.5
Burundi	6.2
Cameroon	10.0
Cape Verde	0.1
Central African Republic	2.6
Chad	1.8
Comoros	0.3
Congo (Brazzaville)	2.2
Congo (Democratic Rep.)	49.5
Côte d'Ivoire	7.3
Djibouti	0.1
Egypt (Arab Rep.)	3.1
Equatorial Guinea	0.4
Eritrea	3.2
Ethiopia	55.0
Gabon	0.8
Gambia	0.7
Ghana	6.1
Guinea	5.5
Guinea-Bissau	0.3
Kenya	18.0
Lesotho	1.3
Liberia	1.3
Libya/GSPLAJ	0.5
Madagascar	9.4
Malawi	8.2
Mali	6.1

TABLE 9.1 *(Continued)*

	Fuelwood production (million tonnes)
Mauritania	0.1
Mauritius	N
Morocco	7.2
Mozambique	17.5
Namibia	1.6
Niger	3.5
Nigeria	80.0
Réunion	N
Rwanda	4.0
São Tomé and Príncipe	0.1
Senegal	2.5
Seychelles	N
Sierra Leone	2.3
Somalia	2.6
South Africa	5.1
Sudan	5.1
Swaziland	0.7
Tanzania	49.0
Togo	2.7
Tunisia	2.4
Uganda	16.5
Western Sahara	N
Zambia	7.1
Zimbabwe	9.8
Total Africa	**441.4**
Bahamas	
Belize	0.1
Canada	4.3
Costa Rica	2.7

TABLE 9.1 (*Continued*)

	Fuelwood production (million tonnes)
Cuba	1.6
Dominica	N
Dominican Republic	3.3
El Salvador	5.7
Guadeloupe	N
Guatemala	11.0
Haiti	4.5
Honduras	6.1
Jamaica	1.2
Martinique	N
Mexico	20.0
Nicaragua	3.2
Panama	1.1
Puerto Rico	N
Trinidad & Tobago	N
United States of America	56.5
Other	N
Total North America	**121.3**
Argentina	3.6
Bolivia	1.2
Brazil	65.0
Chile	13.5
Colombia	11.5
Ecuador	3.1
French Guiana	N
Guyana	0.6
Paraguay	4.9
Peru	5.4
Surinam	0.1
Uruguay	1.2
Venezuela	0.7
Other	N
Total South America	**110.8**
Afghanistan	4.6
Armenia	N
Azerbaijan	N
Bangladesh	24.0
Bhutan	1.2
Brunei	0.1
Cambodia	5.6
China	149.0
Cyprus	N
Georgia	0.1

TABLE 9.1 (*Continued*)

	Fuelwood production (million tonnes)
India	207.0
Indonesia	121.0
Japan	0.9
Kazakhstan	0.2
Korea (DPR)	3.2
Korea (Republic)	0.3
Kyrgyzstan	N
Laos	0.3
Malaysia	6.0
Mongolia	0.1
Myanmar (Burma)	15.0
Nepal	16.5
Pakistan	24.5
Philippines	30.5
Singapore	N
Sri Lanka	7.3
Tajikistan	N
Thailand	24.5
Turkey	4.5
Turkmenistan	N
Uzbekistan	N
Vietnam	24.5
Other	N
Total Asia	**670.9**
Albania	0.3
Austria	2.8
Belarus	0.6
Belgium and Luxembourg	0.4
Bosnia-Herzegovina	0.5
Bulgaria	0.8
Croatia	0.9
Czech Republic	0.4
Denmark	0.4
Estonia	1.1
Finland	2.9
FYR Macedonia	0.5
France	7.1
Germany	2.2
Greece	0.7
Hungary	1.2
Ireland	0.1
Italy	4.0
Latvia	2.6
Lithuania	0.9

(*continued on next page*)

TABLE 9.1 (*Continued*)

	Fuelwood production (million tonnes)
Moldova	0.3
Netherlands	0.1
Norway	0.2
Poland	1.2
Portugal	0.4
Romania	2.7
Russian Federation	15.0
Serbia & Montenegro	N
Slovakia	0.2
Slovenia	0.5
Spain	2.3
Sweden	2.8
Switzerland	0.8
Ukraine	1.3
United Kingdom	0.2
Other	N
Total Europe	**58.4**
Iran (Islamic Republic)	1.4
Iraq	N

TABLE 9.1 (*Continued*)

	Fuelwood production (million tonnes)
Israel	N
Jordan	N
Lebanon	0.3
Syria (Arab Republic)	N
Total Middle East	**1.7**
Australia	2.0
Fiji	N
French Polynesia	N
New Caledonia	N
New Zealand	N
Papua New Guinea	4.0
Samoa	0.1
Solomon Islands	0.1
Vanuatu	N
Other	N
Total Oceania	**6.2**
Total World	**1 410.7**

COUNTRY NOTES

The Country Notes on Wood Fuels reflect the data and comments provided by WEC Member Committees in 2003, supplemented where necessary by information provided for the 2001 and 1998 WEC *Survey of Energy Resources* or derived from published sources.

Albania

Forestry/wood processing

Quantity of raw material available	237 thousand tonnes oil equivalent

Argentina

Forestry/wood processing

Quantity of raw material available	2.4 million tonnes
Direct use from combustion	23 095 TJ

Australia

Forestry/wood processing

Quantity of raw material available	~ 25 million tonnes
Yield of solid fuel (operational)	11 GJ/tonne
Electricity generating capacity	76 500 kW
Direct use from combustion	~ 66 000 TJ

Includes Tumut pulp and paper mill power plants, plus Maryvale pulp and paper and Visy's plant in Brisbane.

Direct combustion assumes 6 million tonnes of firewood used mainly for domestic heating. Calorific value of 11 GJ/tonne assumed.

Belgium

Black liquor/bark

Quantity of raw material available	0.2 million tonnes
Electricity generating capacity	31 000 kW
Electricity generation	585 TJ

Data refer to 1996.

Brazil

Forestry/wood processing[a]

Quantity of raw material available	6.99 million tonnes
Electricity generating capacity	188 213 kW
Electricity generation	4 748 TJ
Direct use from combustion	65 900 TJ
Total energy production	76 618 TJ

Black liquor

Quantity of raw material available	10.9 million tonnes
Electricity generating capacity	502 000 kW
Electricity generation	12 654 TJ
Direct use from combustion	102 828 TJ
Total energy production	130 921 TJ

[a] Wood and wood residues from paper and pulp industry.

Bulgaria

Forestry/wood processing

Direct use from combustion	23 037 TJ

Canada

Forestry/wood processing

Quantity of raw material available[a]	50.6 million tonnes
Solid fuel production capacity	6 240 TJ/yr
Yield of solid fuel	17.865 GJ/tonne
Solid fuel production	3 560 TJ
Electricity generating capacity	1 586 000 kW
Electricity generation	45 014 TJ
Direct use from combustion	548 000 TJ
Total energy production	596 574 TJ

[a] Comprising 17.7 wood waste + 21.9 black liquor + 11.0 cord wood.

Cote d'Ivoire

Firewood and charcoal constitute 60% of the national energy consumption. As well as household consumption, wood fuels are also used in restaurants, ironwork, bakeries, potteries, curing and drying feed.

Croatia

Wood residues	
Quantity of raw material available	0.845 million tonnes

Data refer to 1996.

Czech Republic

Forestry/wood processing	
Quantity of raw material available	2.6 million tonnes
Electricity generation	1 372 TJ
Direct use from combustion	15 338 TJ

Denmark

Forestry/wood processing	
Quantity of raw material available	0.6 million tonnes
Solid fuel production (wood pellets)	2 217 TJ
Electricity generation	801 TJ
Direct use from combustion	6 226 TJ
Total energy production	9 729 TJ

Egypt (Arab Republic)

Forestry/wood processing	
Quantity of raw material available	1.2 million tonnes

Estonia

Forestry/wood processing	
Quantity of raw material available	0.567 million tonnes
Solid fuel production	8 692 TJ

Finland

Forestry/wood processing	
Quantity of raw material available	21 million tonnes
Solid fuel production capacity	3 600 TJ/yr
Solid fuel produced (pellets)	1 800 TJ
Electricity generating capacity	2 000 000 kW
Electricity generation	25 000 TJ
Direct use from combustion	185 200 TJ
Total energy production	210 200 TJ

The present use of bioenergy is dominated by residues and by-products from the forest industry in Finland.

The share of black liquor (and other similar liquors) was 42% in the total use of renewable energy sources (excluding peat) in 2001. Bark, sawdust, etc. also have a significant share (about 24%).

In the main, forestry/wood-processing resources are exploited by co-generation plants producing electricity and heat. Heat energy produced in co-generation plants and heat energy from direct combustion plants are shown as direct use from combustion. Electricity produced mainly in co-generation plants is shown as electricity generation.

Firewood and harvesting residues are excluded.

France

Forestry/wood processing	
Electricity generation	4 772 TJ
Direct use from combustion	341 326 TJ
Total energy production	346 098 TJ

Germany

Forestry/wood processing	
Direct use from combustion	182 442 TJ
Wood waste, etc.	
Electricity generation	639 GWh

Data relate to 2001.

Hungary

Forestry/wood processing

Quantity of raw material available	1.167 million tonnes

Iran (Islamic Republic)

Forestry/wood processing

Quantity of raw material available	0.2 million tonnes

Italy

Forestry/wood processing

Direct use from combustion[a]	785 TJ

Data relate to 2001.

[a] Produced by district heating plants fuelled by residues of wood industries (alpine zones of northern Italy).

Japan

Forestry/wood processing

Quantity of raw material available	1.46 million tonnes
Electricity generating capacity	50 000 kW

Data relate to FY 1999.

Jordan

Forestry/wood processing

Quantity of raw material available	N million tonnes

Korea (Republic)

Forestry/wood processing

Direct use from combustion	3 542.5 TJ

Latvia

Forestry/wood processing

Quantity of raw material available	1.8 million tonnes
Solid fuel production capacity	28 600 TJ/yr
Yield of solid fuel	16 GJ/tonne
Solid fuel production	28 600 TJ
Direct use from combustion	28 600 TJ

Wood residues

Quantity of raw material available	1.3 million tonnes
Solid fuel production capacity	12 600 TJ/yr
Yield of solid fuel	9.7 GJ/tonne
Solid fuel production	12 600 TJ
Direct use from combustion	12 600 TJ

Lithuania

Forestry/wood processing

Quantity of raw material available	2.05 million tonnes
Electricity generating capacity	1 500 kW
Electricity generation	14 TJ
Direct use from combustion	27 481 TJ
Total energy production	27 495 TJ

The Forest Inventory and Management Institute estimates that forested land area accounts for about 30.6% (or about 20 000 km^2) of the country's territory. Total consumption of firewood and wood waste in 2002 was assessed at 3.351 million m^3 (or 656 800 toe). The major share of firewood was used directly by final consumers and the rest for the production of heat in heating plants.

Netherlands

Forestry/wood processing

Direct use from combustion	
Households	5 400 TJ
Industry	1 750 TJ

Paraguay

Forestry/wood processing

Quantity of raw material available	1.433 million tonnes
Direct use from combustion	20 511.3 TJ

Philippines

Forestry/wood processing	
Electricity generation	22 981 TJ

Portugal

Forestry/wood processing	
Quantity of raw material available	6 million tonnes
Biogas production	51 TJ
Electricity generating capacity	360 000 kW
Electricity generation	26 588 TJ
Direct use from combustion	52 111 TJ
Total energy production	78 750 TJ

Romania

Forestry/wood processing	
Quantity of raw material available	2.4 million tonnes
Direct use from combustion	30 000 TJ
Sawdust	
Quantity of raw material available	2.8 million tonnes
Biogas production capacity	N TJ/yr
Biogas production	N TJ
Direct use from combustion	45 000 TJ

In a country where it is reported that more than 6 million hectares are covered by forests, and even larger areas are devoted to agriculture, the biomass potential is estimated to be around 200 PJ/yr. In 2000 fuelwood (as a primary energy resource, including biomass) accounted for 115.84 PJ and in 2001 90 PJ. The majority (89%) of biomass is used for domestic heat production (traditional stoves) in rural areas. Only a small part is used in modern, small and medium-size boilers. The wood-processing industry uses about 11 million m^3/yr. It is reported that the wood industry produces in excess of 3 million tonnes of waste. Additionally, some 560 000 m^3 of sawdust and other wood waste resulting from wood processing are not yet being utilised for energy purposes. Instead they are discarded and thus threaten damage to the environment.

The biomass resources for heat are estimated to come from:

- sorting the industrial wood 12.8 PJ/yr;
- waste from primary processes in wood-processing industry 7.9 PJ/yr;
- waste from secondary processes in wood-processing industry 5.1 PJ/yr;
- waste as bark and sawdust 5.9 PJ/yr.

Serbia & Montenegro

Forestry/wood processing	
Quantity of raw material available	2 500 TJ
Direct use from combustion	3 500 TJ

Slovenia

Wood residues	
Quantity of raw material available	0.94 million tonnes
Yield of solid fuel	10 GJ/tonne
Solid fuel production	9 000 TJ
Electricity generating capacity	8 500 kW
Electricity generation	120 TJ

Data refer to 1996.

South Africa

Wood and forestry residues are used as a source of energy. However, no comprehensive analysis of the use of these wastes has been conducted.

Spain

Forestry/wood processing	
Quantity of raw material available	6.9 million tonnes
Electricity generating capacity	151 623 kW
Electricity generation	3 098.76 TJ

Direct use from combustion	99 708.18 TJ
Total energy production	102 806.94 TJ

Data based on work carried out for the preparation of the Plan for the Promotion of Renewable Energy in Spain, taking into account 1.4 mtoe forest residues and 1 mtoe wood agricultural residues. It is estimated that each tonne is equivalent to 0.35 toe.

Swaziland

Forestry/wood processing

Quantity of raw material available	0.63 million tonnes

Sweden

Energy crops (especially willow)	
Direct use from combustion	1 800 TJ
Black liquor	
Electricity generation	8 800 TJ
Direct use from combustion	126 000 TJ
Total energy production	134 800 TJ
Wood and industrial waste	
Electricity generation	7 000 TJ
Direct use from combustion	110 400 TJ
Total energy production	117 400 TJ

Switzerland

Waste paper, paper sludge

Quantity of raw material available	253 GWh
Electricity generation	26 TJ
Direct use from combustion	551 TJ
Total energy production	577 TJ

Taiwan, China

Forestry/wood processing

Quantity of raw material available	0.62 million tonnes

Black liquor

Direct use from combustion	8 171 TJ

Thailand

Forestry/wood processing

Electricity generating capacity	20 000 kW
Electricity generation	2 522.88 TJ
Direct use from combustion	52 283.2 TJ
Total energy production	54 806.08 TJ

Turkey

Forestry/wood processing

Electricity generating capacity	72 000 kW

United Kingdom*

Wood	
Production	204 ttoe
Direct use from combustion (domestic)	204 ttoe
Wood waste	
Production	266 ttoe
Input to heat generation	66 ttoe
Direct use from combustion (industrial)	200 ttoe

Domestic wood use includes the use of logs in open fires, cooker boilers and other wood-burning stoves. The figure given is an approximate estimate based on a survey carried out in 1989. A survey to provide current information was undertaken for 2001/2002. Unfortunately, the results proved inconclusive and estimates for domestic wood use remain unchanged.

In 1997, the industrial wood figure (which includes sawmill residues, furniture manufacturing waste, etc.) was included as a separate category for the first time. This was due to the availability of better data as a result of a survey carried out in 1996 on wood-fired combustion plants above 400 kW thermal input. A follow-up survey was subsequently carried out for 2000. This survey highlighted that there were fewer sites (174) operating than in 1996 due to the imposition of more stringent emissions control.

* Source: Digest of United Kingdom Energy Statistics 2003

United States of America

Forestry/wood processing

Quantity of raw material available	153 million tonnes

Uruguay

Forestry/wood processing

Quantity of raw material available	N million tonnes
Direct use from combustion	N TJ

Black liquor

Quantity of raw material available	0.05 million tonnes
Electricity generation	80.8 TJ
Direct use from combustion	607.1 TJ

Chapter 10

Bioenergy (other than Wood)

COMMENTARY

COMMENTARY

Introduction

Biomass resources are potentially the world's largest and most sustainable energy resource, comprising approximately 220 billion oven-dry tonnes (odt) (or c. 4 500 EJ) of annual primary production; the annual bioenergy potential is about 2 900 EJ, although only a fraction could realistically be used on a sustainable basis and at competitive prices. Estimates of the future potential contribution of bioenergy range from 67 to 450 EJ per annum.

Bioenergy is a broad term embracing a large range of feedstock. It can be classified into three main categories: (i) woodfuels, (ii) agrofuels, and (iii) fuels derived from urban waste. Bioenergy can also be classified according to a chosen technology route: (i) traditional applications (e.g. firewood and charcoal) and (ii) modern uses (e.g. electricity generation and combined heat and power (CHP)). Modern applications are rapidly replacing traditional uses, in particular in industrial countries, e.g. in Finland and Sweden 15–20% of their primary energy is generated from biomass. Bioenergy is not a transition fuel as it has often been portrayed, but a fuel that will continue to be the prime source of energy for many people for the foreseeable future.

Comparison with the WEC *Survey of Energy Resources 2001* Commentary

It is not possible to do justice in a few pages to such a complex issue as bioenergy. The SER 2001 commentary (Rosillo-Calle, 2001) presented an overview of bioenergy with particular attention to the use of residues, modernisation efforts to upgrade bioenergy and a brief discussion of its potential role in climate change mitigation. It also highlighted the main reasons why bioenergy had become increasingly important in the previous decade; such issues are still largely valid today. The present commentary is slightly different, e.g. the word 'biomass' has been replaced by 'bioenergy', for the sake of clarity; after a brief note on wood fuels, agrofuels and municipal solid waste (MSW) (the last named was excluded from previous commentaries), the text concentrates primarily

2004 Survey of Energy Resources

on non-woody biomass. It also assesses some of the most recent technological, policy and legislative changes, all of which are playing key roles in the successful penetration of bioenergy as a cost-effective and competitive energy option.

Continuing Difficulties with Data and Classification of Bioenergy

Previous commentaries have highlighted the problem of the lack of good-quality data. In recent years, thanks to considerable efforts by some international agencies (e.g. FAO, IEA) and national governments, data have improved significantly, particularly in industrialised countries. In poorer countries there are still serious problems and lack of data continues to hamper sound decision-making. This is largely caused by lack of financial and human resources for adequate data collection and analysis and by the informal nature of traditional bioenergy. Many WEC member countries are unable to produce adequate and reliable data on bioenergy or cannot supply any data at all.

The problem is particularly acute in respect of economic data, which are not readily available or are quoted in a way that makes comparisons very difficult. The inability to fully address the indigenous biomass resource capability and its likely contribution to energy and development is still a serious constraint on the full realisation of bioenergy's potential.

A further constraint is confusion with respect to terminology. FAO has been attempting to address this problem for some considerable time and after many consultations a document is currently being finalised which will hopefully solve some of these problems (see FAO/WE, 2003). It is a difficult problem because bioenergy is an integral part of many cultures. FAO classifies bioenergy into three main groups according to the nature of the biomass: (i) woodfuels; (ii) agrofuels and (iii) municipal by-products (see Fig. 10.1). FAO/WE (2003) contains a more detailed discussion of terminology. Bioenergy can also be classified according to the technological route: (i) traditional bioenergy (firewood, charcoal, residues) and (ii) modern bioenergy (associated with industrial wood residues, energy plantations, use of bagasse, etc.); see, e.g. Goldemberg and Coelho (2003).

This chapter deals primarily with 'bioenergy other than wood'. This term includes agricultural and livestock residues and herbaceous crops grown specifically for energy purposes, but excludes woody biomass (wood and wood-based residues, including sawmill waste and black liquor) and forest plantations grown specifically for energy, all of which are discussed in Chapter 9. However, the implications of large-scale plantations (herbaceous and woody crops) are briefly discussed.

Bioenergy Potential

There have been many attempts to quantify bioenergy potential, but difficulties arise owing to its complex and varied nature: ranging from resource availability to economic, technological, ecological, social, cultural and environmental factors. In addition, there is considerable uncertainty with regard to the potential role of dedicated energy forestry/crops, since residues (all sources) have a much more limited potential. The plantation potential ranges from 100 Mha to over 1 300 Mha. A recent study (IPCC-TAR, 2001) has estimated the global bioenergy potential from plantations at 440 EJ/yr (10.12 billion toe); it assumes that all agricultural lands not needed for food production (1.28 billion ha or 9.70% of total land) would be used for forest plantations.

Biomass resources are potentially the world's largest and most sustainable energy source. Previous commentaries have quoted a potential renewable resource comprising 220 billion odt (or about 4 500 EJ) of annual primary production and an annual bioenergy potential of about 2 900 EJ, although realistically only 270 EJ could be considered available on a sustainable basis and at competitive prices (Hall and Rao, 1999). Various scenarios have estimated the potential contribution from bioenergy for

PRODUCTION SIDE, SUPPLY	MAJOR COMMODITIES	USER SIDE, DEMAND EXAMPLES
Direct Woodfuels	WOODFUELS	**Solid:** Fuelwood (wood in the rough, chips, sawdust, pellets), Charcoal
Indirect Woodfuels		**Liquid:** Black liquor, Methanol, Pyrolitic oil
Recovered Woodfuels		**Gases:** Products from gasification and pyrolisis gases of above fuels
Fuel crops	AGROFUELS	**Solid:** Straw, Stalks, Husks, Charcoal from agrofuels
Agricultural by-products		**Liquid:** Ethanol, Raw vegetable oil, Oil diester, Methanol, Pyrolitic oil
Animal by-products Agroindustrial by-products		**Gases:** Biogas, Producer gas, Pyrolisis gases from agrofuels
Municipal by-products	MUNICIPAL BY-PRODUCTS	**Solid:** Municipal solid wastes (MSW)
		Liquid: Sewage sludge, Pyrolitic oil from MSW
		Gases: Landfill gas, Sludge gas

FIGURE 10.1 Biofuel classification scheme proposed (Source: FAO/WE (2003)).

the period 2025–2050 at between 67 and 450 EJ for the research focus (RF) scenario, and from 28 to 220 EJ for the demand driven (DD) scenario. The share of total final energy demand lies between 7 and 27% (Hoogwijk et al., 2001). Thus, the problem is not availability of biomass resources but the sustainable management, competitive and affordable delivery of energy to those who need it to provide them with modern energy services. This implies that both production and use of bioenergy must be modernised.

Bioenergy is not a transition fuel as it has often been portrayed, but a fuel that will continue to be the prime source of energy for many people for the foreseeable future. For example, an IEA (2002) study concluded: 'Over 2.6 billion people in developing countries will continue to rely on biomass for cooking and heating in 2030… this is an increase of more than 240 million from current use. In 2030 biomass use will still represent over half of residential energy consumption…'.

Current Contribution

Bioenergy is the most important renewable energy (RE) resource and currently provides about 55 EJ, mostly in the form of woody and herbaceous residues. In future, energy plantations are, as stated above, expected to play an increasing role, with their future contribution being increasingly for modern uses.

Agricultural and Livestock Residues

These are currently the main sources of non-wood bioenergy and will continue to be so in the short to medium term. The expected increase of bioenergy consumption, particularly in its modern forms, could have a significant impact not only in the energy sector, but also in the drive to modernise and diversify current agricultural activities, with the consequent development of rural societies. A reliable and affordable supply of energy is a key prerequisite for socio-economic development.

Residues are a large and under-exploited potential energy resource, and offer many opportunities for better utilisation, as a large proportion of them are readily available and represent a good opportunity at low costs. With proper management, residues can lead to greater productivity in agriculture. There have been many attempts to estimate global production and use of residues, e.g. Woods and Hall (1994) estimated it as 93 EJ (c. 2.14 billion toe), although assessments vary considerably.

There are a number of important factors which need to be addressed when considering the use of residues for energy. First, there are many other competing uses, e.g. for animal feed, erosion control, animal bedding, fertilisers (dung), etc. Secondly, there is the problem of agreeing on a common methodology for determining what is and what is not a recoverable residue, e.g. estimates of livestock residues often vary by a factor of five.

Many residues would be difficult to utilise and thus the most reasonable approach would be to concentrate efforts on the more promising, e.g. sugarcane, corn, rice husks, etc. together with plantations.

As an example, more than 300 million tonnes of bagasse is produced worldwide and mostly used as fuel in sugarcane mills. The energy content of 1 tonne of bagasse (50% moisture content) is 2.85 GJ/tc (tonne cane) milled. This excludes barbojo (tops and leaves), which represents the largest part of the sugarcane (55%) and which is currently mostly burned off or left to rot in the fields. In Brazil alone, the cogeneration potential ranges from 4 GW using conventional technologies to 47 GW by 2025 with BIG/GT (biomass-integrated gasification gas-turbine). The importance of sugarcane does not only stem from its potential as an energy crop, but from the many other products that can be obtained simultaneously, e.g. sugar, ethanol and other by-products such as yeast, fertilisers, etc.

The use of livestock residues is more controversial. In view of concerns about environmental and health hazards, it is questionable whether animal manure should be used as an energy source on a large scale, except in specific circumstances. In addition, the variations in quantification are so large that figures are often meaningless. One of the most important developments of the past decade has been the use of poultry litter in combustion plants. Poultry litter is refuse from broiler houses and contains materials such as wood shavings, shredded paper or straw, mixed with droppings. The material has a calorific value of between 9 and 15 GJ/tonne, with a moisture content varying between 20 and 50%, depending on husbandry practices. Currently there is about 150 MW installed capacity worldwide (75 MW in the UK and over 50 MW in the USA) and it is growing rapidly, generating new economic, energy and environmental benefits from a resource largely wasted in the past.

Plantations

The potential of energy forestry/crops has sometimes been overstated. This stems from the assumption that large-scale energy plantations would be established primarily on degraded lands. However, recent studies have questioned this, as a key to producing low-cost energy forestry/crops is good-quality land (http://bioenergy.ornl.gov/reports/fuelwood/chap5.html). This study also shows that plantations aimed at generating electricity can be financially viable only when local conditions are favourable and/or the costs of conventional fuels are high.

Overall, it seems that large-scale energy plantation predictions are unlikely to be achieved; a more likely scenario would be closer to 300–500 Mha, despite the potential availability of land[1]. There are various reasons for this:

- degraded land is less attractive than good-quality land owing to higher costs and lower productivity, although the importance of bringing degraded land into productive use is recognised;
- capital and financial constraints, particularly in developing countries;
- cultural practices, mismanagement, perceived and potential conflict with food production, population growth, etc.;
- productivity would have to increase far beyond what may realistically be possible, although large increases could be achieved;
- increasing desertification problems and the potential impact of climate change on agriculture, which at the present time are too unpredictable;
- emerging energy alternatives (e.g. clean coal technology, wind power, etc.);
- water constraints.

Municipal Solid Waste

Previous commentaries have excluded a discussion on MSW owing to: (i) the nature of MSW, which comprises many organic and non-organic materials; (ii) the difficulties and high costs associated with sorting such material, making it unlikely as a candidate for RE except for disposal purposes; (iii) re-use of MSW was mostly for recycling; (iv) MSW disposal was mostly undertaken in landfill sites or incineration plants.

However, in the past few years important changes have taken place which make MSW more attractive; it is now recognised as an RE resource, although this is still debatable. MSW has been included in this commentary for the following reasons: (i) recognition of its increasing role as a potential energy source: e.g. in the EU about 7% of RE originates from MSW and this share is expected to grow; (ii) stringent environmental legislation in many countries, which has cut pollution from MSW-fuelled plants considerably, thus making it more acceptable from an energy and environmental point of view. For example, in the USA emissions from waste-to-energy (WTE) plants have been cut from 4 260 g toxic equivalent (TEQ) in 1990 to 12 g TEQ in 2000.

It is difficult to give reliable figures on total energy generated from MSW as this varies considerably, depending on the nature of this resource. This is further complicated by differences in the amount of MSW generated in different countries, and between rural and urban dwellers within individual countries (e.g. 314 kg/yr per capita in Japan, 252 kg/yr in Singapore or 170 kg/yr in Brazil). Its composition, and hence its potential as an energy source, varies considerably but generally it contains from 8 to 12 MJ/kg.

Rogner et al. (2001) have estimated the global economic energy potential from MSW at about 6 EJ (c. 138 mtoe). The USA alone produces about 240 million tonnes/yr of MSW. Worldwide, approximately 130 million tonnes of MSW are combusted annually in over 600 WTE plants in 35 countries to produce electricity and steam for heating and other industrial applications (Themelis, 2003).

Yet, despite increasing efforts to deal with this problem, in particular through WTE plants, hundreds of millions of tonnes of MSW still end up in landfills, emitting enormous amounts of GHG. Current RD&D on MSW focuses on ways to minimise waste, material recycling, energy recovery and landfill of the non-biodegradable fractions. The utilisation of the biodegradable fraction of MSW as a bioenergy is, in most countries, intimately linked with waste-management policies and public perceptions.

Transportation Fuels

In the past few years there have been important advances in the field of alternative transportation fuels, primarily ethanol and bio-diesel: these are briefly described below.

[1] This could change if carbon trading (CT) becomes a significant factor.

Ethanol fuel. The past 2 years have seen important changes in this area including: (i) a rapid increase in ethanol fuel production, particularly in the USA (from 31.9 billion litres in 2001 to over 38 billion litres in 2003); (ii) an increasing number of countries running or planning ethanol fuel programmes (see Fig. 10.2); (iii) increasing interest in flexible fuel vehicles (FFVs), including in Brazil: this offers considerable flexibility to both the producers and consumers of ethanol fuel; (iv) disappointing advances in the field of cellulose-based ethanol, with costs remaining a stumbling block; (v) direct electrochemical conversion in fuel cells.

Biodiesel. The production of biodiesel has increased dramatically in the past few years, particularly in the EU. Both the automobile industry and the biodiesel producers have played a major role in developing the technical and normative prerequisites. Blends of up to 20% (B-20) can be used in almost all diesel engines; B-20 does not require any engine modifications and provides the same payload capacity and range as conventional diesel, whilst increasing the engine lubricant consumption significantly. Higher blends can also be used in most modern engines with little or no modifications, although some material compatibility and warranty issues with higher blends have not yet been fully resolved.

The EU leads the world in the production and use of biodiesel, with a current installed capacity of over 2.3 billion litres; biodiesel supplies over 1% of the diesel market. The most important biodiesel producer and consumer is Germany, where production has increased from about 79.5 million litres in 1998 to nearly 1.26 billion litres in 2003[2]. It was expected that by the end of 2003 there would be about 1 700 service stations selling biodiesel in Germany alone (www.ufop.de).

Biodiesel production in the USA has also increased rapidly, mainly as a result of support from government and soybean producers, together with environmental pressures. But this market is still small compared to the EU, at just over 61 million litres in 2001, although the market is potentially very large. For example, the American Biofuels Association (ABA) considers that with government incentives comparable to those provided for ethanol, biodiesel consumption could reach 7.5 billion litres or about 8% of diesel consumption, primarily to power bus fleets, heavy-duty trucks and agricultural vehicles, mostly in blends of about 20%.

The US Department of Energy is currently developing a low-cost biodiesel from mustard seed that could add a further 5–10 billion US gallons (20–38 billion litres), if the research proves successful. The estimated costs are about US$ 1.00/gal (0.26 cents/l).

Technological Developments

Despite the fact that many technologies have reached the demonstration and validation stage and have shown that both small and large-scale applications are possible, technological advances have not been matched by economics. This is partly due to high costs of raw materials, low productivity, a large range of conversion technologies, competition from other energy sources, etc. Nonetheless, technological advances are opening up many new opportunities, and there are already many mature technologies that can meet such criteria, without being necessarily more expensive than fossil fuels (if all costs are internalised).

Combustion technology is already proven, while other more advanced technologies such as gasification are still at the development stage. Current interest centres on CHP and co-firing. Co-firing with fossil fuels has received considerable attention, particularly in the EU and the USA, and is emerging as one of the most important alternatives for large-scale biomass utilisation, if current technical, economical and cultural barriers can be overcome.

The driving force in co-firing is coal and the need to reduce GHG. Coal is abundant and will

[2] Includes plants under construction and due to enter operation in 2003.

Country	Production (2001)	Remarks
Australia	150 million litres/yr (current production – all uses)	Plan for 2-5% blend nationwide by 2010, corresponding to 1–2 billion litres/yr production
Brazil	12.6 billion litres (2002)	Could increase production significantly if carbon trading and exports take off
Canada	240 million litres/yr (current production – all uses)	Target 1.35 billion litres/yr by 2010
China	3 billion litres/yr (current production – all uses)	Plan to produce additional 3 billion litres/yr from maize, sweet sorghum, cassava and sugarcane
EU-15	Over 2 billion litres of ethanol	Plan to increase ethanol significantly. Will need 8.5 - 14 billion litres/yr by 2010. Potentially a large market
India	1.8 billion litres/yr (current production – all uses)	Plan for 5% blend, trial at 300 refuelling stations, initial demand 350 million litres/yr
Mexico	70 million litres/yr (current production – all uses)	Programme to produce 1.3 billion litres/yr
South Africa	385 million litres/yr from coal and gas, 40 million litres from cane	Plan for 12% blend nationwide; plan to increase fermentation ethanol from cane
Thailand	150 million litres/yr (current production – all uses)	Plan for 10% blend, using molasses, cassava, and sugarcane, corresponding to 0.7 billion litres/yr
USA	About 11.2 billion litres (2003)	Large expansion programme; over 16 billion litres by 2005 and c.20 billion litres in 2015. Mostly from maize
Others: Colombia, Cuba, Peru, Central America, Ethiopia, Malawi, etc.	Large potential, ranging from a few million litres to hundred millions	Mostly ethanol from sugarcane to be blended with gasoline in various proportions. Some ethanol-diesel blends. Malawi blends 15% since 1982. Other programmes are at various stages.

FIGURE 10.2 Summary of current ethanol production and plans for bioethanol fuel use.

continue to be used, but needs to meet increasingly stringent environmental goals: biomass offers a good and cheap alternative. Blends of 15–25% biomass with coal have been widely tested and it has been shown that bioenergy can provide about 15% of the total energy input, with only feed intake systems and burner modifications. Current interest is to reach a 40% blend and also in co-firing with natural gas, coal and biomass. In countries with large reserves of coal and abundant biomass resources, co-firing is one of the most important and promising alternatives for using bioenergy on a large scale.

Co-firing in existing coal-fired power plants makes it possible to achieve greater efficiency in converting biomass into electricity, e.g. 33–37% when fired with coal. There are also important environmental benefits, e.g. lower sulphur emissions, about 30% reduction in NO_x (see www.eren.doe.gov/biopower/), in addition to significant savings in plant and infrastructure investment, etc.

The past 2 years have failed to produce any major development in gasification and generally the performance of many plants has been rather disappointing, as they have failed to produce the expected results. A new and promising research area is the production of hydrogen via gasification.

Policies and Legislative Changes

Policies serve as the framework within which objectives and targets are set at the national and regional level to ensure sustainable production, marketing and use of bioenergy. Unfortunately bioenergy is often neglected in political, economic and social agendas. Bioenergy involves many factors, ranging from production, preparation, transportation, and conversion of raw materials for distribution and utilisation to final use, and this complicates matters.

Nonetheless, a growing number of countries have introduced legislation or have some specific policies to deal with the various aspects of bioenergy; generally, however, it does not receive the same benefits as conventional energy sources (Trossero, 2003). Despite the growing number of countries supporting bioenergy, much is still needed to integrate it into mainstream energy systems.

Legislative measures have a key role to play. Without a legal framework, bioenergy will not penetrate the energy market on any significant scale, as recent experience shows. For example, the EU uses a mixture of incentives to promote RE, ranging from government-guaranteed purchase prices for RE electricity (e.g. in Germany, Denmark and Spain) to competitive mechanisms (e.g. in Ireland and the UK). The EU has also approved specific legislation to promote alternative fuels [COM (2001) 547]. The driving force for the support of biofuels in the EU is the Commission's Green Paper: *Towards a European Strategy for the Security of Energy Supply* [COM (2000) 769], which introduced the objective of alternative fuels supplying 20% of the road transport sector by the year 2020.

In the USA the most important pieces of legislation to promote RE, particularly liquid biofuels, include: (i) the Alternative Motor Fuels Act of 1988 (AMFA), which created incentives for the production of vehicles designed to operate on any combination of fuel alcohols and gasoline; (ii) the Energy Policy Act of 1992 (EPAct); (iii) the Clean Cities Program (CCP), designed to encourage communities to coordinate the voluntary acquisition of alternative fuels vehicles (AFVs); (iv) the Clean Air Act Amendments (CAAA) which requires that certain regions use oxygenated, reformulated gasoline in cities where smog levels are high, etc. The law stipulated that a certain percentage of oxygenates must be from renewable sources.

Many countries have also introduced stringent legislation to deal with MSW and hence the growing interest in it as an energy resource, particularly in industrialised countries, e.g. the EU Directorate (EU, 1999/31/EC) requires all biodegradable materials to be reduced by 65% to the 1995 level by 2016; legislation also requires landfilling of combustible material to be phased out by 2010. In the USA, as a result of the introduction in 1995 of the Maximum Available Technology (MACT), led by the Environmental Protection Agency (EPA) of the US Congress, pollution from WTE is currently at the same level as any other comparable plant.

Social Changes

Previous commentaries estimated that bioenergy consumption in rural areas of developing countries (including all types of biomass and all end-uses) was roughly 1 tonne per person per year, and about 0.5 tonne in semi-urban and urban areas (15% moisture, 15 GJ/t); this is still generally valid today. There are many differences: e.g. bioenergy per person per year averages 6.7 GJ in Africa; and 5.6 GJ in Latin America. It seems that while in relative terms, traditional uses of bioenergy might be declining in some parts of the world (e.g. Latin America and Asia), in absolute terms the total amount of bioenergy is increasing, particularly as a modern energy carrier.

Consumption patterns are also changing rapidly. For example, many urban consumers in Africa are shifting to charcoal as their living standards improve; also, many consumers use bioenergy in both its traditional and modern forms. There are many variations due to the large number of factors involved, including availability of supply, climatic differences, population growth, socio-economic development, cultural factors, etc.

The past 2 years have also witnessed a growing interest in biotrade, particularly from northern Europe and Canada. In the past bioenergy was only traded locally; the fact that regions with abundant raw material can trade internationally will open up new possibilities. This is particularly important for ethanol fuel, of which currently only 10% is internationally traded, the rest being commercialised internally. This acts as a barrier to future expansion since supply remains insufficiently elastic[3]. Increased biotrade could bring significant benefits, e.g. through the creation of stable markets, better use of underutilised biomass, income generation, and so forth.

A major challenge still remaining is how best to tackle the problems posed by the traditional uses of bioenergy, e.g. low combustion efficiency and health hazards. For bioenergy to have a future it must provide people with what they want—cheap and convenient fuels for lighting, power, etc. at an affordable price. Considerably more effort is needed to integrate policies, technologies and markets to achieve these goals. Undoubtedly bioenergy, primarily in its modern applications, will continue to play a major role in the future.

Frank Rosillo-Calle
Imperial College, London

[3] For example, Japan is planning to blend about 5% of ethanol with gasoline (this will require about 10 billion litres/yr) but is reluctant to go ahead because currently no country can guarantee to supply that amount. Brazil is keen to supply the Japanese ethanol fuel market but Japan would not wish to rely on a single supply source. Only when a large number of countries are able to supply ethanol, will major consumers feel confident to go ahead; in other words, only when ethanol fuel becomes an internationally traded commodity.

References

FAO/WE, 2003. Bioenergy Terminology, FAO Forestry Department, FAO, Rome.

Goldemberg, J., Coelho, S.T., 2003. Renewable Energy—Traditional vs. Modern Biomass, Energy Policy, in press.

Hall, D.O., Rao, K.K., 1999. Photosynthesis, sixth ed., Studies in Biology, Cambridge University Press.

Hoogwijk, M., den Broek, R., Berndes, G., Faaij, A., 2001. A Review of Assessments on the Future of Global Contribution of Biomass Energy, First World Conference on Biomass Energy and Industry, Sevilla, James & James, London.

IEA, 2002. Energy Outlook 2000–2030, IEA, Paris (www.iea.org).

IEA, 2003. Municipal Solid Waste and Its Role in Sustainability, IEA Bioenergy, International Energy Agency, Paris.

IPCC-TAR, 2001. Climate change 2001: mitigation. In: Mentz, B., et al. (Eds.), Third Assessment of the IPCC, Cambridge University Press.

Rogner, et al., 2001. Energy Resources, World Energy Assessment: Energy and the Challenge of Sustainability; Part II Energy Resources and Technology Options, UNDP, pp. 135–171 (Chapter 5).

Rosillo-Calle, F., 2001. World Energy Council—Survey of Energy Resources, Chapter 10: Biomass (Other than Wood) Commentary.

Themelis, N.J., 2003. An Overview of the Global Waste-to-Energy Industry, Waste Management World, 2003–2004 Review Issue, July–August 2003, pp. 40–47 (See also www.iswa.org).

Trossero, M.A., 2003. Wood Energy and Climate, paper presented at UNECE, Timber Committee, 61st Session of the Economic and Social Council, Geneva, October 2003.

Woods, J., Hall, D.O., 1994. Bioenergy for Development: Technical and Environmental Dimensions, FAO Environment and Energy Paper 13, FAO, Rome.

TABLE 10.1

Bagasse: estimated potential availability—2002

	Bagasse potential availability (thousand tonnes)
Benin	16
Burkina Faso	130
Burundi	65
Cameroon	340
Chad	104
Congo (Brazzaville)	108
Congo (Democratic Republic)	212
Côte d'Ivoire	554
Egypt (Arab Republic)	3 651
Ethiopia	935
Gabon	59
Guinea	82
Kenya	1 751
Madagascar	105
Malawi	850
Mali	104
Mauritius	1 803
Morocco	522
Mozambique	554
Nigeria	22
Senegal	310
Sierra Leone	23
Somalia	652
South Africa	9 019
Sudan	2 424
Swaziland	2 200
Tanzania	608
Togo	16
Uganda	522
Zambia	759
Zimbabwe	1 843
Total Africa	**30 343**
Barbados	147
Belize	386
Costa Rica	1 174
Cuba	11 481
Dominican Republic	1 681
El Salvador	1 551
Guatemala	6 194
Haiti	16

TABLE 10.1 (Continued)

	Bagasse potential availability (thousand tonnes)
Honduras	1 043
Jamaica	571
Mexico	16 538
Nicaragua	1 206
Panama	489
St Kitts and Nevis	65
Trinidad & Tobago	340
United States of America	10 641
Total North America	**53 523**
Argentina	5 477
Bolivia	978
Brazil	76 829
Colombia	8 224
Ecuador	1 614
Guyana	1 079
Paraguay	375
Peru	2 771
Suriname	33
Uruguay	23
Venezuela	1 940
Total South America	**99 343**
Bangladesh	746
China	28 426
India	63 650
Indonesia	7 009
Japan	593
Malaysia	359
Myanmar (Burma)	326
Nepal	359
Pakistan	10 805
Philippines	6 480
Sri Lanka	65
Taiwan, China	567
Thailand	20 987
Vietnam	2 901
Total Asia	**143 273**
Unspecified	851

(continued on next page)

TABLE 10.1 (*Continued*)

	Bagasse potential availability (thousand tonnes)
Total Europe	**851**
Iran (Islamic Republic)	652
Total Middle East	**652**
Australia	18 300
Fiji	1 090
Papua New Guinea	172
Western Samoa	7
Total Oceania	**19 569**
Total World	**347 554**

Sources: Bagasse potential availability based on production of cane sugar published in the *I.S.O. Sugar Yearbook 2002*, International Sugar Organization; Bagasse potential availability conversion factor from United Nations *Energy Statistics Yearbook 2000* (assumes a yield of 3.26 tonnes of fuel bagasse at 50% humidity per tonne of cane sugar produced).

COUNTRY NOTES

The Country Notes on Bioenergy reflect the data and comments provided by WEC Member Committees in 2003, supplemented where necessary by information provided for the 2001 and 1998 editions of the WEC *Survey of Energy Resources*.

Albania

Municipal solid waste	
Quantity of raw material available	405 ttoe

Argentina

Municipal solid waste	
Quantity of raw material available[a]	5.5 million tonnes
Sugarcane bagasse	
Quantity of raw material available	5.9 million tonnes
Ethanol installed capacity	6 237 TJ/yr
Ethanol fuel production[b]	2 525 TJ
Electricity generating capacity	182 061 kW
Electricity generation	131 031 TJ
Direct use from combustion	37 028 TJ
Total energy production	170 584 TJ
Agricultural residues	
Quantity of raw material available	5.4 million tonnes
Electricity generating capacity	81 916 kW
Electricity generation	1 186 TJ

[a] The figure shown corresponds to the Ciudad Autónoma de Buenos Aires and 31 municipalities of the Province of Buenos Aires. This area produces 40% of the country's total waste.
[b] Raw material used: molasses.

Generally, the use of biomass resources takes place in a dispersed way, except for bagasse, mostly used as fuel in the cane sugar mills for electricity and steam production. Part of the bagasse is also used by the paper industry as raw material. Vegetable wastes mainly include harvest residues (straw and stem from maize, wheat, rice, etc.) and from the manufacturing of food products (rice chaff, peanut shells, etc.).

Municipal waste corresponds to the rubbish from the residential, commercial and public service sectors collected by the municipal services and have almost been totally used for ecological landfill.

Lately 'biodiesel' has aroused high expectations, but so far production on an industrial scale has not been achieved.

Australia

Municipal solid waste	
Quantity of raw material available	~ 6.9 million tonnes
Yield of solid fuel	~ 9 GJ/tonne
Electricity generating capacity	103 700 kW
Sugarcane bagasse[a]	
Quantity of raw material available	11.4 million tonnes
Yield of solid fuel	~ 9.3 GJ/tonne
Electricity generating capacity	368 600 kW

[a] Data refer to 1997.

For MSW, generation is 98 700 kW from landfill gas and 5 000 kW from MSW gasification (SWERF plant, Wollongong).

Sugar industry generation includes the Rocky Point sugar mill cogeneration plant, which uses some wood waste in the non-crushing season.

Approximately 80 megalitres/yr ethanol is produced.

Biodiesel production is relatively low and is estimated to be below 20 million litres in 2002.

The Bureau of Rural Sciences has developed a bioenergy atlas for Australia.

Austria

Municipal solid waste	
Quantity of raw material available	1.5 million tonnes
Electricity generation	9 028 TJ

Direct use from combustion	4 435 TJ
Total energy production	13 463 TJ
Other biomass	
Quantity of raw material available	6.0 million tonnes
Biodiesel capacity[a]	1 125 TJ/yr
Direct use from combustion	31 565 TJ
Total energy production	56 623 TJ

Data refer to 2001.
[a] Data refer to 2000.

Belgium

Municipal solid waste	
Quantity of raw material available	1.1 million tonnes
Electricity generating capacity	76 600 kW
Electricity generation	1 765 TJ

Data refer to 1996.

Bolivia

Animal dung	
Direct use from combustion	3 270 TJ
Sugarcane bagasse	
Direct use from combustion	10 458 TJ
Crop residues	
Direct use from combustion	307 TJ

Data refer to 1996.

Botswana

Municipal solid waste	
Direct use from combustion	1 420 TJ

Estimated.

Brazil

Municipal solid waste	
Quantity of raw material available	20.0 million tonnes
Biogas production capacity[a]	1 640 TJ/yr
Yield of biogas[b]	0.0230274 GJ/tonne
Biogas produced[c]	427.38 TJ
Electricity generating capacity	47 631 kW
Electricity generation	1 201 TJ
Total energy production	2 796 TJ
Sugarcane bagasse	
Quantity of raw material available	87.2 million tonnes
Electricity generating capacity	1 287 000 kW
Electricity generation	19 296 TJ
Direct use from combustion	732 489 TJ
Total energy production	777 488 TJ
Rice hulls	
Quantity of raw material available[d]	2.58 million tonnes
Electricity generating capacity	18 225 kW
Electricity generation	488 TJ
Cane juice	
Quantity of raw material available	76.97 million tonnes
Ethanol production capacity	331 834 TJ/yr
Yield of ethanol	2.6 GJ/tonne
Ethanol production	198 530 TJ
Total energy production	200 841 TJ
Molasses	
Quantity of raw material available	10.3 million tonnes
Yield of ethanol	7.5 GJ/tonne
Ethanol production	77 212 TJ
Total energy production	79 717 TJ

[a] Biogas potential production = 71 252 562 m^3/yr.
[b] Medium value (GJ/m^3).
[c] Adopted density 1 kg/m^3 at 15 °C and 101.325 kPa. Efficiency factor = 26%.
[d] Rice hull is equivalent to 22% of total rice crop mass.

Bulgaria

Production of energy sector raw materials could open up new possibilities for forestry and agriculture, contributing to employment in rural areas, increasing the standard of living and incomes through development of energy cultures, promoting RE based on biomass, regional support through co-financing of RE projects, and processing and marketing of agricultural production for RE.

In the long term CHP usage of biomass has the largest potential amongst all forms of RE.

The advantages of biomass usage, based on new technologies, can be seen in the utilisation of biogas. Therefore, a campaign for the promotion and support of decentralised bio-energy installations is of special importance. Such installations could combine different technologies.

Unfortunately so far there are no installations in Bulgaria for biogas generation.

There are at the moment at least two factors stimulating research into biodiesel fuels usage in the area of transport: the decreasing supply of oil and the negative influence of transport vehicles on the environment. It is known that the transport sector has considerable effects on the environment. During the last 10 years its share in global warming has increased from less than 20% to more than 25%.

In Bulgaria basic oils are obtained from sunflowers and, in small amounts, from rape, soya and corn. In earlier years oils were also produced from flax, castor oil, etc. The Bulgarian oil extraction enterprises have an annual production capacity of 750 000 tonnes of oil-bearing seeds, of which on average about 50% is utilised. The Bulgarian enterprises are equipped for reprocessing sunflower oil, but they can also reprocess rape oil. As sunflower oil is preferred for food purposes, the more likely raw material for future production of biodiesel will be rape oil and in small amounts lower-quality sunflower and other plant oils. For the 2000/2001 season, 90 000 decares of rape were sown and harvested, and during the 2001/2002 season 250 000 decares were sown. (Methanol production through the distillation of bio-products from corn, beetroot or potatoes is also a future possibility for the replacement of conventional fuels).

Canada

Municipal solid waste	
Quantity of raw material available	21 million tonnes
Biogas production	9 200 TJ
Electricity generating capacity	85 300 kW
Electricity generation	2 421 TJ
Direct use from combustion	8 820 TJ
Total energy production	20 441 TJ
Crop residues—corn	
Quantity of raw material available	20 million tonnes
Crop residues—cereal grain	
Quantity of raw material available	25 million tonnes
Various—wheat	
Ethanol production capacity	466.4 TJ/yr
Yield of ethanol	7.2 GJ/tonne
Ethanol production	466.4 TJ
Various—corn	
Ethanol production capacity	466.4 TJ/yr
Yield of ethanol	7.8 GJ/tonne
Ethanol production	63.6 TJ

Data refer to 1999.

Côte d'Ivoire

Data concerning the use of biomass energy (apart from wood and charcoal) are unavailable. To resolve this problem, a strategy is being devised to collect data on production and consumption of all forms of biomass.

There is a programme for restructuring the institutional framework of renewable energies and a project concerning the inventory and the evaluation of agricultural and industrial wastes.

Natural biomass, agricultural waste and industrial waste constitute the potential renewable energies for direct use.

Biomass energy in different forms (firewood, charcoal by city dwellers, agricultural and industrial wastes) is consumed by 78% of the population.

The agricultural and industrial energy resources are estimated at more than 4 mtoe/yr. They constitute an important source of energy and essentially come from palm oil, manufactured wood, coffee, rice and sugarcane.

The principal technologies used for the conversion of biomass into energy are carbonisation, gasification and fermentation.

Croatia

Municipal solid waste	
Quantity of raw material available	1.1 million tonnes
Crop residues—wheat straw	
Quantity of raw material available	0.25 million tonnes
Crop residues—maize stalks	
Quantity of raw material available	0.51 million tonnes
Crop residues—barley straw	
Quantity of raw material available	0.03 million tonnes
Crop residues—from fruit growing	
Quantity of raw material available	0.16 million tonnes

Data refer to 1996.

Czech Republic

Municipal solid waste	
Quantity of raw material available	1.6 million tonnes
Biogas production	15 TJ
Electricity generation	702 TJ
Direct use from combustion	4 429 TJ
Agricultural residues	
Quantity of raw material available	4 million tonnes
Biodiesel production	3 000 TJ
Biogas production	61 TJ
Industrial waste	
Quantity of raw material available	1 million tonnes
Electricity generation	14 TJ
Direct use from combustion	5 592 TJ
Sewage sludge	
Biogas production	1 481 TJ

Denmark

Municipal solid waste	
Quantity of raw material available	9.5 million tonnes
Electricity generating capacity	275 000 kW
Electricity generation	5 098 TJ
Heat production	21 164 TJ
Total energy production	31 843 TJ
Agricultural residues—straw	
Quantity of raw material available	2.7 million tonnes
Electricity generating capacity	131 000 kW
Electricity generation	2 235 TJ
Total energy production	13 698 TJ
Agricultural residues—slurry, etc.	
Quantity of raw material available	24 PJ
Biogas production	1 138 TJ
Electricity generating capacity	20 000 kW
Electricity generation	374 TJ
Total energy production	1 490 TJ
Agricultural residues—other veg. waste	
Electricity generating capacity	70 kW
Electricity generation	51 TJ
Total energy production	764 TJ
Fish oil	
Quantity of raw material available	191 TJ
Electricity generation	65 TJ
Total energy production	75 TJ
Sewage sludge	
Quantity of raw material available	3 PJ
Biogas production	330 TJ
Electricity generating capacity	7 000 kW
Electricity generation	111 TJ
Total energy production	823 TJ
Landfill gas and municipal waste gas	
Quantity of raw material available	1 PJ
Biogas production	286 TJ
Electricity generating capacity	5 000 kW

Electricity generation	79 TJ
Total energy production	733 TJ

Egypt (Arab Republic)

Municipal solid waste	
Quantity of raw material available	2.4 million tonnes
Sugarcane bagasse	
Quantity of raw material available	1.4 million tonnes
Ethanol production capacity	456.25 TJ/yr
Biodiesel production capacity	22.83 TJ/yr
Total energy production	479.08 TJ
Cotton stalks	
Quantity of raw material available	1.2 million tonnes
Rice straw	
Quantity of raw material available	3.4 million tonnes
Animal dung	
Quantity of raw material available	6 million tonnes
Biogas production capacity	40 TJ/yr
Yield of biogas	4.1 GJ/tonne
Biogas production	15 TJ
Direct use from combustion	15 TJ
Sewage sludge	
Quantity of raw material available	2.4 million tonnes
Electricity generating capacity	18 000 kW
Industrial waste	
Quantity of raw material available	3 million tonnes
Food processing waste	
Quantity of raw material available	2 million tonnes

Estonia

Municipal solid waste	
Quantity of raw material available	0.569 million tonnes
Biogas production (landfill gas)	107 TJ

Finland

Municipal solid waste	
Quantity of raw material available	0.4 million tonnes
Solid fuel production capacity	N TJ/yr
Electricity generating capacity	50 000 kW
Electricity generation	770 TJ
Direct use from combustion	1 130 TJ
Total energy production	1 900 TJ
Biogas	
Electricity generation	0.1 TJ
Direct use from combustion	0.5 TJ
Total energy production	0.6 TJ
Landfill gas	
Electricity generating capacity	3 000 kW
Electricity generation	40 TJ
Direct use from combustion	150 TJ
Total energy production	190 TJ
Wastewater	
Electricity generating capacity	6 000 kW
Electricity generation	90 TJ
Direct use from combustion	324 TJ
Total energy production	414 TJ

France

Municipal solid waste	
Quantity of raw material available	11.2 million tonnes
Electricity generation	8 867 TJ
Direct use from combustion	27 018 TJ
Total energy production	35 885 TJ
Agricultural residues	
Biogas production	220 TJ
Direct use from combustion	3 208 TJ
Total energy production	3 428 TJ
Biofuels	
Ethanol production	2 415 TJ
Biodiesel production	11 214 TJ
Total energy production	13 629 TJ
Landfill gas	
Biogas production	8 080 TJ
Other waste	
Biogas production	4 667 TJ

The above data relate only to metropolitan France and exclude overseas departments (DOM).

In the DOM, 2002 production from bagasse was 1 224 TJ electricity and 4 442 TJ heat.

Germany

Municipal solid waste	
Biodiesel production capacity	940 000 tonnes/yr
Biogas production capacity	160 MW
Biogas production	9 600 TJ
Electricity generating capacity	852 000 kW
Electricity generation	11 200 TJ
Landfill gas[a]	
Electricity generating capacity	142 MW
Electricity generation	88 GWh
Sewage sludge gas[a]	
Electricity generating capacity	75 MW
Electricity generation	732 GWh
Liquid biofuels[a]	
Plant capacity	500 000 tonnes/yr
Other biogas[a]	
Electricity generating capacity	200 MW
Electricity generation	74 GWh

[a] Data refer to 2001.

Ghana

Agricultural residues	
Quantity of raw material available	
Coconut shell and husk	0.135 million tonnes
Groundnut shells	0.0475 million tonnes
Rice straw and husk	0.120 million tonnes

Data refer to 1990.

Greenland

Municipal solid waste	
Solid fuel production capacity	214 TJ/yr
Yield of solid fuel	10.5 GJ/tonne
Solid fuel production	281 TJ
Direct use from combustion	83 TJ
Waste from fishing industry	
Yield of biodiesel	38.62 GJ/tonne
Biodiesel production	12.33 TJ

Hong Kong, China

Municipal solid waste	
Quantity of raw material available[a]	2.8 million tonnes
Biogas production capacity[b]	350 TJ/yr
Yield of biogas	0.0084 GJ/tonne
Biogas produced[c,d,e]	> 120 TJ
Electricity generating capacity[d]	6 384 kW

[a] Data refers to 2001.

[b] A town gas plant uses landfill gas from the nearby Shuen Wan landfill site as process fuels to crack naphtha to produce town gas.

[c] Biogas is also utilised at sewage treatment works (STW): (1) at Shatin STW to supply a dual fuel 1 120 kW engine for generating electricity (on-site application); (2) at Shatin and Taipo STWs to fuel boiler plant to maintain temperature of sludge digestion process; (3) at Taipo STW to supply a dual fuel engine which drives a process air-blower.

[d] Landfill gas engines with a total capacity of 6 384 kW have been installed in three strategic landfill sites (WENT, NENT and SENT) to generate electrical power for on-site usages. There is an additional 373 kW capacity at two other sites (Tseung Kwan O and Jordan Valley). Four of these sites also use landfill gas for heating in the leachate treatment process.

[e] Data on electricity and heat generated by landfill gas in 2002 are not available.

Hungary

Municipal solid waste	
Quantity of raw material available	0.258 million tonnes
Yield of solid fuel	12.5 GJ/tonne
Solid fuel production	19 908 TJ
Biodiesel production capacity	114 TJ/yr
Yield of biodiesel	38 GJ/tonne

Biogas production capacity	133 TJ/yr
Yield of biogas	23 GJ/tonne
Biogas production	100 TJ
Electricity generating capacity	26 064 kW
Electricity generation	290 TJ
Direct use from combustion	32 141 TJ
Agricultural residues	
Quantity of raw material available	0.175 million tonnes

Iceland

Municipal solid waste	
Direct use from combustion	45 TJ

Indonesia

Sugarcane bagasse	
Quantity of raw material available	6.5 million tonnes
Agricultural residues—rice husk	
Quantity of raw material available	14.3 million tonnes
Agricultural residues—coconut shells	
Quantity of raw material available	1.1 million tonnes
Agricultural residues—coconut fibre	
Quantity of raw material available	2.0 million tonnes
Agricultural residues—palm oil residues	
Quantity of raw material available	8.5 million tonnes

Iran (Islamic Republic)

Municipal solid waste	
Quantity of raw material available	15.33 million tonnes

Ireland

Municipal solid waste	
Electricity generating capacity	14 732 kW
Electricity generation	324 TJ

Israel

Municipal sewage	
Electricity generating capacity	1 000 kW
Electricity generation	28.8 TJ
Industrial sewage	
Electricity generating capacity	500 kW
Electricity generation	9 TJ

Italy

Municipal solid waste[a]	
Quantity of raw material available	3.7 million tonnes
Biogas production[b]	7 970 TJ
Electricity generating capacity	499 545 kW
Electricity generation	30 204 TJ
Direct use from combustion	600 TJ
Total energy production	38 774 TJ
Agricultural residues	
Biodiesel production	3 627 TJ
Biogas production[b]	64 TJ
Electricity generating capacity	7 946 kW
Total energy production	3 691 TJ
Sewage sludge	
Biogas production[b]	55 TJ
Electricity generating capacity	7 772 kW
Farm slurries	
Biogas production[b]	118 TJ
Electricity generating capacity	2 110 kW
Food industry by-products	
Electricity generating capacity	222 000 kW
Electricity generation	11 599 TJ
Crop residues	
Direct use from combustion	39 600 TJ

Data refer to 2001.

[a] Includes biogas from dumping ground.

[b] All used to generate electricity. In order to avoid duplication, the related electricity production is excluded

from electricity generation while electricity generating capacity includes the capacity of biogas-fired power plants.

Japan

Municipal solid waste	
Quantity of raw material available[a]	51 million tonnes
Electricity generating capacity[b]	829 000 kW
Sugarcane bagasse	
Quantity of raw material available[b]	0.2 million tonnes
Electricity generating capacity[b]	27 000 kW

[a] Data relate to FY 1997.
[b] Data relate to FY 1999.

Jordan

Municipal solid waste	
Quantity of raw material available	1.095 million tonnes
Electricity generating capacity	1 000 kW
Electricity generation	19 800 TJ
Direct use from combustion	N TJ

Jordan has adopted a special programme for bioenergy by which pre-feasibility studies for the utilisation of MSW for electricity generation have been prepared since 1993 through cooperation with GEF.

The outcome of these studies resulted in implementing the first biogas project in Jordan and in the region with a capacity of about 1 MW of electricity. This project is owned, operated and maintained by the Jordan Biogas Company (JBCo), and is going to be expanded up to 5 MW by the year 2005. The Greater Amman Municipality is actively working in this field, where several waste treatment projects, including bioenergy power plants, on a commercial/ private finance basis are in hand.

Korea (Republic)

Municipal solid waste	
Electricity generating capacity	12 358 kW
Electricity generation	740.8 TJ
Rice bran/soy oil waste[a]	
Biodiesel generating capacity	231.1 TJ/yr
Biodiesel production	35.4 TJ
Sludge[b]	
Biogas production capacity	3 601.6 TJ/yr
Biogas production	1 693.4 TJ

[a] Used for transportation.
[b] Alcohol, sewage.

Latvia

Municipal solid waste	
Quantity of raw material available	0.28 million tonnes
Ethanol production capacity	208 TJ/yr
Yield of ethanol	26 GJ/tonne
Biodiesel production capacity	93.5 TJ/yr
Yield of biodiesel	37.4 TJ
Biogas production capacity	60.5 TJ/yr
Yield of biogas	7.3 GJ/tonne
Biogas production	36.1 TJ
Electricity generating capacity	2 100 kW
Electricity generation	36.1 TJ

At present Latvia does not sort its municipal waste (0.47 million tons a year) and therefore the solid waste (0.28 million tons a year) is not used for incineration.

In Riga, the capital of Latvia, biogas is obtained from the total unsorted garbage mass, which is used to generate electricity.

The production capacity of ethanol is 8 000 tonnes/yr but that of biodiesel is 2 500 tonnes/yr. There is no legislation in Latvia at the moment on the use of bioethanol and biodiesel fuel. Since the costs of bioethanol and biodiesel fuel are higher than the costs of fossil fuels, they are not used in practice, apart from research and experimentation.

Lebanon

Municipal solid waste

Quantity of raw material available	1.44 million tonnes

Lithuania

Straw

Quantity of raw material available	0.004 million tonnes
Direct use from combustion	105 TJ

Luxembourg

At the present time the government is preparing a study for the evaluation of biomass resources and their exploitation.

Mexico

Sugarcane bagasse

Quantity of raw material available	12.6 million tonnes

Production of bagasse includes non-fuel burning. All fuel use of bagasse occurs in the sugar and pulp and paper industries. The Balance Nacional de Energía shows that in 2002, 84 081 TJ (11.9 million tonnes) of bagasse was consumed by the sugar sector for energy purposes and 212 TJ (30 000 tonnes) in the pulp and paper sector, including autoproduction of electricity. Bagasse consumed for non-energy purposes was about 3 385 TJ (479.8 thousand tonnes).

The production and consumption of biogas is about 301 TJ and the principal use is for electricity generation. The only sectors that produce and consume biogas are food, beverages and tobacco (97 TJ) and the municipal sector (204 TJ). Gross electricity generation from biogas is about 12 GWh.

Monaco

Municipal solid waste

Quantity of raw material available	0.07 million tonnes
Electricity generating capacity	2 600 kW
Electricity generation	26 TJ
Direct use from combustion	72 TJ
Total energy production	98 TJ

Data refer to 1996.

Morocco

Animal dung

Biogas production capacity	4.00 TJ/yr
Yield of biogas	0.56 GJ/tonne
Biogas production	4.00 TJ

Data refer to 1996.

Nepal

Agricultural residues

Quantity of raw material available	1.05 million tonnes
Solid fuel production capacity	0.51 TJ/yr
Yield of solid fuel	12.56 GJ/tonne
Solid fuel production	13.15 TJ
Direct use from combustion	13.15 TJ

Dung

Quantity of raw material available	1.96 million tonnes
Solid fuel production capacity	0.29 TJ/yr
Yield of solid fuel	10.90 GJ/tonne
Solid fuel production	20.08 TJ
Biogas production capacity	0.36 TJ/yr
Yield of biogas	10.90 GJ/tonne
Biogas production	1.33 TJ
Total energy production	21.41 TJ

Netherlands

Municipal solid waste

Electricity generation	10 296 TJ

Direct use from combustion	1 085 TJ
Total energy production	11 381 TJ

Landfill gas

Biogas production	2 763 TJ

Sludge

Biogas production	2 041 TJ

Fermentation

Biogas production	5 632 TJ

New Zealand

Landfill

Quantity of raw material available	744.55 TJ
Electricity generating capacity	9 000 kW
Electricity generation	223.37 TJ

Sewage

Quantity of raw material available	385.05 TJ
Electricity generating capacity	11 340 kW
Electricity generation	115.42 TJ
Direct use from combustion	190.00 TJ
Total energy production	305.42 TJ

Paraguay

Sugarcane bagasse

Quantity of raw material available	0.36 million tonnes

Cane juice

Quantity of raw material available	0.038 million tonnes
Ethanol production capacity	861.6 TJ/yr
Yield of ethanol	1.303 GJ/tonne
Ethanol production	295.4 TJ

Agricultural residues—cotton

Quantity of raw material available	0.285 million tonnes
Electricity generation	9.3 TJ
Direct use from combustion	4 089.9 TJ
Total energy production	4 099.2 TJ

Agricultural residues—other

Quantity of raw material available	0.068 million tonnes
Electricity generation	37.3 TJ
Direct use from combustion	1 022.5 TJ
Total energy production	1 059.8 TJ

Philippines

Municipal solid waste

Electricity generation	6 TJ

Sugarcane bagasse

Electricity generation	6 518 TJ

Crop residues—coconut

Electricity generation	7 046 GWh

Crop residues—rice

Electricity generation	2 934 GWh

Animal

Electricity generation	146 GWh

Poland

Agricultural residues—manure

Biogas production	1 054 TJ

Agricultural residues—straw etc.

Quantity of raw material available	20 million tonnes
Direct use from combustion	25 063 TJ

Industrial waste

Direct use from combustion	13 970 TJ

Other

Direct use from combustion	3 641 TJ

Portugal

Municipal solid waste

Quantity of raw material available	1 million tonnes
Electricity generating capacity	81 000 kW
Electricity generation	1 840 TJ

Romania

Municipal solid waste	
Direct use from combustion	N TJ

Senegal

Municipal solid waste	
Electricity generating capacity	20 000 kW
Agricultural residues—peanut shells	
Electricity generating capacity	22 000 kW
Biomass potential (per annum)	
Peanut shells	197 500 tonnes (221 MW)
Palmetto shells	1 740 tonnes
Sugarcane bagasse	250 000 tonnes (20 MW)
Rice husks	217 212 tonnes
Sawdust	3 000 m^3
Millet/Sorghum/Maize stalks	4 052 900 tonnes
Typha reed	1 000 000 tonnes
Cotton stalks	23 991 tonnes
Peanut haulm	790 617 tonnes

Serbia & Montenegro

Municipal solid waste	
Quantity of raw material available	33 600 TJ
Agricultural residues	
Quantity of raw material available	58 400 TJ
Orchard	
Quantity of raw material available	15 000 TJ
Vineyard	
Quantity of raw material available	6 400 TJ

Singapore

Municipal solid waste	
Electricity generating capacity	135 000 kW
Electricity generation	3 994.68 TJ

Slovakia

Agricultural residues	
Quantity of raw material available	2.4 million tonnes
Solid fuel production capacity	10 904 TJ/yr
Ethanol fuel production capacity	412 TJ/yr
Biodiesel production capacity	216 TJ/yr
Biogas production capacity	560 TJ/yr
Total energy produced	12 092 TJ

Slovenia

Municipal solid waste	
Electricity generating capacity	2 776 kW
Electricity generation	43 TJ

South Africa

Sugarcane bagasse	
Quantity of raw material available	6.974 million tonnes
Electricity generating capacity	105 000 kW

There are a number of sugar mills that use bagasse as fuel for CHP plants. Some bagasse is used as input fibre for paper.

There are two landfill sites that have, in the past, produced landfill gas: one is at Grahamstown in the Eastern Cape, the other near Johannesburg. The current status of these sites is unknown.

Afrox has a site in the Free State that produces helium. The waste gas is enriched in methane using a membrane process and is then used to fuel a generator at the plant. The installed capacity is 72 kW_e. The plant is not fully utilised and produces about 0.6 TJ/yr.

Other biomass estimates:	million litres
Potential sunflower seed oil production	600
Potential ethanol production	
Cassava	3 000
Sugarcane	500
Bagasse	250
Molasses	100

Maize	1 000
Sorghum straw	150
Wheat straw	200

In addition, bagasse, maize, sorghum and wheat residues could be used to provide bioenergy. The potential has not been fully explored.

Spain

Municipal solid waste[a]	
Quantity of raw material available	18.4 million tonnes
Electricity generating capacity	94 100 kW
Electricity generation	2 376.4 TJ
Crop residues—dry fruit shells	
Electricity generating capacity	2 240 kW
Electricity generation	35.77 TJ
Direct use from combustion	1 255.08 TJ
Total energy production	1 290.85 TJ
Crop residues—grape and olive	
Electricity generating capacity	50 900 kW
Electricity generation	1 426.01 TJ
Direct use from combustion	5 950.41 TJ
Total energy production	7 376.42 TJ
Agricultural residues—other[b]	
Electricity generating capacity	54 500 kW
Electricity generation	1 421.89 TJ
Direct use from combustion	63.85 TJ
Total energy production	1 485.74 TJ
Landfill gas and sewage sludge gas	
Biogas production	6 168.41 TJ
Electricity generating capacity	64 687 kW
Electricity generation	1 545.16 TJ
Direct use from combustion	365.34 TJ
Total energy production	8 078.91 TJ
Animal dung[c]	
Biogas production	872.91 TJ
Electricity generating capacity	3 250 kW
Electricity generation	72.93 TJ
Direct use from combustion	236.81 TJ
Total energy production	1 182.65 TJ
Other[d]	
Ethanol production capacity	4 844.13 TJ/yr
Yield of ethanol	26.79 GJ/tonne
Ethanol production	4 844.13 TJ
Biodiesel production capacity	226.09 TJ/yr
Yield of biodiesel	37.68 GJ/tonne
Biodiesel production	226.09 TJ
Electricity generating capacity	22 450 kW
Electricity generation	517.50 TJ
Direct use from combustion	14 621.18 TJ
Total energy production	20 208.90 TJ

[a] Data refer to 1999.
[b] Including cereal straw and other agricultural residues.
[c] Including biogas produced by animal residues and agro-alimentary industries.
[d] Including cereals for the production of ethanol, oils and fats for the production of biodiesel, black liquor and other types of biomass.

Swaziland

Sugarcane bagasse	
Quantity of raw material available	1.32 million tonnes

Sweden

Municipal solid waste	
Quantity of raw material available	2.46 million tonnes
Biogas production	5 000 TJ
Electricity generation	1 100 TJ
Direct use from combustion	20 200 TJ
Total energy production	26 300 TJ
Agricultural residues—wheat	
Ethanol production	1 900 TJ

Switzerland

Municipal solid waste	
Quantity of raw material available	3.01 million tonnes
Biogas production	2 491 TJ
Electricity generating capacity	262 000 kW
Electricity generation	4 910 TJ
Direct use from combustion	9 479 TJ
Total energy production	16 880 TJ
Sludge	
Quantity of raw material available	125 GWh
Direct use from combustion	338 TJ

Other

Quantity of raw material available	744 GWh
Electricity generation	78 TJ
Direct use from combustion	2 197 TJ
Total energy production	2 275 TJ

Taiwan, China

Municipal solid waste

Quantity of raw material available	9.4 million tonnes
Yield of biogas	0.38 GJ/tonne
Biogas production	334 TJ
Electricity generating capacity	265 000 kW
Electricity generation	17 698 TJ
Total energy production	18 032 TJ

Sugarcane bagasse

Quantity of raw material available	0.53 million tonnes
Electricity generating capacity	60 980 kW
Electricity generation	4 142 TJ
Direct use from combustion	2 824 TJ
Total energy production	6 966 TJ

Agricultural residues—rice hulls

Quantity of raw material available	0.3 million tonnes

Agricultural residues—hog manure

Quantity of raw material available	79.2 million tonnes
Biogas production capacity	8 210 TJ/yr
Yield of biogas	0.1 GJ/tonne
Biogas production	173 TJ
Electricity generating capacity	1 700 kW
Electricity generation	115 TJ
Direct use from combustion	753 TJ
Total energy production	1 041 TJ

Thailand

Municipal solid waste[a]

Quantity of raw material available	11.56 million tonnes
Biogas production capacity[b]	51.25 TJ/yr
Yield of biogas	10.98 GJ/tonne
Electricity generating capacity[c]	2 500 kW
Electricity generation	64.8 TJ

Sugarcane bagasse

Electricity generating capacity	67 100 kW
Electricity generation	8 464.26 TJ
Direct use from combustion	156 663.56 TJ
Total energy production	165 127.22 TJ

Agricultural residues—paddy husk

Quantity of raw material available	2.986 million tonnes
Electricity generating capacity	80 100 kW
Electricity generation	2 526.034 TJ
Direct use from combustion	32 506.08 TJ
Total energy production	35 032.114 TJ

Agricultural residues—oil palm shell

Quantity of raw material available	0.151 million tonnes
Direct use from combustion	1 636.95 TJ

Agricultural residues—coconut shell

Quantity of raw material available	0.092 million tonnes
Direct use from combustion	143.44 TJ

Agricultural residues—cassava rhizome

Quantity of raw material available	3.155 million tonnes
Direct use from combustion	29 057.14 TJ

Agricultural residues—corn cob

Quantity of raw material available	0.232 million tonnes
Direct use from combustion	4 185.28 TJ

Agricultural residues—sorghum leaves & stem

Quantity of raw material available	0.22 million tonnes
Direct use from combustion	0.423 TJ

Agricultural residues—palm oil

Quantity of raw material available	4.03 million tonnes
Biodiesel production capacity[d]	213 TJ/yr
Yield of biodiesel	7.91 GJ/tonne
Biodiesel production	71 TJ

Agricultural residues—tapioca	
Quantity of raw material available	16.87 million tonnes
Ethanol production capacity[e]	18.08 TJ/yr
Yield of ethanol	4.5 GJ/tonne
Ethanol production	1.5 TJ
Other—animal waste	
Quantity of raw material available	13.67 million tonnes
Biogas production capacity	405.51 TJ/yr
Yield of biogas[f]	4.56 GJ/tonne
Biogas production[g]	385.23 TJ/tonne
Other—waste water	
Biogas production capacity	994.84 TJ/yr
Yield of biogas	9.6 GJ/tonne
Biogas production	826.06 TJ/tonne

[a] Total quantity of MSW after recycling process (recycle ratio = 18%).
[b] Kampangsan landfill site.
[c] Phuket province.
[d] Capacity of biodiesel plant = 20 000 litres/day.
[e] Capacity of plant = 5 000 litres/day.
[f] Yield of biogas production from swine waste, volatile solid yields from other animal wastes are as follows (GJ/tonne volatile solid): cattle = 6.45; buffalo = 6.01; chicken = 5.08; elephant = 5.06.
[g] 95% of total capacity.

Turkey

Municipal solid waste	
Electricity generating capacity	1 400 kW
Agricultural residues—vegetal waste	
Quantity of raw material available	1.846 million tonnes
Industrial waste	
Electricity generating capacity	19 000 kW
Biogas	
Electricity generating capacity	7 200 kW
Animal waste	
Quantity of raw material available	3.763 million tonnes

United Kingdom*

Poultry litter, meat and bone, straw, farm waste and short-rotation coppice	
Production	355 ttoe
Input to electricity generation	283 ttoe
Direct use from combustion	72 ttoe
Sewage gas	
Production	184 ttoe
Input to electricity generation	130 ttoe
Direct use from combustion	53 ttoe
Electricity generating capacity	96 MW
Electricity generation	397 GWh
Landfill gas	
Production	892 ttoe
Input to electricity generation	879 ttoe
Direct use from combustion	14 ttoe
Electricity generating capacity	472.9 MW
Electricity generation	2 679 GWh
Waste and tyres[a]	
Production	763 ttoe
Input to electricity generation	672 ttoe
Direct use from combustion	91 ttoe
Municipal solid waste combustion	
Electricity generating capacity	278.9 MW
Electricity generation[b]	958 GWh
Other	
Electricity generating capacity[c]	165.7 MW
Electricity generation[d]	870 GWh
Wastes[e]	
Electricity generation	494 GWh

[a] Comprises municipal solid waste, general industrial waste and hospital waste.
[b] Biodegradable part only.
[c] Includes the use of farm waste digestion, waste tyre, poultry litter, meat and bone and straw combustion, and short-rotation coppice.
[d] Includes the use of farm waste digestion, poultry litter combustion, meat and bone combustion, straw and short-rotation coppice.
[e] Non-biodegradable part of municipal solid waste plus waste tyres.

Landfill gas. Landfill gas exploitation has benefited considerably from the Non-Fossil Fuel Obligation (NFFO) and this can be seen from the large rise in the amount of electricity generated since 1992. Further commissioning of landfill gas projects under NFFO will continue to increase the amount of electricity generated from this technology. In 2002, 13 new schemes came on line under NFFO.

Sewage sludge digestion. In all sewage sludge digestion projects, some of the gas produced is used to maintain the optimum temperature for digestion. In addition, many use CHP systems. The electricity generated is either used on site or sold under the NFFO.

Short-rotation coppice. Short-rotation willow coppice development is now becoming well established with demonstration projects underway in Northern Ireland and England. Under Northern Ireland's second Non-Fossil Fuel Renewable Energy order for electricity, two projects were live at the end of 2002.

In England, Project ARBRE in south Yorkshire was contracted under NFFO 3 to generate 10 MW of electricity of which 8 MW were to be exported to the local grid. This project has run into difficulties and has recently (2002/2003) been sold to new owners who are currently evaluating their options on taking the project forward.

Straw combustion. Straw can be burnt in high-temperature boilers, designed for the efficient and controlled combustion of solid fuels and biomass to supply heat, hot water and hot air systems. There are large numbers of these small-scale batch-fed whole bale boilers. The figures given are estimates based partly on 1990 information and partly on a survey of straw-fired boilers carried out in 1993–1994. A 31 MW straw-fired power station near Ely, Cambridgeshire was commissioned in 2000 and has been exporting electricity since September of that year.

Waste combustion. Domestic, industrial and commercial wastes represent a significant resource for materials and energy recovery. Wastes may be combusted, as received, in purpose-built incinerators or processed into a range of refuse-derived fuels (RDF) for both on-site and off-site utilisation.

Nineteen WTE plants were in operation in 2002 burning MSW, RDF and general industrial waste (GIW).

Waste can be partially processed to produce coarse RDF which can then be burnt in a variety of ways. By further processing the refuse including separating off the fuel fraction, compacting, drying and densifying, it is possible to produce an RDF pellet. The pellet has around 60% of the gross calorific value of British coal.

The generation of MSW has been split between biodegradable sources and non-biodegradable sources using information on calorific values of the constituent parts. Approximately 66% of generation from MSW was estimated to be from biodegradable sources.

General industrial waste combustion. Certain wastes produced by industry and commerce can be used as a source of energy for industrial processes or space heating. These wastes include general waste from factories such as paper, cardboard, wood and plastics.

A survey conducted in 2001 has highlighted that although there are six WTE plants burning GIW, these are all MSW facilities. As no sites are solely burning GIW for heat or electricity generation, this feedstock is being handled under the MSW category.

Specialised waste combustion. Materials in this category include scrap tyres, hospital wastes, poultry litter, meat and bone and farm waste digestion. The large tyre incineration plant with energy recovery did not generate in 2002. Although part of waste tyre combustion is of biodegradable waste, because there is no agreed method of calculating the small biodegradable content, all the generation from waste tyres has been included under non-biodegradable wastes when calculating renewables eligible for the Renewables Obligation (RO) and the European Union's Renewables Directive (RD).

One poultry litter combustion project started generating electricity in 1992, a second began in 1993. Both of these are NFFO projects. In addition, a small-scale CHP scheme began generating towards the end of 1990. However, this has now closed due to new emissions regulations. A further NFFO scheme started generating in 1998 at Thetford, Norfolk, and during 2000 a Scottish Renewable Order (SRO) began to generate. During 2000–2002 one of the earlier poultry litter projects was fuelled mainly by meat and bone. A new poultry litter scheme became fully operational in 2001.

There was a farm waste digestion project generating electricity under the NFFO but it ceased to operate post-2002.

Co-firing of biomass with fossil fuels. Co-firing of biomass with fossil fuels is now eligible under the RO, the first time that any RE initiative has included co-firing. As the purpose of this was to enable markets and supply chains for biomass to develop, and not to support coal-fired power stations, the following limits are placed on co-firing:

- only electricity generated before 1st April 2011 will be eligible;
- from 1st April 2006 at least 75% of the biomass must consist of energy crops.

Co-firing of biomass fuel in fossil fuel power stations is not a new idea. Technically it has been examined and proven to various degrees in power stations worldwide. It has not been considered at the large power station scale in the UK until fairly recently but a number of utilities are now investigating use of a range of biomass products at various coal-fired power station sites. The ability of coal station furnaces to cope with the introduction of such biomass is dependent on a number of factors including biomass composition and furnace design. Current trials are planned to look at possible substitution at up to 20% on a thermal basis. However, the scale of implicit fuel preparation and plant transport systems may limit the scope for substitution in addition to coal furnace considerations. The 2002 data for biomass include biomass use at one UK coal-fired power station, although other stations have begun to burn biomass in 2003.

*Source: Digest of United Kingdom Energy Statistics 2003.

United States of America

Municipal solid waste	
Quantity of raw material available	146.2 million tonnes
Biogas production capacity	145 895 TJ/yr
Electricity generating capacity[a]	3 308 000 kW
Electricity generation[a]	72 651 TJ
Agricultural residues—corn stover and wheat straw	
Quantity of raw material available	176 million tonnes
Corn	
Ethanol fuel production capacity	236 834 TJ/yr
Yield of ethanol	2.5 gal/bushel of corn
Ethanol production[b]	171 553 TJ
Other	
Ethanol fuel production capacity	4 916 TJ/yr
Biodiesel production capacity	9 391 TJ/yr
Yield of biodiesel	40.2 GJ/tonne
Biodiesel production	2 053 TJ
Electricity generating capacity[c]	539 000 kW
Electricity generation[c]	9 636.7 TJ

[a] Includes landfill gases.
[b] Includes production from corn and other.
[c] Includes agricultural residues.

Uruguay

Sugarcane bagasse	
Quantity of raw material available	0.05 million tonnes
Electricity generation	24.3 TJ
Direct use from combustion	477 TJ
Agricultural residues—sunflower husks	
Quantity of raw material available	0.01 million tonnes
Direct use from combustion	164.4 TJ
Agricultural residues—rice husks	
Quantity of raw material available	0.21 million tonnes
Electricity generation	8.8 TJ
Direct use from combustion	558.6 TJ

Data refer to 2001.

Chapter 11

Solar Energy

World Energy Council
CONSEIL MONDIAL DE L'ENERGIE

COMMENTARY

Introduction

Renewable energy is the most abundant permanent energy resource on earth. Fig. 11.1 provides, at a glance, a comparison between annual renewable energy (solar and its most important indirect forms, biomass, wind, hydro) and total non-renewable resources (oil, gas, uranium, coal).

This commentary is solely concerned with solar radiation—direct (beam) radiation and circumsolar and diffuse (sky) radiation—which is, as shown in Fig. 11.1, the earth's prime energy resource.

In fact, taking the best guesses and only to give an order of magnitude, the energy content of the annual solar radiation which reaches the earth and its atmosphere is 2 895 000 EJ, compared to the total non-renewable energy resources of 325 300 EJ (oil, 8 690 EJ; gas, 17 280 EJ; uranium, 114 000 EJ; coal, 185 330 EJ). The energy content of other major renewables is estimated as 1 960 EJ (hydro, 90 EJ; wind, 630 EJ; photosynthetic storage/biomass, 1 260 EJ), a very small fraction of annual solar radiation. Current world primary energy consumption is about 425 EJ/yr.

Today, 80% of worldwide energy use is based on fossil fuels. Several risks are associated with their use. Energy infrastructures—power plants, transmission lines and substations, and gas and oil pipelines—are all potentially vulnerable to adverse weather conditions or human acts. During summer 2003, one of the hottest and driest European summers in recent years, the operations of several power plants, oil and nuclear, were put at risk owing to a lack of water to cool the condensers. World demand for fossil fuels (starting with oil) is expected to exceed annual production, very likely within the next two decades. Shortages of oil or gas can initiate international economic and political crises and conflicts. Burning fossil fuels releases emissions to the environment, with effects at the local (e.g. benzene), regional (aerosols, short-lived gases) and global (greenhouse gases) level.

Concerns regarding present energy systems are therefore growing because of the inherent risks connected with security of supply and potential international conflicts, and on account of the potential damage they can do to the

2004 Survey of Energy Resources

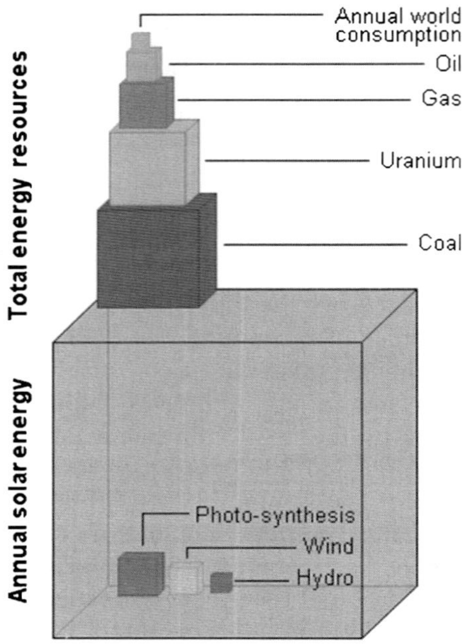

FIGURE 11.1 Order of magnitude of energy sources on earth (Source: Lomborg, 2001).

natural environment in many and diverse ways. World public opinion, international and national institutions, and other organisations are increasingly aware of these risks, and they are pointing to an urgent need to fundamentally transform present energy systems onto a more sustainable basis.

A major contribution to this transformation can be expected to come from solar radiation, the prime energy resource. In several regions of the world we can see the seeds of this possible transformation, not only at the technological level, but also at policy levels, as for example, analysing policies and plans put in place by world leaders in renewable energy like the European Union and those member states with a particular commitment to solar energy development.

Fig. 11.2 shows an 'exemplary path' to 2050/2100 studied by the German Advisory Council on Global Change, which points, in the long term, to a major contribution to global energy consumption from solar energy.

This scenario, which will be published in full in the first half of 2004, is based on the recognition that it is essential to turn energy systems towards sustainability worldwide, both in order to protect the natural life-support systems on which humanity depends and to eradicate energy poverty in developing countries. It foresees that in a hundred years the world could turn again to solar and renewable energy, which until well into the 19th century was the world's sole supply; wood, charcoal and straw were still the prevailing sources of energy everywhere, except for a few European countries. Of course, this new solar era can be envisioned because of the tremendous scientific and technological advances made during the last centuries.

By 2100 oil, gas, coal and nuclear, as shown in Fig. 11.2, should cover less than 15% of world energy consumption while solar thermal and photovoltaic should cover about 70%. Key elements of this long-term scenario are the energy efficiency and energy intensity policies that will make the contribution of renewable and solar energy a substantial factor. Those policies will deeply transform the building, industry and transport sectors, increasing their reliance on renewable energy resources.

The transition towards this possible future has already started. In the following paragraphs an attempt will be made to show this by reviewing the state of the art regarding *Solar Radiation Resource Assessment* and the status and rate of growth of the major solar energy technologies, their technical and market maturity as well as institutional and governmental policies and approaches to promote their integration into the world's energy systems. Buildings are already the largest human fabric intercepting solar radiation and, therefore, potentially, a driving sector for the development of solar energy use.

Solar Radiation Resources

Solar is the world's most abundant, permanent source of energy: Fig. 11.1 depicts the amount of solar radiation intercepted annually by the whole of the earth's surface. The amount

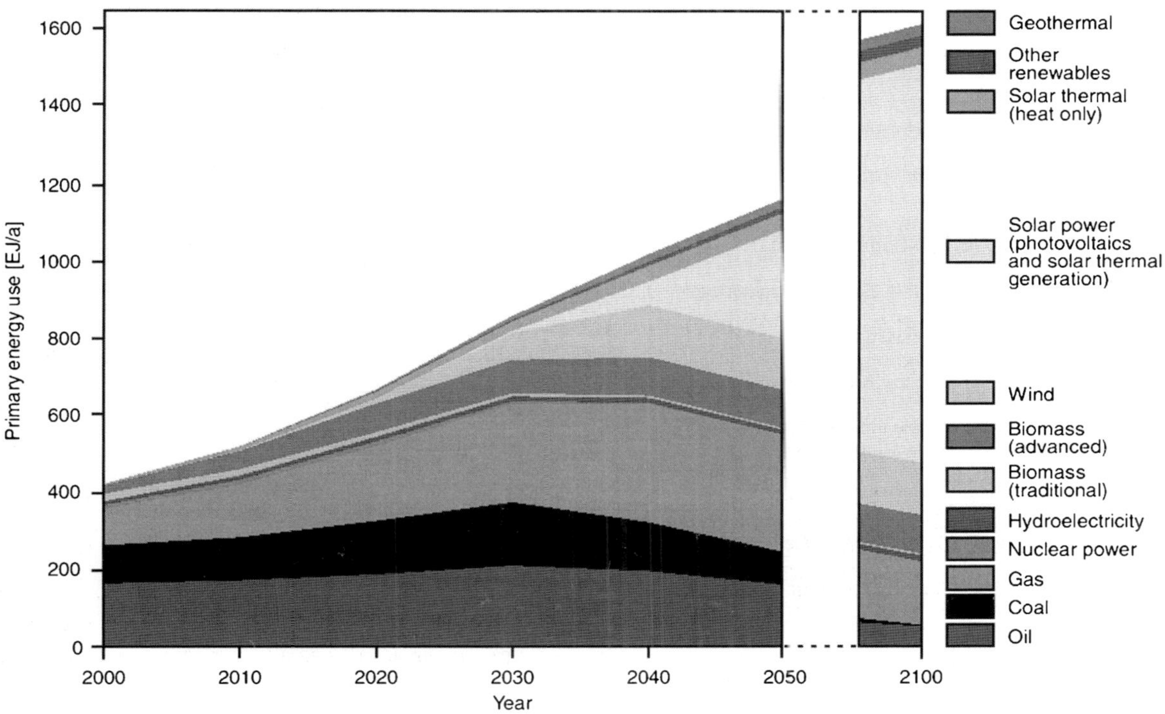

FIGURE 11.2 Transforming the global energy mix: The exemplary path to 2050/2100 (Source: WBGU, 2003).

of solar radiant energy falling on a surface per unit area and per unit time is called irradiance. The average extraterrestrial irradiance or flux density at a mean earth–sun distance, known as the *solar constant*, normal to the solar beam on the outer fringes of the earth's atmosphere, is 1 366 W/m^2, according to a recent estimate. When this power density is averaged over the surface of the earth's sphere, it is reduced by a factor of 4. A further reduction by a factor of 2 is due to losses in passing through the earth's atmosphere. Thus, the annual average horizontal surface irradiance is approximately 170 W/m^2. When 170 W/m^2 is integrated over 1 year, the resulting 5.4 GJ that is incident on 1 m^2 at ground level is approximately the energy that can be extracted from one barrel of oil, 200 kg of coal, or 140 m^3 of natural gas.

However, the flux changes from place to place. Some parts of the earth receive up to about 40% more than this annual average. The highest annual mean irradiance of 300 W/m^2 can be found in the Red Sea area, and typical values are about 200 W/m^2 in Australia, 185 W/m^2 in the United States and 105 W/m^2 in the United Kingdom. These data show that the annual solar resource is almost uniform (within a factor of about 2), throughout almost all regions of the world. It has already been shown that economically attractive applications of solar energy are not limited to just the sunniest regions. Northern European countries offer good examples of this.

Fig. 11.3 gives the potential of solar energy of 1% (minimum) or 10% (maximum) of unused land in different regions of the world. For all regions in Fig. 11.3 the minimum solar potential is estimated at 1 575 EJ/yr and the maximum is 49 837 EJ/yr. The lowest estimate is more than three times the current global primary energy consumption of about 425 EJ/yr.

However, the important figure for any area is the ratio between the solar input and local energy

Region	Unused land (Gha)	Assumed for solar energy* (Mha)		Solar energy potential (EJ/yr)	
	Available	Min.	Max.	Min.	Max.
NAM (North America)	0.5940	5.94	59.4	181.1	7 410
LAM (Latin America & the Caribbean)	0.2567	2.57	25.7	112.6	3 385
AFR (Sub-Sahara Africa)	0.6925	6.93	69.3	371.9	9 528
MEA (Middle East & North Africa)	0.8209	8.21	82.1	412.4	11 060
WEU (Western Europe)	0.0864	0.86	8.6	25.1	914
EEU (Central & Eastern Europe)	0.0142	0.14	1.4	4.5	154
FSU (Newly independent states of the former Soviet Union)	0.7987	7.99	79.9	199.3	8 655
PAO (Pacific OECD)	0.1716	1.72	17.2	72.6	2 263
PAS (Other Pacific Asia)	0.0739	0.74	7.4	41.0	994
CPA (Centrally planned Asia & China)	0.3206	3.21	32.1	115.5	4 135
SAS (South Asia)	0.1038	1.04	10.4	38.8	1 339
WORLD TOTAL	3.9331	39.33	393.3	1 575.0	49 837
Ratio to the current primary energy consumption (425 EJ/yr)				3.7	117

* The maximum corresponds to 10% of the unused land; the minimum corresponds to 1% of the unused land

FIGURE 11.3 Minimum and maximum solar energy potential from unused land (Source: Climate Change 2001: Working Group III: Mitigation).

consumption, which varies from less than 100 in some energy-intensive countries to over 10 000 in some developing countries.

The energy delivered by the sun is both intermittent and changes during the day and with the seasons. To evaluate the efficiency of solar energy systems, the 'standard sun', the flux of about 1 000 W/m^2 that the sun can deliver to a surface area on earth directly facing the sun, is used and which, consequently, is rated in terms of 'peak watts' (output under a 1 kW/m^2 illumination).

In a period of rapidly growing deployment of solar energy systems, it is imperative that solar resource parameters and their space/time specificity are well known by solar energy professionals, planners, decision makers, engineers and designers. Because these parameters depend on the applications (flat solar thermal collectors, solar thermal power plants, photovoltaic, window glass, etc.), they may differ widely, and might be unavailable for many locations, given that irradiance measurement networks or meteorological stations do not provide sufficient geographically time-site specific irradiance coverage. This coverage is especially useful because it allows assessment of the output of a solar system in relation to the technical characteristics of the system, local geography and energy demand. It therefore allows a better assessment of the feasibility of a solar energy application and of its value.

Today, there is a proliferation of data and solar maps, several easily accessible on the Internet. Users can also customise maps, for example, developing regional maps from national maps, incorporating other data such as the location of various infrastructures and natural features, by using geographic information system (GIS) technology.

A few examples of organisations and databases concerned with solar energy resources found on the Internet are included in the references at the end of this commentary.

Light, Heat, Electricity and Fuels from Solar Energy

The energy content of solar radiation can be used directly or converted into other energy forms useful in our daily lives—light, heat, electricity, fuels and other applications—by solar technologies.

Some are better known and popular, such as photovoltaic and solar thermal collectors, others are less known, such as the solar detoxification of contaminated waters or solar distillation. The value of the less familiar uses is often underestimated or even ignored. For example, 1 m^2 intercepting a flux of 1 000 W/m^2 of solar energy could deliver over 600 W of heating energy, if the solar radiation were delivered directly into a building through a square metre of glass. In comparison 1 m^2 of photovoltaic technology can deliver 100 AC watts of peak electricity and a solar thermal collector 300 W for heating domestic water. However, no one solar technology can be declared to be more important than any other—each plays a role. Their diversity makes solar an important option to power different energy systems in all countries of the world.

The role of an individual solar technology can be enhanced by integrating it with one or more other technologies that are either energy

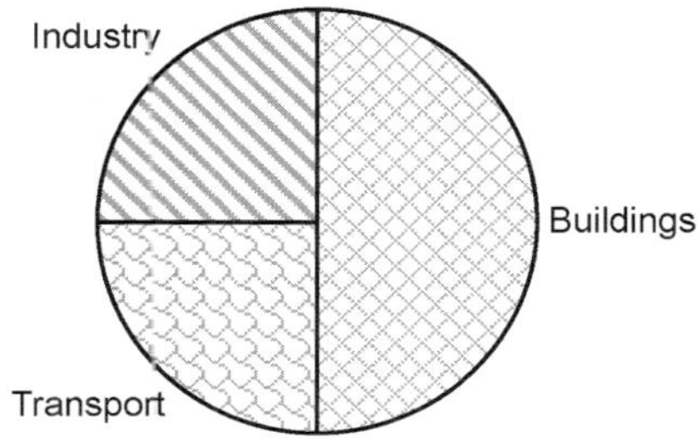

FIGURE 11.4 Energy consumption in buildings, industry and transport.

efficient or solar powered. Replacing electric hot water boilers with solar heat and other conventional electrical appliances (light bulbs, refrigerators, washing machines, etc.) with innovative energy-efficient models, can reduce electricity demand and increase the significance of, e.g. the contribution of photovoltaic electricity to the whole energy budget.

Buildings offer excellent examples of this approach. They are the modern world's main and most widespread technological systems, and the most direct expression of a people's culture. Solar technologies can make a substantial contribution to the energy budget of a modern building, and consequently to the world's energy use. Buildings should be seen as complex energy systems and as the largest collectors of solar energy.

In everyday life, residential, commercial and industrial buildings need lighting, heating, cooling and electricity. Fig. 11.4 is a simplified illustration of the shares of energy consumption in three different sectors—buildings, industry and transport—in the industrial nations.

In industrialised countries, from 35 to 40% of total primary energy consumption is used in buildings. However, if we take into account the energy used to manufacture materials and the infrastructure to serve buildings, buildings' share of total primary energy consumption can be around 50%. In Europe, 30% of energy use is for space and water heating alone, representing 75% of total building energy use. Fig. 11.5 shows how the building energy budget can be changed from conventional buildings to a passive solar house standard. The heat consumption has

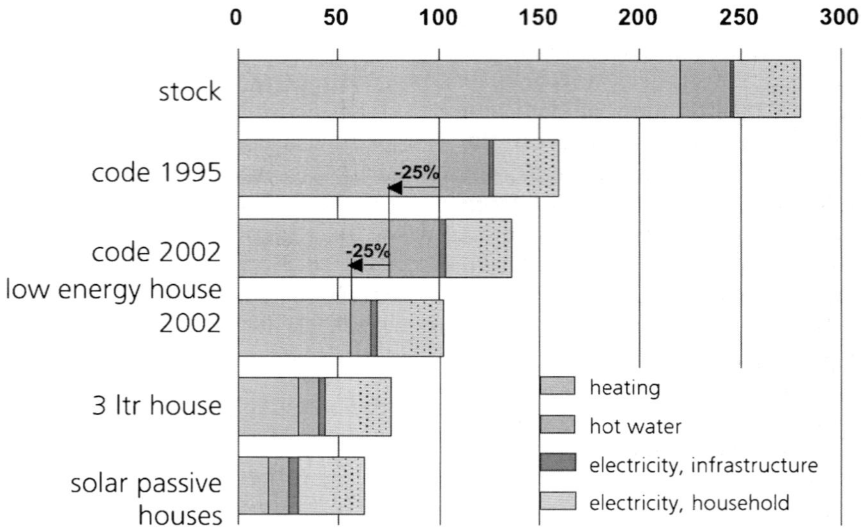

FIGURE 11.5 Specific energy demand for conventional buildings and passive house standard in Germany (kWh/m^2/yr) (Source: Eurec 2002).

also been significantly reduced by adopting a design which allows the low winter sun to penetrate the south side of the building but provides shade in the high summer sun. For such houses, consumption can be reduced by up to 75% or more.

This result is the outcome of buildings that have been continually transformed over the last few years. A growing number of new energy-related elements are being incorporated in the roofs and walls—the so-called skin or envelope. Insulation, special glass and smart windows are being utilised to reduce energy demand whilst active solar heating and electricity-generation systems are supplying the energy needed to operate the building.

The design approach has been changed to a 'whole building' approach, which sees the various problems and solutions as a whole and tackles them in an integrated and intelligent way right from the start of the design process, when every choice is decisive. It moves beyond the simple concept of 'energy saving' or 'solar energy' and deals with them at the same time by involving multidisciplinary teams to develop new forms of integration for all the above-mentioned solar technologies. This approach allows for new buildings to be designed and built yielding 30–70% improvements in energy efficiency, with construction costs 2–10% higher.

These new approaches are initiating a revolution in the building construction industry, which could become the main driver for the application of technologies for the production of electricity and heat from solar radiation. Globally, about 8–10 million new buildings are constructed every year, most of them in developing countries. Large areas of these countries do not have access to gas for water heating or to an electricity grid, thus making solar energy an attractive alternative. Even if a tiny fraction of these buildings were solar, the implications on the solar and energy industry could be enormous, not only from a technological point of view but also from a cultural point of view. It would be a contributory factor to changing the way conventional sources of energy and solar energy are thought about.

The solar revolution in the building industry will not happen, however, by itself. It will require new regulation and building codes that make binding energy-saving measures and the integration of new energy-efficient and solar technologies in buildings. Regulation has been the secret behind several success stories in the use of solar thermal collectors: for example, the 1980 regulation in Israel requiring every new building with a height of less than 27 m to have a solar thermal system on its roof. Similar regulations adopted over the last few years by several large and small towns (e.g. Barcelona in Spain) have stimulated a significant growth in solar thermal installations.

Aesthetics and architectural aspects may also play a crucial role in revolutionising building construction. There is a view that the integration of solar collectors should be consistent with the design of the roof and facade of a building, as if integration should be synonymous with invisibility. According to other views, integration should enhance the aesthetic of a building without hiding the solar energy systems.

Because buildings do not exist in isolation, the 'whole building' approach can be extended to further buildings or blocks of buildings, as in the photovoltaic application shown in Fig. 11.6. This depicts Cosmotown Kiyomino SAIZ, a complex of 79 homes built by the Hakushin company, with the Kubota Corporation supplying a roof-integrated 3 kW photovoltaic power generation system for each house.

FIGURE 11.6 The Japanese Cosmotown Kiyomino SAIZ housing development.

This illustration also underlines an argument, often raised against solar energy utilisation: namely land usage. Solar energy is often seen to be a 'dispersed' source of energy compared with concentrated fossil fuels and nuclear energy. This argument is misleading because the solar energy systems installed on walls and roofs in Kiyomino do not use land additional to that used for the construction of the buildings themselves. Moreover, land usage for fossil fuel infrastructures for transportation, distribution and waste storage can be considerable.

The extension of solar energy use from a block of solar buildings to an entire city could become possible. There are several cities around the world that are working in this direction, aiming at greater use of solar energy within the context of a long-term plan for sustainable urban development. Such projects focus on cities as complete systems, in which passive solar heating and cooling, daylighting, solar photovoltaic, and solar thermal technologies are integrated.

The markets for these last two technologies have been growing rapidly over the last decade, especially in industrialised countries. Annual growth of solar energy for hot water and electricity production increased at more than 32% per annum in the period 1971–2000, albeit from a very low base. However, the contribution of solar radiation to the world total primary energy supply is still very, very small, at about 0.04%.

In the following paragraphs the most widely used solar systems for the production of electricity, heat and fuels are reviewed.

Solar Photovoltaic Systems (PV)

The emerging markets for PV power generation systems include a large variety of applications ranging from a few watts to megawatts. The International Energy Agency identifies four primary applications for photovoltaic power with a rated power of 40 W or more:

- off-grid domestic systems that provide power to households and villages that are unconnected to the grid;
- off-grid non-domestic systems (these were the first commercial applications of PV, to power equipment installed far from the grid for telecommunications, water pumping, warning lights and environmental data recording);
- grid-connected distributed PV systems to supply power to a building or other load that is also connected to the grid;
- grid-connected centralised systems that are designed to provide an alternative to conventional centralised power generation or to strengthen utility distribution systems.

By end-2002 a cumulative total of about 1 500 MW of PV power had been installed globally, with about 400 MW installed during 2002. Most of this capacity represents grid-connected distributed systems, integrated in residential, commercial and industrial building envelopes (walls, roofs, glass).

Whilst interpreting this result, it should be noted that the PV growth is currently taking place mainly in a few industrialised countries. By end-2002, the IEA-PVPS[1] countries had installed capacity totalling 1 350 MW, of which Japan accounted for 637 MW, Germany 295 MW and the USA 212 MW.

These three leading countries have the most important national programmes for the installation of grid-connected PV on roofs. With regard to PV in Japan most of the supporting measures provide a direct capital subsidy for the installation cost of the equipment. In Germany the successful 'feed-in tariff' incentive was confirmed at the end of December 2003 by the German Parliament. The regulation will guarantee that every operator of a photovoltaic installation will receive a fixed basic reimbursement of €0.457 for each kWh fed into the public grid. An additional €0.117 will be guaranteed for photovoltaic installations on roofs of up to 30 kW and €0.093 for installations over 30 kW. Facades will attract a further payment of €0.05. Each of these payments will last for 20 years. In addition, the German Government has cancelled the former limitation of promoting up to a total of 1 000 MW. The situation is different in the US, where the 'Million Roofs' solar programme is voluntary; a large variety of sub-programmes are promoted at State level and include commercially built homes with PV as a standard feature (Solar Advantage Homes Program promoted by The Sacramento Municipal Utility District).

Regarding the total installed PV power per capita, Japan is first in the world with 5.00 W per capita, followed by Switzerland with 3.80 and Germany with 3.37.

PV is also expanding rapidly in developing countries, where the newly installed capacity accounted for 2–3 MW in 2002.

The PV industry has grown rapidly in recent years and increased by 44% in 2002. The value of PV sales amounted to about US$ 3.5 billion in 2002 and is projected to grow to more than US$ 27.5 billion in 2012.

Fig. 11.7 shows cell and module production and production capacity in the year 2002. Japan is the leading producer, followed by Europe and the USA.

Fig. 11.8 specifies the volume of PV production (for the top 10 manufacturers), showing each company's share of the world total, from 1999 to 2002.

Although PV is a fast-growing sector, it remains very small compared to other new renewable electrical energy markets, such as wind. In addition, PV is also still the most expensive of the solar technologies for electricity production.

However, while measuring the value of PV in cost per kWh produced, the many features of PV should not be undervalued. Its versatility allows it to have many stand-alone applications, significantly reducing the overall project cost by avoiding the expense of grid connection. The value of PV is also important when used

[1] International Energy Agency Photovoltaic Power Systems (IEA-PVPS) Programme: Australia, Austria, Canada, Denmark, Finland, France, Germany, Israel, Italy, Japan, Korea, Mexico, the Netherlands, Norway, Portugal, Spain, Sweden, Switzerland, the United Kingdom, the United States of America.

Unit: MW	Cell production	Cell production capacity	Module production	Module production capacity
Japan	244	361	260	405
Europe	134	226	133	217
USA	121	177	81	148
Rest	21	37	8	22
TOTAL	**520**	**801**	**482**	**792**

FIGURE 11.7 Cell and module production and production capacity in 2002 in the IEA-PVPS countries.

to meet some of the most fundamental needs of people living in developing countries, such as water pumping or food and medicine refrigeration.

The cost of PV should therefore be seen in a medium/long-term perspective as the dramatic falls of past years continue. Today, the wholesale costs for PV are estimated at about an average of US$ 4.10/W for modules. With respect to PV off-grid systems, their cost is greater than that of grid-connected systems because they require storage batteries and associated equipment.

	PRODUCTION								GROWTH	
	1999		2000		2001		2002		2002/2001	
	(MW)	(%)	(MW)	(%)	(MW)	(%)	(MW)	(%)	(MW)	(%)
Sharp (Japan)	30.0	14.9	50.4	17.5	75.0	19.2	123.1	21.9	48.1	64.0
BP Solar (UK)	32.5	16.1	41.9	14.6	54.2	13.9	73.8	13.1	19.6	36.2
Kyocera (Japan)	30.3	15.1	42.0	14.6	54.0	13.8	60.0	10.7	6.0	11.1
Shell (UK-NL)	22.2	11.0	28.0	9.7	39.0	10.0	57.5	10.2	18.5	47.4
Sanyo Electric (Japan)	13.0	6.5	17.0	5.9	19.0	4.9	35.0	6.2	16.0	84.2
AstroPower (USA)	12.0	6.0	18.0	6.3	26.0	6.7	29.7	5.3	3.7	14.2
RWE Solar (Germany)	10.0	5.0	14.0	4.9	23.0	5.9	29.5	5.3	6.5	28.3
Isofoton (Spain)	6.1	3.0	9.5	3.3	18.0	4.6	27.4	4.9	9.3	51.8
Mitsubishi Electric (Japan)	-	-	12.0	4.2	14.0	3.6	24.0	4.3	10.0	71.4
Photowatt (France)	10.0	5.0	14.0	4.9	14.0	3.6	17.0	3.0	3.0	21.4

FIGURE 11.8 The world's top 10 photovoltaic manufacturers (PV News, Vol. 22, No. 3, March 2003).

In 2002 the system cost in the off-grid sector varied from about US$ 10 to US$ 18/W. For grid-connected systems the lowest reported price was US$ 4/W, even if a typical price was US$ 7/W.

Current trends show that technological developments, subsidies, manufacturing scale-up, mass production techniques, volume sales and other factors will certainly bring the costs down further. PV could probably be expected to become one of the cheaper options for electricity production.

Technological research and development continue for all PV components and systems, from PV cells to inverters, batteries, battery charge controllers, switchgear, cabling and mounting structures. Improvements and cost reduction are reported for all of them.

Solar Thermal Systems

The use of energy in the form of heat is one of the largest items in the energy budget. Solar thermal technologies can supply heat for domestic hot water (DHW) and space heating in residential, commercial and institutional buildings, for swimming pool heating, solar-assisted cooling, solar-assisted district heating, for industrial process heat and for desalination. The most important developments are in the installation of solar thermal systems for hot water (< 100 °C) which have proved to be efficient and reliable.

More than 100 million m^2 of solar thermal collector area were estimated to be in operation in the world at the end of 2002. The newly installed area during 2002 was more than 10 million m^2. China leads with an estimated market at 5.5 million m^2 of collector area per year, mostly evacuated tube collectors.

It must be noted that using the measure of square metres of collector area installed and in operation is a simplification necessary to estimate and compare markets. However, solar thermal applications can be very different in terms of technique (glazed collectors that include flat-plate and evacuated tube collectors, unglazed collectors for heating swimming pool water, air collectors, etc.), solar yields and costs.

The unglazed collector market for swimming pool heating is dominant in Australia (with 2 million m^2 in operation), USA and Canada (with 15 million m^2). This market is between 15 and 20 times bigger than the unglazed market in Europe. Solar thermal systems with glazed and vacuum collectors for DHW and space heating cover most of the European solar thermal market.

Fig. 11.9 provides a general overview of the most important solar thermal collector markets (newly installed collectors) and the area in operation in 2002. China, Europe (with Austria, Germany and Greece leading) and the USA account for 84% of the annual market and 75% of the area in operation.

Solar thermal collector penetration has been stronger in countries where there are both national or local long-term policies and supporting measures. For example, to counter a slowdown of the solar thermal collector industry in 2002, the German Government raised the incentives for solar water heating (SWH) systems from €92 to €125 per m^2 of collector surface in February 2003 and thus contributed to improving the market in 2003. Locally, city regulations that require the installation of solar thermal collectors to supply hot water for buildings have stimulated a rapid growth of such installations, as in the case of the city of Barcelona in Spain.

The estimated annual solar thermal energy production from the collector areas in operation depends on the solar radiation available and the solar thermal technology used. For example, in Austria, estimated annual solar yields are 350 kWh/m^2 for flat-plate collectors, 550 kWh/m^2 for vacuum collectors and 300 kWh/m^2 for unglazed collectors. Estimated annual yields for flat-plate collectors are 700 kWh/m^2 in Australia, 400 kWh/m^2 in Germany and 1 000 kWh/m^2 in Israel.

The estimated annual global solar thermal collector yield of all recorded systems is around

	Total home market (newly installed glazed collectors only) m² (2002)	In operation (glazed, vacuum and unglazed) m² (2002)	Planned to be in operation m² (year)
China	5 500 000	40 000 000	230 000 000 (2015)
Europe	1 196 540	12 300 000	100 000 000 (2010)
USA *	961 453	11 128 293	n.a.
Turkey *	630 000	6 482 000	15 000 000 (2010)
Israel *	390 000	3 500 000	n.a.
Japan *	314 000	7 360 296	n.a.
Australia *	75 000	3 200 000	n.a.
India*	50 000	550 000	n.a.

*** 2001**

FIGURE 11.9 Estimated solar thermal collector areas in major world markets: newly installed, in operation and planned.

23 000 GWh. Europe accounts for approximately 4 600 GWh.

Costs can also be very different: in Greece, a DHW thermosiphon system for one family unit of 2.4 m^2 collector and 150 l tank costs €700 (VAT included); in Germany, a system for the same purpose costs around €4 500 (4–6 m^2 and 300 l tank).

The success of solar thermal systems for the production of DHW has provided motivation for other possible applications for solar heat. In recent years, there have been several significant developments in the field of solar combi-systems (combination of solar water and heating system). These have occurred in the field of large-scale solar heating systems for multi-family houses and housing developments as well as for cooling and production of heat for industrial processes. Such applications have a great potential but instances are still uncommon.

Solar Thermal Power Plants

Solar thermal power plants use direct sunlight, so they must be sited in regions with high direct solar radiation, as those shown in Fig. 11.10.

Among the most promising areas are the South-Western United States, Central and South America, Africa, the Middle East, the Mediterranean Countries of Europe, Iran, Pakistan, and the desert regions of India, the former Soviet Union, China and Australia.

Direct solar radiation can be concentrated and collected by a range of concentrating solar power (CSP) technologies to provide medium to high temperature heat. This heat is then used to operate a conventional power cycle. CSP falls into three categories: parabolic troughs, power towers and heat engines.

Today, total installed CSP capacity is 364 MW. Most of this power (354 MW) is from the nine California SEGS (solar electricity generating systems) power plants, connected to the Southern California grid. SEGS were installed during the period 1984–1991, and range in size from 14 to 80 MW_e, with over 2 million m^2 of parabolic troughs. The trough technology is the most mature solar thermal power technology and has demonstrated a maximum peak efficiency of 21% in terms of

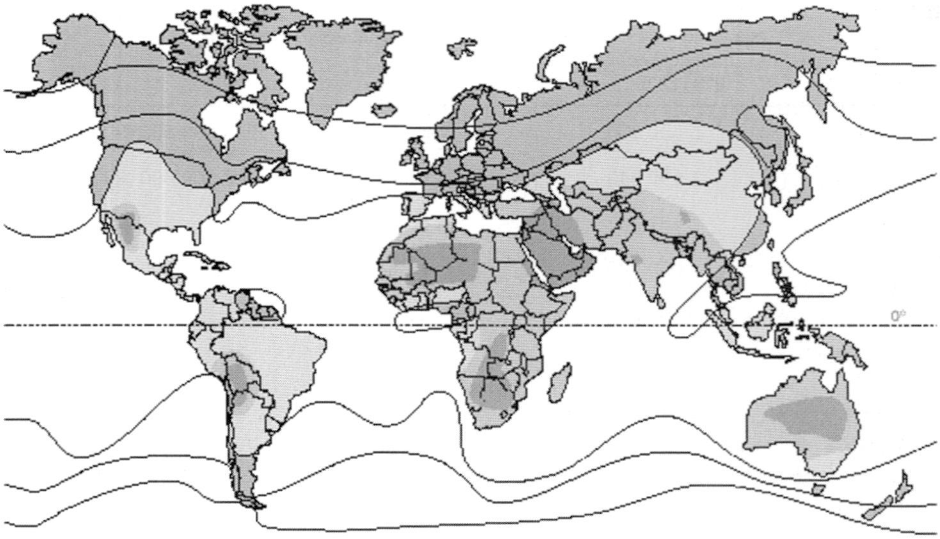

FIGURE 11.10 Regions of the world appropriate for concentrating solar power (CSP) (Source: European Commission).

conversion of direct solar radiation into grid electricity. California SEGS supply 800 million kWh/yr to the local grid at a cost of about US 10–13 cents/kWh.

The remaining CSP installed capacity is from central receiver or tower systems (10 MW in the USA, 2 MW at the Plataforma Solar de Almería (PSA), Spain; plus test plants) and from parabolic dishes (200 kW).

Until they were surpassed by the PV system installed capacity a few years ago (1997/1998), the California SEGS were the only plants producing a substantial amount of electricity from solar radiation.

Today, interest in CSP is picking up again, and projects totalling about 1 000 MW are being planned or have already started in Algeria, Australia, Egypt, Greece, India, Israel, Italy, Mexico, Morocco, Spain and the USA. A new generation of trough systems are under development, some at a conceptual or research level, others as prototype and demonstration plants.

Trough technology with direct steam generation is under experimentation at the PSA on Spain's Mediterranean coast.

The Fresnel Principle Solar Collector is being developed by the Belgian company Solarmundo (in 1999 the largest prototype of a Fresnel Collector, with a total reflector area of 2 500 m^2, was erected) and by the Australian company Solar Heat and Power. In 2003 Solar Heat and Power started installation of a 24 000 m^2 pilot CLFR (compact linear Fresnel reflector) array attached to the coal-fired 2 000 MW Liddell power station in the Hunter Valley, New South Wales.

Solargenix Energy (formerly Duke Solar) of the USA signed a long-term contract with Sierra Pacific Resources in January 2003 to supply 50 MW of electricity generated by solar power using parabolic troughs from a plant to be located in Eldorado Valley, near Boulder City, Nevada. This new plant should provide experience of new systems and lead to an expansion of solar thermal power both elsewhere in the USA and also worldwide.

Since 2000 an innovative approach in trough technology has been promoted by ENEA (The Italian Agency for New Technology, Energy and the Environment). ENEA is proposing substantial innovations in SEGS plants, among these, the use of molten salt as the thermal fluid and two thermal storages, hot and cold, to allow temperatures up to 550 °C to be reached and for the plant to be operated for 24 h a day. Starting in 2005, it is planned that a first large demonstration plant, called Archimede, will be built adjacent to an ENEL (the biggest Italian utility) gas combined power plant (Fig. 11.11). The solar plant will increase the ENEL plant's capacity by 20 MW.

The experience gained with the operational plant and those currently under construction should assist cost-reducing developments and the further revival of solar thermal electric systems on a worldwide basis. According to a study by Greenpeace and ESTIF (European Solar Thermal Industry Association), starting from the current level of just 364 MW, the total installed capacity is expected to exceed 5 000 MW by 2015, with additional capacity rising at almost 4 500 MW/yr by 2020. One long-term scenario foresees that by 2040, CSP's contribution would be at a level of almost 630 000 MW.

FIGURE 11.11 Archimede plant as envisioned at the site of ENEL power plant in Priolo Gargallo (Sicily, Italy).

The most promising countries in terms of governmental targets or potentials for CSP are Spain, the United States, Mexico, Australia and South Africa.

Solar Energy Storage Systems

As a result of solar energy's intermittent nature, the growth in worldwide usage will be constrained until reliable and low-cost technology for storing solar energy becomes available.

The sun's energy is stored on a daily basis by nature through the process of photosynthesis in foodstuffs, wood and other biomass. The storage of energy from intermittent and random solar radiation can be done artificially, by using energy storage technologies (thermal storage, chemically charged batteries, hydro storage, flywheels, hydrogen, compressed air), some well known and widely applied, others still under development.

Thermal storage for solar heat and chemically charged batteries for off-grid PV systems are the most widely used solar energy storage systems today. The most likely long-term candidate for solar energy storage is hydrogen. Many think that it will be both a primary fuel and an 'energy carrier' (as is electricity today) of the future. Its development is also expected to be supported by its potential for transforming transportation and stationary energy systems. There are more than a dozen processes that can be used to produce hydrogen, as shown in Fig. 11.12.

Steam reforming (SMR) Thermal cracking Partial oxidation (POX) Coal gasification (CG) Biomass gasification (BG) Electrolysis Grid (coal, nuclear) Solar photovoltaic Solar thermal power	Wind Photochemical Photo-electrochemical Thermochemical Solar thermal Nuclear Biological production Thermal decomposition

FIGURE 11.12 Hydrogen production technologies (Source: ASES, Renewable Hydrogen Forum, National Press Club, Washington, DC, October 2003).

A number of photobiological and electrochemical ways are being explored with a view to producing hydrogen from solar radiation. It can be produced by splitting water molecules into hydrogen and oxygen, using electricity or heat generated from solar radiation (photovoltaic, solar thermal power, concentrated solar heat). Hydrogen and oxygen can be converted back into electricity, heat and water in fuel cells. Fuel cells are thus an efficient component of the hydrogen cycle for solar energy storage.

Currently, there are hundreds of prototypes and demonstration systems in operation around the world that use the intermittent solar source to produce hydrogen and then continuous power. These systems are expected to benefit mainly off-grid sites in remote areas. However, they will also be useful for on-grid sites as well, for example, in ensuring power stabilisation and reducing expensive peak energy costs for buildings. Producing hydrogen using PV in a building will allow the building to become a hydrogen factory. Excess electricity or hydrogen can be supplied to the grid or to community refuelling stations.

The largest hydrogen factories envisioned would be built in the sunnier countries of the world, where large PV and CSP plants could produce hydrogen remotely, which would then be used on site or alternatively delivered as power and heat on demand, where needed.

Conclusion

Energy saving, energy efficiency and multidisciplinary and system approaches are required in order to broadly increase the use of solar energy.

Even if individual solar technologies for daylight, electricity, heat and fuels have made tremendous progress over the last years, the greatest challenge is to integrate them in an efficient and orderly way into energy infrastructures and the environment.

Visible changes are already taking place in the building construction industry, in particular with passive solar heating, cooling and daylighting designs. Roofs and walls are already becoming the potentially largest area in which photovoltaic and thermal systems to collect solar radiation can be integrated. Hydrogen, as the long-term candidate for solar energy storage, promises to overcome the bottleneck from solar radiation intermittency.

The overall challenge to using solar energy is, however, increasingly not so much technological as cultural and political. Only strong public policy and political leadership can move forward the application of solar and other renewables.

This is clearly shown by those countries whose Governments have established firm goals for the penetration of renewable energy into primary energy and electrical energy production or have adopted specific policy mechanisms that have achieved great success in promoting the installation of systems for producing electricity from renewables. Examples are the successful feed-in laws schemes adopted in several European countries, for instance, Austria, Germany and Spain; the Renewables Portfolio Standard (RPS) adopted by several of the American states, which ensures that a minimum amount of renewable energy is included in the portfolio of electricity production; and city ordinances requiring solar systems to be used for the production of hot water for residential and commercial buildings.

Appropriate policy measures have shown that solar applications can be boosted with many positive side effects, from the creation of new industries, new jobs and new economic opportunities, to the protection of the environment.

Cesare Silvi
ISES Italia

References

Aitken, D., 2003. Transitioning to a Renewable Energy Future. ISES International Solar Energy Society White Paper, Freiburg, November 2003.

American Solar Energy Society (ASES), 2003. Renewable Hydrogen Forum, National Press Club, Washington, DC, October 2003.

ESRA European Solar Radiation Atlas (up to the Ural mountains and including neighbouring Mediterranean territories): http://www.helioclim.net/esra/

European Solar Thermal Industry Federation (ESTIF), 2003. Sun in Action II A Solar Thermal Strategy for Europe, April 2003.

Gordon, J., 2001. Solar Energy—The State of the Art—ISES International Solar Energy Society Position Papers, ISES and James & James, London.

Greenpeace and European Solar Thermal Industry Association (ESTIA), 2003. Solar Thermal Power 2020 Exploiting the Heat from the Sun to Combat Climate Change.

International Energy Agency (IEA), 2002. Renewables in Global Energy Supply, an IEA Fact Sheet, November 2002.

International Energy Agency (IEA), 2003. Trends in Photovoltaic Applications, Survey Report of Selected IEA Countries Between 1992 and 2002. Report IEA-PVPS T1-12:2003, August 2003.

Lomborg, B., 2001. The Skeptical Environmentalist, Cambridge University Press, Cambridge.

National Solar Radiation Database (NSRDB) (United States): http://rredc.nrel.gov/solar/old_data/nsrdb/

PV News, Vol. 22, No. 3, March 2003.

Satel-Light The European Database of Daylight and Solar Radiation: http://www.satel-light.com/index.html

SoDa Integration and exploitation of networked Solar radiation Databases for environment monitoring: http://www.soda-is.com/

Solar Radiation Modeling and GIS Databases: http://iamest.jrc.it/pvgis/pv/solmod/solmod0.htm

WBGU German Advisory Council on Global Change, 2003. World in Transition—Towards Sustainable Energy Systems. Summary for Policy Makers.

Table Notes

At this point in time, the quantification of solar energy in terms of installed capacity and annual output of electricity and heat presents extraordinary difficulties, which are probably greater than those encountered with any other source of energy. The combination of comparatively newly developed technologies, rapid market growth and widespread, virtually worldwide, diffusion (often at the level of individual households, many in remote rural areas) makes comprehensive enumeration extremely difficult, if not impossible. This means that any aggregate data on a national level can be no more than indicative of the situation.

Table 11.1 is confined to data on photovoltaic generating capacity, as available from the following sources:

- WEC Member Committees, 2003;
- *Trends in Photovoltaic Applications*, August 2003, IEA Photovoltaic Power Systems Programme;
- National Sources.

The data covered in Table 11.1 constitute a sample, reflecting the information available in particular countries: they should not be considered as complete, or necessarily representative of the situation in each region. For this reason, regional and global aggregates have not been computed.

TABLE 11.1

Solar energy: installed photovoltaic capacity at end-2002

	Installed capacity (kW_p)
Africa	
Algeria	1 200
Côte d'Ivoire	250
Egypt (Arab Republic)	3 000
South Africa	10 000
Swaziland	57
North America	
Canada	10 000
Mexico	16 160
United States of America	212 200
South America	
Argentina	4 700
Brazil	13 500
Uruguay	N
Asia	
China	43 300
Hong Kong, China	110
India	63 500
Japan	636 842
Korea (Republic)	5 418
Nepal	1 661
Singapore	60
Sri Lanka	14
Thailand	5 651
Turkey	400
Europe	
Austria	9 000
Bulgaria	5
Croatia	12
Czech Republic	10
Denmark	1 600
Finland	3 052
France	17 241
Germany	294 700
Greece	330
Hungary	30
Italy	22 000
Latvia	3
Luxembourg	1 000
Netherlands	26 326
Norway	6 384
Portugal	1 668
Romania	36
Spain	20 503
Sweden	3 297
Switzerland	19 400
United Kingdom	4 136

TABLE 11.1 (*Continued*)

	Installed capacity (kW_p)
Middle East	
Iran	177
Israel	503
Jordan	184
Lebanon	5
Oceania	
Australia	39 130

Note:
(1) The data shown for France include French Overseas Departments (DOM).

COUNTRY NOTES

The Country Notes on Solar Energy have been compiled by the editors. In addition to national, international, governmental publications/web sites and direct personal communications, the following publications have been consulted:

- *Photovoltaic Power Systems Programme, Annual Report 2002*, International Energy Agency;
- *Trends in Photovoltaic Applications: Survey Report of Selected IEA Countries Between 1992 and 2002*, International Energy Agency—Photovoltaic Power Systems Programme, September 2003;
- *Sun in Action II—A Solar Thermal Strategy for Europe*, European Solar Thermal Industry Federation, 2003;
- *Renewable Energy World*, James & James (Science Publishers) Ltd.

Information provided by WEC Member Committees has been incorporated as available.

Albania

The Albania-EU Energy Efficiency Center (EEC) is carrying out a pilot project on the exploitation of photovoltaic systems for pumping irrigation and potable water. The average solar radiation for some of the main counties ranges from 9 813 kJ/m^2 in January to 24 101 kJ/m^2 in August and the average number of hours with sun (measured at four meteorological stations) over the period 1951–1990 ranges from 2 406 to 2 685.

The National Energy Agency and EEC have carried out a number of studies for installing solar panels in both the residential and service sectors. Based on these studies, the EEC has obtained small grants from various donors and has installed 15 solar panel systems. With EEC promotion through awareness campaigns, Albanian citizens have started installing solar panels for hot water. If the solar panel systems in Albania were to be similarly developed to those in Greece, the potential production of hot water would amount to 360 GWh$_t$ (or 75 MW$_t$ of installed capacity). These figures correspond to a total surface area of solar panels of 300 000 m^2 (or 0.3 m^2 per family), while the solar panel penetration in countries such as Israel and Greece is in excess of 0.45 m^2 per family.

Algeria

Algeria is well endowed with solar insolation and interest in the usage of the resource began in 1962 with the establishment of the Solar Energy Institute.

The Sahara, 86% of the land area, receives an average insolation of 3 500 h/yr, the high plateaux (10% of the land area) receive 3 000 h/yr and the coastal region (4% of the land area and home to the majority of the population) receives 2 650 h/yr. The average solar energy received in the three regions is 2 650, 1 900 and 1 700 kWh/m^2/yr, respectively.

Whilst at the present time the share that solar power contributes to the overall supply of energy is small, it has proved invaluable for the rural electrification of isolated settlements, especially in the south of the country. To date, 20 villages (representing 1 000 households) have been electrified using a PV system with a working installed capacity of approximately 500 kW$_p$. PV systems have also been employed for telecommunications, street lighting and water pumping.

A second PV programme is in hand to electrify 800 households in 16 villages.

At the beginning of 2003 NEAL (New Energy Algeria) joined the IEA Implementing Agreement on Solar Power and Chemical Energy Systems (SolarPACES). This 14-member-country Agreement conducts R&D in the field of CSP and chemical energy systems. Such studies on CSP (using solar-troughs, -towers and -dishes) have furthered the harnessing of solar-powered generation.

As part of its pledge to expand electricity generation capacity, Algeria plans to increase the share of solar-powered generation to 5% by 2010. The country would like to supply Spain and Italy with electricity from solar-powered plants. To this end, plans are being formulated for two 1.2 GW undersea cables.

Argentina

Argentina's PAEPRA (Programa de Abastecimiento Eléctrico a la Población Rural de Argentina) was established in 1995 and the PERMER (Proyecto de Energía Renovable en el Mercado Eléctrico Rural) project was subsequently designed to support it.

The main development goal of PERMER is to improve the quality of life of rural inhabitants who have not been reached by the Electric Transformation Programme. It will aim to achieve this objective through:

- provision of an electricity service that meets the basic needs of lighting and social communication, with decentralised supply sources based on technologies mainly using renewable resources;
- promotion of the participation of the private sector in the provision of this supply so as to achieve the sustainability of the project;
- strengthening the institutional capacity of regulation agencies with reference to the implementation and use of renewable energies;
- improvement of information on sources of renewable energies existing in the country.

The project is financed by a US$ 30 million loan from the International Bank for Reconstruction and Development (IBRD), a US$ 10 million donation from the Global Environment Facility (GEF), and contributions from the provinces, service concession companies, service users (through tariffs) and the Ministry of Education (for schools), plus a minimum contribution from the National Treasury.

Investigations have found that more than 2 million rural inhabitants and 6 000 public services (schools, health care centres, water services, police, civil registries, etc.) lack an electricity supply. In order for these rural sites to be supplied with electricity, a combination of PV, wind, micro water turbines and/or diesel generators will be utilised.

The Electrification of the Dispersed Electric Market (MED), foreseen in the PERMER project, will supply solar home systems (SHS) for stand-alone installations but the method of generation for community schemes will be chosen from the range of renewables available and according to the lowest cost at the time. Private companies, established in the relevant Argentine provinces, will implement the project.

The 2002 economic crisis, and in particular, the devaluation of the Argentine currency, seriously affected the performance of the project. By end-2002 the project was in abeyance because of an interruption in the payments from the World Bank. However, they resumed at the end of January 2003 although it was necessary to reprogramme the project's activities. Negotiations took place with the IBRD during which the original agreement was amended so that it could take account of the actual economic situation at any time. With these modifications in place, it is expected that the project will regain its normal progress.

Australia

Solar PV is one of the best established renewable technologies in Australia, with three decades of technology and market development. The overall market expanded by 27% in 2002, with the grid-connected segment growing 95%, cell production doubled (to 20.5 MW) and production capacity trebled (to 30.5 MW). In addition, a new concentrator system manufacturer commenced operation, with a 5 MW facility and 0.2 MW of production in 2002. Over 70% of Australian PV cell production was exported.

Australian researchers continue to lead the world in PV cell development and commercialisation. Several new technologies are on the verge of commercial production. A new generation of university and trade-trained graduates are entering the job market and are set to enhance community acceptance and use of PV. The Australian PV industry is beginning the development of a PV Roadmap, to coordinate and focus its activities over the coming decade. The Roadmap will cover research priorities, industry infrastructure needs and customer requirements and will set industry targets for deployment. It will facilitate the introduction into the market of several new Australian PV manufacturers, with new PV technologies and new market aims.

Public expenditure on PV research, demonstration and market development totalled AUD 20.6 million in 2002. This was matched by private company expenditure. The major portion of public expenditure (73%) was for market-based incentive programmes: there has been a clear trend in public support from R&D to market incentives over the past 6 years.

The Centre for PV Engineering at the University of New South Wales (UNSW) continues its world-leading research into high-efficiency wafer and thin-film silicon cells. Other areas of research include buried contact cells, silicon light emission, silicon-based quantum wells and superlattices, new energy up- and down-conversion concepts and energy collection using optical-frequency antennas.

BP Solar significantly increased both its mono- and poly-silicon cell efficiencies by installation of new plasma-enhanced chemical vapour deposition (PECVD) silicon nitride systems on its production lines. BP continues its development of automated production equipment.

Pacific Solar is developing and commercialising a thin-film PV technology called Crystalline Silicon on Glass (CSG), based on initial research at the UNSW. In addition to its R&D on CSG modules, Pacific Solar has developed and commercialised its own module inverters and roof mounting systems. CSG module manufacture is scheduled for 2005.

The Centre for Sustainable Energy Systems at the Australian National University (ANU), in conjunction with energy utility Origin Energy, has developed a new thin-film PV technology to be known as 'Sliver cells'. Origin Energy plans to commence construction of a pilot plant to commercially demonstrate the potential of the technology in 2003. The plant is to be constructed in Adelaide, and is being designed to be expandable to approximately 10 MW p.a. capacity.

The ANU team is also developing parabolic trough and paraboloidal dish PV concentrator systems, and a Combined Heat and Power Solar System.

Murdoch University is developing methods of producing low-cost silicon from a number of new sources for both wafer-based and thin-film silicon solar cells.

Sustainable Technologies International (STI) has, after many years of research, commenced the world's first pilot production of titania dye sensitised solar tiles and panels. Demonstration systems are being installed. Solar Systems Ltd continues development and commercialisation of its PV tracking concentrator dishes for off-grid community power supplies or end-of-grid applications. Current systems achieve 500 times concentration and use air or water cooling. System efficiencies of 20% have been achieved. The systems are currently based on silicon cells, but work is continuing on development of non-silicon devices, which are expected to achieve 40% efficiency.

PV Solar Energy Pty Ltd has developed and demonstrated a new PV roof tile, based upon a versatile extruded aluminium frame. The tile uses a new low-cost pluggable PV junction box, developed by Tyco Electronics and monocrystalline solar cell laminates.

The aim of the Photovoltaic Rebate Program (PVRP) is to encourage the development and use of building-integrated PV. Over 950 systems were installed in 2002, amounting to 1.2 MW. Forty per cent of customers, accounting for 48% of installed capacity, were on grid-connected buildings and a total of AUD 5.8 million was allocated in grants. Since the start of the programme in 2000, 3 760 systems, amounting to 4 MW, have been installed and grants of AUD 21.3 million have been provided. PVRP is funded by the Australian Government, with administration by the State Governments. An initial amount of AUD 31 million was allocated over 4 years, with grants of AUD 5 000 per kW provided, to a maximum of AUD 7 500 per residential system and AUD 10 000 per community building system. A further allocation of AUD 5.8 million has been announced and the programme has been extended to 2005. Grants will be reduced to AUD 4 000 for residential systems and 8 000 for community systems.

The aim of the Remote Renewable Power Generation Program is to increase the use of renewable energy for power generation in off-grid areas, to reduce diesel use, to assist the Australian renewable energy industry, to assist in meeting the infrastructure needs of indigenous communities and to reduce greenhouse gas emissions.

Each State has established a slightly different programme, to meet the specific needs of local off-grid applications. However, in general, the target groups are indigenous and other small communities, commercial operations, including pastoral properties, tourist facilities and mining operations, water pumping and isolated households that operate within diesel grids or use direct diesel generation. One megawatt of PV has so far been installed under this programme. Although it is not PV specific, almost all systems installed to date include some PV and PV makes up 92% of installed capacity. The overall programme has funds up to AUD 264 million allocated to it, of which AUD 12 million have been allocated to date.

Core funding for this programme is provided to the States by the Australian Government, on the basis of diesel fuel excise collected in the years 2001/2002 to 2003/2004 from diesel fuel used by public generators not connected to main electricity grids. Grants of up to 50% of the capital cost of renewable energy systems are available for diesel replacement. The programme is administered by the State Governments, with additional funding provided by some States. The programme will extend to 2009/2010, though some States are likely to expend their allocation before then.

A specific allocation of AUD 8 million has been made to the Aboriginal and Torres Strait Islander Commission (ATSIC) for the Bushlight Program to assist with the development of industry capability and local understanding of renewable energy systems in indigenous communities.

Annual PV sales in Australia increased by 27% in 2002, stimulated by government grant programmes for rooftop applications and off-grid diesel replacement. Australia's established off-grid PV market in industrial, agricultural and commercial applications such as telecommunications, signalling, water pumping, electric fences and cathodic protection continued to dominate the PV market, accounting for 64% of the 2002 applications and 58% of cumulative installations. The on-grid market grew strongly during 2002, with sales increasing 95% over 2001 levels, and will hopefully be able to maintain momentum with the lower grant levels available from 2003 to 2005. Grid-connected applications for household, community and commercial buildings accounted for 14% of installations. Grid systems now account for 11% of PV installed in Australia.

As natural gas became the preferred fuel for water heating, the market for solar water heaters went into decline during the 1980s. However, with attention turning to renewable energy, the market saw some stabilisation in the 1990s. By end-2001 it was estimated that a total of 3.198 million m^2 solar collectors had been installed, of which 1.198 million m^2 were flat plate. The estimated thermal energy production in 2001 totalled 1 436.6 GWh.

Austria

There is no programme aimed specifically at the promotion of solar power in Austria but rather at renewable energy in general. Small renewable energy technologies (excluding large hydropower and biomass) benefit from two separate support schemes:

- The section of one scheme relating to other renewable technologies (non-small hydropower) requires that electricity suppliers must source a minimum of 1% (rising to 4% by 2008) of their electricity from such applications.
- The second renewable support scheme is the feed-in tariff system. Utilities are obligated to purchase power from selected renewable technologies at above-market tariffs that are determined by the Government. These tariffs were originally set by each individual Land, but legislation passed in July 2002 transferred this responsibility to the Federal Government so that now the feed-in tariffs can be made consistent across the country. This move will allow renewable resources to be used more efficiently throughout the country, providing the same level of renewable generation at lower overall cost.

It has been reported that by 2001 over 2.3 million m^2 of thermal solar collectors had been installed in Austria, which represents about 0.3 m^2 per capita—one of the highest ratios in Europe. However, the main thrust of the solar energy sector has been the photovoltaic market. An increase of more than 45% in installed capacity occurred during 2002, reaching a level of approximately 9 MW by the end of the year. Although the initial installations were stand-alone systems, in recent years the emphasis has been on grid-connected systems, so that they now represent more than two-thirds of the overall installed capacity.

Austria saw its first PV manufacturing plant open in 2002, with production of standard and semi-transparent crystalline silicon panels.

A 2001 Government study determined that by 2010 electricity generation could potentially be supplied by 0.1 TWh solar PV and heat generation by 2.2 TWh solar thermal.

Botswana

Currently, solar energy is mainly used for water heating, village electrification, powering telecommunication systems, water pumping and desalination.

The National Photovoltaic Rural Electrification Programme, funded by the Government, started in 1997. The overall goal is to provide photovoltaic energy to rural households and institutions on affordable financial terms. Since its inception the Programme has made over 300 installations throughout the country mostly for households use, although a few are used for income-generating activities such as shops and poultry farms.

The Botswanan and the Japanese Governments (through the Japanese International Cooperation Agency) are jointly implementing a PV Master Plan project which will initiate an institutional framework as well as strategies to optimise the implementation of solar electrification.

Brazil

The resource potential is available in two publications: Atlas Solarimétrico do Brasil—Banco de Dados Terrestre, UFPE, 2000 and Atlas de Irradiação Solar do Brasil, LABSOLAR-UFSC & DGE-INPE, 1998. However, the methodologies used in the Atlases are different. The former is a model based on ground station information and the latter uses a model based on satellite data.

The total photovoltaic power installed in Brazil is estimated to be between 12 and 15 MW_p, 50% of the projects are for telecommunications systems and the other 50% for rural energy systems, but output data are not collected.

The Brazilian Government established the Programme for Energy Development of States and Municipalities (PRODEEM—Programa de Desenvolvimento Energético de Estados e Municípios) in December 1994. The aim of the Programme is to provide energy for the basic social demands of poor communities where they are isolated from conventional connected systems. PRODEEM primarily uses photovoltaic systems and supplies only community schemes (generation for schools, health clinics, water pumping, public lighting systems, etc.).

Bulgaria

Of all the renewable energies available to Bulgaria, solar energy provides the greatest potential. Whilst almost the whole country is suitable for direct transformation of solar energy, both in terms of heating and electricity generation, some areas are especially suitable for the location of larger capacity industrial systems.

The first 10 kW photovoltaic system is presently being prepared for installation.

The Energy Efficiency Agency has determined that there are many possibilities for the utilisation of solar energy in Bulgaria. Studies have shown that the country can be divided into three zones according to the solar insolation received, namely:

Zone A—encompasses regions in the south east, part of the southern Black Sea coastal region and the valleys of the rivers Struma, Mesta and Maritza. The amount of sunshine is over 2 200 h/yr and the total solar radiation received on a horizontal surface is greater than 1 600 kWh/m^2.

Zone B—encompasses regions in the Danube plain, the Dobrudja region, the Trace lowland, west Bulgaria, the Balkan hollow fields and Stara Planina mountain regions. The amount of sunshine varies from 2 000 to 2 200 h/yr and the total solar radiation received on a horizontal surface varies from 1 500 to 1 600 kWh/m^2.

Zone C—encompasses the remaining part of the country's territory but mainly the mountainous regions, where the amount of sunshine is less than 2 000 h/yr and the total solar radiation received on a horizontal surface is less than 1 500 kWh/m^2.

Canada

In recent years the Canadian PV industry has grown steadily, serving mainly the domestic off-grid market and the export market; the grid-connected PV sector has not displayed the same rate of development. There are approximately 150 organisations actively promoting PV power. Many of them are members of the Canadian Solar Industries Association or Énergie Solaire Québec.

By end-2002 the PV installed capacity had risen to 10 MW from 1 MW in 1992. This growth has been sustained in an unsubsidised market owing to the popularity of the technology's ability to serve remote areas of the country. The applications have been utilised for transport route signalling, navigational aids, isolated residential buildings, telecommunications and remote sensing and monitoring.

Canada is now working to reduce greenhouse gas emissions (as part of the commitment in its ratification of the Kyoto Protocol in December 2002). In order to increase the awareness of PV and, additionally, to bring about cost reductions, several federal departments have joined with industry and other interested parties to complete projects within the Technology Early Action Measures Program, the MicroPower-Connect initiative and the expanded Renewable Energy Deployment Initiative (REDI) program for On-Site Generation at Federal Facilities.

To achieve these objectives, there have been several industrial developments:

- ATS Automation Tooling System has announced a project to commercialise a next-generation solar cell technology.

- ARISE Technologies Corporation and Cook Homes have joined forces to market PV solar homes to the residential sector.
- Carmanah Technologies Inc. is developing improved LED lighting and solar products with partners, BC-Hydro and BCIT.
- Xantrex Technology Inc. is launching a customer-financing programme for renewable energy products.

With a view to promoting grid-connected systems in the medium to long term, the Government will invest in R&D for building-integrated PV technology and support both the development of a technical guideline for the interconnection of small power supplies and demonstrations of PV on high-profile buildings throughout the country.

China

It is estimated that two-thirds of the country receives solar radiation energy in excess of 4.6 $kWh/m^2/day$; the western provinces are especially well endowed.

By early 2002 it was thought that as many as 7 million Chinese households situated in isolated areas still did not have access to electricity. There was no grid connection in these areas and the future installation of a grid was considered infeasible. To rectify this situation the Government has developed many schemes under The Brightness Program.

The country has been working with the World Bank/Global Environment Facility (GEF) on a plan (Renewable Energy Development Project) to install 10 MW of PV in Solar Home Systems (SHS) over 5 years. Furthermore, with a view to providing power to isolated areas, the Chinese Government is instituting a massive electrification programme based on renewable energy technologies and valued at more than 1.8 billion RMB. Globally, it is one of the most ambitious. Supported by the Beijing Jikedian Renewable Energy Development Centre, the US and German renewable energy programmes and the Institute for Sustainable Power (ISP), the governmental National Development and Reform Commission (formerly the State Development Planning Commission) is implementing The Township Electrification Program (Song Dian Dao Xiang).

The Program will form part of the national strategic programme and also be linked with the United Nations Development Program (UNDP)/-GEF 'Capacity Building for the Rapid Commercialisation of Renewable Energy' project. Links to the World Bank/GEF Renewable Energy Development Project and the Global Village Energy Partnership are also a possibility.

Song Dian Dao Xiang, initiated in 2001 with installation completed just 20 months later in June 2003, had the aim of installing power systems in 1 061 townships (representing about 1 million people). Where available, systems utilised the hydro resource but the majority were PV, PV-diesel or PV-wind hybrid, totalling 20 MW. The townships participating in the projects were located in the Provinces of Tibet, Inner Mongolia, Sichuan, Gansu, Qinghai, Xinjiang, Yunnan, Hunan, Shaanxi and Chongquing.

It has been reported that the second phase of the Program is planned to run from 2005–2010 with the intention of electrifying a further 20 000 villages.

Private enterprise has also been involved in bringing power to the villages: examples are the 2001 Shell and Sun Oasis collaboration to supply SHS for up to 78 000 houses in Xinjiang Province and the 2003 Shell project to install centralised systems for 1 300 houses in nine villages in Xinjiang and Yunnan Provinces.

After Song Dian Dao Xiang is completed, it has been said that China will have the greatest number and density of installed hybrid village power systems in the world.

It is reported that the country is the world's largest manufacturer of Domestic Hot Water (DHW) systems and also the largest user. Chinese expert opinion provided in a private communication for the present *Survey* suggests that there are in the region of 26 million m^2 of solar water heaters installed in China.

The State Economic and Trade Commission plans for renewable energy to supply 2% of total energy supply by 2015, incorporating an average annual 17% growth for solar thermal systems.

Côte d'Ivoire

There is a plentiful supply of solar insolation in Côte d'Ivoire, with the potential estimated at between 1 500 and 1 800 kWh/m^2/yr. The Government, the private sector and some non-governmental charitable organisations have been active in promoting the exploitation and use of solar energy. The applications foreseen include pumping water in villages, hospital refrigeration units in rural areas, the use of solar thermal energy to dry feeds, etc.

At the present time an official project, financed by the Spanish Government, to install solar PV in 105 villages is about to come to fruition.

The current research programme is concentrating on the restructuring of the framework for the use of renewable energies. A major difficulty is the non-availability of renewable energy data in Côte d'Ivoire.

Denmark

Although in general the Danish Government considers renewable energy to be of utmost importance to the national energy plan, in recent years the growth in the installed solar capacity has not been rapid. This is largely due to the fact that despite the Danish Energy Authority subsidising R&D in the energy field via the Energy Research Programme (Energiforskningsprogrammet—EFP) to the sum of DKK 40 million in 2002 (DKK 70 million expected in 2003), the effort to promote PV has been somewhat fragmented. However, it was hoped that during 2003, this situation would be rectified and a national PV strategy would be formulated.

As part of the plan to increase the deployment of building-integrated systems, a 4-year nationwide solar cell project, SOL 1000, is being implemented. The programme is intended to demonstrate low-cost and architecturally acceptable integration of PV technology on the existing housing stock.

Despite a buoyant market for solar thermal systems during the 1990s, the Government's apparent lack of interest has resulted in a generally low level of deployment throughout the country: at end-2001 it was estimated that 55 m^2/1 000 inhabitants solar collectors had been installed. However, Samsø, the Danish Renewable Energy Island, has a 2 500 m^2 district heating scheme along with many small domestic systems and Skive Kommune in Jutland intends to equip as many public buildings as possible with solar systems.

Egypt (Arab Republic)

The Egyptian Government incorporated renewable energy into its energy planning programmes at the beginning of the 1980s and published a solar atlas of the country as early as 1991. This shows that the average global radiation ranges from 5.4 kWh/m^2/day on the Mediterranean coast to 7.1 kWh/m^2/day in Upper Egypt.

Until recent years the emphasis of the Ministry of Electricity and Energy's New and Renewable Energy Authority (NREA) work had been on the demonstration and testing of projects rather than their market commercialisation. The technologies of both solar thermal (domestic solar water heaters—DSWH, solar industrial process heat—SIPH, solar thermal electricity generation—STEG) and PV power were thoroughly tested.

However, DSWH have been produced locally since the early 1980s and over 400 000 m^2 have now been installed, mostly in new cities and tourist villages. A Government policy has made it mandatory that all new communities must have DSWH installed and the annual production of collectors is now in excess of 25 000 m^2.

Egypt's industrial sector accounts for a very substantial percentage of total energy use and much work has been undertaken in the field of SIPH. Following two successful low-temperature pilot projects (using flat-plate collectors) in the food and textile industries, a pharmaceutical company has been chosen for a pilot project to generate saturated steam at 175 °C/8 bar by utilising solar parabolic trough technology. The project was expected to be operative by end-2003 and to contribute a significant energy saving.

In the mid-1990s various governmental STEG studies concluded that there was considerable scope for the Integrated Solar Combined Cycle System (ISCCS) using concentrated solar technology with a conventional gas turbine combined cycle. Feasibility studies were undertaken and with GEF/World Bank support granted, the first 127 MW ISCCS plant with a 31 MW solar component is expected to be in operation by 2006. It is expected that this plant will be the first of a series of solar/fossil fuel hybrid power plants. It is planned that by 2010 about 750 MW will be installed. In a longer time-frame, the NREA envisages that by 2017, 4 050 MW of STEG plants will be installed, producing 25.5 TWh/yr of electricity for internal consumption and export.

Whilst the PV applications of water pumping, desalination, refrigeration, village electrification, etc. remained part of the testing processes, the uses for telecommunications, navigation lights and advertising boards have all been implemented. Total capacity currently stands at about 3 MW_p.

Finland

The seasonal variation of solar radiation in Finland is large: 90% is obtained between the months of March and September. The solar insolation in southern Finland is 1 000 kWh/m^2/yr on a horizontal surface.

Although larger applications are used in remote areas for telecommunication base stations, weather stations, etc., Finland's domestic market is dominated by SHS for summer houses which average approximately 50–100 W_p. Currently, the installed PV capacity totals just over 3 MW of which dwellings account for about 90%. However, the recent growth of PV in building-integrated systems is expected to continue. The total amount of solar collectors for DHW and heating is about 10 000 m^2.

Whilst the PV sector remains fairly modest, the Ministry of Trade and Industry's Action Plan contains a national target of 40 MW_p installed PV capacity by 2010, with a corresponding 2025 level of 500 MW_p.

France

By end-2002 France (including its overseas departments (DOM)) had a total installed PV capacity of 17 241 kW_p of which 10 437 kW_p were off-grid domestic, 4 862 kW_p were off-grid non-domestic and 1 942 kW_p were grid-connected distributed. The annual average increase in capacity between 1992 and 2002 was just over 25% but to date, owing to the country's heavy reliance on nuclear power, renewable energy in total has provided only a small share of total energy supply. In 1999, however, the body charged with promoting renewable energy, ADEME (Agence de l'environnement et de la maîtrise de l'énergie), instigated a development programme and signed a 7-year contract with the Government. The programme is intended to promote renewable energies with target capacities to be attained by 2006. France's overseas departments and Corsica, where the high cost of electricity is the main consideration, will benefit most.

At end-2001 there were 556 000 m^2 of solar thermal panels installed in France, giving an annual heat output of about 750 TJ.

The objective of ADEME's Plan Soleil 2000–2006 is to boost the capacity of solar thermal power in metropolitan France to accord with sustainable development. It is planned to achieve this objective by a combination of financial incentives, professional training, quality control and publicity campaigns.

Germany

By a factor of 11 Germany has the highest level of installed PV capacity amongst the European members of the IEA-PVPS. At end-2002 capacity stood at 295 MW as compared with the next biggest country (the Netherlands with 26 MW).

This achievement has been brought about after the Federal Ministry of Environment (BMU) estimated that out of a technical potential of 450 GWh annual generation from all renewable energies, PV could contribute as much as 19%. Between 1992 and 2002 the average annual increase was nearly 48%, driven in the later years by the highly successful 100 000 Rooftops Solar Electricity Programme which came into force at the beginning of 1999 (running to 2003). Additionally, the Electricity Feed Law of 1991 was replaced in April 2000 by the Renewable Energy Sources Act (EEG) which guarantees a feed-in tariff for PV.

Since third quarter 2002 the BMU has been charged by the Federal Government with responsibility for renewable energy and R&D is conducted under the '4th Programme on Energy Research and Energy Technology'. Within this programme, the 'Way Paving Programme Photovoltaic 2005' has been formulated with three main goals:

- cost-reduction for PV cells and modules by decreasing production costs and by increasing cell and module efficiencies;
- cost-reduction, technical optimisation and removing other obstacles that prevent the use of PV in different types of buildings;
- PV for decentralised, grid-independent electricity supply.

The Second Amendment of the Renewable Energies Act came into force on 1 January 2004. It is expected to promote even further growth in the German solar industry in the coming years. The 2004 subsidies available for bringing a solar power installation into operation on buildings will be between €0.54 and €0.574/kWh. An additional bonus of €0.05/kWh will be paid for facade installations. The new law has abolished the upper limit for subsidies, providing large installations with €0.457/kWh.

Solar thermal technology is also expected to benefit from a market incentive programme. The German Solar Industry Association has stated that it expects growth to be in double figures again in 2004.

In January 2004, it was announced that Shell Solar and Gesellschaft für Solarenergie are to build the world's largest solar power station, south of Leipzig. The station will be built on the site of a former lignite mine ash deposit and will comprise some 33 500 solar modules with a total output of 5 MW. The grid-connected station is expected to supply power sufficient for 1 800 households and to be operational in July 2004.

Greece

Despite the existence of a very high Hellenic potential for solar energy applications and the beginning of their deployment in the mid-1970s, major applications have so far been restricted to SWH collectors. There has been a negligible market for large-scale hot water systems in the commercial sector (hotels, hospitals and swimming pools) and an even smaller penetration of large systems for hot water, air-conditioning and space heating in industry.

In order to promote solar energy the Hellenic State accommodates a very favourable taxation environment for solar applications but individual consumers' purchases are mainly limited to SWH collectors, because of the high production cost of photovoltaic applications.

It is envisaged that large-scale mirror-type solar collectors, etc. will, in the next 5–10 years, be developed so that electricity generating capacity utilising solar energy can be substantially increased.

Hong Kong, China

The annual mean daily global solar radiation for Hong Kong is 14.46 MJ/m^2 but a variety of factors inhibit this resource from being utilised: the density of population and high-rise buildings creates difficulties in siting installations and furthermore, the typhoons the country is subject to raise costs associated with the anchorage of such installations; the comparatively low demand for heating energy; the lack of an indigenous solar equipment manufacturing industry necessitates the importation of high-priced (versus savings achieved) solar equipment; the relatively slow return on investment in solar installations is incompatible with the demand for fast payback terms by the building industry.

However, the findings from a Consultancy Study on the Potential Applications of Renewable Energy in Hong Kong commissioned by HKSAR Government suggested that the resource potential for PV power in Hong Kong is 5 944 GWh/yr. This study report can be found at http://www.emsd.gov.hk/emsd/e_download/wnew/stage1_report.pdf.

Because Hong Kong is largely urbanised, most of the large solar thermal applications have, to date, been government projects installed in rural areas or new towns. Schemes to heat water either directly or indirectly have been installed in a swimming pool, slaughterhouse, prison, hospital, military camp and public bathhouses.

The country has been successful in deploying PV applications to supply electrical power to the monitoring equipment of automatic weather stations in remote locations and approximately 60% of the battery-operated aids to navigation are now powered by solar PV.

Two universities are currently carrying out research on building-integrated PV (BIPV) systems as the Government believes this concept could be incorporated into the facades, roofs and shading devices of buildings. Two grid-connected building-integrated PV systems with a total capacity of 73 kW have already been installed.

In the coming 3 years planned solar energy projects include (a) a solar hot water supply system with 80 m^2 of collector panels and (b) various PV projects with a total installed capacity of 1 800 kW.

There is a hybrid renewable energy system on a remote island that includes a 3.3 kW PV system plus a 400 l solar thermal hot water system (about 1 kW equivalent).

India

The Ministry of Non-Conventional Energy Sources (MNES), working in conjunction with the Indian Renewable Energy Development Agency (IREDA) continues to promote the utilisation of all forms of solar power as part of the drive to increase the share of renewable energy in the Indian market. This promotion is being achieved through R&D, demonstration projects, government subsidy programmes, programmes based on cost recovery supported by IREDA and also private sector projects.

India has a good level of solar radiation, receiving the solar energy equivalent of more than 5 000 trillion kWh/yr. Depending on the location, the daily incidence ranges from 4 to 7 kWh/m^2 with the hours of sunshine ranging from 2 300 to 3 200 per year. Solar thermal and solar photovoltaic technologies are both encompassed by the Solar Energy Programme that is being implemented by the MNES. The Programme, regarded as one of the largest in the world, plans to utilise India's estimated solar power potential of 20 MW/km^2, and 35 MW/km^2 solar thermal. The country has also developed a substantial manufacturing capability, becoming a lead producer in the developing world.

Within the overall drive towards renewable energy, the MNES conducts separate programmes for both solar thermal and solar PV.

- The Solar Thermal Programme covers solar water heating (with an enormous potential of 140 million m^2), solar cooking, solar air heating and solar buildings.

The commercial sector (hotels, hostels, hospitals and other large institutions) have favoured solar thermal applications; to date, water heating systems with a total collector area of 680 000 m^2 have been installed.

Five types of solar cookers have been developed:

- cardboard solar cooker—low-cost, portable, one or two dishes at a time;
- box solar cooker—small, four dishes at a time, intended for small families, to date 530 000 have been sold;
- dish solar cooker—fast cooking device for homes and small establishments, for 10–15 people, to date some 500 have been installed;
- community solar cooker for indoor cooking—large, automatically tracked parabolic reflector standing outside kitchen through an opening in the north wall with a secondary reflector further concentrating the rays on to the bottom of the black-painted cooking pot, for 40–50 people, to date in the region of 60 have been installed;
- solar steam cooking system—large, automatically tracked parabolic reflectors, coupled in a series and parallel combination, generating steam for use in community kitchens, for thousands of people, usually installed in conjunction with a conventionally fuelled boiler; the world's largest solar cooking system installed at Tirumala in Andhra Pradesh, has the capacity to provide food for 15 000 people per day, to date six such systems have been installed.

Solar air heating technology has been applied to various industrial and agricultural processes (e.g. drying/curing, regeneration of dehumidifying agents, timber seasoning, leather tanning) and also for space heating; many types of solar dryers have been developed for use in different situations, to date 5 000 m^2 have been installed.

Solar buildings have been promoted by the MNES in an effort to increase energy efficiency; the state government in Himachal Pradesh is actively promoting the incorporation of passive solar design into building design, in excess of 20 buildings either already exist or are under construction.

Aditya Solar Shops (showroom-cum-sales and service centres) are being established nationwide; in the beginning the MNES plan was to promote the establishment of the shops in major cities but their success has led to a project for at least one to exist in each district throughout the country.

- Solar Photovoltaic Programme (SPV) has been promoted by the MNES for the past two decades, aimed particularly at rural and remote areas. By end-2002 more than 1.03 million SPV systems representing 107 MW_p had been deployed (of which 46 MW_p had been exported). Domestically, the technology has been applied to home lighting (9.1 MW_p), street lighting (3.5 MW_p), solar lanterns (4.9 MW_p), water pumping (6.6 MW_p), telecommunications (16.3 MW_p), stand-alone power plants (3.8 MW_p) and others (16.8 MW_p). In excess of 800 000 lighting systems had also been installed.

Following the success of the MNES' country-wide SPV demonstration and utilisation programme during the period of the Ninth Plan, it is planned, with certain modifications, to continue it during the Tenth Plan (2002–2007).

Of the approximately 80 000 villages not currently connected to the grid, about 18 000 are too remote ever to be considered. The MNES has the objective that by 2010 they will all have access to power from renewable energy sources, with the Tenth Plan electrifying 5 000 of them.

In a country where agriculture is a major component of the economy, the SPV Water Pumping Programme will continue to promote the large-scale use of PV powered pumping systems for farmers.

Encouraged by the MNES, a strong R&D base and an indigenous manufacturing sector (ranging from silicon material to solar cells, PV modules, complete systems and power plants) have been developed, enabling efficiencies to improve, production yields to rise and costs to be reduced.

The Ministry is also implementing a programme for water-pumping windmills, small aerogenerators and wind-PV hybrid systems to enable the huge Indian wind resource to be harnessed in conjunction with the solar power available. These applications will be fully researched and demonstrated prior to deployment in remote areas.

- The Solar Grid Power Programme comprises two components: the Solar Thermal Power Programme and the Solar PV Power Programme.

Using World Bank/GEF and German Government (through KfW Bankengruppe) finance, it is planned to set up a 140 MW Integrated Solar Combined Cycle Power Plant in Mathania, Rajasthan. The plant will consist of a 35 MW solar component based on parabolic trough collectors and a 105 MW combined cycle unit running on re-gasified LNG.

The grid-connected Solar PV projects are designed to provide voltage support in remote sections of weak grids, and for installation on public buildings in urban centres for peak load shaving and in island locations where they can be substituted for a hydrocarbon fuel. To date 31 projects with an aggregate capacity of 2.5 MW have been installed in 10 States and 3 Union Territories. A further seven projects totalling 550 kW are under construction.

Indonesia

The archipelago of Indonesia comprises over 17 000 islands (according to the latest count using satellite mapping) of which approximately 6 000 are inhabited. Difficulties in extending the national grid across the islands to the widely dispersed population meant that in 1995 only about 58% of the country's 62 000 villages were electrified. Historically, areas that could not be supplied with conventional electricity from the national grid have relied upon hydro-electricity and stand-alone diesel generators to power mini-grids, or used kerosine for lighting.

Indonesia's situation close to the equator and its annual average insolation level make it highly suitable for the installation of solar energy devices, especially for the huge rural population and in remote areas.

Both solar thermal and solar PV applications have been installed throughout the country. Most of the solar thermal components are of Indonesian origin and have been installed for domestic water heating in urban areas and for the use of agriculture or crop drying in rural areas. Although there is now some indigenous assembly of PV modules, most solar PV systems are imported and used for small, remote applications, such as SHS, PV pumps and repeater stations.

The Indonesian Agency for the Assessment and Application of Technology (BPPT) coordinates the Fifty Megawatt Programme under which it is planned to install PV electricity in 1 million homes by 2005.

Iran (Islamic Republic)

With its richness of oil and gas reserves, Iran has little incentive to develop its solar resource. However, with an ever-increasing demand for electricity, the Iran Electric Power Industry includes in its research the possibilities of renewable energy. To this end the Ministry of Energy has a Renewable Energy Department.

The daily average solar radiation ranges from 2.8 kWh/m^2 in the south-east to 5.3 kWh/m^2 in the central region. The first PV application, installed in 1993, was followed by several more in 1998, but all projects are small and are used for street lighting, agricultural pumping and for electricity generation at a border post. Solar water heaters have been installed in small numbers.

Whilst the utilisation of solar power to produce electricity has, in the past, been comparatively rare, research continues on the use of solar energy combined with thermal power. It has been reported that 101.25 MW of solar energy projects are planned in conjunction with combined cycle steam power plants.

Israel

With an annual incident solar irradiance of approximately 2 000 kWh/m^2 and few natural energy resources, Israel has pioneered the use of solar energy. However, whilst the 1980 law requiring the installation of solar water heaters has had a dramatic effect, PV activity remains largely in the realm of academia.

The 1980 'Solar Law' is an amalgam of different legislative measures, all designed to lay down national standards and regulations. The Planning and Building Law requires the installation of solar water heaters for all new buildings (including residential buildings, hotels and institutions, but not industrial buildings, workshops, hospitals or high-rise buildings in excess of 27 m), dictating the size of the installation required for a particular type of building; the Land Law governs solar installations in existing multi-apartment buildings and the Supervision of Commodities and Services Law provides governmental supervision of the quality of installations and their guarantees.

At the present time about 80% of Israel's residential buildings have solar thermal systems, the vast majority of which are utilised for water heating. It has been reported that the use of solar collectors saves the country in the region of 600 000 toe/yr. By end-2001 the total collector area in operation was estimated to be 3.5 million m^2 and the energy produced, 3.5 million MWh (based on an annual incident solar irradiation of 2 000 kWh/m^2 and a 50% average system efficiency factor).

Although the Israel Electric Corporation is required to purchase electricity from private producers, there are no incentives for PV systems. The extensive national grid precludes the same penetration by PV as has been enjoyed by solar water systems. There is no PV module manufacturing capability within the country and currently most activity is concentrated on maintaining the technical excellence that has been achieved through academic research. However, during 2002 PV-operated cameras for vehicle number plate recognition were installed for use on Israel's first toll road. Additionally, there are instances of PV being used for lighting, irrigation, pumping, refrigeration and in parking ticket machines.

At end-2002 there were 503 kW_p of installed PV power, of which 283 kW_p was off-grid domestic, 200 kW_p was off-grid non-domestic, 6 kW_p was grid-connected distributed and 14 kW_p was grid-connected centralised.

In November 2002 the Government passed a resolution stating that by 2007 at least 2% of total electric energy must be generated from renewable energy and by 2016, it must rise to 5%.

Italy

Since the early 1980s, the main thrust of solar energy in Italy has been photovoltaic, the development of which has ranged from research on materials and devices and experimentation for grid and non-grid applications to the dissemination of such technology through various incentive programmes.

By end-2002 the total PV capacity installed was approximately 22 MW_p of which about 6.3 MW_p was for rural electrification, 5.3 MW_p for off-grid domestic applications, 6.7 MW_p for on-grid centralised systems and 3.7 MW_p for on-grid distributed.

The Roof-Top Program of the Ministry of Environment has provided financial support and thus a strong incentive to the growth of PV installations.

Currently, it is estimated that in the region of 300 000 m^2 of solar water heaters are installed.

In November 2003 ENEL (the largest Italian utility) and ENEA (the Italian Agency for New Technology, Energy and Environment) announced their collaboration on the 'Archimede' project. ENEL's existing Priolo Gargallo gas combined cycle power plant located in Sicily will be expanded with the addition of a solar plant to be constructed alongside. The innovative technology, albeit inspired by the 3rd century BC mathematician, will use parabolic mirrors to concentrate and accumulate the power from the sun during daylight hours. With the use of a new fluid based on low-cost fused salts and capable of allowing high temperatures to be reached, the process will use the thermal energy collected to produce vapour and thus electricity during a 24-hour cycle. It is estimated that there will be a saving of 12 500 toe per annum and a reduction of 40 000 tonnes of CO_2 emissions.

Japan

As one of the 20 members of the Implementing Agreement on Photovoltaic Power Systems (IEA-PVPS) Japan had the highest installed PV capacity (636 842 kW_p) at end-2002, which was more than double that of the next highest country: Germany. This level represents an annual increase of 42% between 1992 and 2002.

Of the 636.8 MW_p total capacity, 0.6 MW_p was for the off-grid domestic market, 72 MW_p for off-grid non-domestic, 561.3 MW_p for grid-connected distributed and 2.9 MW_p for grid-connected centralised.

Following the Japanese Government's Sunshine Project, launched in answer to the problems created by the oil crises of the 1970s, and its subsequent New Sunshine Program, launched in 1993 as a way to efficiently overcome barriers related to new energy, the '5-Year Plan for PV Power Generation Technology R&D (FY2001–FY2005)' was launched in 2001. The programme aims at a range of developments, in addition to establishing technology for PV to be available for approximately the same generation cost as the household electricity rate.

The 1997 New Energy Law led to The Total Primary Energy Supply Outlook in 1998 which specified that the target for installed PV was to be 5 000 MW by 2010. In 2001 this target was reduced to 4 820 MW. The Ministry of Economy, Trade and Industry (METI) is charged with promoting the measures necessary to achieve this target. The 'Renewable Portfolio Standard' Law introduced during 2002 requires energy suppliers to use a certain percentage of renewable energy.

In addition to the main demonstration programmes ('PV Field Test for Industrial Use' and 'Demonstrative Development of Centralised Grid-Connected PV Systems') both started in FY 2002, METI also began in FY 2002, three implementation programmes ('Residential PV System Dissemination Programme', 'Introduction and Promotion of New Energy at the Regional Level' and 'Financial Support for Entrepreneurs Introducing New Energy').

The Japanese Government is approaching the development of its PV industry in an holistic manner—not only is the METI involved but also the following Ministries: Land, Infrastructure and Transport; Posts and Telecommunications; Education, Culture, Sports, Science and Technology; Environment. Local governments and municipals and the utilities are also involved in the cause.

The PV industry has grown rapidly in recent years with annual output of PV cells increasing from 50 MW in 1998 to in excess of 200 MW in 2002. As a result of this growth, the Japan PV Energy Association has announced the 'Self-sustainable PV Industry Vision: the Genesis of PV Industry for Energy and Environment'—a roadmap to promote PV up to 2030.

Off-grid non-domestic PV systems without governmental support are being implemented for the use of telecommunications, traffic signs, telemetering, ventilation and lighting.

The production and deployment of solar hot water systems began more than 50 years ago and the market developed during the ensuing three decades. The oil crises of the 1970s fostered further growth but in the late 1990s stagnation set in, not least because of the Government's termination of low-interest loans. At end-2001 it was estimated that a total of 7.360 million m^2 glazed collectors had been installed, of which 7.219 million m^2 were flat plate collectors and 0.141 million m^2 were vacuum collectors.

Jordan

Several studies and surveys on the utilisation of solar energy have been made for a number of locations in the country through the so-called 'Phoebus Project'. Their outcome was encouraging and indicated that the utilisation of solar energy for remote applications is feasible.

The use of solar energy for thermal applications, especially for generating electricity, is technically possible, owing to the high solar energy intensity (which exceeds 2 000 kWh/m^2/yr at some locations) and other favourable factors. However, from an economic point of view, such power stations are not yet viable and will need additional support to be able to compete with conventionally generated power.

Jordan lies in the so-called earth–sun belt area and has a high potential of solar energy, with annual averages of global solar energy of about 1 800 kWh/m^2/yr.

The main use of solar energy is for domestic water heating, with approximately 30% of houses having such installations; to supply this market, more than 25 manufacturers are producing locally designed solar water heater systems.

In addition, 100 PV systems are used in remote areas throughout the country. These systems cover various applications, such as water pumping, telecommunications, schools and others, with a total capacity of 184 kW$_p$.

Kenya

Kenya receives a plentiful supply of solar radiation, averaging between 4 and 6 kWh/m^2/day.

Although Kenya Power and Lighting's Rural Electrification Programme (begun in 1985) has resulted in grid power being made available to a small percentage of rural households, only 50% of the urban and 5% of the rural population have access to a public supply of electricity.

With the establishment of indigenous equipment manufacturers, sales outlets, installers and after-sales service agents, in excess of 150 000 PV-based SHS have been sold. Current sales number over 20 000 per year.

Potentially a very large market for PV systems exists in Kenya, but to date implementation has tended to be restricted to those sections of society that receive a regular income. The systems are mostly used to power TVs, radios and lights. Additionally, Solarnet (an NGO) has instituted the Solar for Schools Project which is aimed at off-grid rural boarding schools.

Korea (Republic)

Despite solar PV having a large potential in Korea, by end-2002 the total capacity installed was only 5 410 kW$_p$. Historically, the high cost of PV installations has discouraged the market and in order to reverse this situation, the Government has launched 'Solar Land 2010', a programme of the '2nd General National Energy Plan', 2002–2011.

The programme aims to increase the level of interest in PV by means of a range of projects, e.g. the installation of 3 kW PV systems in 30 000 houses by 2010; the commercialisation of thin-film cells, thereby achieving a cost reduction; subsidising both 20% of total installation cost and the difference in the electricity price between PV and conventionally generated electricity, etc.

Latvia

The Latvian global solar radiation varies between 900 and 1 100 kWh/m^2/yr and although the amount of sunshine the country receives is only about 1 200 h/yr, solar power is being utilised to good effect.

PV systems have been installed in lighthouses and lightships and solar collectors have been installed in two schools. In 2002 the solar thermal project, 'Thermal Solar Collectors at

Aizkraukle Secondary School Nr. 2' was the largest solar collector project in the Baltic States.

Lebanon

At the present time with its widespread electrical grid connections, no real market exists in Lebanon for photovoltaic applications. Very few PV systems have been installed and those that do exist are in extremely remote areas—some have been installed as a personal initiative, others as demonstration projects.

However, SWH has been gaining ground for both residential and commercial applications. It is estimated that approximately 100 000 m^2 had been installed by 2000.

Lithuania

The total annual potential of solar energy in Lithuania is assessed at 1 000 kWh/m^2 and the technical potential is about 1.5 TWh per annum. At present the total area of solar collectors that are used for production of heat is about 800 m^2 and total heat production in 2002 was 400 MWh.

Mexico

Although Mexico's average solar energy resource is estimated at 5 kWh/m^2/day, by end-2002 the country still did not possess a national PV programme. The existing installations have mainly been provided by poverty alleviation programmes operated by municipal authorities.

During 2002, nearly 1 200 kW PV capacity was installed (an increase of 14% over 2001) of which 594 kW was for rural electrification, 237 kW for telecommunications, 237 kW for off-shore oil platforms and 120 kW for water pumping and cathodic protection. The Mexican PV industry was expecting the 2003 market to double with respect to the previous year.

By end-2002 a cumulative total of 16 161 kW PV capacity had been installed of which 12 943 kW represented off-grid domestic capacity, 3 208 kW off-grid non-domestic and 10 kW grid-connected distributed.

At end-2002 there were 448 000 m^2 of flat plate solar collectors installed, mainly used for water heating for various purposes.

The Federal Government has declared renewable energy to be of national interest in its Mexican National Energy Programme 2001–2006. It is planned to encourage the participation of the private sector and to expand the scope of PV rural electrification.

Netherlands

The national government's change in its renewable energy policy, the political uncertainty of 2002 and the current contraction of the Dutch economy have all contributed to a slowing in the growth of the PV market in the Netherlands.

The 2001 change of policy regarding renewable energy resulted in the PV R&D Programme being replaced by a more general 'Renewables in The Netherlands' Programme. Additionally, the target of 1 500 MW_p PV in 2020 was also abandoned. As PV does not perform well in terms of output, production and installation costs when compared with wind and biomass, the sector has lost the impetus it once had.

Subsidies and support schemes have been altered in recent years and although the governance of the most important regulation for the private consumer market (the EPR—Energie Premie Regeling) was moved from the Ministry of Economic Affairs to the Ministry of Housing, it has continued. On 1 January 2003 it was transformed from a fiscal measure to a subsidy scheme. The EPR has proved popular with the domestic market and many thousands of customers have taken the opportunity to purchase a small (less than 600 W_p) PV system. Moreover, bringing the regulation of the EPR under the Ministry of Housing increases the possibilities for encouraging property developers to install PV applications.

At end-2002 a total of 26 326 kW_p PV had been installed, of which 4 632 kW_p was off-grid, 19 214 kW_p was grid-connected distributed and 2 480 kW_p was grid-connected centralised.

Development of the solar thermal sector began in the mid-1970s, mainly concentrating on the installation of large projects. Following a significant decline in the mid-1980s, the Dutch Government introduced a subsidy scheme at the beginning of the 1990s in order to stimulate the market. In 1994 the first Long-Term Agreement for the Implementation of Solar Hot Water Systems came into existence. The signatories to the 4-year Agreement were the Dutch Government, the solar industry and the energy utilities. It proved to be effective and in 1998 the Agreement was extended for a further 4 years.

The deployment of solar systems has been successful in the house building industry, owing to the encouragement by local authorities. By 2001 nearly 15% of all new residential dwellings were supplied with a DHW system and in total, 269 112 m^2 (of which 201 877 m^2 were flat plate collectors, 2 000 m^2 were vacuum collectors and 65 235 m^2 were unglazed collectors) were in use. The volume of solar thermal installed stabilised at about 30 000–35 000 m^2/yr during the period 2001–2003.

Restraints on the national government budget have led to an uncertain situation regarding the continuation of the EPR scheme in 2004. The discussion is continuing between Dutch Ministers and Parliament.

Norway

The majority of Norway's commercial solar market consists of off-grid PV-systems. At end-2002 a total of about 100 000 systems had been installed, mostly in recreational cabins. The panels, used for re-charging batteries for lighting, etc., are typically 50–60 W in size.

In addition, the Norwegian coastal service has installed some 2 200 solar beacons along the coast. It is planned that all off-grid lighthouses will be thus supplied in the future.

The EU Project, 'PV Nord', is designed to demonstrate and evaluate PVIB (PV in buildings) in Norway, Sweden, Denmark and Finland. At least one project is being conducted in Norway but generally the political willingness to support PV in the country is limited and electricity from solar panels will hardly represent a great share of Norwegian electricity generation in the foreseeable future.

Poland

The Centre of Photovoltaics (PV Centre) is based at the Warsaw University of Technology. One purpose of the Centre is to act as the link between the different bodies involved in the market for PV (academia, research institutions, industry and consumers).

To date there have been few applications of PV in Poland but in December 1999 the PV Centre installed and monitored the first stand-alone PV system for a pedestrian crossing. This application is expected to grow now that the General Polish Road Authority has become involved.

In December 2000 the PV Centre installed the first building-integrated thin-film system in the country. Since then other installations have been completed and the Centre continues to report to the Government with a view to encouraging the wider dissemination of PV in the country.

Portugal

The utilisation of the Portuguese solar resource has, to date, been low. There has been an historical dependence on imported fossil fuels, which has resulted in the country having the highest GDP energy intensity in the EU.

In order to provide a more integrated approach to energy supply and demand and achieve a higher energy efficiency, one objective of the Government's 2001 E4 Programme (Energy Efficiency and Endogenous Energies) is to promote the development of the renewable energies for power and thermal generation and thereby reduce CO_2 emissions.

At end-2002, only 1 668 kW_p (of which 76% were stand-alone applications and 24% were grid-connected) of PV capacity was installed. Installed solar thermal collectors total some 200 000 m^2, based on a sustainable annual market of about 7 000 m^2.

By 2010 it is planned that 50 MW of PV and 1 million m^2 of solar thermal (based on a sustainable annual market of 150 000 m^2) will be installed.

In addition to Portugal's first building-integrated grid-connected PV system (12 kW_p) developed by the National Institute for Engineering and Industry Technology, the world's largest PV power plant (64 MW_p) is currently under development—this plant alone, once completed, will exceed the planned 2010 capacity.

Romania

Solar energy is studied as part of two Government-financed research projects included in the National Research Programs: MENER (Energy and Environment Research) and AMTRANS (Buildings and Transport Research).

It has been determined that, taking into account the abundant wind and solar resource available to Romania, there is considerable scope for the development of hybrid systems. There are more than 80 000 households, together with schools, farms and isolated buildings, which are not grid-connected and could benefit from the installation of such applications.

Amongst the research and demonstration projects co-financed by the EU and the Romanian Government to date are:

- hybrid solar wind systems: 5 kW PV;
- PV systems for social objectives in remote places: 5 kW PV;
- PV/thermal power plant: 1.6 kW;
- building-integrated PV and grid-connected PV applications: Solar Amphitheatre 10 kW (grid);
- grid-connected demonstration system developed by Icemenerg—5kW (grid);
- other rural electrification: 3.9 kW;
- telecommunication and other: 5 kW.

With regard to thermal applications: about 10% of the collectors installed during 1980–1990 (approximately 800 000 m^2) are working. This results in an annual production of approximately 140 TJ. Five TJ have been installed in recent years by international and Romanian companies (Viessmann, Velux, Corina Gealan, Alfa-Bit, Solarterm).

Russian Federation

With its vast size, Russia necessarily receives a very substantial amount of solar radiation, but the geographical diversity of the country means that the resource is not uniformly available. The average solar radiation in the southern regions is about 1 400 kWh/m^2/yr whilst the remote northern areas receive about 810 kWh/m^2/yr.

The regions with the best potential comprise the North Caucasus, regions bordering the Black Sea and the Caspian, and the southern parts of Siberia and the Far East. Areas below or near latitude 50°N have particularly favourable solar radiation. The resource is extremely seasonal: at 55°N it ranges from 1.69 kWh/m^2/day in January to 11.41 kWh/m^2/day in July.

Although it has been estimated that the gross potential, the technical potential and the economic potential for solar energy are 2.3 trillion tce, 2 300 million tce and 12.5 million tce, respectively, Russia's enormous indigenous fossil fuel reserves have meant that historically, little attention has been paid to the renewable energies. However, with about 10 million people having no access to an electricity grid and most rural settlements having no centralised heat supply, the possibilities for off-grid solar energy or hybrid applications are huge.

Singapore

Most solar PV installations are dispersed stand-alone industrial/commercial systems, e.g. car park sign illumination, street lighting, emergency building lighting, bus stop lighting.

Grid-connected applications total approximately 10 kW installed capacity. With no government-driven incentive programme in prospect for grid-connected PV, a significant PV market in Singapore is unlikely. Experimental projects will occur sporadically, and industrial/commercial installations will continue where they are thought to be economically viable.

South Africa

In 1994 the newly elected Government of National Unity launched its Reconstruction & Development Programme and thereby accelerated the trend for PV installations. In the same year Eskom launched a huge electrification programme and by 1999 the national level of household electrification had risen from 36 to 68%.

The next phase of the project was to address the powering of the large dispersed rural population. It was not feasible to provide grid connection in isolated areas which were much more suited to having off-grid PV.

Receiving an annual average global solar radiation of approximately 5.5 kWh/m^2/day, the country is well placed for utilising this resource but at the present time there is no official register or survey of the actual installed capacity. SWH for domestic use and for the heating of swimming pools is becoming more fashionable although again the market penetration is unknown.

Spain

Despite an abundant solar resource, the development of the solar thermal market has historically experienced a low rate of growth. Continuing high costs have been blamed whereas the decreasing costs for solar PV energy have stimulated the market over the years.

The Instituto para la Diversificación y Ahorro de la Energía's (IDAE) Plan for the Promotion of Renewable Energy in Spain was approved by the Spanish Government on 30 December 1999 and covers the period 2000–2010. This sets out the programme for the development of low-temperature solar-thermal energy, high-temperature solar-thermal energy and solar-PV:

- low-temperature solar-thermal energy: the potential is estimated at 26.5 million m^2 of solar panels; by 2010 it is expected that 4.5 million m^2 of solar panels will have been installed in the existing residential sector, new homes, apartment blocks and hotels;
- high-temperature solar-thermal energy: Spain has built up a particular expertise with regard to high-temperature systems, having conducted much research at the Almeriá Solar Platform in the south-east of the country. It is now envisaged that by 2010, 200 MW of solar or solar-hybrid power will be installed, with an annual output of 413 GWh;
- solar-PV: the potential resource is estimated at 2 300 MW. The following targets have been set for 2010—20 MW stand-alone installations plus 50 MW of installations less than 5 kW and 65 MW of installations greater than 5 kW.

At the beginning of 2003 a 1.2 MW$_p$ solar PV plant covering 70 000 m^2 was opened near the town of Tudela in the Navarre region. The location of the site receives 1 600 kWh/m^2/yr solar radiation; whilst the central section of the 12 602 PV panels is connected to the grid the remaining 'distributed' area will be used for research on a variety of PV technologies and types of panel.

Swaziland

Comprehensive data for solar energy are being established through a 5-year project which started in 2001. The project aims at establishing the wind and solar regime in the country. Data from the measuring stations are being collected for validation. Currently, insolation is estimated at 5 kWh/m^2/day.

Although during the first half of the 1990s, several solar projects were installed in rural locations, they mostly failed. One success has been a project started in 2000 under UNESCO's Solar Village Programme. The village of Mphaphati was chosen for the establishment of a demonstration project at the primary school. It included providing lighting for three classrooms and a security system, the installation of audio-visual aids, SHS for the teacher's houses and water pumping for the school's vegetable garden. Subsequently, a solar-powered cellular payphone has been installed in the village shop and a rural cinema established.

Sweden

With its electricity generation currently dependent on nuclear and hydro, Sweden's market for solar energy is negligible. As in Norway and Finland, the main application of PV is in the domestic off-grid sector, where installations are sited in remote cabins, campers, caravans and boats. A system for the promotion of renewable energy through tradable electricity certificates was launched in 2003, but prices were probably too low to promote a high demand for PV.

By end-2002 only 3 297 kW$_p$ PV had been installed, of which nearly 80% were for the off-grid domestic market. It is thought that an increase in the incidence of solar PV will occur over time but the technology will probably not be utilised for large-scale electricity power generation within the next 5–10 years.

Likewise, the market for solar thermal systems has not been strong and by end-2001 it was estimated that a total of 192 157 m^2 had been installed, of which 156 522 m^2 were flat plate collectors, 1 704 m^2 were vacuum collectors and 33 931 m^2 were unglazed collectors.

Switzerland

Following the Government's national programme *Energy 2000*, launched in 1990, *Swiss-Energy* (also 10 years) is a further programme for the promotion of renewable energy and more efficient use of energy.

Switzerland has a dedicated national PV programme which focusses not only on all aspects of RD&D, but also the promotion of the technology and its market deployment. By end-2002 the campaign 'Solar electricity from the utility' had resulted in more than 130 utilities providing solar electricity to customers. Approximately 50% of the population have access to solar-generated electricity and in excess of 30 000 customers purchase 5 GWh per annum.

Thailand

The solar map for Thailand created in 1999 by the Department of Energy Development and Promotion (DEDP) shows that the entire country's average energy from sunlight amounts to 18 MJ/m^2/day and that in the April–May period the majority of regions record their highest levels (20–24 MJ/m^2/day). Thus, there is considerable scope for the development of applications utilising solar energy.

The Government's Renewable Energy Development Plan includes provision for research into PV battery-charging systems for a Non-electrified Rural Villages Project.

During 2003 the School of Renewable Energy Technology (SERT, located at Naresuan University) completed the first phase of its Energy Park. The Park, designed to provide a centre for exhibition, demonstration, research and marketing of solar energy technology, will be completed in 2005.

At the present time approximately 5.5 MW of stand-alone PV and grid-connected applications have been installed. Their siting is mostly in isolated rural areas where they are used at solar cell battery-charging stations, for pumping water in villages and as navigation aids and communication repeaters. Demonstration projects also exist for PV/wind/diesel hybrid systems for power generation in national parks and wildlife sanctuaries.

It has been reported that the target is for 20 MW of PV systems to be installed by 2006, of which 8 MW is expected to be stand-alone systems for remote areas (water pumping, mini-grids for schools and health centres, and traffic lights), 3 MW for residential and commercial buildings (own consumption and for the return of excess electricity back to the utilities) and 9 MW will be for grid support.

Many indigenous companies have expressed interest in creating PV-related businesses and one, Thai Photovoltaics Ltd, is reported to be establishing the country's first solar cell manufacturing plant using thin-film technology.

Turkey

Turkey's utilisation of its significant solar radiation resource is largely in the form of solar thermal collectors. The market was initiated during the 1970s in response to the growth of the tourism industry and the need for plentiful hot water. The country's energy supply difficulties and the political and economic uncertainties of the 1980s provided further impetus to market development. Although deployment has been extensive—it is estimated that there are in the region of 8 million m^2 of flat plate collectors installed—the sector has not demonstrated a high degree of advanced technology. Turkish customers have historically preferred simple, inexpensive installations, albeit that this approach has led to problems of utilisation and maintenance.

It is expected that the solar thermal market will continue to grow in Turkey by means of increasing the number of collectors installed on roofs and also possibly through large-scale projects such as greenhouse heating during the winter months in the agricultural areas of southern Turkey.

United Kingdom

In 1990 the British Government introduced the first Non-Fossil Fuel Obligation Renewable Order (NFFO) for England and Wales (and subsequently the four further Orders). Scotland has had its own three Orders (SRO) and Northern Ireland two (NI NFFO). The Orders were designed to impose an Obligation on electricity suppliers (the Renewables Obligation) to deliver a specified proportion of their supplies from electricity generated from specified sources of renewable energy or to purchase Renewables Obligation Certificates or to make a buyout payment.

In November 2001 the Performance and Innovation Unit (PIU, part of the Cabinet Office) announced how £100 million of additional funding for renewables was to be spent (e.g. grants for small-, medium- and large-scale PV installations) and in early 2002 published a review of energy policy. This included recommendations on the way in which renewables could be further utilised beyond 2010. The Government responded to the recommendations in the report in an Energy White Paper published in February 2003. This not only endorsed the stated aim for 10% of generated electricity to come from renewable energy sources by 2010 but strengthened the aspiration by a target of doubling the renewables' share of electricity to 20% by 2020.

Although the commitment is present, the UK had only 4 136 kW_p installed PV capacity by end-2002, representing less than 0.5% of the total capacity installed in the 20 IEA-PVPS member countries. However, three key programmes (research and development; field tests and demonstrations and participation in international programmes) supported the installation of 1.4 MW during the year (2001 had seen an increase of only 0.8 MW). The demonstration programme aims to deliver 3 000 domestic PV systems and 140 larger (non-domestic) systems over a 3-year period. The field tests are aimed at challenging technical and institutional barriers. In the period 2002/2003 to 2005/2006 the Government is making a total of £348 million available to support the programme and to increase the momentum necessary to achieve the 2010 target.

Solar collectors for heating water are used in the UK to a limited extent. In 2001 it was estimated that they contributed 78.5 GWh for heating swimming pools, while the solar contribution to DHW supply in 2002 was estimated at 55.2 GWh.

United States of America

The US solar resource is far in excess of all projected energy demand in the mid-term. Solar insolation levels vary from less than 400 W/m^2 to over 700 W/m^2, depending on latitude, climate (primarily average cloud cover), terrain, and application (i.e. using a fixed-angle collector as compared to a sun-tracking collector). However, the country has approximately 9 million km^2 of land area and currently the primary limitation to further utilisation of the vast solar resource is the high cost of solar power relative to conventional and other renewable resources.

Solar energy has thus far primarily competed in niche markets such as residential hot water supply, remote power generation, consumer electronics, and limited grid-tied applications to offset utility peak power requirements. To supply more than a relatively small fraction of total energy needs, expensive solar power technologies will require the use of expensive energy storage.

Planned capacity additions are 108 MW for central station solar thermal, 7 MW for central station solar PV, and 475 MW for dispersed solar PV. The goal of the Solar Energy Technologies Program (Solar Program) is to increase the supply of renewable energy through cost reductions in solar technology. The Department of Energy, working with industry, will achieve these cost reductions by increasing photovoltaic cell efficiency, system reliability and manufacturing capability, and by reducing the production cost of PV, CSP, and SWH systems.

The Solar Program currently develops PV and CSP systems for electricity, and solar heating systems for domestic and commercial hot water production. Electricity generation can be accomplished on scales ranging from kilowatts to megawatts and can be used by residential and commercial buildings, manufacturing plants, and electric utilities. The Solar Program also conducts research and development on related balance-of-system components such as direct current and alternating current power inverters, battery charge controllers and storage systems, as well as commercial and industrial facility lighting through fibre optic solar lighting technologies.

Uruguay

Uruguay has a well-developed electricity grid with about 95% of the population having a connection. Where systems have been installed, the principal applications for the utilisation of solar power have been for water heating and for PV, mainly for lighting.

Chapter 12

Geothermal Energy

World Energy Council
CONSEIL MONDIAL DE L'ENERGIE

COMMENTARY

Geothermal energy is heat energy from the depths of the earth. It originates from the earth's molten interior and from the decay of radioactive materials in underground rocks. The heat is brought near the surface by crustal plate movements, by deep circulation of groundwater and by intrusion of molten magma, originating from a great depth, into the earth's crust (see Fig. 12.1). In some places the heat rises to the surface in natural streams of steam or hot water, which have been used since prehistoric times for bathing and cooking.

Zones of high heatflow may be located close to the surface where convective circulation plays a significant role. Deep circulation of groundwater along fracture zones brings heat to shallower levels, collecting the heatflow from a broad area and concentrating it into shallow reservoirs. By drilling wells, this heat can be tapped to supply pools, greenhouses and power plants.

The quantity of this heat energy is enormous; it has been estimated that over the course of 1 year, the equivalent of more than 100 million GWh of heat energy is conducted from the earth's interior to the surface. Yet geothermal energy tends to be relatively diffuse, which makes it difficult to tap. If it were not for the fact that the earth itself concentrates geothermal heat in certain regions—typically regions associated with the boundaries of tectonic plates (see Fig. 12.2) then geothermal energy would be essentially useless.

Geothermal resources are renewable within the limits of equilibrium between offtake of reservoir water and natural or artificial recharge. Within such an equilibrium the energy source is renewable for a long period of time. At other sites the resource lifetime, if not recharged, may be limited to several decades. In any case, if it is not technically renewable, the global geothermal potential represents a practically inexhaustible energy resource. The issue is not the finite size of the resource, but the availability of technologies able to tap the resource economically.

Nature of the Geothermal Energy Resource

On average, the temperature of the earth increases by about 3 °C for every 100 m in depth. This means that at a depth of 2 km, the temperature of the earth is about 70 °C, increasing to 100 °C at a depth of 3 km, and so on. However, in some places, tectonic activity allows hot or molten rock to approach the earth's surface, thus creating pockets of higher temperature resources at easily accessible depths.

2004 Survey of Energy Resources

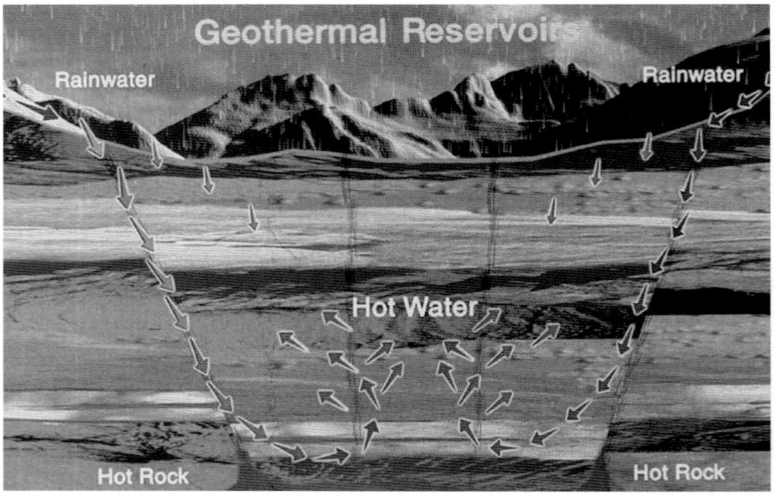

FIGURE 12.1 A representative geothermal reservoir.

The extraction and practical utilisation of this heat requires a carrier which will transfer the heat towards the heat-extraction system. This carrier is provided by geothermal fluids forming hot aquifers inside permeable formations. These aquifers or reservoirs are the hydrothermal fields. Hydrothermal sources are distributed widely but unevenly across the earth. High-enthalpy geothermal fields occur within well-defined belts of geologic activity, often manifested as earthquakes, recent volcanism, hot springs, geysers and fumaroles. The geothermal belts are associated with the margins of the earth's major tectonic or crustal plates and are located mainly in regions of recent volcanic activity or where a thinning of the earth's crust has taken place.

FIGURE 12.2 World map showing lithospheric plate boundaries (Source: U.S. Geological Survey).

One of these belts rings the entire Pacific Ocean, including Kamchatka, Japan, the Philippines, Indonesia, the western part of South America running through Argentina, Peru, Ecuador, Central America, and western North America. An extension also penetrates across Asia into the Mediterranean area. Hot crustal material also occurs at mid-ocean ridges (e.g., Iceland and the Azores) and interior continental rifts (e.g., the East African Rift, Kenya and Ethiopia).

Low-enthalpy resources are more abundant and more widely distributed than high-enthalpy resources. They are located in many of the world's deep sedimentary basins, e.g. along the Gulf Coast of the United States, western Canada, in western Siberia, and in areas of central and southern Europe, as well as at the fringes of high-enthalpy resources.

There are four types of geothermal resources: hydrothermal, geopressured, enhanced geothermal systems (formerly Hot Dry Rock, HDR), and magma. Although they have different physical characteristics, all forms of the resource are potentially suitable for electric power generation if sufficient heat can be obtained for economical operation.

Hydrothermal Resources

These are the only commercially used resources at the present time. They contain hot water and/or steam trapped in fractured or porous rock at shallow to moderate depths (from approximately 100–4 500 m). Hydrothermal resources are categorised as vapour-dominated (steam) or liquid-dominated (hot water) according to the predominant fluid phase. Temperatures of hydrothermal reserves used for electricity generation range from 90 °C to over 350 °C, but roughly two thirds are estimated to be in the moderate temperature range (150°–200 °C). The highest quality reserves contain steam with little or no entrained fluids, but only two sizeable, high-quality dry steam reserves have been located to date, at Larderello in Italy and The Geysers field in the United States.

Recoverable resources available for power generation far exceed the developments to date. Many countries are believed to have potential in excess of 10 000 MW_e which would fulfil a considerable portion of their electricity requirements for many years (e.g. the Philippines, Indonesia and the USA).

Important low-enthalpy hydrothermal resources are not necessarily associated with young volcanic activity. They are found in sedimentary rocks of high permeability which are isolated from relatively cooler near-surface groundwater by impermeable strata. The water in sedimentary basins is heated by regional conductive heat flow. These basins (e.g. the Pannonian Basin, in eastern Europe) are commonly hundreds of kilometres in diameter at temperatures of 20°–100 °C. They are exploited in direct thermal uses or with heat pump technology.

Geopressured Resources

Geopressured geothermal resources are hot water aquifers containing dissolved methane trapped under high pressure in sedimentary formations at a depth of approximately 3–6 km. Temperatures range from 90° to 200 °C, although the reservoirs explored to date seldom exceed 150 °C. The extent of geopressured reserves is not yet well known world-wide, and the only major resource area identified to date is in the northern Gulf of Mexico region where large reserves are believed to cover an area of 160 000 km^2. This resource is potentially very promising because three types of energy can be extracted from the wells: thermal energy from the heated fluids, hydraulic energy from the high pressures involved, and chemical energy from burning the dissolved methane gas.

Enhanced Geothermal Systems—EGS (Formerly Hot Dry Rock Resources)

These resources are accessible geologic formations that are abnormally hot but contain little or no water. The EGS potential is 200 GW in the USA and 60 GW in Europe. The basic concept in HDR technology is to form a man-made geothermal reservoir by drilling deep wells (400–5 000 m) into high-temperature, low-permeability rock and then forming a large heat-exchange system by hydraulic or explosive

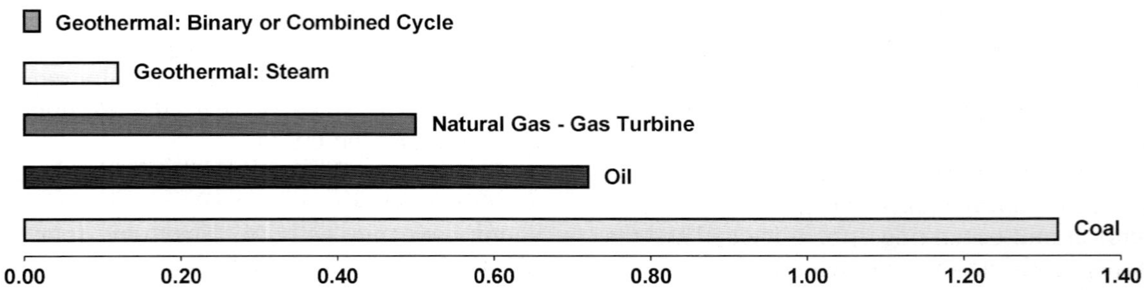

FIGURE 12.3 Relative emissions (kgCO_2/kWh) for different energy resources.

fracturing. Injection and production wells are joined to form a circulating loop through the man-made reservoir, and water is then circulated through the fracture system.

Ecological and Environmental Impact

Geothermal energy use has a net environmental impact. Geothermal power plants have fewer and more easily controlled emissions than any similar-sized fossil fuel power plants (see Fig. 12.3). Direct heat uses are even cleaner and are practically non-polluting when compared to conventional heating.

There are other environmental advantages to geothermal energy, such as the fact that power plants using geothermal energy require far less land area than other energy resources (see Fig. 12.4). Another advantage, which differentiates geothermal energy from other renewables is its continuous availability 24 hours a day, all year round.

Mainstream Technologies

Geothermal energy has been used for centuries for bathing, therapeutic utilisations and water heating. Only in the 20th century has geothermal energy been harnessed on a large scale for other purposes, such as space heating, industry and electricity generation. The range of potential methods for utilising any geothermal resource depends mostly on the temperature of the resource.

Direct Heat Use

Lower-temperature geothermal resources occur in many world regions. They can provide useful energy for space and water heating, district heating, greenhouse heating, warming of fish ponds in aquaculture, crop drying, etc. (for the shares of the main direct heat uses, see Fig. 12.5). Geothermal fluids are generally pumped through a heat exchanger to heat air or a liquid in direct use, although the resources may be used directly if the salt and solid contents are

Technology	Land occupied in m^2/GWh/yr for 30 years
Coal (including mining)	3 642
Solar thermal	3 561
Photovoltaics	3 237
Wind (turbines and roads)	1 335
Geothermal	404

FIGURE 12.4 Land uses for different energy technologies.

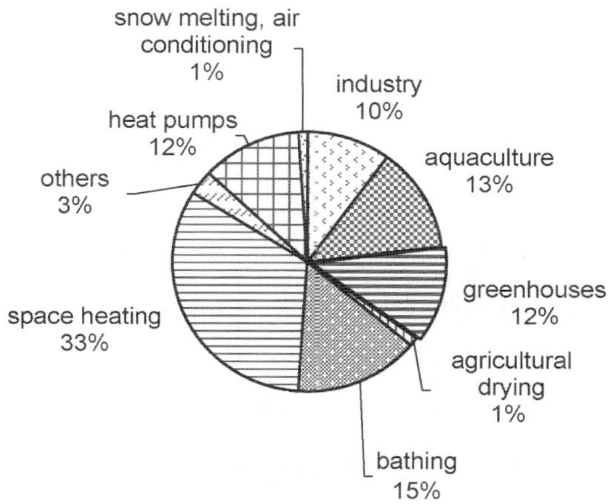

FIGURE 12.5 Direct-heat uses.

low. In comparison with geothermal electricity production, direct use has several advantages, such as higher energy efficiency (50–70%), generally the development time is shorter and less capital investment is involved.

Geothermal heat pump (GHP) technology can use geothermal sources of 20 °C or less. GHP can move heat in either direction; in winter heat is removed from the earth and delivered to the home or building—heating mode, while in summer heat is removed from the home or building and delivered for storage to the earth–air conditioning mode.

Geothermal Power Generation Technologies

There are several types of energy-conversion processes for generating electricity from hydrothermal resources. These include dry steam and flash steam systems, which are traditional processes; binary cycle and total flow systems, which are newer processes with some significant advantages.

Dry Steam Plants. Conventional steam-cycle plants are used to produce energy from vapour-dominated reservoirs. As is shown in Fig. 12.6, steam is extracted from the wells, cleaned to remove entrained solids and piped directly to a steam turbine. This is a well-developed, commercially available technology, with typical unit sizes in the 35–120 MW_e capacity range. Recently, in some places, a new trend of installing modular standard generating units of 20 MW_e has been adopted. In Italy, smaller units in the 15–20 MW_e range have been introduced.

FIGURE 12.6 Dry steam power plant (Source: Geothermal Education Office).

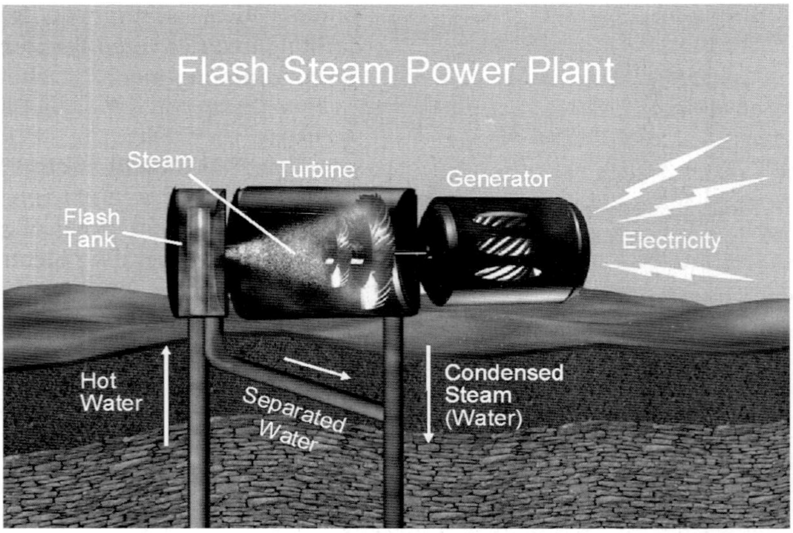

FIGURE 12.7 Flash steam power plant (Source: Geothermal Education Office).

Flash Steam Plants. More complex cycles are used to produce energy from liquid-dominated reservoirs which are sufficiently hot (typically above 160 °C) to flash a large proportion of the liquid to steam. As shown in Fig. 12.7, single-flash systems evaporate hot geothermal fluids to steam by reducing the pressure of the incoming liquid and direct it through a turbine. In dual-flash systems, steam is flashed from the remaining hot fluid of the first stage, separated and fed into a dual-inlet turbine or into two separate turbines. In both cases, the condensate may be used for cooling while the brine is re-injected into the reservoir. This technology is economically competitive at many locations and is being developed using turbogenerators with capacities of 10–55 MW_e. A modular approach, using standardised units of 20 MW_e, is being implemented in the Philippines and Mexico.

Binary-Cycle Plants. Operating experience over the years has confirmed the advantages of binary geothermal plants, not only for low-enthalpy water-dominated resources but also for high-enthalpy resources with high aggressive brine or brine with high non-condensible content. The systems deliver sustainable zero-pollution energy and avoid a long-term depletion of the resource (which is 100% reinjected).

- *Low-Enthalpy Resources (100°–160 °C)*. For low-enthalpy resources, binary plants based on the use of organic Rankine cycles (ORC) are utilised to convert the resource heat to electrical power (see Fig. 12.8). The hot brine or geothermal steam is used as the heat source for a secondary (organic) fluid, which is the working fluid of the Rankine cycle.

In the early 1980s, in order to increase the power output from a given brine resource by increasing the thermal cycle efficiency, a supercritical cycle using isobutane was developed, as well as a cascade concept. The supercritical cycle may be slightly more efficient than the cascading cycle, but the cascading system has the advantage of lower operating pressures and lower parasitic loads in the cycle pumps. For example, at a power plant in Southern California, a three-level arrangement was employed and resulted in increased efficiency or power output gain of about 10% over that achievable with a simple ORC. In all of the above arrangements, a modular approach was employed so that

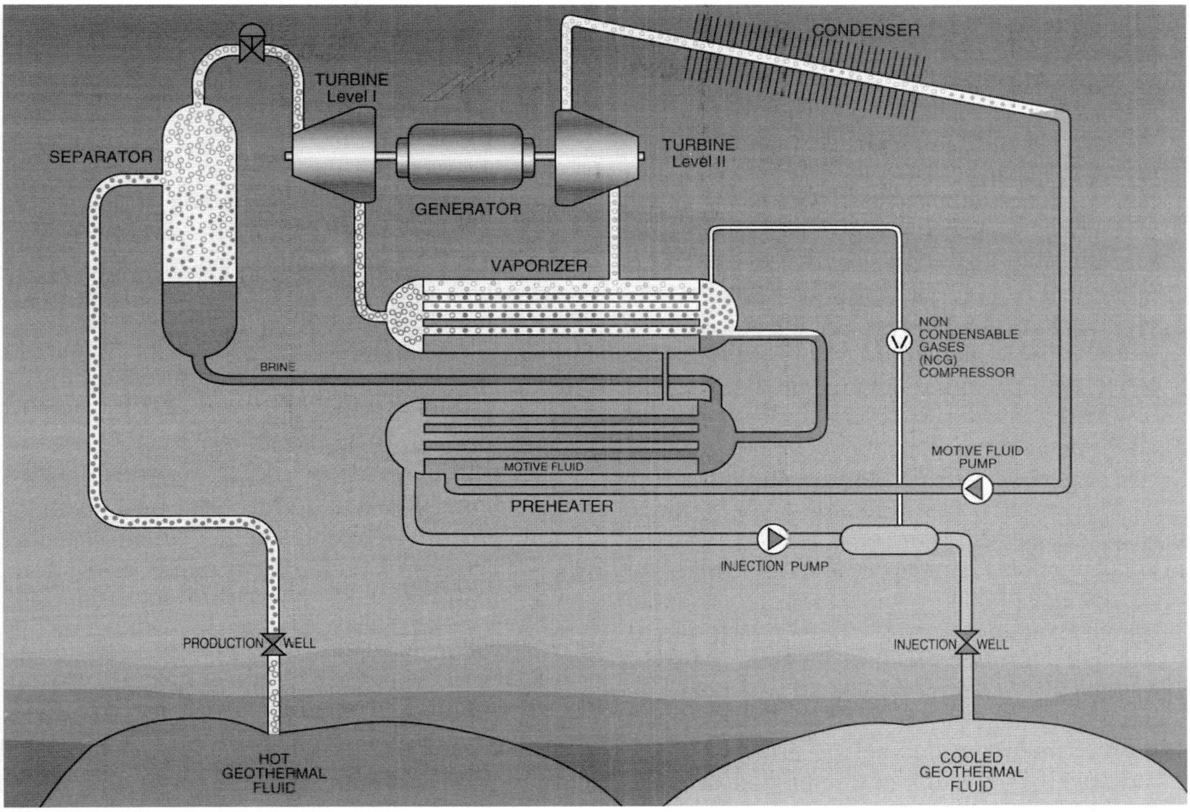

FIGURE 12.8 Air-cooled binary plant (Source: Ormat).

high plant availability factors of 98% and above were achievable.

- *Moderate-Enthalpy Resources (160°–190 °C).* For moderate-enthalpy, two-phase resources with steam quality between 10 and 30%, binary plants are also efficient and cost-effective. Furthermore, when the geothermal fluid has a high non-condensible gas (NCG) content, even higher efficiency can be obtained than with condensing steam turbines.

This binary two-phase configuration is used in the São Miguel power plant in the Azores Islands (Fig. 12.9). Separated steam containing NCG is introduced in the vaporiser heat exchanger to vaporise the organic fluid. The geothermal condensate at the vaporiser exit is then mixed with the hot separated brine

FIGURE 12.9 Two-phase binary geothermal power plant on São Miguel Island, Azores (Source: YRE International Mission in Azores—FEE Portugal).

to provide the preheating medium of the organic fluid.

Since the onset of silica precipitation is related to its concentration in the brine, dilution of the brine with the condensate effectively lowers the precipitation temperature at which silica crystallises. This lower temperature added 3.5 MW of heat to the cycle representing 20% of the total heat input. This additional heat is utilised at the same thermal efficiency as the remaining heat, owing to the nature of the combined steam-brine cycle. Since the cycle efficiency is about 17%, this low-temperature heat produces about 600 additional kW.

The second way to improve the utilisation of the resource is the use of a regenerative cycle by the addition of a recuperator heat exchanger between the organic turbine and the air-cooled condenser, since the organic vapour tends to superheat when the vapour is expanded through the turbine. In this case the recuperator reduces the amount of heat that needs to be added to the cycle from the external source, thereby reducing the brine flow rate required. This results in a reduction of about 7% in the total heat input required to produce the design level of power output.

- *Geothermal Combined Cycle Plants.* To best utilise a steam-dominated resource, a Geothermal Combined Cycle is used where the steam first flows through a back-pressure steam turbine and then is condensed in the organic turbine vaporiser. The condensate and the brine are used to preheat the organic fluid as in the two-phase binary configuration above.

  This concept was first used in 1989 in repowering a back-pressure steam plant in Iceland, then with a 30 MW plant in Hawaii in 1992, followed by a 125 MW plant in the Philippines and a 60 MW plant in New Zealand.

Experimental Plants. At an experimental stage are several other systems aiming to utilise the geothermal resource in an efficient manner:

- Trilateral cycle—out of the binary total flow systems, the most well conceived is the trilateral cycle, which has been partially tested.
- Absorption and Absorption/Regenerative cycles—of the various cycles proposed, the most advanced system is the Kalina cycle, which was tested on an energy recovery plant. One geothermal plant has been tested recently in Iceland and one is planned in the USA. A demonstration is yet to be made in a geothermal power plant to prove the practicality of the concentration variations, the high pressure of the system, and other factors.
- Field-tested Systems
  - The total flow steam cycle (bi-phase), although conceptually elegant and theoretically efficient, did not attain sustained commercial operation in its prior trials, mainly because of clogging in nozzles.
  - The direct heat exchanger usage encountered serious problems of fouling and excessive hydrocarbon fluid loss.
  - Hybrid system: this is a complex design combining internal combustion engines with heat recovery from the hot brine and exhaust. Tests have yet to demonstrate the validity of the concept.

Recent Developments

Current Uses and Commercial Status

Electricity from geothermal energy has been generated in Italy for almost 100 years. Until 1974, the total installed capacity for converting geothermal energy into electricity was only about 770 MW_e (in Italy, Japan, New Zealand, the United States, and Mexico). Following the second oil shock, the worldwide installed capacity achieved its highest growth of 17.2% per year. The number of geothermal power-producing countries increased from 10 to 17. Recent emphasis has been placed on power production using the liquid hydrothermal resource since power production with dry steam has been commercially viable for several decades. At the end of 2002 a total of over

8 200 MW_e geothermal electricity generating capacity was operational worldwide, in more than 20 countries. Substantial market penetration has thus far occurred only with hydrothermal technology.

Going into some detail, the seven countries with the largest electric power capacity are: the USA with over 2 000 MW_e is first, followed by the Philippines (1 931 MW_e); five other countries (Mexico, Italy, Indonesia, Japan and New Zealand) have capacities (in 2003) ranging from 440 to 950 MW_e each. These seven countries represent over 90% of the world capacity and about the same percentage of world output, amounting to about 51 000 GWh.

The strong decline in the USA in recent years, due to over-exploitation of the giant Geysers field has been partly compensated by important additions to capacity in several countries: Indonesia, Mexico, Italy, New Zealand, Costa Rica, and Russia. Newcomers in the electric power sector recently (2003) are Papua New Guinea and Germany.

Direct use or non-electric applications of geothermal energy have enjoyed modest growth. The statistics are a reflection of limited data and it is likely that there are many direct-use applications not reflected in the figures. Most data reflect the situation as evaluated in 2000. An update of data is currently in preparation for the 2005 World Geothermal Congress and it is expected that this will reveal a decrease in the growth rate for all sectors (Geothermal Resources Council Bulletin, July–August 2003). However, Table 12.1 shows that three leading countries, the US (5 366 MW_t), China (2 814 MW_t) and Iceland (1 800 MW_t) cover 59% of the world capacity, which has reached 17 000 MW_t. Out of about 60 countries with direct geothermal plants, besides the three mentioned above, Turkey, Italy, Switzerland, Sweden, France, Canada, Germany, Japan and New Zealand have sizeable capacity.

Heat Pump Applications

During the last decade a number of countries have encouraged individual house owners to install ground-source heat pumps to heat their homes in the winter and (as needed) to cool them in summer. Financial incentive schemes have been set up, commonly funded by the governments and electric utilities, as the GHP reduce the need for peak power and thus replace new electric generating capacity. The US Government Heat Pump Consortium estimates that there were 750 000 GHP units installed in the USA (in 2002), which reduced electricity demand by some 1 900 MW. In 2002 the European Union countries installed more than 50 000 GHP units (Geothermal Heat Pump Consortium, October 2003).

R&D

Several demonstration projects are under development in the framework of *Enhanced Geothermal Systems*, with the participation of the USA, the European Union, Germany, France, Switzerland, Italy, Japan and Australia. HDR projects are under development at Fenton Hill (USA), Soultz-sous-Forêts (France) and Hunter Valley (Australia).

The Future

The growth rate of the geothermal energy market is not limited by a lack of resources. During the early oil crises, intensive investigations led to the discovery of many geothermal reservoirs for electricity generation, some of which are in operation, while about 11 000 MW_e of proven resources are not yet tapped. In the near future, the growth rate will most probably be 3–4% annually, as has been the case during the past few years.

The report prepared by the Geothermal Energy Association shows that geothermal resources using today's technology have the immediate potential to support between 35 450 and 72 390 MW of electrical generation capacity Using enhanced technology, currently under development, the geothermal resource could support between 65 580 and 138 130 MW of electrical generation capacity. Assuming a 90% availability factor, which is well within the range experienced by modern geothermal power plants, this electric capacity could produce over a trillion kWh of electricity annually.

Worldwide the report indicates that geothermal power could serve the electricity needs of 865 million people, or about 17% of the world's population. It identifies 39 countries which could be 100% geothermal powered, mostly in Africa, Central and South America and the Pacific.

However, if the environmental impacts of energy use are internalised, then the real value of geothermal technology, including its superior environmental characteristics and local resource features, will be taken into account and the geothermal market will become more profitable. As a result, there will be enhanced geothermal exploration and R&D. The growth rate should then reach 6–7% and more. This outlook should encourage the development of other geothermal resources. The EGS and geopressured technologies may reach maturity around 2010.

Lucien Bronicki
Michael Lax
Israeli WEC Member Committee

References

Bronicki, L., 1998. Eighteen Years of Field Experience with Innovative Geothermal Power Plants, World Energy Council, Houston, TX.
Bronicki, L., 2002. Geothermal Power Stations, third ed., Encyclopedia of Physical Science and Technology, vol. 6., pp. 709–718.
Fridleifsson, I.B., 2000. Geothermal Energy for the Benefit of the People Worldwide, web site: www.geothermie.de
Gawell, K., Reed, M., Wright, P.M., 1999. Preliminary Report: Geothermal Energy, the Potential for Clean Power from the Earth, Geothermal Energy Association.
GeothermEx, 1998. Peer Review of the Hot Dry Rock Project at Fenton Hill, New Mexico, a report for the Office of Geothermal Technologies, US Department of Energy.
Grassiani, M., Krieger, Z., 1999. Advanced Power Plants for Use with Hot Dry Rock (HDR) and Enhanced Geothermal Technology, ORMAT, Yavne, Israel.
Huttrer, G.W., 2000. The Status of World Geothermal Power Generation 1995–2000, Proceedings, World Geothermal Congress, pp. 23–27.
Lund, J.W., 2000. World status of geothermal energy use, overview 1995–1999, Proceedings, World Geothermal Congress, pp. 4105–4111.
Lund, J.W., Freeston, D.H., 2000. World direct uses of geothermal energy 2000, Proceedings, World Geothermal Congress 2000, pp. 1–21.
Nakicenovic, N., Grübler, A., McDonald, A., 1998. Global Energy Perspectives, Cambridge University Press, Cambridge.
U.S. DOE, 1998, Strategic Plan for the Geothermal Energy, NREL, U.S. Department of Energy, Office of Geothermal Technologies.
Williams, S., Porter, K., 1989. Power Plays–Profiles of America's Independent Renewable Electricity Developers, Investor Responsibility Research Center, Washington, DC.
World Energy Council, 1994. New Renewable Energy Resources, Chapter 4, Kogan Page.
World Energy Council, 2000a. World Energy Assessment, United Nations Development Programme, United Nations Department of Economic and Social Affairs.
World Energy Council, 2000b. Energy for Tomorrow's World–Acting Now! World Energy Council Statement.

Table Notes

The data shown in Table 12.1 reflect as far as possible those reported by WEC Member Committees in 2003.

When not available from WEC Member Committees, data were drawn from the *Proceedings of the World Geothermal Congress*, Kyushu & Tohoku, Japan, 28 May–10 June 2000, International Geothermal Association, and the Geothermal Resource Council, together with advice from Michael Lax of the Israeli WEC Member Committee. National statistical sources have also provided a small amount of data.

Installed electricity generating capacity in the USA (2 002 MW_e) reflects the level reported by J. Lund at the end of 2002. This level is somewhat lower than that published by the US DOE/EIA and reported by the US WEC Member Committee (2 216 MW_e). The difference is attributable to the treatment of downrated capacity.

The direct use of geothermal energy is not only inherently difficult to quantify but in some instances can be subject to constraints on reporting for reasons of confidentiality, etc. The statistics shown for both capacity and output should therefore be treated as, at best, indicative of the situation in a particular country. In many cases the figures have not been updated since the last World Geothermal Congress (2000). As far as possible, direct use includes the capacity and output of geothermal (ground-source) heat pumps.

Annual capacity factors have been calculated on the basis of end-year capacity levels, as average-year data were not available. In general, therefore, the factors shown will tend to be understated. The capacity factors of 1.00 given for direct use of geothermal energy in certain countries reflect the assumptions made in the surveys consulted.

TABLE 12.1

Geothermal energy: electricity generation and direct use at end-2002

	Electricity generation			Direct use		
	Installed capacity (MW_e)	Annual output (GWh)	Annual capacity factor	Installed capacity (MW_t)	Annual output (TJ)	Annual capacity factor
Algeria				100	1 588	0.50
Kenya	57	447	0.90	1	11	0.25
Tunisia				20	173	0.28
Total Africa	**57**	**447**	**0.90**	**121**	**1 772**	**0.46**
Canada				378	1 022	0.09
Costa Rica	145	984	0.78			
El Salvador	161	940	0.67			
Guadeloupe	4	25	0.67			
Guatemala	28	175	0.71	3	108	1.00
Honduras				1	18	0.76
Mexico	853	5 398	0.72	164	3 920	0.76
Nicaragua	35	275	0.90			
United States of America	2 002	13 357	0.76	5 366	25 006	0.15
Total North America	**3 228**	**21 154**	**0.75**	**5 912**	**30 074**	**0.16**

TABLE 12.1 (*Continued*)

	Electricity generation			Direct use		
	Installed capacity (MW_e)	Annual output (GWh)	Annual capacity factor	Installed capacity (MW_t)	Annual output (TJ)	Annual capacity factor
Argentina	1	N		26	450	0.55
Chile				N	7	0.55
Colombia				13	266	0.63
Peru				2	50	0.65
Venezuela				1	14	0.63
Total South America	**1**	**N**		**42**	**787**	**0.60**
China	29	100	0.39	2 814	31 406	0.35
Georgia				250	6 307	0.80
India				80	2 516	1.00
Indonesia	807	6 238	0.88	7	43	0.19
Japan	547	3 431	0.72	258	5 836	0.72
Korea (Republic)				51	1 076	0.67
Nepal				1	22	0.66
Philippines	1 931	10 248	0.61	3	40	0.38
Thailand	N	2	0.61			
Turkey	18	81	0.53	820	15 757	0.61
Total Asia	**3 332**	**20 100**	**0.69**	**4 284**	**63 003**	**0.47**
Austria	1	4	0.48	58	420	0.23
Belgium				4	108	0.87
Bulgaria				107	1 638	0.48
Croatia				114	550	0.15
Czech Republic				13	130	0.33
Denmark				3	84	0.81
Finland				300	3 200	0.34
FYR Macedonia				81	511	0.20
France				326	4 914	0.48
Germany				397	1 568	0.13
Greece				100	350	0.11
Hungary				250	3 600	0.46
Iceland	202	1 433	0.81	1 800	24 700	0.44
Italy	862	4 660	0.62	680	8 916	0.42
Lithuania				41	681	0.53
Netherlands				11	58	0.17
Norway				6	32	0.17
Poland				55	274	0.16
Portugal	16	105	0.75	2	3	0.25
Romania				191	2 128	0.35
Russian Federation	73	300	0.47	307	6 131	0.63
Serbia & Montenegro				86	2 415	0.89
Slovakia				75	340	0.14

(*continued on next page*)

TABLE 12.1 (*Continued*)

	Electricity generation			Direct use		
	Installed capacity (MW_e)	Annual output (GWh)	Annual capacity factor	Installed capacity (MW_t)	Annual output (TJ)	Annual capacity factor
Slovenia				103	1 080	0.33
Spain				70	1 051	0.47
Sweden				377	4 129	0.35
Switzerland				547	2 387	0.14
United Kingdom				3	36	0.38
Total Europe	**1 154**	**6 502**	**0.64**	**6 107**	**71 434**	**0.37**
Israel				63	1 714	0.86
Jordan				153	1 541	0.32
Total Middle East				**216**	**3 255**	**0.48**
Australia	N	1	0.68	10	295	0.90
New Zealand	448	2 715	0.69	308	7 081	0.73
Total Oceania	**448**	**2 716**	**0.69**	**318**	**7 376**	**0.74**
Total World	**8 220**	**50 919**	**0.71**	**17 000**	**177 701**	**0.33**

COUNTRY NOTES

The Country Notes on Geothermal Energy have been compiled by the editors together with contributions from Michael Lax of the Israeli WEC Member Committee. A wide range of sources have been consulted, including national, international and governmental publications/ web sites. Use has also been made of direct personal communications.

Argentina

Argentina is in the forefront of South American utilisation of geothermal resources. High-temperature geothermal heat exists in the western region, along the Andes range, and moderate to low-temperature thermal fields have been identified in other parts of the country.

Since a 670 kW binary-cycle pilot power plant at Copahue went off-line in 1996, the emphasis in recent years has been on the development of direct uses of geothermal power. At present there are 134 direct-use projects with an installed capacity of 25.7 MW_t.

Australia

In recent years the geothermal situation in Australia has changed. The expansion of activity has been caused by the successful demonstration of binary hydrothermal power plants, the commercial success of ground-source heat pumps and also increasing government support of initiatives to reduce greenhouse gas emissions.

Geothermal energy is used directly by the numerous hot water bathing pools throughout the country and a district heating scheme in western Victoria. A hot water spa in the latter region received local government approval in 1999. Ground and water-source heat pumps have increased in popularity throughout the country, with at least 2 000 installations in place (in 2000); an expansion in the market of 50% per annum is expected, including commercial-size hot water systems for drying fruit and vegetables.

Australia has also been found to have a very significant HDR resource, particularly in the centre of the country, extending into the north-eastern corner of South Australia and the south-western corner of Queensland. Research aimed at evaluation of HDR began in 1994.

There are only two experimental geothermal electric power sites in Australia. A 20 kW plant in Mulka (South Australia) operated for three and a half years in the late 1980s. An eight-fold scale-up of Mulka was commissioned at Birdsville (Queensland) in 1992 (150 kW) and ran until end-1994. After environmental considerations dictated a change in the working fluid, and also after a change of ownership, the plant was put back on line for demonstration in mid-1999.

Austria

There has been a certain amount of development of Austrian geothermal resources since 1995. The aggregate installed capacity of 58 MW_t is utilised for direct applications such as district heating, spa heating, bathing, swimming and the heating of greenhouses.

A 500 kW_e binary power plant at Altheim was brought into operation in 2001. Later that year a second 250 kW_e binary power plant was installed at Rogner Hotel, Bad Blumau. The annual output is about 4 GWh.

In addition, it has been reported that there are in the order of 19 000 heat pump installations throughout the country, with an estimated total capacity of 228 MW.

Bulgaria

Bulgaria's theoretical potential for geothermal energy usage has been evaluated at 14 387 TJ/yr, but up to now the installed heat capacity is only 107 MW.

In 1998 the Geological Institute of the Bulgarian Academy of Sciences, under the auspices of the Ministry of Environment and Water, completed an assessment of the geothermal resources in 162 fields, 103 of which are under the protection of the Ministry of Public Health. As a result it was concluded that:

- the water temperature of the reservoirs evaluated ranges from 20°–100 °C;
- in around 33% of the reservoirs the water temperature ranges from 20°–30 °C;
- in around 43% of the reservoirs the water temperature ranges from 40°–60 °C;
- the total flow rate of the sub-thermal ($T < 20$ °C) and thermal waters ($T > 20$ °C) is up to 4 600 l/s, within which 3 000 l/s is the average flow rate for thermal waters.

Geothermal heat sources could be used for the production of CHP, while district heating systems could be constructed on the basis of a combined supply of geothermal energy and a classical energy resource.

Canada

It has been demonstrated from research undertaken since 1974 that Canada has a plentiful and widespread geothermal potential. The abundance of hydro-electric resources and inexpensive fossil fuels have, however, proved disincentives to large-scale development. Resources of high-temperature geothermal energy have been established but to date none have been utilised. Rather it has been applications utilising the low-temperature resources that have come to fruition. A continuing development of the Meager Creek Geothermal Project, located some 170 km north of Vancouver, aims to develop the site potential of some 110 MW_e.

Direct utilisation of geothermal energy has followed four routes (GHPs, aquifer thermal energy storage, energy from mine waters and hot spring resorts) and provides an estimated total installed capacity of 378 MW_t.

China

With fast economic growth and increasing environmental concerns, the utilisation of geothermal energy in China increased by 12% per annum during the 1990s. Studies have identified more than 3 200 geothermal features, of which some 50 fields have been investigated and explored. High-temperature resources are mainly concentrated in southern Tibet and western parts of Yunnan and Sichuan Provinces, whereas low-medium temperature resources are widespread over the vast coastal area of the south-east, the North China Basin, Songliao Basin, Jianghan Basin, Weihe Basin, etc.

The primary development has been in the growth of geothermal energy used directly. In 1998 it was reported that there were in excess of 1 600 sites being used for installations as diverse as drying, fish farming, irrigation and earthquake monitoring. However, the main emphasis has been on the expansion of installations for space heating, sanatoria and tourism.

The development of geothermal power generation has been, by comparison, relatively slow, owing to the large hydro-electric resources in those provinces with high-temperature geothermal resources (Tibet and Yunnan). The largest power complex is located at Yangbajing (Tibet). China's aggregate capacity is approximately 30 MW_e, generating 100 GWh annually.

Costa Rica

The Central American volcanic belt passes through Costa Rica, evidenced by numerous volcanoes and geothermal areas. The fields of Miravalles, Tenorio and Rincón de la Vieja are located in the north-western part of the country and have been studied in detail.

To date, Costa Rica's geothermal resources have only been utilised for electric power generation. A 55 MW_e single flash condensing unit was commissioned in March 1994 at Miravalles, followed soon afterwards by an additional

5 MW_e backpressure unit. A second 55 MW_e condensing unit came on stream in 1998, and subsequently (in 2000) another 29.5 MW_e backpressure unit increased the installed generation capacity of the Miravalles field to 144.5 MW_e. A further 18 MW_e unit (unit 5) is under construction.

Exploration work on the slopes of the Rincón de la Vieja volcano at the Las Pailes and Borinquen geothermal fields is ongoing.

El Salvador

Like Costa Rica, El Salvador lies on the Central American volcanic belt and thus there is a plentiful geothermal resource. The main emphasis has been on using the resource for power generation and although a potential exists for the direct use of geothermal, it has not yet been developed.

Geothermal energy accounts for over 24% of El Salvador's electricity output. In 2002, power generation from the Ahuachapán and Berlín geothermal facilities was 940 GWh.

Of the 161 MW_e of geothermal capacity currently installed in El Salvador (95 MW_e at Ahuachapán, and 66 MW_e at Berlín), only about 119 MW_e is reported to be actually available (63 MW_e at Ahuachapán and 56 MW_e at Berlín).

Geotérmica Salvadoreña (GESAL) estimates that it can increase effective capacity in the next 3 years by about 50 MW_e with the addition of one new 28 MW_e condensing unit at Berlín; relocation of the 10 MW_e wellhead power plant from Berlín to the Cuyanausul Geothermal Field; and a 12 MW_e upgrade of the Ahuachapán installations. GESAL is also exploring the possibilities of expanding operations at other Central American geothermal fields.

Two other prospective geothermal areas are San Vicente in the centre of the country and Chinameca in the east; each has an estimated capacity of 50 MW_e. Future studies are also planned for Santa Rosa de Lima and Obrajuelo Lempa.

Ethiopia

Ethiopia is one of the African countries possessing geothermal potential. Considerable resources of both high- and low-enthalpy geothermal have been located in the Ethiopian Rift Valley and in the Afar depression. Exploration that began in 1969 has, to date, revealed the existence of 24 prospects having about 700 MW_e potential.

In mid-1998 the 8.5 MW_e Aluto-Langano geothermal plant became operational. Aluto-Langano became the first geothermal power plant in Africa to use integrated steam and binary power technology. The plant has recently become inactive, owing to resource problems.

In addition to the Aluto-Langano geothermal field, the other areas under exploration are Corbetti in the Lakes District and Tendaho in the Central Afar region.

France

There are only low-enthalpy geothermal resources in metropolitan France; high-enthalpy geothermal resources are found only in France's overseas departments.

Although the first French geothermal district heating plant was constructed in 1969 in the Paris region, the main development of geothermal energy began following the oil crises of the 1970s. The resources are found in two major sedimentary basins: the Paris Basin and the Aquitaine Basin in the southwest. Other areas (Alsace and Limagne) have geothermal potential but it cannot be so readily utilised. At end-1986 there were 74 plants in operation but by end-1999 this number had fallen to 61, of which 41 were in the Paris region, 15 in the Aquitaine Basin and 5 in other regions. The installed capacity is mainly used for space heating (97%) but also greenhouse heating (2%) and fish and animal farming (1%). In addition, since the 1980s some of France's low-enthalpy resources have been utilised by the installation of heat pumps. At the present time several thousand plants exist, mainly for collective or individual building heating.

Since the 1980s the French authorities have supported research into the potential of HDR.

Georgia

Geothermal resources are prevalent throughout the area of the South Caucasus and are utilised intensively in Georgia. It has been reported that the country's considerable reserves are being particularly efficiently used in the Tbilisi field. The installed capacity effectively available for direct heat applications has been estimated to be in the region of 250 MW_t.

Germany

Germany does not possess high-enthalpy steam reservoirs. Its geothermal resources are located in the North German sedimentary basin, the Molasse Basin in southern Germany and along the Rhine graben.

Germany's 2000 Renewable Energy Law (REL) mandates a doubling of renewable power in the country's electricity market to 10% by 2010. The REL sets specific tariffs for each renewable energy, based upon its real cost. Electricity from geothermal plants would receive €0.08–0.10/kWh. An outcome of the REL is the first German geothermal power plant (200 kW_e), inaugurated at Neustadt-Glewe in 2003.

At end-1999 total installed capacity for direct use of geothermal energy stood at 397 MW_t, of which 55 MW_t represented 27 major centralised plants and 342 MW_t small decentralised earth-coupled heat pumps and groundwater heat pumps.

Greece

So far, geothermal energy has encountered severe opposition from the local population (chiefly the inhabitants of the islands involved) because of the lack of an appropriate introduction and public relations policy by PPC, the Greek state power company. By mid-2002, this had been remarked upon by the Minister of Development himself. As a result, appropriate legislative means and procedures, incorporated into a new law, have been adopted quite recently, specifically for the promotion of geothermal energy and the liberalisation of the energy market, including electricity production and supply and natural gas.

High-enthalpy geothermal fields occur in the islands of Milos and Nisyros, which are located in the South Aegean volcanic arc. There is currently no electricity generation from these two fields because of the opposition of the local people. Deep drilling for high enthalpy was carried out in the 1980s on both islands.

The estimated high-enthalpy geothermal potential of the island of Milos amounts to 27–180 MW_e with another 120 MW_e considered as probable reserves. PPC attempted to install a prototype electrical generating unit of about 2 MW_e in the mid-1980s, but the whole project was eventually stopped because of operational problems (mainly due to inadequacies in the system's desalination technique) and consequently opposition from the local population. However, it is anticipated that if local opposition problems can be overcome an overall installed electrical generating capacity of about 40–50 MW_e is quite feasible.

For the island of Nisyros the estimated high-enthalpy potential amounts to 50–100 MW_e—it is envisaged that electrical generating capacity of about 20 MW_e could be installed.

There is a plan by PPC to exploit geothermal energy for electricity generation on the island of Lesvos (Mitilini), where an exploration project exists. Drillings have so far pointed to a potential electrical generating capacity of about 7–8 MW_e.

Low-temperature geothermal fields occurring in structurally active sedimentary basins have a big potential. A small proportion of this heat resource is currently used in greenhouse heating, balneology, fisheries and desalination.

Guadeloupe

The double flash plant at La Bouillante in the French Overseas Department of Guadeloupe is at present the only example of the island's geothermal energy being utilised for electricity production. The plant was commissioned in 1985 but was closed between 1991 and 1996.

The French Agency for Environment and Energy Management (ADEME) contributed to the development of the Bouillante high-enthalpy field by supporting 20% of the cost of drilling new wells.

In July 2002, the original power plant (GB1) was hooked up to BO-5, one of the wells drilled for the new 10 MW_e plant (GB2), replacing the BO-2 well that had reached the end of its useful life. This hook-up increased GB1's average net capacity to 4 MW_e at end-2002, with the prospect of an increase to 5 MW_e when further new wells are connected. Industrial production from GB2 was expected to start around the end of 2003.

Guatemala

Guatemala's Instituto Nacional de Electrificación (INDE) has five geothermal areas for development. All five areas (Zunil, Amatitlán, Tecuamburro, San Marcos and Moyuta) lie in the active volcanic chain in southern Guatemala. INDE has conducted investigative work and development of geothermal power since 1972 and to date 58 MW_e has been proved, with a further 398 MW_e estimated as potential additional capacity.

The first geothermal power plant in the country was constructed in the Amatitlán area; electricity production from a 5 MW_e back-pressure plant began in November 1998. Subsequently the unit was dismantled. Eventually, expansion of the field and construction of a 25 MW_e combined cycle plant are envisaged. The Amatitlán field also supports the direct use of geothermal energy, in the form of using steam for drying concrete blocks and a fruit dehydration plant.

A second 24 MW_e geothermal plant (in the Zunil I field) has been in commercial operation since September 1999. Following INDE's exploratory drilling work, a contract was signed with Orzunil I for the private installation and operation of the plant. Until 2019 the company will buy steam from INDE and sell power to the national grid. Exploratory drilling in the Zunil II field has shown that it possesses 50 MW_e potential.

Hungary

Hungary possesses very considerable geothermal resources and it has been estimated that the country has the largest underground thermal water reserves and geothermal potential (low and medium enthalpy) in Central Europe.

To date, there has been no utilisation of geothermal energy for the production of electricity. The principal applications of geothermal power used directly are greenhouse heating (63.5%), space heating (23%), industrial process heat (0.5%), and other uses (13%). It has been reported that geothermal heat pumps represent an additional 3.8 MW_t and four spas supply a further 14.2 MW_t.

In the mid-1990s the Hungarian Oil and Gas Company (MOL) began a programme to promote the development of geothermal energy. Three pilot projects have been studied, two of which involve cascaded use of geothermal heat for electricity production and subsequent direct applications.

Iceland

Geothermal energy resulting from Iceland's volcanic nature and its location on the Mid-Atlantic Ridge has been utilised on a commercial scale since 1930. The high-temperature resources are sited within the volcanic zone, whilst the low-temperature resources lie mostly in the peripheral area.

The total annual primary energy consumption in Iceland is 499 GJ per capita, which is higher than that in any other country. Geothermal energy provides about 55% of the total primary energy supply, while the share of hydropower is 17%, that of oil 25% and that of coal 3%. The principal use of geothermal energy in Iceland is for space heating: about 87% of all energy used for house-heating comes from geothermal resources. A total of 26 municipally-owned geothermal district heating systems are located in Iceland, the largest of which is in Reykjavík, serving 177 000 people.

Whilst 77% of the direct use of geothermal heat is used for space heating, 8% is used for industrial process heat, 6% for swimming pools, 4% for greenhouses, 3% for fish farming and 2% for snow melting. Total installed capacity for direct use was 1 800 MW_t at end-2002, giving an annual output of 24 700 TJ.

In recent years there has been an expansion in Iceland's energy-intensive industrial sector. To meet an increased demand for power, the capacity of geothermal plants has grown rapidly from 50 MW_e and currently stands at over 200 MW_e. Geothermal electricity generation was 1 433 GWh in 2002.

Two co-generation power plants are in operation. The Svartsengi energy plant has a capacity of 200 MW_t for hot water production and 46 MW_e for electricity generation, of which 8.4 MW_e comes from binary units using low-pressure waste steam. At the Nesjavellir energy plant there is an installed capacity of 200 MW_t for hot water production and 90 MW_e for electricity production. Two conventional geothermal power plants are in operation in Iceland, Krafla (60 MW_e) and Namafjall (3 MW_e).

A binary plant of the Kalina type has been installed at Husavik. In generating 2 MW_e, the geothermal water is cooled from 120° to 80 °C; it is then used for district heating in the locality.

Indonesia

The islands of Indonesia possess enormous geothermal resources: geological surveys have identified as many as 244 prospects, of which 70 are specified as high-temperature reservoirs with an estimated total resource potential of nearly 20 000 MW_e. Of this potential about 49% is in Sumatra, 29% in Java-Bali, 8% in Sulawesi and 14% in other islands.

A very small amount of geothermal energy is used directly for bathing and swimming, all instances being in West Java.

The financial crisis that hit Indonesia towards the end of 1997, and the resultant adverse effect that it had on power sector demand and growth, resulted in some delay to the development of geothermal energy. By December 2002, however, the country had increased its geothermal electric power generation capacity to 807 MW_e. This figure includes currently operating facilities with a capacity of 330 MW_e at Gunung Salak, 140 MW_e at Kamojang, 145 MW_e at Darajat, 110 MW_e at Wayang Windu, 2 MW_e at Sibayak, 20 MW_e at Lahendang and an additional 60 MW_e at Dieng.

In the future the Government plans to significantly alter the fuel mix of electricity generation by increasing the use of coal, geothermal energy and hydro power and thus reducing the use of oil and gas.

Iran (Islamic Republic)

So far two exploration wells have been drilled to a depth of 3 000 m for geothermal utilisation. The initial data indicate that at least 100 MW of electricity could be produced.

Italy

Italy is one of the world's leading countries in terms of geothermal resources. The high-temperature steam-dominated reservoirs lie in a belt running through the western part of the country from Tuscany to Campania (near Naples). Commercial power generation from geothermal resources began in Italy in 1913 with a 250 kW unit. Subsequently the main emphasis has been on the production of power rather than on direct use of the heat.

Following the limited development of resources during the first half of the 20th century, it was the second half that saw rapid growth. By end-1999, total Italian installed geothermal capacity stood at 621 MW_e. The growth continued in recent years and by the end of 2002 Italy had 862 MW_e of geothermal power plants installed.

In addition to the Italian country report presented at the World Geothermal Congress 2000, a detailed analysis of direct uses was also presented. This analysis found that several large geothermal fish farms (approximately 110 MW_t), larger hotels and balneological spa uses (in the Abano district and on the island of Ischia) had been excluded from the country report.

Italian direct uses (excluding balneological/swimming pool use) can be conservatively estimated at about 680 MW_t with a production of approximately 9 000 TJ/yr.

Japan

Japan has a long history of geothermal utilisation, both direct and for power generation. The first experimental power generation took place in 1925, with the first full-scale commercial plant (23.5 MW_e) coming on-line at Matsukawa, in the north of the main island of Honshu, in 1966. Following each of the two oil crises, development of Japan's geothermal resources was accelerated and by end-1984, 314.6 MW_e capacity had been commissioned. Growth continued and unit size decreased as technological improvements occurred. By end-1999, installed capacity stood at 546.9 MW_e (consisting of 19 units at 17 locations). The existing plants are all located in the Tohoku region of northern Honshu and on the southern island of Kyushu.

The country's power generation potential from geothermal is estimated to be in the order of 2 500 MW_e. The planned government deregulation of the electricity sector, bringing about lower medium and long-term electricity costs, is expected to result in geothermally generated power becoming uncompetitive.

Direct use of geothermal hot water has a long tradition in Japan, where enjoyment of natural baths is a national recreation. There is widespread usage of geothermal heat for purposes other than bathing (which accounts for 11% of capacity): space heating (including hot water supply) 51%; greenhouse heating 13%; snow melting 12%; fish breeding 9%; air conditioning (cooling) 2% and industrial process heat and other 1% each. The quantification of direct use capacity is particularly difficult in Japan. It is thought to be considerably larger than the reported figures.

Hot spring water above 15 °C is widely available, thus there is little demand for heat pumps.

Beginning in 1980, the New Energy and Industrial Technology Development Organization (NEDO) initiated 52 surveys to evaluate those areas with the most promising geothermal potential for power generation. Starting in FY 1999, NEDO has operated a Project on Geothermal Power Generation Development. This scheme provides subsidies for drilling geothermal exploration/production/re-injection wells in addition to the installation of specified equipment at sites where surveys and construction work have already progressed. Subsidies are also available for part of the costs of installing binary generating equipment at geothermal sites. The scheme had a budget of 1.106 billion yen for FY 2003.

Jordan

Jordan is blessed with many thermal water resources spread along the Rift, in addition to thermal wells on the Eastern Plateau.

Several studies on the evaluation and assessment of these resources have been conducted by the Natural Resources Authority (NRA), and the results have shown that the thermal resources are of the low-enthalpy type (< 150 °C), and that their optimum use is for space heating, greenhouses and fish farming.

Kenya

Kenya possesses substantial geothermal resources at Olkaria near Lake Naivasha (about 80 km north-west of Nairobi) and at other locations in the Rift Valley.

The first geothermal unit came into operation at Olkaria in July 1971, with an initial installed net capacity of 15 MW_e. Two more 15 MW_e units were added, so that by end-1999 capacity had risen to 45 MW_e. The geothermal power output was increased by 12 MW_e in 2000 when the first two stages of Kenya's first private geothermal plant were installed at Olkaria III. The additional stage of Olkaria III (38 MW_e) is under development. The 64 MW_e Olkaria II power plant was commissioned in November 2003, bringing the country's installed capacity to 121 MW_e.

Geothermal capacity totalling 173 MW_e is planned to be in operation by 2005 and a total of 576 MW_e by 2017. In order to attract sufficient investment funds to achieve this goal, the restructuring of the power industry must continue. It is expected that in the future the power industry will be a partnership of the private and public sectors.

A minimum amount of geothermal energy is used for direct heat. For the time being flowers are being grown on an experimental basis, but it is intended that this should become a commercially viable operation.

Mexico

Reflecting the country's location in a tectonically active region, Mexico's geothermal manifestations are particularly prevalent in the central volcanic belt, as well as in the states of Durango, Chihuahua, Baja California and Baja California Sur. Development has, in the main, been concentrated on electric power production, although there is some utilisation of geothermal power for direct purposes.

In 2003 Mexico celebrated its 30th anniversary of uninterrupted geothermal electric operation. The last 30 years have witnessed a number of achievements in developing the country's geothermal potential, including bringing online the Los Azufres (1982), Los Humeros (1990) and Las Tres Vírgenes (2001) geothermal fields. Mexico's geothermal-electric capacity grew from 37.5 MW_e to 853 MW_e during the period—a figure that jumped to 953 MW_e when the 100 MW_e Los Azufres II was commissioned in the summer of 2003.

Geothermal heat used directly is predominantly utilised for bathing and swimming. Of the reported 164.19 MW_t installed capacity (end-1999), virtually 100% was located in resorts throughout the volcanic zone. Minimal amounts of direct heat (if any) are utilised for space heating, greenhouse heating, agricultural drying, timber drying and mushroom breeding.

New Zealand

New Zealand is exceptionally rich in geothermal fields, as well as a large number of other geothermal features. Substantial capacity exists for both the generation of geothermally produced power and also for geothermal energy used directly.

The first geothermal power plant came into operation at Wairakei, north of Lake Taupo (North Island) in November 1958, with an initial capacity of 69 MW_e. The second stage of development, which added a further 123 MW_e of capacity, began operation in October 1963. Wairakei was the second geothermal power station to be built in the world and the first to tap a hot pressurised water resource. Owing to an initial very rapid run-down in field pressure, the maximum output achieved from the station was 173 MW_e. In 1983 all high-pressure turbine/generator units were decommissioned, owing to the decline in high-pressure steam output from the field. The current installed capacity of Wairakei is 162 MW_e, with an additional 15 MW_e binary capacity planned to be in service by 2005.

Between 1966 and 1990 three more power plants were commissioned within the central

North Island's Taupo Volcanic Zone (TVZ), in the localities of Reporoa and Kawerau. Their combined capacity (one back pressure unit, 3 binary units and 4 combined-cycle units) amounted to 130 MW_e.

Between 1996 and 1999 four plants were commissioned: the 55 MW_e McLachlan plant (Taupo locality), a 25 MW_e combined-cycle plant at Rotokawa (Taupo locality), the 12 MW_e Ngawha binary plant (Northland locality, about 245 km north of Auckland) and a 55 MW_e combined-cycle at Mokai (Taupo locality). The Rotokawa Extension plant, adding another 6.5 MW_e to the existing plant, was completed in 2002, bringing the geothermal installed capacity to almost 448 MW_e.

Potential generation capacity from the geothermal resources of the TVZ has been conservatively estimated at 2 000 MW_e.

At end-1999 installed capacity for direct heat uses stood at 307.9 MW_t. The main user of direct heat is at Kawerau, where a 210 MW_t plant generates clean process steam for various procedures within a pulp and paper mill operation. Geothermal steam at other locations is also used for agricultural drying (10% of direct-heat capacity), bathing and swimming (9%), space heating (7%) and fish and animal farming (6%).

Nicaragua

Nicaragua is the Central American country with the greatest geothermal potential, in the order of several thousand megawatts. Reserves that can be estimated with a higher degree of confidence total about 1 100 MW_e. Medium- and high-temperature resources are associated with volcanoes of the Nicaraguan Depression, which parallels the Pacific Coast.

Geothermal exploration began at the end of the 1960s, focussing on the Momotombo and San Jacinto-Tizate geothermal fields. Studies increased after 1973, at a time when the oil crisis had a large impact on Nicaragua's economy. Geothermal electricity production started at Momotombo in 1983.

Exploitation of geothermal power in the Momotombo area, located at the foot of the volcano of the same name, began when the first 35 MW_e single-flash unit was commissioned in 1983. A second 35 MW_e unit was added in 1989. Gross output of electricity reached a peak of 468 GWh in 1992 but subsequently fell away to a low of 121 GWh in 1998 owing to over-exploitation of the field and lack of re-injection.

In 1999, ORMAT secured a 15-year contract to improve electricity output at Momotombo. Since then, the company has drilled four deep wells (OM-51 to OM-54), and of these, only OM-53 was a good producer (9–11 MW). A number of wells have been cleaned of scale and chemical inhibition systems installed. About 80% of waste geothermal fluids are being injected back to the reservoir and a new reservoir management plan has been implemented. Since May 2002, these efforts have increased and stabilised the electrical output of the flash plant at about 29 MW_e. In November 2002, a 7.5 MW_e binary energy converter came online, raising generation capacity at Momotombo to about 35 MW_e. The field now has 12 production wells, and four injection wells.

A Geothermal Master Plan for Nicaragua was completed in November 2001. It assessed the geothermal resource potential of identified fields and prospects in the country. At present, concessions for geothermal exploration and/or exploitation are in place for Momotombo, San Jacinto-Tizate and Casita-San Cristóbal. Apart from these, the three most promising geothermal prospects are:

- El Hoyo-Monte Galán. Located west of Momotombo, this field has an estimated capacity of 200 MW for 30 years;
- Managua-Chiltepe. This area is located about 15 km NW of Managua, Nicaragua's capital, and has an estimated capacity of up to 150 MW for 30 years;
- Masaya-Granada-Nandaime. This area, which includes several volcanoes and geothermal prospect areas, is near the northwestern shore of Lake Nicaragua; it may produce 200 MW_e for 30 years.

To date all geothermal energy has been used for power generation but the Government, with support from the European Union and the UN Economic Commission for Latin America and the Caribbean, will conduct a geothermal rural electrification and direct application pilot project in the areas of the Cosigüina Volcano and Ometepe Island. Low-enthalpy fluids will be investigated for use in grain-drying, fish farming and heating greenhouses.

Papua New Guinea

Although there are numerous volcanic islands scattered along the western Pacific Rim, the first geothermal commercial development occurred only recently on the island of Lihir, off the northeast coast of Papua New Guinea. In 2003, Lihir Gold Ltd commissioned a 6 MW_e geothermal power plant, providing the nearby gold mine with 10% of its power needs. Lihir is among 40 identified geothermal resources in PNG; 38 are unexplored.

Philippines

The Philippines archipelago is exceptionally well endowed with geothermal resources. Today the country is the world's second largest user of geothermal energy for power generation.

The geothermal plants in the Philippines are generating about one-fifth of the national electricity supply from six fields, in which there are 11 areas in production. The fields, spread throughout the islands, are at Mak-Ban (Luzon), Tiwi (Luzon), Tongonan (Leyte), Palinpinon (Negros), Bac-Man (Luzon) and Mindanao (Mindanao). Operations began in 1979 with 278 MW_e and grew steadily until the mid-1980s, when installed capacity reached 894 MW_e. Further capacity was not added until 1993, after which is grew rapidly again to reach 1 909 MW_e by 2000. The end-2002 capacity is reported by the Philippines WEC Member Committee to have been 1 931 MW_e.

Three new geothermal areas at Mt. Labo (Luzon), Northern Negros (Negros) and Cabalian (Leyte) are presently in an advanced stage of development.

Within the terms of the Philippine Energy Plan, the Government is planning to increase geothermal capacity by 526 MW_e by 2008. Output would increase to 13 865 GWh but the geothermal contribution would fall to 18.5% (from the current 23%) owing to the use of natural gas for power generation.

Poland

Poland has substantial resources of geothermal energy, but not at high temperatures. The available resource ranges from 50° to 100 °C.

At present there are installations in operation at Bańska (near Zakopane), Pyrzyce (near Szczecin) and Mszczonów (near Skierniewice).

Geothermal water is mainly used for heating purposes; geothermal heat is not utilised for electricity generation.

Portugal

The limited geothermal resources in mainland Portugal have been developed for direct use, whereas geothermal occurrences in the Azores islands are utilised for the production of electricity as well as being used directly.

There are about 50 natural low-enthalpy occurrences spread throughout the mainland.

Twelve areas with potential for developing geothermal electricity generation have been identified on the islands of Faial, Pico, Graciosa, Terceira and São Miguel in the Azores. At the present time the installed geothermal power capacity of São Miguel binary dual-phase power plant is 16 MW_e.

As the estimated potential of the Ribeira Grande field is in the range of 80 MW_e, it is envisaged that an additional 24–30 MW_e capacity could be constructed by 2010, thereby meeting 40–45% of the electrical demand of the island.

São Miguel also has an installed 'direct use' capacity of 1.5 MW_t (end-1999) using geothermal energy for direct heat. Six small greenhouses use the 90 °C waste water from a nearby geothermal power plant in order to grow experimental crops.

Romania

The relevant geological formations and locations are the following:

- Triassic limestones and dolomites in the central part of the Western Plain, in the Bihor County (Oradea—Bors area);
- Pliocene (Pannonian) sands and sandstones in the Western Plain (eastern part of the Pannonian Depression) in the Satu Mare, Bihor, Arad and Timis counties;
- Senonian sandstones in the foredeep of the Southern Carpathians in the Getic Depression;
- Late Jurassic (Malm)—Early Cretaceous (Neocomian) carbonate rocks in the central and eastern part of the Moesian Platform.

Past and present drilling activity has included the drilling of almost 200 exploration wells with definite geothermal indications, of which 60 have become production wells and six have become injection wells.

Reinjection of waste geothermal fluid is only performed in highly-fractured carbonate reservoirs, in the Bors area in order to keep the reservoir pressure and free flow of the wells, and in the Oradea and North Bucharest areas as a means of waste disposal. With regard to the sand and sandstone reservoirs, due to the argillaceous minerals, several full-scale reinjection experiments have failed, as a consequence of an abnormal increase in the injection pressure, quite similar to that observed in the majority of tests performed in analogous environments in other parts of the world. Since for most fields reinjection is not yet economically feasible, the flow rates (either in pumping or in free-discharge wells) have to be restricted.

Russian Federation

Geothermal resources have been identified in several areas of the Federation: the Northern Caucasus (Alpine and Platform provinces), Western Siberia, Lake Baikal and, most significantly, in Kamchatka and the Kuril Islands. It has been estimated that the high-temperature resources defined to date in the Kamchatka Peninsula could ultimately support generation of 2 000 MW_e or more. However, at the present time Russia's energy sector is based on fossil fuels and the exploitation of hydroelectric and nuclear power, and therefore the contribution from geothermal energy is tiny. Over the past 30 years there has been some development of high-temperature resources for power generation, but the main thrust of Russian geothermal utilisation has been, and continues to be, for direct purposes.

Investigations into using geothermal energy for power generation in Kamchatka began in 1957, and in 1966 a 4 MW_e single-flash plant was commissioned at Pauzhetka. It was enlarged to 11 MW_e in 1980; in 1999 a 12 MW_e geothermal power plant was put into operation at Verkhne Mutnovsky, followed in 2002 by the inauguration of the 50 MW_e Mutnovsky geothermal power plant.

At end-1999 installed capacity for direct use amounted to more than 300 MW_t. The heat is used mainly for space and district heating but also for a range of agricultural purposes (greenhouses, soil heating, fish and animal farming, cattle-breeding), for various industrial processes (manufacturing, wool washing, paper production, wood drying, oil extraction) and for spas and recreational bathing.

Although there is much scope for the installation of heat pumps in Russia, their use is presently at an early stage of development.

Slovakia

In the Košice Basin there is a confirmed source of geothermal water at 130 °C. It is proposed to utilise the 100 MW_t output for heating the town of Košice, with a production of 600 GWh/yr, operational from 2005/2006.

Sweden

The only reported use of the geothermal energy resource in Sweden is from heat pumps. It has been estimated that by 1998 in the region of 55 000 had been installed, with an aggregate capacity of 377 MW_t.

Switzerland

Switzerland's installed capacity for utilising geothermal energy has grown rapidly in recent years and the country now ranks among the world leaders in direct-use applications (there is no geothermal-based electricity). There are two main components to Switzerland's geothermal energy: the utilisation of shallow resources by the use of horizontal coils, borehole heat exchangers (BHE), foundation piles and groundwater wells, and the utilisation of deep resources by the use of deep BHEs, aquifers by singlet or doublet systems, and tunnel waters. In virtually all instances heat pumps are the key components.

At end-1999 there were in the region of 21 000 ground-source heat pumps installed throughout the country, representing about 500 MW_t. The remaining approximately 50 MW_t of capacity was utilised for bathing and swimming (17 locations, 25 MW_t), space heating (20 MW_t), air conditioning (five locations, 2.2 MW_t) and snow melting (0.1 MW_t).

Following successful drilling to tap deep aquifers for a district heating network at Riehen (on the border with Germany and operational since 1995), the network was extended and thus became the first example of cross-border geothermal utilisation.

There remains substantial room for growth in Switzerland's geothermal sector. The annual growth-rate for heat pumps is estimated at 15% and the Government is actively supporting research and development into geothermal energy.

Thailand

Investigations of geothermal features in Thailand began in 1946 and subsequently more than 90 hot springs located throughout the country were mapped. However, it was not until 1979 that systematic studies of the resources began.

A small (0.3 MW_e) binary-cycle power plant was installed at Fang, in the far north near the border with Myanmar. Since commissioning in December 1989, this sole Thai geothermal plant has operated successfully, with an 85–90% availability factor. In addition, the Electricity Generating Authority of Thailand (EGAT) is using the 80 °C exhaust from the power plant to demonstrate direct heat uses to the local population. The exhaust is being used for air conditioning, cold storage and crop drying. A further example of utilising the heat directly is a public bathing pond and sauna that have been constructed by the Mae Fang National Park.

Geothermal systems at San Kampaeng, Pai and nine other locations are reported to be under further investigation, but to date Thailand's national programme on geothermal energy has still not been firmly established and no other developments have occurred.

Turkey

A significant factor in Turkey's high geothermal potential is the fact that the country lies in the Alpine-Himalayan orogenic belt. Geothermal exploration began during the 1960s, since when about 170 fields have been identified. Although some of this number are high-enthalpy fields, 95% are low-medium enthalpy resources and thus more suited to direct-use applications.

At end-1999, geothermal installed capacity for direct uses totalled 820 MW_t, of which 392 MW_t provided the space heating and thermal facilities of 51 600 residence-equivalents, 101 MW_t provided heating for 45.4 ha of greenhouses and 327 MW_t was utilised for bathing and swimming (194 spas).

Following research undertaken in 1968 into using geothermal resources for the production of electricity, a 0.5 MW_e pilot plant was installed in 1974 in the Kizildere field (near Denizli in south-western Turkey). In 1984 the 20.4 MW_e single-flash Kizildere geothermal power plant came into operation. In addition to electricity generation, the plant has an integrated liquid CO_2 and dry ice production factory that utilises the geothermal fluids. Its capacity was reported by the Turkish WEC Member Committee to have been 17.5 MW_e at end-2002.

To date, at least four other geothermal fields with electric power generating potential have been discovered and studied to varying degrees.

United States of America

The USA possesses a huge geothermal resource, located largely in the western half of the country. Research has shown that geothermal energy has been used in North America for many thousands of years but the first documented commercial use was in 1830 in Arkansas. In 1922 an experimental plant began generating electricity in California but, proving to be uneconomic, it soon fell into disuse. Another 38 years were to pass before the first large-scale power plant began operations at The Geysers, north of San Francisco, California. The USA is the world's largest producer of electricity generated from geothermal energy.

Only California, Nevada, Hawaii and Utah utilise geothermal energy for power generation; investigative studies undertaken in Oregon during the early 1990s proved to be unsuccessful. However, the 1990s saw dramatic change in the geothermal power industry: plants came on line, plants were retired, there were changes of ownership (resulting, in some cases, in operational efficiencies), etc. By end-2002 total effective capacity stood at 2 002 MW_e.

Although no central agency tracks all US geothermal wells, the Department of Energy quotes generation from geothermal energy as 13 357 GWh in 2002, representing 0.3% of total US electricity production. At The Geysers a project for injecting recycled wastewater into the reservoir has become the world's first wastewater-to-electricity system.

Geothermal heat suitable for direct utilisation is far more widespread through the US, ranging from New York State in the east to Alaska in the west. At end-1999 a total of 566 MW_t installed capacity was used for fish and animal farming (129 MW_t), greenhouse heating (119 MW_t), bathing and swimming (107 MW_t), district heating (99 MW_t), space heating (83 MW_t), agricultural drying (20 MW_t), industrial process heat (7 MW_t), and snow melting (2 MW_t). In addition, it is estimated that 45 000 heat pumps have lately been installed annually, resulting in a total capacity of some 5 366 MW_t at end-2002. Apart from a decline in industrial process heat, direct uses of geothermal energy continue to expand. The heat pump market is expected to continue to grow strongly, to reach an estimated 1.5 million units in service by 2010.

The goal of the Geothermal Technologies Program is to improve the performance and reduce technology and market entry costs of geothermal energy to competitive levels. Focussed R&D is the key strategy to improving performance and reducing costs by:

- reducing the cost of drilling;
- reducing the capital cost of geothermal surface systems;
- doubling the exploration success rate;
- increasing the amount of economic geothermal resources.

More specifically, the Program conducts R&D in remote sensing, geophysical and geochemical techniques, well bore integrity and lost circulation, hard rock bit technology, advanced drilling concepts, acid resistant cements, high-temperature electronics, reservoir modelling and analysis, fracture analysis, advanced heat rejection, enhancement of air-cooled condensers, improved materials, power plant process monitoring, and removal of non-condensable gases. The Program also conducts cost-shared projects with the US geothermal industry to locate new geothermal resources and expand the resources at existing geothermal fields.

Chapter 13

Wind Energy

World Energy Council
CONSEIL MONDIAL DE L'ENERGIE

COMMENTARY

Introduction and Résumé

Wind energy continues to be one of—if not the—fastest growing energy technologies, with an average growth over the last 5 years of more than 30% p.a. Europe continues to be the 'home' of wind power, with 75% of the globally installed capacity.

A major development in recent years has been that offshore installations have started to take off. If offshore developments prove positive, a huge new wind resource will be available. This will spur a technological development of even bigger wind turbines, up to 5 MW.

With the increasing installation of wind turbines, electricity systems face new challenges. In Western Denmark more than 20% of the annual electricity consumption is now delivered from wind turbines and the installed wind turbine capacity exceeds the minimum demand. This gives a need for new concepts for system control.

The increasing installation of wind turbines leads to a downward trend in the price of the turbines and the cost of generation. This follows from conventional industrial theory and can also be observed in practice. Thus, the reduced costs can be an important factor in promoting the diffusion of wind power globally. Other important factors are the reduced dependence on fossil energy resources and the absence of emissions from wind turbine operations.

A Global Resource

Wind resources are available in most parts of the world. With today's technology and price level, an annual average wind speed above 5 m/s is necessary to make wind power feasible. Even if one allows for 90% of the technical potential to be impossible to harvest (owing to conflicts of interest with other forms of land use), the potential wind resource is several times higher than the global electricity demand.

A limiting factor for the use of wind energy is that very windy areas tend to be sparsely populated. Patagonia is an excellent example of a very good wind resource but very little demand for electricity. Also large parts of Africa have a good wind resource available but not the infrastructure to tap it.

The development of wind power has thus primarily taken place in Europe, where the wind resource is good, electricity demand high and

2004 Survey of Energy Resources

the willingness (and ability) to pay for a cleaner form of energy exists.

Continued Growth

The global growth rate of new wind turbine installations has been close to 40% p.a. on average over the period 1996–2001[1]. The growth in annual installations declined from this level to about 28% in 2002, giving a total global installed wind turbine capacity of 31 400 MW by the end of 2002.

Most of the installed capacity of wind turbines is concentrated in a few countries, with Germany alone accounting for 38% of the global installed capacity. In the next four places come Spain, the USA, Denmark and India. More than 80% of the world's wind capacity is installed in these countries.

Similarly, the supply of wind turbines is dominated by a few companies, with the five major companies supplying around 75% of production in 2002. The tendency to concentration has been reinforced by a merger at the end of 2003 of the world's number 1 and number 3 suppliers (the Danish companies Vestas and NEG Micon, respectively), giving the new company a global market share of more than 35%.

Traditionally, specialised companies have dominated wind turbine manufacturing, but recently General Electric entered the wind industry through the purchase of Enron Wind Corporation's assets. An earlier attempt by ABB to enter the scene through an internal development failed. Nevertheless, the increasing size of the market for wind turbines will most likely make it more interesting for the traditional manufacturing companies in the years to come.

The Standard Technology

Traditionally, the majority of the world's wind turbines have three glass-reinforced plastic blades. The power train includes a low-speed shaft, a step-up gearbox and an induction generator, either four- or six-pole. There are numerous other possibilities, however. Carbon fibres are becoming an alternative blade material with the ever-increasing length of the blades. Variable-speed machines have become the most common. Variable speed brings several advantages—it means that the rotor turns more slowly in low winds (which keeps noise levels down), it reduces the loadings on the rotor and the power conversion system is usually able to deliver current at any specified power factor. One major manufacturer builds direct-drive machines, without a gearbox. These are usually of the variable-speed type, with power-conditioning equipment.

Towers are usually made of steel and the great majority are of the tubular type. Lattice towers, common in the early days, are now rare, except for very small machines in the range 100 kW and below.

As the power in the wind increases with the cube of the wind speed, all wind turbines need to limit their power output in very high winds. There are two principal means of accomplishing this, with pitch control on the blades or with fixed, stall-controlled blades. Pitch-controlled blades are rotated as wind speeds increase so as to limit the power output and, once the 'rated power' is reached, a reasonably steady output can be achieved, subject to the control system response. Stall-controlled rotors have fixed blades, which gradually stall as the wind speed increases, thus limiting the power by passive means. These dispense with the necessity for a pitch control mechanism, but it is rarely possible to achieve constant power as wind speeds rise. Once peak output is reached the power tends to fall off with increasing wind speed, and so the energy capture may be less than that of a pitch-controlled machine. Pitch control is now the favoured technology for larger wind turbines.

Trend Towards Bigger Turbines

Whereas a wind turbine in the 1980s would have a size of 25–75 kW, modern-day wind

[1] Wind Force 12—a publication by EWEA and Greenpeace.

turbines are typically in the order of 1 MW. The size of wind turbines is increasing continuously for a number of reasons. The wind speed is more stable and less influenced by turbulence as the height of the hub increases. The lack of available sites for wind projects in major markets like Germany and Denmark gives an incentive to increase size. Finally, the development of larger onshore wind turbines can be seen as part of the positioning for the expected boom in the offshore market (see below).

However, a number of different constraints are limiting the size of wind turbines. As size increases, the forces working on the mechanical parts of the turbines such as shaft, gear and bearings increase. This leads to the need for larger size components and gives a more than linear increase in the weight of the whole construction. Another constraint is the limitations involved in transporting large components such as rotor blades by road. This is all the more true because wind turbines are often erected in remote areas where the infrastructure is not very well developed.

The largest prototype established as of January 2004 is a 4.5 MW Enercon 112 turbine erected in 2002. During 2004, 5 MW prototypes are expected from Prokon Nord and from REpower—both German companies. These projects bring the rotor diameter of the wind turbines to well over 100 m.

Offshore Wind Parks—The New Frontier

The first major offshore wind park was the 160 MW Horns Rev wind park located in the North Sea and owned by the Danish utility, Elsam. Horns Rev consists of 80 × 2 MW Vestas wind turbines and was commissioned in 2002. In 2003 a similarly sized wind park was commissioned in the Baltic Sea near Nysted by the other Danish utility, Energy E2. At the end of 2003 approximately 550 MW of offshore wind parks will be established worldwide, of which 393 MW will be in Denmark.

FIGURE 13.1 Wind farm at Rejsby Hede, Denmark (Source: Elsam).

The principal prospects for offshore wind power are, however, elsewhere. Through two rounds of bidding the British Crown Estate has offered leases to develop a total of 29 offshore wind projects with a total installed capacity of up to 8 600 MW in the waters around the UK. Before construction can start, the projects need to obtain planning consent. By January 2004, 11 projects representing 1 200 MW had been approved.

Similarly, a huge amount of project development activity is ongoing in German waters in the Baltic Sea and the North Sea. A number of other European countries such as the Netherlands, Belgium, Ireland and Sweden have programmes for offshore wind parks. The planning horizon for offshore projects has been quite long, however, and the real breakthrough has yet to be seen.

Offshore development promises a number of benefits over onshore wind parks. The wind conditions tend to be more stable offshore, giving a higher load factor (up to 40%) for the wind turbines and thus a higher production per installed MW. Each wind park can be larger offshore as more space is available and there are no close neighbours. The size of individual projects may go as high as 1 000 MW and thus similar to that of many conventional power stations. Finally, an offshore location permits larger wind turbines to be erected, since some of the logistical constraints facing onshore wind parks do not apply when ship transport is an option.

A disadvantage of offshore development is the need for much stronger foundations, owing to the combined effect of wind and waves. Also grid connection will typically be more expensive, since dedicated cables reaching 20–50 km offshore will often be necessary.

Integrating Large-Scale Wind Generation—A Major Challenge

Traditionally, wind power has been a distributed source of electricity generation with relatively small units dispersed in the grid, and has typically been connected to the lower voltage levels—below 60 kV. Being a marginal source of production, wind turbines have also generally been exempt from delivering the same system services as conventional power stations—i.e. voltage control, frequency control and the ability to survive prolonged periods with low grid voltage.

The image of wind power as a marginal source of production is gradually changing as some areas have an increasing share of wind power in the generation mix. An extreme example is the grid in the western part of Denmark. Fig. 13.2 provides some salient data.

Fig. 13.2 shows that the installed wind capacity is in the order of twice the minimum load on the system. This means that even though wind power on average only supplies 20% of the energy, it will frequently happen that wind power alone supplies the whole of the demand in the area. Only strong interconnections with neighbouring areas have made it possible to balance this situation.

With the major part of the wind connections at the lower voltage levels in the grid, a new management strategy for the grid is also necessary. Power flows in the grid have traditionally been unidirectional from the central production facilities to the consumers. With a high degree of distributed generation, power may flow in both directions. The lower voltage levels of the grid become both distribution and collection systems. A more active management of the lower voltage levels of the grid is thus necessary to reduce the risk of failures. Since the investments in the grid are typically very long term, it is a serious task for system operators to monitor the development and to adapt to changes in the production mix.

The development of offshore wind parks can, to some extent, ease the situation for the system operator. Offshore wind parks are larger and thus connected at higher voltage levels, and the larger size makes it feasible to put stricter demands on the wind parks to behave as conventional power stations. They should be able to regulate production within the limits of the wind, e.g. when the wind speed is high enough for generation, it can contribute to frequency control and voltage control.

In brief, the good news from the Western Danish system is that integration of large proportions of wind energy in the generation mix has proved possible, without compromising security of supply. An optimal use of wind power in larger proportions will, however, demand a new management philosophy for the grid.

Minimum load	Maximum load	Installed wind capacity	Installed decentralised CHP	Installed conventional capacity
1 189 MW	3 685 MW	2 371 MW	1 621 MW	3 107 MW

FIGURE 13.2 Western Danish electricity grid.

Cost Reductions Over Time

The costs of electricity from wind power have been gradually decreasing over time. A number of different effects are involved in the cost reductions:

- larger scale wind turbines decrease the cost per installed kW, owing to economies of scale;
- more efficient manufacturing as the industry matures;
- better siting and adaptation of turbines to local conditions;
- reduced maintenance costs.

On aggregate, the cost reductions have amounted to around 15% each time the total installed capacity of wind turbines has doubled. The present rate of growth in wind capacity equals a doubling each 3–5 years. Investment costs in 2001 were around US$ 765 per kW. The kWh cost depends heavily on the wind conditions on the site. With a capacity factor of 25% this gives costs of US 3.6 cents/kWh. If the historical cost reductions are continued, the investment cost will be down to US$ 550 per kW by 2010 and with continued improvements in the capacity factor to 28% the specific costs will be down to US 2.6 cents/kWh.

Such cost developments will make wind turbines very competitive with the long-term marginal costs of other electricity technologies from a purely financial point of view. The present very low interest rates also favour capital-intensive technologies such as wind power.

Including Externalities

Environmentally speaking, wind power can be said to be a very 'honest' technology. You can hear and see the environmental issues directly. Technically speaking, it is visual intrusion and noise that constitute the most serious environmental impacts of a wind farm. In particular, the visual impact of wind parks is, in practice, the major limiting factor on the space available for wind power development.

On other environmental issues such as the greenhouse effect, acidification and other forms of air or water pollution, wind power is significantly cleaner than conventional fossil fuel energy sources. With carbon constraints being valued through emission trading systems in several industrialised countries and through project-based mechanisms like Joint Implementation and the Clean Development Mechanism (in countries in transition and developing countries, respectively), the competitiveness of wind power will be enhanced.

The increased focus on security of supply and independence from imported fossil fuels may further stimulate demand for wind power.

Outlook: Promising

Not even the most committed promoters of wind power technology believe that the growth rates of recent years will continue at the same high rates. But even with more modest growth rates, wind power may become a significant part of world electricity supply during the next 10 years (Fig. 13.3).

	2007	2010	2012	2020
IEA World Energy Outlook 2002		55 000 MW		112 000 MW
BTM World Market Update 2002	83 000 MW		177 000 MW	
Wind Force 12 by EWEA and Greenpeace		234 000 MW		1 261 000 MW

FIGURE 13.3 Projections of global installed wind capacity.

In all projections Europe will continue to be the 'home' of wind turbines. One of the reasons is that for years to come it will still be necessary to have schemes integrating the value of environmental benefits in the tariff, in order to promote the implementation of wind turbines on a larger scale.

Søren Varming
ECON Analysis, Denmark

TABLE 13.1

Wind energy: installed generating capacity and annual electricity output at end-2002

	Installed capacity (MW_e)	Annual output (GWh)
Cape Verde Is.	3	5
Egypt (Arab Rep.)	68	235
Kenya	N	N
Morocco	54	140
Somalia	N	N
South Africa	N	N
Tunisia	19	50
Total Africa	**144**	**430**
Canada	236	449
Costa Rica	62	259
Guadeloupe	8	20
Jamaica	N	N
Mexico	6	11
Neth. Antilles	12	30
USA	4 685	12 000
Total N. America	**5 009**	**12 769**
Argentina	26	73
Brazil	22	54
Chile	2	5
Uruguay	N	N
Total S. America	**50**	**132**
China	468	1 000
Cyprus	2	4
Hong Kong, China	N	N
India	1 702	3 700
Indonesia	1	2
Japan	415	598
Korea (Rep.)	13	15
Nepal	1	2
Philippines	N	N
Sri Lanka	3	4
Taiwan, China	3	6
Thailand	N	N
Turkey	19	48
Total Asia	**2 627**	**5 379**
Austria	139	200
Belgium	44	75
Bulgaria	N	N
Czech Republic	6	2
Denmark	2 889	4 877
Estonia	2	3
Finland	43	63
France	148	264
Germany	12 001	16 800
Greece	349	654
Hungary	2	1
Ireland	137	426
Italy	788	1 600
Latvia	22	12
Luxembourg	16	25
Netherlands	693	1 200
Norway	97	76
Poland	58	61
Portugal	195	400
Romania	N	N
Russian Federation	7	15
Spain	4 825	9 792
Sweden	340	600
Switzerland	5	5
Ukraine	46	75
United Kingdom	552	1 450
Total Europe	**23 404**	**38 676**
Iran (Islamic Republic)	11	34
Israel	7	14
Jordan	2	3
Lebanon	N	N
Total Middle East	**20**	**51**
Australia	105	340
New Caledonia	3	5
New Zealand	36	151
Total Oceania	**144**	**496**
Total World	**31 398**	**57 933**

TABLE 13.1 (Continued)

Notes:

(1) The data shown largely reflect those reported by WEC Member Committees in 2003, supplemented by national and international published sources, in particular: the American Wind Energy Association, the European Wind Energy Association, *IEA Wind Energy Annual Report 2002*, *Windpower Monthly* and *Wind Directions*.

(2) In many instances, output in 2002 has been estimated by the editors.

COUNTRY NOTES

The Country Notes on Wind Energy have been compiled by the editors. In addition to national Wind Energy Associations' web sites and government publications/web sites, numerous national and international sources have been consulted, including the following publications:

- *IEA Wind Energy Annual Report 2002*, International Energy Agency;
- *Wind Directions*, Magazine of the European Wind Energy Association;
- *Renewable Energy World*, James & James (Science Publishers) Ltd;
- *Energy Briefing Reports*, Norton Rose.

Information provided by WEC Member Committees has been incorporated as available.

Albania

Historically, wind energy has been used in isolated areas by traditional windmills and for water pumping; however, in recent years attention has turned to utilising the resource for power generation. The installations range from a few kilowatts to 2 MW.

Meteorological stations have provided information for the areas of Durres, Kryevidh, Xarre, Bulqize and Milot but the data are approximate and nationally, incomplete. This lack of accurate statistics is a bar to attracting the necessary investment to the sector.

Although an objective of the European Union in the next 20 years is to secure 20% of its electricity from wind, it is considered that, given the Albanian resource, only 4% of its electricity can be generated thus by 2020 (some 400 GWh/yr). However, if priority is given to the construction of 20 coastal wind turbines adjacent to water pumping stations, the areas lying beside the Adriatic can be safeguarded from flooding.

Pumping stations located in the coastal lowlands take around 30 GWh/yr (some 0.7% of domestic power generation). Studies conducted by the National Agency of Energy have shown that these areas have a sufficient wind resource for them to be considered as suitable sitings for turbines. Average annual wind speed is around 4–6 m/s at a height of 10 m (with an average annual energy density of 150 W/m^2).

It has been predicted that given a total investment of US$ 150 million (financed over 20 years), a 20 wind turbine project providing an additional 400 GWh/yr generation could be expected by 2015. This assumption is based on each installation totalling 9 MW and comprising turbines of 1.5, 1.0 and 0.6 MW capacity.

Algeria

A wind map of the country has been compiled; it demonstrates that the average wind speed ranges from 2 to 6 m/s. The Adrar region in west central Algeria has been shown to hold the best wind potential, of up to 6 m/s.

The contribution of wind energy is not high in the Algerian energy balance and currently there are no planned projects. At the present time the resource is harnessed to good effect in isolated sites and the main applications are for water pumping, especially in the high plains. One research programme is assessing the technical feasibility of electricity generation at Tamanrasset, in the south of the country.

Argentina

Despite the whole country experiencing a good wind resource, the greatest concentration occurs in the Patagonia region. The resource of this area is considered among the largest and best in the world. The potential, according to some studies, is calculated at 300 000 MW and the mean speed above 8 m/s.

The most desirable wind resource applications are utilising it for electricity generation, mechanical pumping for water troughs for cattle, sheep, pigs, etc. and irrigation.

Electricity generation is accomplished by cooperatives, which supply the electricity to the public.

With reference to mechanical pumping, there are three sizes of water pumping mills on the indigenous market, rated according to the diameter of the rotor. The calculation of power and energy output is based on an average mill. According to the data provided by the manufacturers, the mean power is 1 hp, equivalent to 0.75 kW and the average use is 8 hours/day, 330 days/yr.

The PERMER Project is designed to assist the development of renewable energies in rural markets. For a full description and the current status of the programme, see the Argentinian country note in Chapter 11.

Australia

The development of the wind energy sector got off to a slow start in Australia. The resource had been used historically for water pumping in isolated locations but there existed no comprehensive wind industry. The situation began to change at the end of the 1980s (when the first 20 kW grid-connected turbine was installed in Victoria) and gathered momentum during the 1990s. By end-1999, total installed capacity stood at just over 10 MW (wind-diesel hybrid and grid-connected schemes) and the Australian Wind Energy Association (AusWEA) had just been formed. One of the main objectives of the AusWEA, which has as its members virtually all the major Australian wind developers and wind turbine manufacturers, is to raise awareness of wind as an energy resource and promote it as such.

The Renewable Energy (Electricity) Act of 2000 established the Mandatory Renewable Energy Target (MRET) which came into effect in April 2001. This piece of legislation and the establishment of the AusWEA are considered to have been fundamental in transforming the country's wind industry. By end-2002 about 105 MW capacity had been installed and by end-2003, the total had risen to 197 MW. Some of the projects are innovative research wind-diesel hybrid projects, utilising large wind penetrations and including energy storage or unique installation characteristics. The projects of note are:

- Denham, Western Australia—3 × 230 kW turbines, an area of high wind penetration and up to 70% of energy provided by wind-diesel system using innovative control technology, fully controllable turbines and flywheel energy storage;
- Exmouth, Western Australia—3 × 20 kW Australian turbines with tilt-up towers designed to be lowered in cyclonic wind conditions;
- Huxley Hill Expansion, King Island, Tasmania—an additional 2 × 850 kW turbines to supplement the existing 750 kW of wind power, incorporating energy storage using Vanadium Redox batteries and inverters to allow very high wind penetrations with a stable power supply;
- Mawson Base, Antarctica—3 × 300 kW turbines designed for use in up to 250 km/h winds, first base in Antarctic to have significant electricity from wind energy.

At the present time a total of 2 700 MW of additional capacity has been identified (spread across all states other than the Northern Territory). Some is at the planning stage, some has received approval and some is out to tender and under construction.

The synergy from combining wind and hydropower is being fully explored in Tasmania where very substantial wind energy potential has been identified. The key to unlocking this opportunity has been securing development approval for the AUD 500 million Basslink interconnector, currently being constructed by the UK company National Grid Transco. The undersea cable across Bass Strait will connect Tasmania to the national electricity market via Victoria allowing Tasmania to supply peaking power at premium prices to Victoria at a time of high demand in both summer and winter. It will also enable the export of new renewable energy from wind farm developments to the national electricity market.

There is currently (early 2004) much discussion regarding the MRET. Originally, the target was defined as a 2% increase (to 12.5%) in the 10.5% share of renewables in 1997. However, this was subsequently changed to a flat target of 9 500 GWh by 2010. Owing to an unexpected growth in overall electricity demand, the absolute level of 9 500 GWh is estimated to represent less than a 1% increase in the share of renewables by 2010. It is felt within the wind industry that if this level is not increased, the necessary incentives for investment will fade away and the presently planned capacity will not all be realised. The industry has its own target of installing 5 000 MW by 2010. In January 2004 an MRET Review Panel passed its findings to the Federal Government for consideration.

Austria

Although 70% of Austria's electricity is produced by renewable energy, hydropower supplies the majority share and to date the other renewable energies have played a minority role. With a view to changing this balance and taking into consideration the technologies available, wind power—largely available during the winter months—could complement hydro—at its lowest during the winter.

The first wind measurements were conducted in the late 1980s, discussions regarding feed-in tariffs began in 1991 and the first funding programmes commenced in 1994. These actions brought about the first wind turbines in 1994.

The new electricity laws (ELWOG) which were in effect between 1999 and 2002 stipulated that wind, water, biomass and solar must be offered to the market at a minimum share of 1% as at 1 October 2001, 2% as at 1 October 2003, 3% as at 1 October 2005 and 4% as at 1 October 2007.

The Eco-electricity law (Ökostromgesetz) of July 2002 came into effect on 1 January 2003. The law's objective is to raise the share of renewables (wind, solar, geothermal, hydropower, wave and tide, biomass, biofuels, etc.) to 78.1% of Austrian electricity consumption by 2010 and the non-hydro portion to 4% by 1 January 2008. ELWOG allowed each of the country's states to set its own feed-in tariff whereas Ökostromgesetz has set a flat rate wind tariff of €0.078 per kWh for projects receiving permits by 31 December 2004, guaranteed for 13 years (the ELWOG conditions apply to old turbines).

After growing slowly during the second half of the 1990s, Austria's total installed wind capacity stood at 139 MW at end-2002. However, during 2003, capacity virtually tripled to 415 MW, including the official opening of Europe's highest wind park. Situated at 1 900 m, the Tauernwindpark Oberzeiring consists of 11 turbines; its total 19.25 MW capacity is expected to produce 40 GWh/yr.

Belgium

With its heavy reliance on nuclear power, Belgium has been slow to deploy wind turbines. However, the planned closures of its seven nuclear stations (between 2015 and 2025) have prompted discussion on how renewable energies can take their place. At the present time renewable energy accounts for less than 1% of Belgium's electricity generation.

A demonstration wind farm (21 turbines of 200 kW each, one turbine of 400 kW and one turbine of 600 kW) was established on the eastern pier of Zeebrugge harbour in 1986. Another 600 kW turbine was added in 1999. Apart from the contribution the wind farm makes to the local electricity generation, the main purpose of it has been to publicise and popularise the existence of wind power.

Although in the late 1990s installed wind capacity began to creep upwards, there has proved to be a distinct lack of opportunity for onshore wind farms in Belgium. Attention has turned to the possibilities offshore. Electrabel, the Belgian energy company, has been granted permission to construct the first offshore wind farm in the area known as 'Vlakte van de Raan'. The company has indicated that by end-2004 the entire 100 MW installation will be generating electricity.

By end-2002 it was reported that Belgium had 44 MW installed wind capacity, more than a tripling of the 2000 capacity. By end-2003 it had risen to about 68 MW and it has been forecast that by 2007 the national capacity could rise to 335 MW.

Botswana

The relatively low average wind speeds (2–3 m/s) prevailing in Botswana restrict wind energy applications to water pumping. It is estimated that there are 200 windmills in operation, mostly located in the southern part of the country.

Brazil

According to the 'Atlas do Potencial Eólico Brasileiro' (Brazilian Wind Atlas) the gross wind resource potential is estimated to be about 140 GW. However, only a portion of that amount could be effectively transformed into wind power projects—in the long term, in the region of 30 GW installed power.

The Brazilian Government launched PROINFA—Alternative Sources for Energy Incentive Program, a national programme designed to promote the use of wind, biomass and micro-hydro. The first and second phases of the programme foresee a total of 4.15 GW of wind energy by the end of 2014.

The main application of wind energy in Brazil is for installed capacity to be grid connected. At the present time the following nine projects, totalling 22.075 MW have been installed, generating some 54 GWh/yr:

City	State	Capacity (kW)
Fernando de Noronha	Pernambuco	75
Aquiraz	Ceará	10 000
São Gonçalo do Amarante	Ceará	5 000
Gouveia	Minas Gerais	1 000
Palmas	Paraná	2 500
Fernando de Noronha	Pernambuco	275
Fortaleza	Ceará	2 400
Bom Jardim da Serra	Santa Catarina	600
Olinda	Pernambuco	225

Bulgaria

To date the use of the Bulgarian wind resource has been for traditional purposes only. For example, during the Russian–Turkish War for the liberation of Bulgaria, in excess of 300 wind mills were put to use grinding seed in the regions of Dobrudja and the Black Sea.

Research has shown that the country's wind energy potential will allow development of the resource along the Black Sea coast and in the mountainous areas.

At a level of 10 m, the country can be schematically grouped into three wind energy zones.

Zone of 'Little Dimensional' Wind Energy

Little dimensional wind energy means wind energy equipment (WEE) having an installed capacity from several kilowatts to several dozens of kilowatts. Included would be stand-alone multi-blade WEE for PV hybrid systems, water pumping, grinding seed, etc. The locations best suited for the expansion of small dimensional wind energy are situated in areas where the energy flow density exceeds 100 W/m^2. They include the Danube plain and Thracia, the Struma and Mesta river valleys and the high plains of Western Bulgaria.

Zone of 'Middle Dimensional' Wind Energy

Middle dimensional wind energy means WEE, usually three-bladed turbines, with an installed capacity from several tens of kilowatts to several hundred kilowatts. The most suitable locations include the Black Sea coast and the Dobrudja tableland, one section of the bank of the Danube and up to 1 000 m in the mountainous regions. Here, the energy flow density is between 100 and 200 W/m^2.

Zone of 'Big' Wind Energy

Big wind energy means WEE, usually one-, two- or three-bladed turbines, with an installed capacity from several hundred kilowatts to several megawatts. These installations are usually grid-connected wind farms. The height of the pivot shaft varies from 50 to 100 m but could be higher depending on the length of the blade. The wind resource which would be enough for the development of such capacity covers regions with energy potential exceeding 200 W/m^2. This zone includes open mountain ridges and summits above the 1 000 m line, as well as coastal promontories such as Cape Kaliakra and Cape Emine.

Although there are no wind turbines operating at the present time and no mechanism in place with which to encourage them, it has been reported that a private developer has proposed a 1.3 MW project. Construction of the Peak Murgash installation is planned for 2005.

Canada

Canada's wind energy potential is estimated to be far in excess of its current use of electricity (about 500 TWh). While wind energy will never supply all of the country's electrical requirements, it is not unreasonable to expect this clean, non-polluting, renewable energy source to supply up to 20%. Canada has the ability to manufacture utility-scale wind turbine components such as blades, towers and nacelles but there are no manufacturers of generators, gear-boxes and control systems nor are there any comprehensive wind turbine manufacturing facilities. However, small wind turbines (under 100 kW) are manufactured in Dartmouth, Nova Scotia and Guelph, Ontario.

Wind–diesel projects in remote northern Canadian and Alaskan locations have demonstrated that wind energy can reduce the high costs associated with transporting diesel fuel to these remote sites.

There is a significant rural Canadian, and potentially huge international, market for small non-electric wind turbines for pumping water and aerating ponds. Canada's federal Wind Power Production Incentive (WPPI) provides a production incentive of CDN$ 0.01 per kWh to qualifying wind turbines.

At end-2002, Canada had 236 MW of wind generation plant installed, producing about 449 GWh—enough to supply about 56 000 typical Canadian homes. If this electricity is used to displace coal-generated electricity, it avoids discharging about 449 000 tonnes of carbon dioxide into the atmosphere annually. At end-2003, installed capacity had risen to 322 MW.

Utility-scale wind turbines are installed in Alberta, Saskatchewan, Ontario, Quebec, Prince Edward Island, Nova Scotia and the Yukon and wind farms in Alberta, Saskatchewan, Ontario, Quebec and Prince Edward Island.

Canada's largest wind plant is located in the Gaspé region of Quebec. The Le Nordais project currently has 133 × 750 kW turbines producing 100 MW of electricity at two locations—Cap Chat and Matane. The Pincher Creek area of Alberta is home to a number of large-scale wind projects and is expected to see even more growth in the near future.

China

In China, the power of the wind has been used for water pumping for many hundreds of years but by the 1960s the traditional multi-bladed mechanical windmills had been overtaken by the advent of low-cost diesel engines. It is only in recent years with the country's rapid economic growth that attention has turned again to the wind. In order to power the country's enormous electricity generation requirement and to fulfil the Government's Tenth Five Year Plan (2001–2005) objective of increasing the use of renewable energy resources to account for 5% of total output, wind turbines are now a feature of the Chinese landscape.

By end-2002, the share of renewables in the energy mix was approaching the 5% target and the World Bank estimated that a total of 18 000 MW new capacity (small hydro, wind, biomass, solar and geothermal) would be necessary over a period of 10 years to maintain the level of 5%.

The provinces of Inner Mongolia, Gansu and northern Xinjiang and the southern and eastern coastal areas are well blessed with wind energy and it has been estimated that China's potential could be as high as 160 000 MW. At the present time a small fraction of this potential is being utilised: as at end-2002 total installed capacity was in the region of 460 MW. The Tenth Five Year Plan includes a planned additional 1 192 MW which includes 3–5 wind farms of about 100 MW each, some of which would be located in offshore waters.

In addition to grid-connected or stand-alone wind turbines, the Government's Township Electrification Program (Song Dian Dao Xiang) is installing hybrid PV-wind systems in an effort to electrify the large areas of the country without access to power.

Costa Rica

Costa Rica is reputed to have a better wind regime than California and some of the highest average wind speeds in the world. In addition to using the country's geothermal and biomass resources, the Government is demonstrating its commitment to the utilisation of its wind resource in an effort to develop sustainably and reduce GHG emissions.

In 1993 the Costa Rican Government issued a tender for a 20 MW (30 × 660 kW) grid-connected wind plant near the town of La Tejona. The project was designed for the installation of between 40 and 100 turbines on two parallel ridges to the northwest of Lake Arenal. However, many problems were encountered which delayed the project until the late 1990s. It was not until September 2001 that the turbines were shipped and installation could begin.

A further project, also near Lake Arenal, financed by private and public loans, various banks and the Danish International Development Agency, has been developed. The 24 MW Tierras Morenas wind farm sells approximately 70 000 MWh/yr electricity to the Instituto Costarricense de Electricidad, the state-owned national electric utility, under a 15-year power purchase agreement.

At the present time Costa Rica is the only country in the Central American isthmus to have wind parks connected to the electrical grid. By end-2002, installed wind energy capacity totalled 62 MW and by end-2003 it had increased to 69 MW.

Côte d'Ivoire

Although Korhogo in the north has wind speeds of 2.8 m/s and San Pedro in the south has 3.6 m/s, studies have shown that the country's overall wind potential is not high enough to allow for a great exploitation: the average wind speed was 1.8 m/s in 2003. However, the Government is studying which technology can be used to utilise wind energy at low speeds.

Denmark

With the utilisation of wind energy featuring in each Danish energy strategy, the country has made use of its wind resource since the early 1980s. The installed wind turbine capacity grew slowly but steadily until the mid-1990s when growth became very rapid. This situation continued to end-2002, when it totalled some 2 900 MW. At that point further onshore expansion ceased, owing to a substantial rise in the investment risks taken by the turbine owners selling production on the electricity market. These were caused by a set of complicated regulations and a reduced environmental premium paid to wind power.

Future extension of Danish wind power is likely to take place in the form of offshore wind turbines. Already the country has the world's largest offshore wind farm: by end-2002 all 80 turbines of the Horns Rev installation in the North Sea had been commissioned. The 160 MW farm 14–20 km from the Danish coast is expected to produce 600 MWh annually.

Horns Rev is the first of five farms in a demonstration programme for offshore wind power. During 2003, a sister farm (Nysted) was installed in the Baltic Sea and the Government was expected to invite tenders for an additional three farms. The Government's Action Plan for Energy calls for 4 000 MW of offshore wind power to be established by 2030.

In addition to supplying the home market, Denmark is a major supplier of wind turbines to the world. During 2002, Germany, the global leader in wind capacity, was its largest customer, but the USA, Spain, the Netherlands, Italy, Australia, Greece, Canada and Norway also made substantial purchases.

With the highly significant role that Denmark plays in the world wind industry, R&D is of the utmost importance. The most important areas of research are:

- influence from local wind pattern;
- construction and design of wind mills;
- control and regulation;
- integration with the grid;
- environmental aspects.

One important aspect of R&D is the reduction in the operational and maintenance costs involved with offshore installations.

Egypt (Arab Republic)

Egypt is endowed with an excellent wind energy potential, especially in the Red Sea coast area where a capacity of 20 000 MW could be achieved, as the annual average wind speed is around 10 m/s.

The Wind Atlas for the Gulf of Suez is the result of an investigation of the climatic wind conditions of the area. The investigation, which began in March 1991, was undertaken by the New & Renewable Energy Authority (NREA) and the Risø National Laboratory of Denmark, under the sponsorship of the Egyptian and Danish Governments.

Since 1992, 5 MW wind capacity has been in service at Hurghada and in 2001 a 63 MW wind farm started operating at Zafrana (both on the Red Sea coast).

A further 82 MW of capacity is scheduled to be operative in two stages, late December 2003 and mid-June 2004.

An additional 70 MW, planned in cooperation with Spain, is at the tendering stage and 120 MW in cooperation with Japan is in preparation. It is planned that the capacity will be operational in 2005 and 2006, respectively.

The country's expansion plan contains a target of around 800 MW capacity by 2010, representing 3% of total national capacity.

Finland

During 2002 the Finnish Action Plan for Renewable Energy Sources was updated and the target for the use of renewable energies was set to be at least 50% higher by 2010 than the level in 1995, with 3% of this increase to come from wind power. Thus, the national target for wind power production by 2010 is a capacity of 500 MW and electricity production of 1 100 GWh/yr.

The techno-economic and environmental potential for wind power in Finland is estimated to be 5 400–7 600 MW. At the present time

10 MW of planned wind power capacity is projected.

In 2002, state aid for wind energy was about 33% for the capital investment and €0.0069 per kWh for electricity produced.

France

Despite having a considerable wind resource, France has not been dedicated to developing either the wind industry in particular or renewable energy in general. However Eole 2005, a programme to promote wind power, was introduced in February 1996. The aim was to achieve between 250 and 500 MW installed capacity by 2005 but by end-2002 the total had only reached 150 MW and the country lagged behind Germany, Spain and Denmark by many hundreds of megawatts.

In February 2000, the 'Electricity Law' had come into force and the legislation provided the means by which the French electricity market was open to competition. Previously Electricité de France (EDF) had both sought tenders for wind installations and subsequently decided which would be selected. The Law thus effectively brought Eole 2005 to an end.

One of the conditions of the Electricity Law is that a grant of authorisation to install any wind capacity (up to 4.5 MW) must demonstrate that the capacity is compatible with the 'Programmation pluriannuelle des investissements de production d'électricité' (PPI). The PPI relates to the staging of investments in electricity generation over a period of years with a view to promoting private sector energy production. The first PPI was by an 'Arrêté' (Order) of 7 March 2003 and declared the objective of installing 2 000–6 000 MW wind capacity between 1 January 2003 and 1 January 2007 (including 500–1 500 MW from offshore wind farms).

Early in 2002, the French industry minister had announced a plan for 10 000 MW of wind power by 2010. At that time the President of the French Renewables Association (SER) was quoted as saying that he was 'optimistic' but by end-2003 only 215 MW had been installed.

Germany

The 'Electricity Feed-in' law (*Stromeinspeisungsgesetz*) was the progenitor of German wind power development in the early 1990s: installed capacity rose from 60 MW at end-1990 to 1 120 MW at end-1995. Thereafter, the 80% average annual growth demonstrated during those years was not matched but nevertheless the growth rate between 1996 and 2002 was still a very significant 41% per year.

The increasing utilisation of its wind resource is helping Germany move towards one of its energy policy goals: to double the share of renewable energy in total electricity generation by 2006. The aim of the Federal Government is for there to be 500 MW offshore wind turbines by 2006 and 3 000 MW by 2010.

Germany quickly became the world leader in wind energy, a position which it has maintained. By end-2002 wind capacity had grown to 13 759 turbines and 12 001 MW and the average rated power per turbine had reached 1 395 kW. By end-2003, there had been a further increase to 15 387 turbines representing 14 609 MW and an average rated power of 1 553 kW.

Wind turbines are sited throughout the German Länder but the three coastal states of Lower Saxony, Schleswig-Holstein and Mecklenburg-West Pomerania currently have approximately 48% of the installed wind power.

In recent years attention has turned to offshore projects. To assist their implementation the Government's *Future Investment Program* (ZIP) came into force in 2001. The Program includes an analytical tool for the offshore data collected from research platforms.

Additionally, the Environment Ministry has proposed that the German Renewable Energy Law be revised so that offshore projects would receive a payment (€0.091 per kWh) for their electricity output over 12 years instead of the current 9 years. At the present time the payment is restricted to those installations which start operating before 2006 but an amendment would extend this date to 1 January 2008. Further amendments would reduce payments to some onshore wind turbines.

To date, the number of offshore schemes in receipt of a permit is small but the first data-gathering platform, north of the island of Borkum, started operating at end-August 2003. It is expected that the first phase of a 12-turbine farm will be located adjacent to the platform and that construction will begin during 2004. In December 2002 the Federal Maritime and Hydrographic Agency (BSH) approved the Buergerwindpark Butendiek (80 × 3 MW), to be sited west of the island of Sylt. In December 2003, it was announced that a second research platform was being planned for siting in the Baltic Sea.

Greece

It has been estimated that Greece's very substantial wind resource has the potential to generate 3 000 MW. The areas of high potential are the Aegean islands, Southern Euboea, Eastern Peloponnese and Thrace. The wind power penetration in the autonomous grid of Crete is, at greater than 10%, amongst the highest in the world (with strongly increasing trends). However, the windiest areas tend to be sparsely populated and to have inadequate transmission facilities.

During the 1990s, deployment was slow with capacity only growing from about 19 MW in 1992 to 40 MW in 1998.

Until the late 1990s the majority of the wind power capacity was owned by the Public Power Corporation (PPC). The Liberalisation of the Electricity Market Law together with the EU Directive for Greece to supply 20.1% of its electricity from renewables by 2010 have helped to provide the impetus that the development of the wind sector needed.

By end-2002, installed capacity stood at 349 MW (virtually meeting the Greek Ministry of Development's target of 350 MW). During the course of the year, the Regulatory Authority for Energy (RAE) approved applications for 407 MW capacity for the interconnected mainland and 188 MW for the islands of Evia, Andros and Tinos. In line with the EU Directive, the Ministry of Development has set a new target for wind energy of more than 1 500 MW installed capacity for 2010.

Through a scheme of government incentives a number of local authorities and private investors are developing further wind parks, selling excess electricity to the PPC grid at rates fixed by the Law. PPC has overall operational responsibility for these installations.

In February 2003 the largest private wind farm (59 turbines, 55 MW) started operating in Thrace and in May, 25 applications for a total of 220 MW wind power units were approved by RAE. By end-2003 installed capacity had reached 424 MW (772 turbines) and in excess of 47 MW was under construction. Electricity generation from wind in 2003 totalled 850 GWh.

Wind power research is carried out by a number of public bodies (universities, the Centre for Renewable Energy Resources—CRES) and, to a lesser extent, the PPC. CRES is currently conducting R&D activities into renewable energy-hydrogen projects and the hybrid renewable energy scheme of PPC (a combination of wind power with pumped storage).

Hong Kong, China

The findings from a Consultancy Study on the Potential Applications of Renewable Energy commissioned by the HKSAR Government suggested that the wind resource potential in Hong Kong is about 14 000 GWh/yr.

There are no wind power projects planned for 2004 but a 12-month territory-wide wind measurement programme to collect site-specific data was to be carried out from late 2003. There was also a university programme to study wind energy resources starting mid-2003.

One 2.5 kW wind turbine has been installed as part of a hybrid system to supply electrical power to a dormitory on a remote island and eight small wind power generators (rated from 140 to 500 W) have been installed to supply power to the monitoring equipment of automatic weather stations in remote locations.

Hungary

A project to map Hungary's wind energy resource is currently in progress.

India

The Indian wind power programme was initiated in 1983–1984 and a *Wind Energy Data Handbook* published in 1983 by the Department of Non-conventional Energy Sources (now the Ministry of Non-conventional Energy Sources, MNES) served as a data source for early government initiatives. In 1985 an extensive Wind Resource Assessment was launched, which also signalled the beginning of concentrated development and harnessing of renewable sources of energy and, more specifically, of wind energy. The Assessment has now become the world's largest such programme and to date six volumes of the *Handbook on Wind Energy Resource Survey*, containing a huge volume of accumulated wind data, have been published. It is being implemented through the state Nodal Agencies, the Field Research Unit of the Indian Institute of Tropical Meteorology and the Centre for Wind Energy Technology. An Indian wind atlas has also been proposed.

Initial estimates of the Indian wind resource had put it at 20 000 MW (at the micro level) but studies have revised this figure to just over 45 000 MW (at 50 m hub height). Potential locations with abundant wind have been identified in the following 10 states: Andhra Pradesh, Gujarat, Karnataka, Kerala, Madhya Pradesh, Maharashtra, Orissa, Rajasthan, Tamil Nadu and West Bengal. Taking into account the limitations of the existing grid, it has been estimated that the 10 states have an exploitable technical potential of just under 13 000 MW.

In terms of currently installed wind turbine capacity, India ranks fifth in the world behind Germany, USA, Spain and Denmark. At end-2002 the figure stood at 1 702 MW, of which 63 MW represented demonstration projects and 1 639 MW commercial projects. Tamil Nadu possessed 53% of the commercial plants. By early 2003, commercial installed capacity had already grown to approximately 1 870 MW.

The demonstration projects, which began in 1985, are being implemented in areas not already possessing projects but where commercial developments could follow. It was expected that demonstration projects in Karnataka (two 2 MW) and Kerala (one 2 MW) would commence during 2003.

Use is being made of wind-diesel hybrid projects where an area is dependent on diesel fuel. The first phase of such a project (100 kW) has been completed and the second phase (400 kW) is being implemented. Hybrid projects are being developed for the Lakshadweep and Andaman & Nicobar Islands.

The Indian Renewable Energy Development Agency (IREDA) has played a significant role in the promotion of wind energy, attracting bilateral and multilateral financial assistance from world institutions and the private sector. The Centre for Wind Energy Technology (C-WET), based in Tamil Nadu, acts as a technical focal point for wind power development in India.

Iran (Islamic Republic)

The Iranian WEC Member Committee reports that 11 MW of installed wind power capacity has been generating electricity for the past 3–4 years.

A total of about 119 MW additional capacity is reported as planned for installation as follows:

- 90 MW within the next 1–2 years by the Atomic Energy Agency;

- 64 MW within the next 1–2 years by the Ministry of Energy;
- 23 MW within a year by TAVANIR (a government body of the Ministry of Energy).

Ireland

Ireland's prevailing south-westerly winds from the Atlantic Ocean give a feasible wind resource that has been estimated to be as high as 179 GW, or some 40 times the country's current generating capacity. However, the accessible resource is about 2 190 MW and, in reality, the practicable resource is estimated to be 812 MW.

This abundant wind supply began to be utilised, albeit rather poorly, in the early 1980s with several demonstration schemes. The detailed investigations that followed included the establishment of the Irish Wind Atlas and, in the mid-1990s, the Government's Alternative Energy Requirement (AER I) competition. AER I awarded 15-year fixed-price power purchase agreements.

To date, there have been three programmes offering contracts for wind power projects: AER I offered contracts for 46 MW of capacity, AER III offered 90 MW and AER V offered 240 MW. By end-2002, total installed capacity had reached 137 MW. The results of the AER VI competition were announced in July 2003 and included for the first time contracts for two 25 MW offshore wind demonstration projects.

The Government's Renewable Energy strategy, as contained in the 1999 Green Paper and subsequently the 2000 National Climate Change Strategy, specifies a target of an additional 500 MW of installed renewable electricity generating capacity to be in place in the period 2000–2005. The country is also working with the 2001 EU Directive of meeting 13.2% of its electricity generation from renewables by 2010. Of the renewable energy technologies available, it is considered that by 2005 wind energy will make the greatest contribution and will also account for a large part of the 2010 target.

During 2003, six wind installations were installed, bringing Ireland's total to 31 with a combined capacity of 191 MW. Previously, the largest wind farms had been of 15 MW but by end-2003, the 10 × 2.5 MW turbines installation at Kingsmountain, County Sligo was officially opened.

It was announced in June 2003 that the Irish company, Airtricity and US company, GE had agreed to jointly build the world's largest offshore wind farm. The 200 turbine, 520 MW farm is planned for the area of the Arklow Sandbank, with completion envisaged in about 2008. Early in 2004, seven 3.6 MW turbines had been erected and were approaching operational status.

Italy

The Italian wind resource is most prolific in the southern regions of Campania, Puglia and Molise and on Sardinia, Sicily and the minor islands. Since 1998 Government policies concerning renewable energies have established that wind power plants are particularly favoured.

Total installed wind capacity increased steadily after 1995, and dramatically so in 2001, but generally, deployment has been a slow process. It is thought that problems have arisen owing to difficulties in dealing with a new incentive scheme based on Green Certificates and opposition to the installation of wind turbines from the anti-wind environmental groups and from at least two of the Italian regions.

The Government has enacted a considerable amount of legislation supporting the introduction of renewable energies into the national energy balance. The Italian White Paper for the exploitation of renewable energy sources (RES) states that the targets on wind energy are 700 MW by 2002, 1 500 MW by 2006 and 2 500 MW for 2008–2012. To achieve the stated longer term targets, an addition of around 200 MW in each year would be necessary.

During 2001, 263 MW was added to wind capacity but in 2002 only 106 MW. However, by end-2002 the target for that year had been surpassed, with capacity reaching 788 MW.

There are some large projects under evaluation by local authorities, with the proposed sites for new wind farms mostly in the south/central Apennine region. It is thought that at the present time market development is regaining momentum.

Japan

The Japanese Government instituted its Sunshine Project in answer to the problems created by the oil crises of the 1970s. In 1993, as a way of efficiently overcoming barriers related to new energy, the New Sunshine Program (NSS) was launched; it has been conducted under the aegis of the Agency of Industrial Science and Technology (AIST) in the Ministry of Economy, Trade and Industry (METI) and has included a renewable energy R&D programme that has directed development of wind power in Japan.

Between 1990 and 1994 the New Energy and Industrial Technology Development Organization (NEDO) carried out a wind resource measurement study, and in 1995, together with the Government, started a promotional policy with subsidy programmes.

The Law on Special Measures for Promotion of Utilisation of New Energy (New Energy Law) came into force in mid-1997 and proved to be the encouragement needed for development of the wind sector. Some large-scale wind farms (ranging from 20 to 30 MW) were installed and from the low national capacity prior to 2000, there was significant growth, resulting in an end-2002 figure of 415 MW.

As a result of the UN Climate Change Conference in Kyoto in 1997, Japan agreed to reduce its output of GHG by 6% from 2008 to 2012, compared to the 1990 level. In order to meet this target, the Government amended its previous objective of 300 MW wind power (by 2010) to 3 000 MW in its latest Primary Energy Supply Plan.

April 2002 saw the Government passing further legislation (the Renewables Portfolio Standard—RPS) so that the renewable energy contribution to total electricity supply (1.1% by 2010) will be met.

It was announced at the end of 2003 that trials of Japan's first offshore wind installation (2 × 600 kW turbines) had started at Setana, southern Hokkaido. It is expected that electricity will be generated in early 2004.

Jordan

Studies on Jordan's wind potential have been conducted over a period of years and have shown that the country has a rich wind energy resource. The average annual wind speed exceeds 7 m/s in some areas. A wind atlas has been prepared based on an assessment of the available resource which demonstrates the existence of a potential for several hundred megawatts of wind-power installations.

Twelve wind measuring systems have been installed at promising sites and long-term collection of wind data (speed and direction) and their evaluation is being implemented.

There are two operational wind farms in Jordan: Al-Ibrahimyya, with a capacity of 320 kW (4 × 80 kW), established in 1988 in co-operation with a Danish firm and considered as a pilot project; the other, in Hofa, has a capacity of 1 125 kW (5 × 225 kW), established in 1996 in co-operation with the German Government under a programme called Eldorado. Both wind farms are fully operated and maintained by Central Electricity Generating Company (CEGCo).

The Ministry of Energy and Mineral Resources (MEMR) has issued a call for proposals for the development of a 75–90 MW wind IPP project. It has been reported that technical and financial offers have been received from international companies and that their evaluation was being undertaken during 2003.

Wind energy is also used for water pumping, using a locally manufactured mechanical windmill.

Kenya

At the present time several factors hinder the development of wind energy in Kenya: an inadequate promotion strategy, little information regarding the resource and the non-availability of appropriate technology. Nevertheless, the potential, at its best in the arid and semi-arid northern regions, the coastal area and to some extent in the vicinity of Lake Victoria, has begun to be utilised and in 2002 contributed 200 MWh to the national electricity grid.

Korea (Republic)

Following the country's wind power demonstration project on Cheju Island in 1995, Korea has developed large-scale wind farms to harness its indigenous energy resource. The Government provides subsidies for part of the cost of installation and allows the electricity generated (the majority of which originates from the farms) to be traded at preferential prices.

Latvia

Latvia has favourable conditions for exploiting wind energy: the average yearly velocity of winds blowing over the western coasts of the Baltic Sea is 5.7 m/s.

The tradition of using wind energy in Latvia was revived in 1989, when the first wind power plant (WPP), with a capacity of 16.0 kW, was installed. Initially, the WPPs operated in the off-line mode, with their output used for heat production. The first two plants, on the north-west coast of Latvia in the Ainazi region and with a total capacity of 1.2 MW, were connected to the power grid in 1995.

Three WPPs (Nordex turbines) with the total power of 3.0 MW were installed on the coast in the neighbourhood of Ventspils between 1999 and 2002.

Also during this period, construction of a wind park consisting of 33 directly driven WPPs of the E-40 type (Enercon turbines) with a total capacity of 22 MW began in the Liepaja district—representing the largest such park in the Baltic States. The cost of construction is estimated to be US$ 22 million. Generation began in December 2002.

The sole purchaser of wind-generated energy is the state joint-stock company, Latvenergo.

To date, the only legal instrument in Latvia governing the policy of energy efficiency improvement has been the Law on Energy accepted by the Latvian Government in 1998. According to this law, licensed utilities dealing with electric energy distribution are to purchase, within a licensed area, the surplus of electric energy from small hydro and wind power plants with a capacity up to 2.0 MW, in addition to solar plants planned to be in operation by 1 January 2005. This surplus, intended for sale, is that remaining after local consumption and corresponds to the electricity parameters established in the state.

For 8 years after the start of a plant becoming operational, this extra energy may be sold at a price equalling double the tariff on average sales of electrical energy. The highest purchase tariffs on electric energy are set by the Government for the monopolistic company, Latvenergo. After 8 years the purchase price for surplus electricity will be equal to the average sales tariff on electric energy.

At the present time about 100 MW of WPP is planned for installation over a period of 15 years.

The Institute of Physical Energetics has undertaken research into the construction of directly driven low-power WPPs with the aim of producing inexpensive and optimally designed plants.

Lebanon

There is a high wind energy potential in Lebanon. The country's location between a long western coastline and the Syrian desert to the east results in strong winds both in summer and winter. The best sites are thought to be in Akkar in the north and in the far south, but as yet a wind atlas has not been compiled. Further study is required prior to sites being selected. To date, the turbines installed have been on a small and experimental scale (0.1–10 kW).

Mexico

Mexico's estimated wind potential is about 5 000 MW, most of it located in the south of the country. The Comisión Federal de Electricidad operates 2 MW of the total installed capacity, 1 MW is operated by autoproducers and the other 3 MW are small wind power generators and wind water pumps. Currently, the cost of investment in wind power installations is around US$ 1 000 per kW installed, and the electricity generation cost is between US 5 and 11 cents/kWh.

Morocco

In 1986 the Centre for Renewable Energies (CDER) published a wind atlas for Morocco, albeit it did not cover the entire country. However, it established that the regions with the best wind resources are located in the extreme north on the Mediterranean coast between Tangiers and Tétouan, in the north-eastern Atlas mountains in the area of Taza and along the Atlantic seaboard in the regions of Essaouira and Tarfaya. The wind potential is estimated to be approximately 6 000 MW.

In 1991 the CDER launched a programme to evaluate the available wind resource with a view to completing the atlas. The programme has three phases: the first (1991–1994) concentrated on a study of the Mediterranean coastal area; the second (1995–2000) looked at the north-eastern provinces and the southern coastal areas and the third (2001–2010) is examining the Atlas and Rif mountainous areas.

In addition to various pilot plants being installed, a 3.5 MW demonstration plant was first established in the Tangiers-Tétouan Wilaya prior to the first utility-sized wind park. Abdelkhalek Torres is a 50.4 MW (84 × 600 kW turbines) installation near to Tétouan which was brought into service at the end of August 2000. The Moroccan Office national de l'électricité (ONE) is committed to purchasing the electricity produced (about 226 GWh/yr) for a period of 19 years.

The ONE is planning a 60 MW wind farm to be located near Essaouira. Pre-qualification offers for technological assistance were issued in November 2002 and it is expected that following construction during 2004, the plant will enter service in 2005.

Two further farms are planned: one in the region of Tangiers with a capacity of 140 MW and the other in the area of Tarfaya with a capacity of 60 MW.

Netherlands

During 2001 the Dutch Government changed its policy and set new renewable energy targets in order to comply with its obligations under the Kyoto Protocol.

The new targets are: renewable energy in the overall energy supply: 5% in 2010; 10% in 2020 and electricity generation from renewable energy: 6% in 2005; 9% in 2010. The goal for wind power capacity has been set at 1 500 MW in 2010.

In 2001 renewable energy had only a 1.3% share of overall energy consumption and 2.8% of electricity generation and it was felt that without further action future targets could not be met. Taking this into account, government policy attached a higher priority to those renewable energies that it was felt could make the greatest contribution: offshore wind and biomass.

In 1995 a record number of new wind turbines were installed (in excess of 100 MW), but generally, the 1990s were a decade of capacity growing steadily but unspectacularly. It was not until 2002 that 165 new turbines representing 216 MW were installed bringing the end-2002 total to 693 MW. Forty turbines, with an aggregate capacity of 15 MW, were taken out of service during the year. It was reported in early 2004 that capacity had already reached a total of 927 MW.

In the light of the greater possibilities offered by installing wind turbines offshore, together with experience from Novem's (the Netherlands Agency for Energy and the Environment) 1997 feasibility study for a Near Shore Wind Farm (NSW), a target of 7 500 MW (of which 6 000 MW would be offshore) was set for 2020.

Early in 2002 the consortium Noordzeewind (Nuon Renewables and Shell Wind Energy) was chosen to build a 100 MW demonstration NSW off the coastal area of Castricum and Egmond aan Zee. The project will have a life span of 20 years, at the end of which the plan is to dismantle it. It is hoped during that time, the experience gained will greatly assist the development of further offshore installations. It is expected that the first power will be supplied to the grid in 2005.

As at early 2004, two near-shore projects (totalling 16 MW) are operating.

New Zealand

A wealth of indigenous renewable energy (in particular hydro and geothermal) already supplies about 30% of total energy demand and about 70% of electricity supply. However, owing to its location, New Zealand also has a good wind resource and interest in its harnessing has been increasing.

The first demonstration generator (Brooklyn) was installed in 1993 and subsequently two commercial wind farms (Hau Nui, 3.5 MW and Tararua, 32 MW) were commissioned, in 1997 and 1999, respectively.

The Energy Efficiency and Conservation Act 2000 required the implementation of a National Energy Efficiency and Conservation Strategy (NEECS). One of the mechanisms developed within the Government's Climate Change Programme (to help implement the NEECS targets) is the Projects Mechanism under which credit is given for the displacement of some thermal generation and hence CO_2 emission. Some wind projects are thus expected to become viable earlier than otherwise possible.

Norway

Norway's electricity production is virtually entirely based on hydropower but as there are physical limitations to new schemes, attention has turned to wind energy, albeit with some major obstacles to overcome (financing, public acceptance, etc.).

Although the country has a tremendously high wind resource, at end-2002 wind turbines, installed along the coast, totalled only 97 MW. Moreover, as the electricity generated from wind was only about 76 GWh in 2002, it represented a very minor part of the total electricity generation of 130.6 TWh.

The first wind turbine project was installed in 1986 (55 kW), after which wind capacity grew slowly at first and then accelerated in the later 1990s, to reach 17 MW by end-2001. In September and October 2002, two wind farms of 40 MW each were installed on the island of Smøla and near the town of Havøysund, close to the North Cape.

The latest White Paper on Norwegian energy policy states that the Government will stimulate the use of renewable energies through a comprehensive development programme. The goal for wind energy is the installation of 1 000 MW, generating about 3 TWh/yr, by 2010. This implies in excess of 100 MW new wind energy every year in the period until 2010. The generation costs for these plants are estimated at about 0.25–0.35 NOK/kWh, depending on local conditions.

During 2003, Statkraft (Norway's largest producer of electricity) announced three projects as follows:

- Hitra Wind Park, South Trøndelag County, 56 MW, generating approximately 150 GWh from the last quarter 2004;

- Kjøllefjord Wind Park, Lebesby municipality, Finnmark County, 40 MW, generating approximately 155 GWh from the last quarter 2005;
- Smøla Wind Farm, 110 MW expansion to existing 40 MW, generating approximately 450 GWh from the last quarter 2005.

Poland

Poland is reported to have excellent wind conditions: as much as 75% of the country is considered as favourable and about 5% as very favourable. The average speed varies between 5.5 and 7.0 m/s at a height of 50 m.

At the present time Poland's installed wind capacity totals about 58 MW. There are three wind farms (Barzowice, 5 MW; Cisowo, 18 MW; Zagorze—operational at end-2002—30 MW) and additionally, a number of small single turbines. It is expected that two further wind farms (Skrobotowo, 30 MW and Tymien 50 MW) will become operational during 2005.

The WEC Polish Member Committee reports that according to the 'Assumptions of the Energy Policy' prepared by the Polish Ministry of Economy, a very rapid development of installations is forecast. Under the 'Optimistic' Scenario, it is estimated that wind power capacity will reach 600 MW by 2010.

There are also major plans to develop Poland's offshore wind potential.

Portugal

Despite Portugal's long history of exploring the seas in sailing ships and of using its wind resource for milling corn and pumping water, the country has been slow in utilising it for the production of electricity. The technical wind potential is estimated to be approximately 700 GWh/yr.

The Atlantic archipelagos of the Azores and Madeira both have a high wind energy potential and it was in these islands that the first wind parks were established at the end of the 1980s/beginning of the 1990s.

At the end of the 1990s the legislation relating to renewable energies became more favourable. The European objective is for at least 39% of Portugal's electricity consumption to be generated by renewable sources in 2010 and although the country has recourse to several new technologies, it is thought that wind energy will take a major share.

By end-2002 195 MW of wind capacity had been installed, mostly on the mainland but also in the Azores and Madeira. Further development in all three areas is planned—during 2003 General Electric announced firstly, an agreement with Enernova (a subsidiary of Electricidade de Portugal) to supply turbines totalling 31 MW for three wind farms in northern Portugal and secondly, the supply of turbines totalling 24 MW for two wind farms, one in the north of the country and the other north of Lisbon. Both projects were expected to begin producing electricity by January 2004.

Romania

A national wind map compiled during the 1990s showed that the areas with the best resource are located in the mountains (generally in complex terrain), with altitudes over 1 500 m and wind speeds of 6–9 m/s, and in the Black Sea coastal area, where the wind speed at a standard height of 10 m varies from 7.5 m/s on the continental shelf (with water depth less than 10 m) to 5 m/s on the coast.

Although most of Romania's exploitable wind potential is in the coastal areas and the technical potential is estimated at approximately 2 000 MW (producing about 4.5 TWh/yr), only R&D pilot wind units have operated to date. Several factors have contributed to the slow progress on the wind front: the lack of a regulatory framework, or of any incentives designed to encourage renewable energies and the fact that nuclear power and large hydro plants have, historically, been viewed favourably by the Government. However, it is thought that wind farm projects are likely to be developed on the Black Sea Coast. A 24.5 MW wind farm has been proposed for a 14 km-long site at Constanta.

Russian Federation

Russia has used its high wind resource for many hundreds of years, mainly mechanically for water pumping. However, despite an enormous potential, commercial, large-scale utilisation has never occurred and development has generally been restricted to agricultural uses in areas where a grid connection was infeasible. The areas of greatest resource are the regions where the population density is less than 1 person per km^2.

The coastal areas of the Pacific and Arctic Oceans, the vast steppes and the mountains are the areas of highest potential. In 1935 the wind resource was estimated at 18 000 TWh for the USSR as a whole. More recently, estimates suggest that the European part of Russia has a gross wind energy resource of 29 600 TWh/yr (37%) and the Siberian and Far East part, 50 400 TWh/yr (63%). The technical resource for each is reported to be 2 308 and 3 910 TWh/yr, respectively.

It has been suggested that large-scale wind energy systems can be applied in various locations in Siberia and the Far East (east of Sakhalin Island, the extreme south of Kamchatka, the Chukotka Peninsula in the Magadan region, Vladivostok), the steppes along the Volga river, the northern Caucasus steppes and mountains and the Kola Peninsula where the resource is particularly favourable and there is an existing power infrastructure and major industrial consumers. Additionally, offshore wind parks could be considered in some of these areas, especially in the Magadan region and in the Kola Peninsula where existing hydropower stations could be used to compensate for the intermittent wind power.

In the years following World War II, with the low fuel prices and in an effort to increase the size of generators and thus their efficiency, firstly large-scale thermal and then nuclear power stations were constructed. There was some development in the transition to wind turbines with electrical generators but in the 1995 *New Energy Policy of Russia*, the non-traditional and renewable energies were relegated to playing a small role in servicing small and isolated customers. Even though the Russian Ministry of Fuel and Energy's recent *Energy Strategy for Russia for the Period to 2020* acknowledges that the non-conventional renewable sources of energy (NCRSE) are environmentally clean, socially desirable in isolated communities and can replace the use of locally sourced fossil fuels, a low priority is still accorded. The report estimates that the share of all NCRSE will increase from 0.1 to 1% by 2020 but does not quantify the shares of each technology.

Slovakia

In August 2003 it was reported that Slovakia's first wind farm was being constructed 80 km north of Bratislava. The 2.4 MW project, jointly funded by the Slovak Government and the EU Phare programme and constructed by a German company, is expected to generate a total of 3.6 MWh annually.

Additionally, a 17-turbine installation destined to be constructed in the east of the country was given approval in early 2003.

Although progress has been slow in the past owing to a low price for conventionally generated electricity and a lack of incentives to encourage generation from renewable energies, there are several other projects at various stages of development.

South Africa

Eskom's 'South African Bulk Renewable Electricity Generation Project' is designed to explore the possibilities of using renewable energies, not least wind power, for generation.

Early in 2003 Eskom erected three wind turbines near Cape Town. The three are of different capacities and designs and are essentially a research exercise.

Feasibility studies have shown that the wind measurements are favourable in the area to the north-east of Cape Town. A private company is planning to erect a small wind farm near the town of Darling. The licence for this venture has not yet been granted, due to an incomplete Environmental Impact Assessment.

Wind energy is used extensively in farming communities for pumping water.

Spain

Estimates have shown that the country has a technical wind potential of 15.1 GW, which has provided the wherewithal for an ambitious wind energy policy. From a capacity of just 75 MW in 1994, the end-2002 position was over 6 000% higher at 4 825 MW. By then Spain was second in terms of global installed power, lying behind Germany and ahead of the USA.

The Special Regime of the Electrical Sector Act contained regulations assuring the electricity producers they would have guaranteed access to the grid, with the price per kWh reflecting a bonus above the electricity sale price.

A further 1 377 MW (an increase of 28.5%) capacity was added during 2003, bringing the total to 6 202 MW. However, this was insufficient to retain second place and by the end of the year, the USA had overtaken Spain behind Germany.

Almost all of the Spanish autonomous communities possess wind capacity, from Valencia on the east coast with 20 MW (at end-2003) to Galicia in the north-west with 1 579 MW; only three mainland regions have none.

The Program for Promotion of Renewable Energies (PPER) was devised in 1999 by the Diversification and Energy Saving Agency (IDAE) and was a response to the law on the Electricity Sector which states that by 2010 at least 12% of energy demand must be met from renewables. The original target of the PPER was for wind energy to reach 8 974 MW. The Electricity and Natural Gas Plant: Transmission Grid Development 2002–2011 subsequently amended this figure to 13 000 by 2011.

Sri Lanka

A large section of the Sri Lankan population is without access to electricity and whilst hydropower provides the majority of the generated power, this dependence is vulnerable to drought. In order to increase electricity coverage as well as to satisfy the rapidly growing demand for power, much extra capacity will be required.

In June 2003 a USAID-funded solar and wind mapping survey was presented to the Sri Lankan Government. The survey, conducted by the US National Renewable Energy Laboratory, identified various locations along the north-western coast and the central hill areas for further exploratory work.

At the present time a 3 MW pilot wind project is operating at Hambantota in the south of the country and various small village projects for powering computers, televisions and radios are being implemented. Further projects have been proposed for Bundala, Kirinda and Palatupana.

It has been reported that the Ceylon Electricity Board ultimately hopes to have 200 MW grid-connected wind capacity in the south-eastern quarter of the island.

Swaziland

A regime for utilising the wind resource is not well established in the country but a project aimed at compiling comprehensive wind and solar data started in 2001. Measuring equipment was installed at five sites: two in the high veld, one in the middle veld and two in the low veld. The country's average wind speed is estimated to be 4 m/s.

At the present time wind energy is used for individual water pumping schemes and a country-wide picture of capacity does not exist.

Sweden

Although Sweden was one of the early pioneers in modern wind power development, embarking on a wind energy programme in 1975, deployment has been steady, albeit fairly slow, in the intervening period.

In 2002 the Parliament decided to set 10 TWh as the planning target for electricity production from wind power (providing investment is sufficient) by 2015. During 2003 Green Certificates were introduced, which are accredited to the wind power producers in addition to the power price.

By end-2002 installed capacity totalled 340 MW; generation at 600 GWh represented a doubling of output since 1998.

Sweden has utilised its offshore waters to site wind turbines since 1990, following with installations in 1998, 2000 and 2001. There are several further projects at the planning/feasibility stage: Eurowind is reported to have received permission for a project with 48 × 1.5 MW turbines to be located in Öresund Sound and Vattenfall has undertaken a feasibility study for an offshore project consisting of 3–4 MW turbines west of Karlskrona. The city of Karlskrona is planning for around 100 large-MW turbines offshore.

Ukraine

The wind power potential in Ukraine, whilst very large overall (estimated at some 30 TWh/yr), is considerably higher in the south than in the northern areas. It is considered technically feasible and advisable to use 15–19% of this inherent wind energy.

The first wind plants were installed in 1993 and in 1994 the Cabinet Council of Ukraine passed a special ordinance, *Regarding the Windplant Construction in Ukraine*, as a result of which a construction fund was established by means of a 0.75% increase to electricity bills. Further legislation (the 1996 National Energy Programme, the Integrated Governmental Program for Energy Conservation and the Integrated Wind Plant Construction Program in Ukraine by year 2020) followed.

In 1999 it was reported that a total of 98 turbines were either operating or under construction, representing some 300 MW, and a further 57 (of type USW 56-100) were being assembled. Even then the obsolescent USW 56-100 turbines (the product of a Ukraino-American joint venture) were being replaced by improved German-designed 600 kW turbines which provided an opportunity for a greater penetration of wind power within Ukraine.

The Government envisages that by 2010, overall electric power from wind installations and isolated windmills will total 5.71 TWh/yr, providing 2.5% of Ukraine's total energy consumption.

United Kingdom

To ensure the diversity of electricity generating capacity, the UK Government instituted the Non-Fossil Fuel Obligation Orders (NFFO) for England and Wales and for Northern Ireland (NI NFFO) and the Scottish Renewables Obligation (SRO). The orders were collectively known as the Renewables Obligations. Four Orders were made in England and Wales (1990, 1991, 1995, 1997), two in Scotland (1994, 1997) and two in Northern Ireland (1994, 1996).

The Utilities Act (2000) made substantial changes to the regulatory system for electricity in Great Britain. The Act replaced the NFFO, but contained provisions for the preservation of NFFO contracts for the rest of their term.

Under the Act the NFFO were replaced by the Renewables Obligation and Renewables (Scotland) Obligation, coming into force in April 2002. The Government has imposed an obligation on suppliers that a specified proportion of the electricity supplied must be generated from renewable sources. The obligation is supported by a system of tradable 'Green certificates' (e.g. a supplier which is unable to fulfil its obligation itself can do so by purchasing a certificate from a supplier which has over-achieved). The target began at 3% in 2003 and will rise gradually to 10% by 2010 and thence to 15% by 2015 (announced at the beginning of December 2003). The Government's White Paper published at the beginning of 2003 suggested that by 2020, the contribution could rise to 20%; wind energy would be expected to be the largest player.

Additionally, the Climate Change Levy was introduced in April 2001. It is a levy charged on businesses and the public sector with the objective of reducing carbon emissions.

During 2002, 86 wind turbines representing 88 MW were installed, bringing the aggregate capacity to 552 MW. By end-2003 a further 103 MW brought the total to 655 MW. At the present time the UK has two offshore wind installations totalling 63.8 MW: Blyth off the coast of north-east England and North Hoyle off the coast of North Wales.

In February 2004, a survey carried out by the British Wind Energy Association (BWEA) showed that 22 projects had been confirmed (representing 474 MW) for installation during 2004. Of the 22, 21 are onshore developments and one will be an offshore wind farm at Scroby Sands, off the East Anglian coast. The BWEA also predicts that by end-2005 total installed capacity will have reached in excess of 1 600 MW.

United States of America

The Energy Information Administration (EIA) estimates that the raw wind resource potential of the US is in excess of 3 000 GW. This estimate excludes offshore areas, areas with poor wind potential (average annual wind speeds less than 7 m/s), areas with specific legal or technical restrictions on development for wind use (such as areas with high slope, environmentally restricted areas and urban areas), and areas greater than 20 miles from existing transmission lines. However, most of the land included in this estimate is likely to be precluded from wind development for economic reasons not explicitly accounted for in the estimate, such as high land costs, rough terrain, lack of site access, aesthetic or environmental limitations, the need to upgrade or expand existing transmission capacity to accommodate remote wind capacity, or the need to provide energy storage or back-up generation to maintain grid reliability.

In the context of the National Energy Policy, published in May 2001, the renewable energies (including wind) play a significant role in the diversification of the future national energy supply and the transition to clean, affordable energy sources. The Department of Energy's (DOE) Office of Energy Efficiency and Renewable Energy has its own compatible mission, to 'strengthen America's energy security, environmental quality, and economic vitality through public–private partnerships that bring clean, reliable and affordable energy production and delivery technologies to the marketplace'.

In order to allow wind energy to compete without disadvantage in serving the nation's energy needs, the mission of the DOE's Wind Energy Program is to overcome barriers—energy cost, energy market rules and infrastructure, and energy sector acceptance—through technology R&D, technical support, and collaborative efforts. The program's R&D focus is on low wind speed technology for large and small wind turbines to enable economically competitive wind power development in widespread moderate wind resource areas. Specific efforts include: public/private partnerships for low wind speed technology R&D through advanced component and full turbine prototype development and testing; activities to facilitate integration of wind energy into power delivery systems; and complementary R&D with outreach (States, utilities, industry, and other stakeholders) to identify and remove institutional barriers to wind power development.

The American wind power industry has shown remarkable progress, increasing by an average 34% each year between 1981 and 2002. By end-2002 capacity stood at 4 685 MW (27 states) and by end-2003 (by which time it had overtaken Spain to be second in the world), 6 374 MW (30 states). In recent years the high growth rates have been achieved not least because of the combined effects of declining wind energy costs, available financial incentives and the high costs of competing energy sources. In order to sustain this momentum, the wind industry hopes that the wind energy production tax credit (PTC) which expired at end-2003 will be extended for a further 3 years.

In order to achieve the American Wind Association's target of 6% of electricity generation from wind energy by 2020, an annual growth rate of about 18% is necessary.

Uruguay

At the present time wind energy is utilised for water pumping, etc. in rural areas isolated from the electricity grid. A pilot plant to study the feasibility of grid-connected wind-generated electricity was installed in March 2000.

Chapter 14

Tidal Energy

World Energy Council
CONSEIL MONDIAL DE L'ENERGIE

COMMENTARY

The Tides

The tides are cyclic variations in the level of the seas and oceans. Water currents accompany these variations in sea level which, in some locations, such as the Pentland Firth to the north of the Scottish mainland, can be extreme.

Small tidal 'mills' were used in Southern England, France and in Orkney, which lies to the north of the Pentland Firth, in the Middle Ages. Tidal flows in bays and estuaries offered the potential to drive cereal-grinding apparatus in areas that were too low lying to allow the use of conventional water wheels. In the 20th century the tides were seriously re-examined as potential sources of energy to power industry and commerce.

The explanation of the existence of tides represented one of the greatest challenges to early oceanographers, mathematicians and physicists. It was not until Newton developed his theories of gravitation and the mechanics of motion that a satisfying theory emerged to explain at least some of the properties of the tides. The physics of the 'Newtonian Tidal Theory', which is sometimes referred to as 'Equilibrium Tidal Theory', gives a partial description of tidal behaviour for an abstract planet Earth entirely covered by water, and is outlined in most introductory texts on oceanography (Bearman, 1997).

This theory suggests the establishment of 'bulges' in the fluid surrounding the Earth (Fig. 14.1).

The Earth of course rotates and the two tidal 'bulges', in order to maintain their position with respect to the Moon, have to travel round the Earth at the same rate as the Earth's rotation. The Moon rotates around the Earth, actually about the centre of mass of the Earth–Moon system, every 27.3 days in the same direction that the Earth rotates every 24 h. Because the rotations are in the same direction, the net effect is that the period of the Earth's rotation, with respect to the Earth–Moon system, is 24 h and 50 min. This explains why the tides are approximately an hour later each day.

The equilibrium theory can be extended to include the influence of the Sun. It is possible to consider the establishment of solar 'bulges' in the Earth's oceans as well as the lunar 'bulges'. When these approximately superimpose at the full moon and the new moon, large *spring tides* occur. At the half-moon stage of the lunar cycle, the solar and lunar bulges are 90° out of phase and small *neap tides* occur.

In effect, the tides represent the terrestrial manifestation of the potential and kinetic energy fluxes present in the Earth–Moon–Sun system. These fluxes are complicated by the presence of continents and other landmasses, which modify the form and phase of the tidal wave. This results

2004 Survey of Energy Resources

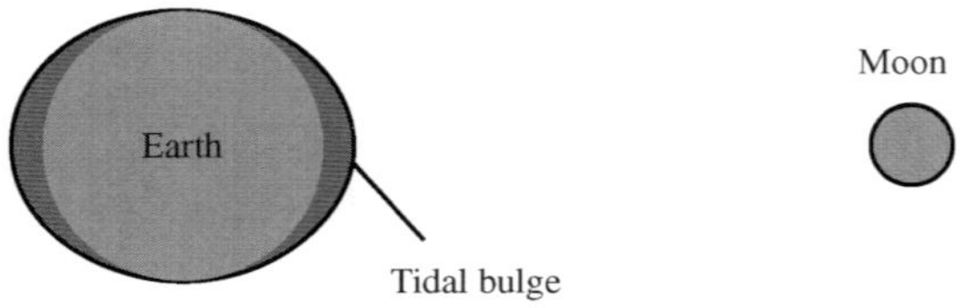

FIGURE 14.1 The tidal bulge.

in some regions of the world possessing substantially higher local fluxes than others. The Bay of Fundy in Canada and the Bristol Channel between England and Wales are two particularly noteworthy examples.

Harnessing the Energy in the Tides

There are two fundamentally different approaches to exploiting tidal energy. The first is to exploit the cyclic rise and fall of the sea level using barrages and the second is to harness local tidal currents in a manner somewhat analogous to wind power.

Tidal Barrage Methods

There are many places in the world in which local geography results in particularly large tidal ranges. Sites of particular interest include the Bay of Fundy in Canada, which has a mean tidal range of 10 m, the Severn Estuary between England and Wales, with a mean tidal range of 8 m and Northern France with a mean range of 7 m. A tidal barrage power plant has, indeed, been operating at La Rance in Brittany since 1966 (Banal and Bichon, 1981). This plant, which is capable of generating 240 MW, incorporates a road crossing of the estuary. It is currently undergoing a 10-year refurbishment.

Other operational barrage sites are at Annapolis Royal in Nova Scotia (20 MW), the Bay of Kislaya, near Murmansk (400 kW) and at Jiangxia Creek on the East China Sea (3.2 MW). Schemes have been proposed for the Bay of Fundy and for the Severn Estuary, but have never been built.

Principles of operation. Essentially the approach is always the same. An estuary or bay with a large natural tidal range is identified and then artificially enclosed with a barrier. This would typically also provide a road or rail crossing of the gap in order to maximise the economic benefit. The electrical energy is produced by allowing water to flow from one side of the barrage to the other, through low-head turbines, to generate electricity.

There are a variety of suggested modes of operation. These can be broken down initially into single basin schemes and multiple basin schemes. The simplest of these are the single basin schemes.

Single basin tidal barrage schemes. These schemes, as the name implies, require a single barrage across the estuary. There are, however, three different methods of generating electricity with a single basin. All of the options involve a combination of sluices which, when open, can allow water to flow relatively freely through the barrage and gated turbines, the gates of which can be opened to allow water to flow through the turbines to generate electricity (Fig. 14.2).

Ebb generation mode. During the flood tide, incoming water is allowed to flow freely through sluices in the barrage. At high tide, the sluices are closed and water retained behind the barrage. When the water outside the barrage has fallen sufficiently to establish a substantial head between the basin and the open water, the basin water is allowed to flow out though low-head turbines and to generate electricity.

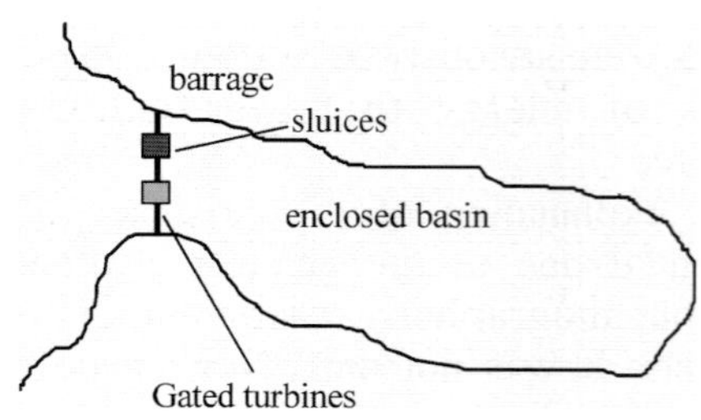

FIGURE 14.2 Hypothetical tidal barrage configuration.

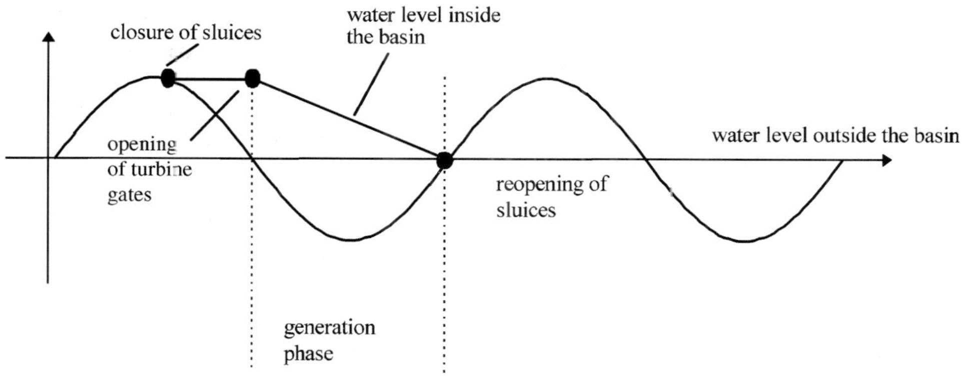

FIGURE 14.3 Water levels in an ebb generation scheme.

The system can be considered as a series of phases. These can be represented as follows to show the periods of generation associated with stages in the tidal cycle (Fig. 14.3).

Typically the water will only be allowed to flow through the turbines once the head is approximately half the tidal range. This method will generate electricity for, at most, 40% of the tidal range.

Flood generation mode. The sluices and turbine gates are kept closed during the flood tide to allow the water level to build up outside of the barrage. As with ebb generation, once a sufficient head has been established the turbine gates are opened and water can, in this case, flow into the basin generating electricity (Fig. 14.4).

This approach is generally viewed as less favourable than the ebb method, as keeping a tidal basin at low tide for extended periods could have detrimental effects on the environment and shipping. In addition, the energy produced would be less as the surface area of a basin would be larger at high tide than at low tide, which would result in rapid reductions in the head during the early stages in the generating cycle.

Two-way generation. It is possible, in principle, to generate electricity in both ebb and flood. Unfortunately computer models do not indicate that there would be a major increase in the energy production. In addition, there would be additional expenses associated in having a requirement for either two-way turbines or a double set to handle the two-way flow.

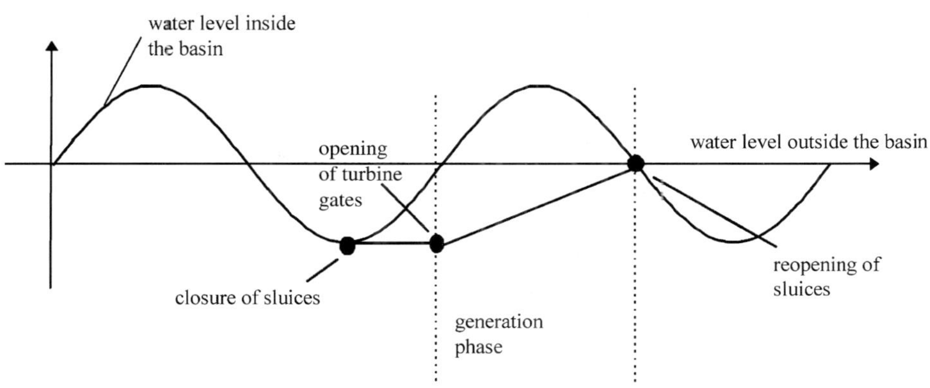

FIGURE 14.4 Water levels in a flood generation scheme.

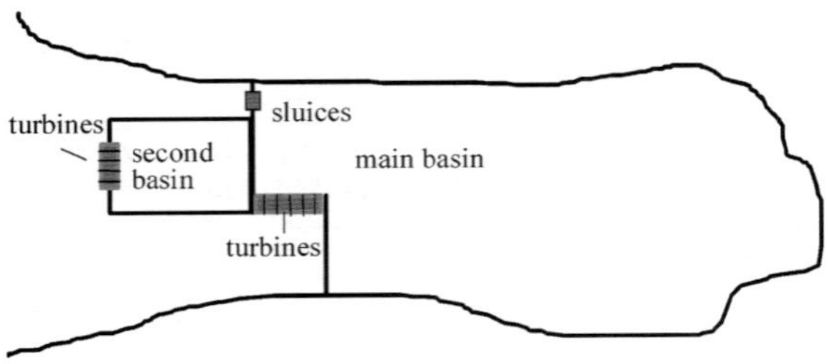

FIGURE 14.5 Hypothetical two basin system.

Advantages include, however, a reduced period with no generation and the peak power would be lower, allowing a reduction in the cost of the generators.

Double basin systems. All single basin systems suffer from the disadvantage that they only deliver energy during part of the tidal cycle and cannot adjust their delivery period to match the requirements of consumers. Double basin systems have been proposed to allow an element of storage and to give time control over power output levels (Fig. 14.5).

The main basin would behave essentially like an ebb generation single basin system. A proportion of the electricity generated during the ebb phase would be used to pump water to and from the second basin, to ensure that there would always be a generation capability.

It is anticipated that multiple basin systems are unlikely to become popular, as the efficiency of low-head turbines is likely to be too low to enable effective economic storage of energy. The overall efficiency of such low-head storage, in terms of energy out and energy in, is unlikely to exceed 30%. It is more likely that conventional pump storage systems will be utilised. The overall efficiency of these systems can exceed 70% which is, especially considering that this is a proven technology, likely to prove more financially attractive.

Possible sites for future tidal barrage developments. Worldwide there is a considerable number of sites technically suitable for development, although whether the resource can be developed economically is yet to be conclusively determined (Boyle, 1996). These include, and this is not a definitive list (Fig. 14.6):

Tidal Current Technology

Principles and history. Presently the development of tidal barrage schemes has been limited. This has been partly a result of the very large capital costs of such systems, associated with the long construction times and fear of environmental impact.

Many engineers and developers now favour, however, the use of technology which will utilise the kinetic energy in flowing tidal currents. The most thoroughly documented early attempt to prove the practicality of tidal current power was conducted in the early 1990s in the waters of Loch Linnhe in the Scottish West Highlands (www.itpower.co.uk). This scheme used a turbine held mid-water by cables, which stretched from a seabed anchor to a floating barge, as shown in Fig. 14.7.

The mid to late 1990s was primarily a time of planning and development as far as tidal current

Site	Mean tidal range (m)	Barrage length (m)	Estimated annual energy production (GWh)
Severn Estuary (UK)	7.0	17 000	12 900
Solway Firth (UK)	5.5	30 000	10 050
Bay of Fundy (Canada)	11.7	8 000	11 700
Gulf of Khambhat (India)	6.1	25 000	16 400

FIGURE 14.6 Possible sites for future tidal barrage development.

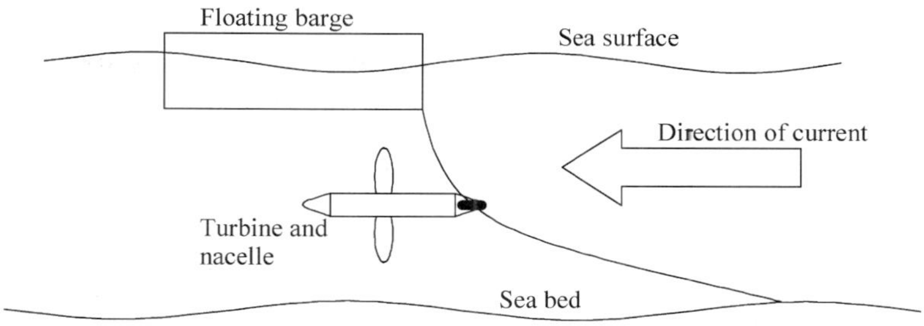

FIGURE 14.7 Schematic diagram of the Loch Linnhe System.

power was concerned, and it was not until the beginning of the 21st century that further systems became ready to test. In 2000 a large vertical axis floating device (the Enermar project www.pontediarchimede.com) was tested in the Strait of Messina between Sicily and the Italian mainland. There are now at least two major prototype systems being tested in UK waters ('SeaFlow' www.marineturbines.com and 'Stingray' www.engb.com) and one in Norway (Hammerfest Strøm www.tidevannsenergi.com). Others are known to be in development. Fig. 14.8 shows an artist's impression of the SeaFlow system, which has been installed in the Bristol Channel between England and Wales. If any of these systems are shown to be effective and to offer opportunities for commercial development, then a new tidal current industry can be expected to develop over the next decade.

FIGURE 14.8 Artist's impression of the SeaFlow system.

The physics of the conversion of energy from tidal currents is very similar, in principle, to the conversion of kinetic energy in the wind. Many of the proposed devices have, therefore, an inevitable, though superficial, resemblance to wind turbines. There is no total agreement on the form and geometry of the conversion technology itself. Wind power systems are almost entirely horizontal axis rotating turbines, as shown schematically in Fig. 14.9 and utilised in the Loch Linnhe project.

In horizontal axis designs, the rotational axis is parallel to the direction of the water flow. Many developers favour this geometry for tidal conversion. Vertical axis systems, such as that shown schematically in Fig. 14.10, in which the axis of rotation is perpendicular to the direction of current flow, have not been rejected.

It is of interest to note that Enemar used a novel Kobold vertical axis turbine.

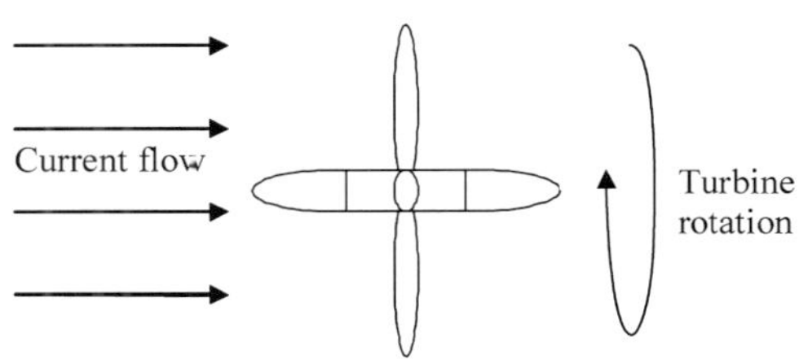

FIGURE 14.9 Schematic diagram of a horizontal axis turbine.

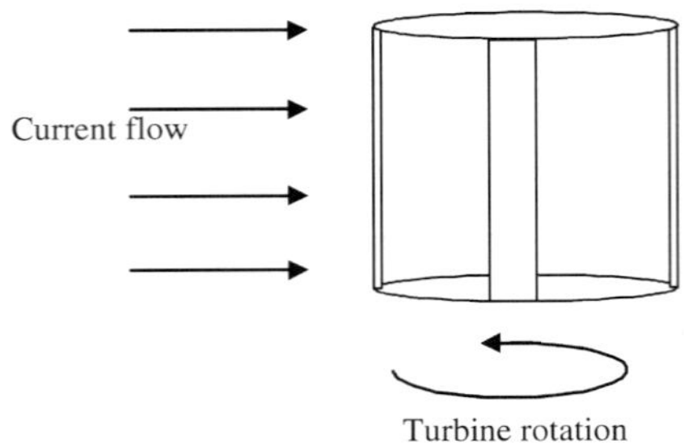

FIGURE 14.10 Schematic diagram of a vertical axis turbine.

The environmental drag forces on any tidal current energy conversion system are very large, when compared with wind turbines of the same capacity. This poses additional challenges to the designer. Designs exist for devices which are rigidly attached to the seabed or are suspended from floating barges, such as the early Loch Linnhe device. It is generally accepted that fixed systems will be most applicable to shallow water sites and moored systems for deep water. There may be exceptions to this, however.

Energy available in tidal currents. The density of seawater, at approximately 1 023–1 025 kg/m^3, is substantially greater than that of air. This, combined with the rapid currents experienced in many sites across the world, suggests that very high energy density fluxes will be encountered. Fig. 14.11 shows the anticipated relationship between current speed and incoming energy density flux.

Spring current speeds exceeding 3 m/s are encountered in many locations, which suggests that the exploitable resource will be considerable. The challenge for designers is to ensure that their system can convert energy efficiently during as much of the tidal cycle as possible, whilst being robust enough to survive the inevitable forces imparted on the system by the currents themselves and by any wave action which might be present. The difficulty of this task should not be overestimated. In practice, the efficiency of conversion of a practical system is likely to be limited to approximately 35%.

It would be wrong to consider the tidal resource purely in terms of the maximum spring current. A channel of width 1 km and depth 50 m in which the spring peak is 3 m/s and the neap peak is 1 m/s would experience a maximum energy flux approaching 700 MW. The mean flux would, however, be under 120 MW. Research is now being conducted to determine how much energy can actually be extracted from a tidal channel, without severely altering the underlying hydraulic nature of the flow and damaging the resource. Preliminary results suggest that in a simple channel 10% of the raw flux can be extracted safely. In a sea loch, however, the proportion could be much higher.

Development options for tidal currents. The environment in which tidal devices will operate is very different from that experienced by wind turbines and there are some rather difficult problems associated with installation, survivability

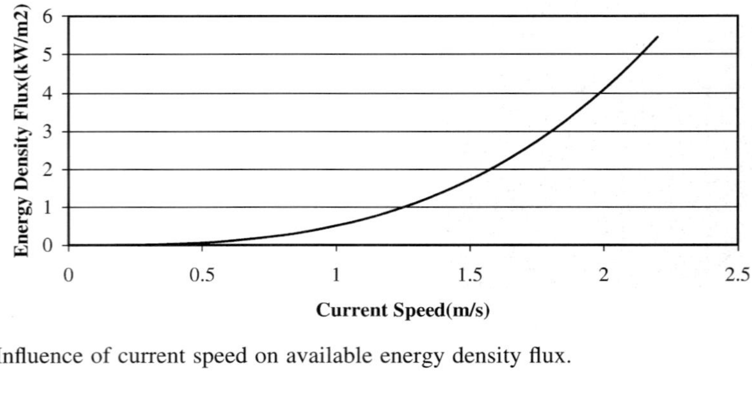

FIGURE 14.11 Influence of current speed on available energy density flux.

and maintenance, which need to be solved before true commercial exploitation can be achieved. Proposed development options often involve the use of dedicated installation and maintenance vessels, which suggests that tidal currents might only be economically developed in large sites, where major hardware can be installed, justifying the use of expensive infrastructure.

Small sites could perhaps be developed, however, using technology which can be installed and maintained using less expensive techniques. The Sea Snail, which has been developed by the Robert Gordon University, and which can be installed using a small seagoing tug, could be an option. This seabed located device is held to the seabed using variable position hydrofoils which generate substantial down force and thus reduce the need to use a large amount of ballast (Fig. 14.12).

Many industrial, commercial and public bodies have suggested that there is a high degree of synergy between the development of a tidal current generation industry and the offshore oil and gas industry. This offers the intriguing prospect of a new renewable industry developing in partnership with the petroleum industry and could, perhaps, result in accelerated development, as a result of the availability of expertise and technology, which would otherwise have to be developed from scratch.

Unlike the wind, tides are essentially predictable as they derive from the astronomic processes discussed earlier in this commentary. Wind power systems are dependent upon random atmospheric processes, which result in it being difficult to integrate large wind power developments into strategic electricity distribution networks. The predictability of the tides will make this integration much easier.

FIGURE 14.12 Launch preparations for the Sea Snail in Orkney, Scotland.

The Future of Tidal Power

The high capital costs associated with tidal barrage systems are likely to restrict development of this resource in the near future. What developments do proceed in the early 21st century will most likely be associated with road and rail crossings in order to maximise the economic benefit. In a future in which energy costs are likely to rise, assuming that low-cost nuclear fusion or other long-term alternatives do not make unexpectedly early arrivals, then tidal barrage schemes could prove to be a major provider of strategic energy in the late 21st century and beyond. Under some local conditions, small-scale barrages might also prove attractive. The technology for tidal barrage systems is already available and there is no doubt, given the experience at La Rance, that the resource is substantial and available.

In the near future it is likely that tidal current systems will continue to appear in experimental form in many places around the world. If these schemes prove successful, then the first truly commercial developments may appear in the first decade of the 21st century. Tidal current systems may not presently have the strategic potential of barrage systems but, in the short term at least, they do offer opportunities for supplying energy in rural coastal and island communities. In the longer term, massive sites such as the Pentland Firth, which has been estimated to have the potential to support a 16 GW capacity development, could become strategically important.

Ian Bryden
Robert Gordon University,
Aberdeen, Scotland

References

Banal, M., Bichon, A., 1981. Tidal Energy in France. The Rance Tidal Power Station—some results after 15 years in operation. Proceedings of the Second International Symposium on Wave and Tidal Energy, Cambridge.
Bearman, G. (Ed.), 1997. Waves, Tides and Shallow Water Processes, The Open University, UK.
Boyle, G. (Ed.), 1996. Renewable Energy, The Open University, UK.

COUNTRY NOTES

The Country Notes on Tidal Energy have been compiled by the editors, drawing upon a wide range of sources. National, international, governmental publications/web sites have all been consulted.

Canada

Embayments at the head of the Bay of Fundy between the maritime provinces of New Brunswick and Nova Scotia have some of the largest tidal ranges in the world. The most promising prospects for tidal power have centred on two sites in this region: the Cumberland Basin (an arm of Chignecto Bay) and the Minas Basin (both at the head of the Bay of Fundy). However, the only commissioned tidal power plant is located at Annapolis Royal, further down the Bay. The 20 MW plant came into operation in 1984: the barrage was primarily built to demonstrate a large-diameter rim-generator turbine. Annapolis uses the largest Straflo turbine in the world to produce more than 30 million kWh per year.

In view of the large tidal energy resource of the two basins, estimated to be 17 TWh per year, different options for energy storage and integration with the river hydro system have been explored. At present there is little prospect of any development going ahead.

China

The south-eastern coastal areas of Zhejiang, Fujian and Guangdong Provinces are considered to have substantial potential for tidal energy. China's utilisation of tidal energy with modern technologies began in 1956: several small-scale tidal plants were built for pumping irrigation water. Thereafter tidal energy began to be used for power generation. Starting in 1958, 40 small tidal plants (total capacity 12 kW) were built for the purpose of generating electricity. These were supplemented from around 1980 by much larger stations, of which the 3.2 MW Jiangxia and the 1.3 MW Xingfuyang schemes were the largest. The majority of the early plants have been decommissioned for a variety of reasons, including design faults, incorrect location, etc. Currently there are seven tidal power stations (plus one tide flood station) with a total capacity of 6 MW.

Since the end of the 1970s emphasis has been placed on optimising the operations of existing plants to improve their performance. Additionally, a feasibility study for a 10 MW level intermediate experimental tidal power station has been undertaken.

France

Relatively few tidal power plants have been constructed in the modern era. Of these, the first and largest is the 240 MW barrage on the Rance estuary in northern Brittany. The 0.8 km long dam also serves as a highway bridge linking St. Malo and Dinard. The barrage was built as a full-scale demonstration scheme between 1961 and 1966 and has now completed 37 years of successful commercial operation. Annual generation is some 640 million kWh.

Originally the barrage was designed to generate on both flood and ebb tides; however, this mode of operation proved to be only partially successful. The barrage is now operated almost exclusively on ebb tides, although two-way generation is periodically instigated at high spring tides.

In 1988 the plant became fully automated, requiring the integration of complex operational cycles imposed by variable heads, and the necessity for continuous regulation of the turbines to optimise energy conversion. A 10-year programme for refurbishing its 24 turbines was begun in 1996, on the plant's 30th anniversary.

Despite its successful operation, no further tidal energy plants are planned for France, which is now dominated by generation from nuclear stations.

India

The main potential sites for tidal power generation are the Gulf of Kutch and the Gulf of Khambhat (Cambay), both in the western state of Gujarat, and the Gangetic delta in the Sunderbans area of West Bengal, in eastern India.

The tidal ranges of the Gulf of Kutch and the Gulf of Khambhat are 5 and 6 m, respectively, the theoretical capacities 900 and 7 000 MW, respectively, and the estimated annual output approximately 1.6 and 16.4 TWh, respectively.

Following a feasibility study for a 3 MW tidal power plant at Durgaduani in the Sunderbans area, a detailed project report is now being drawn up. If the project proceeds, the West Bengal Renewable Energy Development Agency, with MNES assistance, will take it up.

Russian Federation

Design studies for tidal power development have been conducted in Russia since the 1930s. As part of this work, a small pilot plant with a capacity of 400 kW was constructed at Kislogubsk near Murmansk and commissioned in 1968. The success of this installation led to a number of design studies for much larger tidal plants at sites in the north and east of the country: Lumbov (67 MW) and Mezen Bay (15 000 MW) in the White Sea, Penzhinsk Bay (87 400 MW) and Tugur Bay (6 800 MW) in the Sea of Okhotsk. Eventually the Tugur station emerged as the only feasible major scheme. Preliminary design work began in 1972 but the timescale for further development work remains uncertain.

United Kingdom

The large tidal range along the west coasts of England and Wales provides some of the most favourable conditions in the world for the utilisation of tidal power. If all reasonably exploitable estuaries were utilised, annual generation of electricity from tidal power plants would be some 50 TWh, equivalent to about 15% of current UK electricity consumption.

Of six identified sites with mean tidal ranges of 5.2–7.0 m, feasibility studies have been completed for two large schemes: Severn estuary (8 640 MW) and Mersey estuary (700 MW) and for smaller schemes on the estuaries of the Duddon (100 MW), Wyre (64 MW), Conwy (33 MW) and Loughor (5 MW). A governmental programme on tidal energy (1978–1994) concluded that given the combination of high capital costs, lengthy construction periods and relatively low load factor (21–24%), none of these schemes was regarded as financially attractive. A future UK tidal energy programme could include construction of a small-scale scheme primarily to demonstrate the technology and its environmental effects, before progressing to very large schemes on the scale of the Severn.

Chapter 15

Wave Energy

World Energy Council
CONSEIL MONDIAL DE L'ENERGIE

COMMENTARY

Introduction

Work on wave energy began in earnest during the 1970s as a response to the emerging oil crises. There were several government-sponsored programmes throughout the world, particularly in Japan, Norway and the UK. These programmes advanced the technology considerably and their achievements were impressive. Nevertheless, the failure of these programmes to deliver economic supplies of electricity from wave energy left the technology with a credibility problem that has been hard to overcome.

Since the mid-1990s, there has been a resurgence of interest in wave energy, led mainly by small companies. Their endeavours have progressed the technology so that there are now a number of different devices that have been built or that are under construction at this moment around the world. Hence, the next few years will be very interesting for wave energy, as these full-scale prototypes provide the in-service experience required to develop a more mature technology.

These initiatives have been accompanied by government or university-funded activities in numerous countries (including Australia, China, Denmark, India, Indonesia, Iran, Japan, Korea, Mexico, Portugal, Russia and the UK) as well as developments in international organisations such as the European Union, which has funded the development of several devices as well as a Thematic Network on wave energy (Wavenet[1], 2003a), and the International Energy Agency, which has started an Implementing Agreement on ocean energy systems (IEA).

This represents a dramatic change from 10 years ago, when wave energy research was limited to a few universities. It is due, in part, to improvements in device concepts and technology transfer from the offshore oil and gas industry, which has resulted in a reduction in the expected generating cost from the high values documented in the early 1990s (Thorpe, 1992), such that some of the current prototypes (if successful) could prove competitive with other forms of renewables (e.g. PV and offshore wind-Wavenet, 2003a).

The Resource

Wave energy can be considered as a concentrated form of solar energy. Winds are generated by the differential heating of the earth and, as they pass over open bodies of water, they

[1] A name in brackets, whether of an individual or company indicates that this is listed in the references.

2004 Survey of Energy Resources

transfer some of their energy to form waves. The amount of energy transferred, and hence the size of the resulting waves, depend on the wind speed, the length of time for which the wind blows and the distance over which it blows (the 'fetch'), so that the original solar power levels of typically $\sim 100\ \mathrm{W/m^2}$ can be transformed into waves with power levels of over 1 000 kW per metre of wave crest length. Most waves are generated between 30° and 60° in latitude, where the strongest winds blow, but there is also an attractive wave climate within ±30° of the equator (owing to the regular trade winds at these latitudes) and in the high southern latitudes, because of the circumpolar storms. Waves lying within or close to the areas where they are generated, 'storm waves,' produce a complex, irregular sea. These waves will continue to travel in the direction of their formation even after the wind dies down. In deep water, waves lose energy only slowly, so they can travel out of the storm areas with minimal loss of energy as regular, smooth waves or 'swell'. These can persist at great distances from the point of origin. Therefore, coasts with exposure to the prevailing wind direction and long fetches tend to have the most energetic wave climates, for instance the western coasts of the Americas, Europe, Southern Africa and Australia/New Zealand (See Fig. 15.1).

The global wave power resource in deep water (i.e. 100 m or more) is estimated to be $\sim 10^{12}-10^{13}$ W, (Panicker, 1976). As the waves move to shallower waters they lose energy, (e.g. the UK wave power levels at 20 m water depth are about one-third of those in deep water) but detailed variation of sea-bed topography can lead to the focussing of wave energy in concentrated regions near the shoreline called 'hot spots.' Outside the tropics, storms are usually more intense and frequent during winter, which results in wave power levels being higher in that season. Therefore, wave energy provides good seasonal load-following for those regions where peak electricity demand is produced by winter heating and lighting requirements (e.g., northern Europe, western Canada and north-west USA).

The economically exploitable resource varies from 140–750 TWh/yr for current designs of devices when fully mature (Wavenet, 2003a) and could rise as high as 2000 TWh/yr (Thorpe, 1999), if the potential improvements to existing devices are realised.

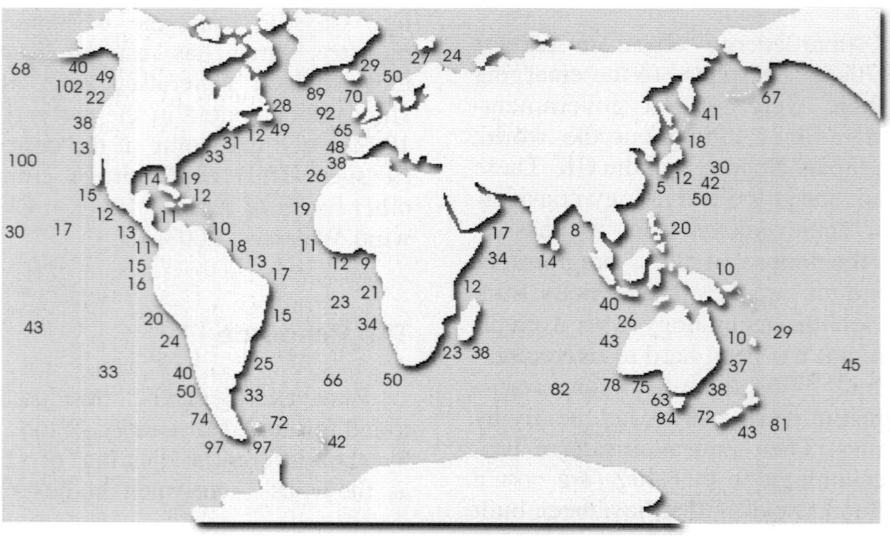

FIGURE 15.1 Average annual wave power levels as kW/m of wave front.

Types of Wave Energy Technology

There are several comprehensive reviews of wave energy (Brooke, 2003; Wavenet, 2003a; Clément et al., 2002). These show that many wave energy devices are at the R&D stage, with only a small range of devices having been tested or deployed in the oceans. Of these, the main types currently deployed or likely to see deployment in the near future are described below.

Oscillating Water Column

The OWC comprises a partially submerged structure forming an air chamber, with an underwater aperture. This encloses a volume of air, which is compressed as the incident wave makes the free surface of the water rise inside the chamber. The compressed air can escape through an aperture above the water column, which leads to a turbine and generator. As the water inside falls, the air pressure is reduced and air is drawn back through the turbine. Both conventional (i.e. unidirectional) and self-rectifying air turbines have been proposed. There have been numerous sea-bed mounted devices deployed world-wide (Fig. 15.2).

The latest developments of this technology include:

- A bottom-mounted, shoreline 500 kW scheme deployed by Wavegen in Scotland (Fig. 15.3);
- A buoyant, nearshore 500 kW OWC under construction for Australia (Fig. 15.4).

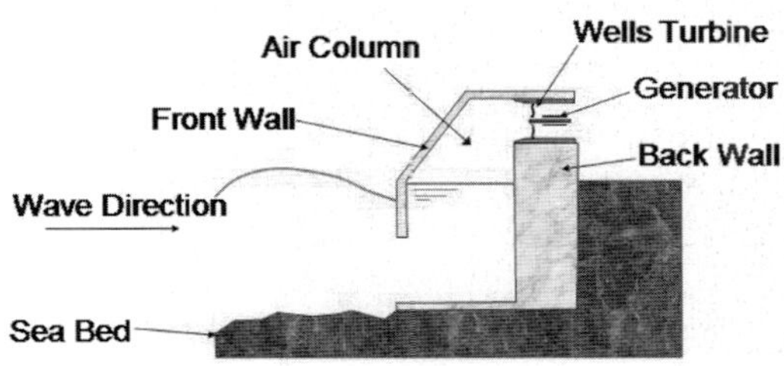

FIGURE 15.2 Oscillating Water Column (schematic).

FIGURE 15.3 OWC (Source: Wavegen).

The Pelamis

The Pelamis is a series of cylindrical hollow steel segments that float in 50 m or more water depth and which are connected to each other by hinged joints (Fig. 15.5).

The device is loosely moored and points into the waves. As waves run down the length of the device, the segments move with respect to each other and actuate hydraulic cylinders incorporated in the joints to pump oil to drive a hydraulic motor/generator via an energy-smoothing system. The device is approximately 120 m long and 3.5 m in diameter with a continuously rated power output of 0.75 MW. At the beginning of 2004, the first full-scale Pelamis converter had been completed and was awaiting deployment (OPD) (Fig. 15.6).

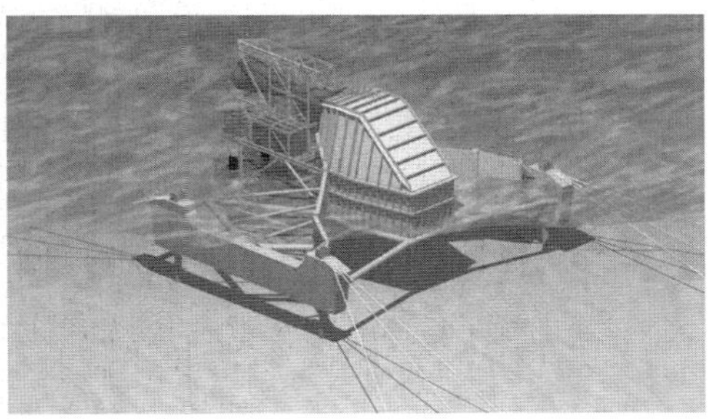

FIGURE 15.4 OWC (Source: Energetech).

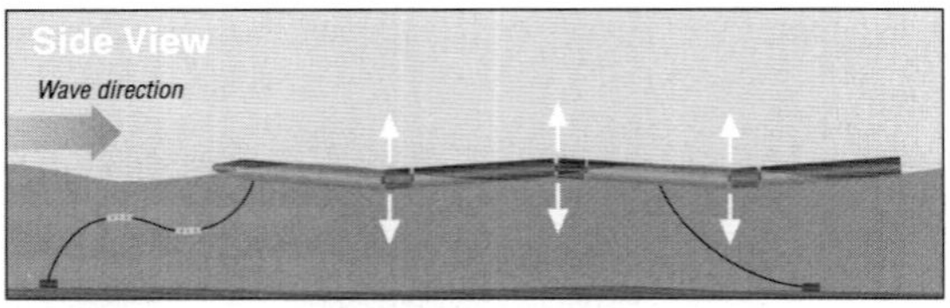

FIGURE 15.5 The Pelamis Wave Energy Converter (schematic) (Source: Ocean Power Delivery).

The Wave Dragon

This device uses a pair of curved reflectors to gather waves to overtop a central ramped reservoir, from which the water is allowed to return to the sea via low-head turbines. A quarter-scale prototype (58 m wide × 33 m long) rated at 20 kW has been deployed in a Danish inlet. The full-size device (estimated to have a generating capacity of ~24 MW) is large, with a 'span' across the reflector arms of 227 m (Fig. 15.7).

The Archimedes Wave Swing

This consists of a cylindrical, air-filled chamber (the 'Floater'), which can move vertically with respect to the cylindrical 'Basement', which is fixed to the sea-bed. The air within the 10–20 m diameter Floater ensures buoyancy. However, a wave passing over the top of the device, alternatively pressurises and depressurises the air within the Floater, changing this buoyancy. This causes the Floater to move up and down with respect to the Basement and it is this relative motion that is used to produce energy using a novel, linear electrical generator. A 2 MW Pilot scheme has been built for Portugal but it has suffered problems in being launched (A.W.S., Fig. 15.8).

The McCabe Wave Pump

This device consists of three narrow steel pontoons that are hinged together and point into the oncoming waves. The key aspect of the scheme is the damper plate attached to

FIGURE 15.6 Artist's impression of the Pelamis Wave Energy Converter (Source: Ocean Power Delivery).

FIGURE 15.7 The Wave Dragon (Source: Wave Dragon).

the central pontoon, which increases the inertia of the central pontoon (by effectively adding mass), thus ensuring that it stays relatively still. Therefore, the fore and aft pontoons move relatively to the central pontoon by pitching about the hinges. Energy is extracted from the rotation about the hinge points by linear hydraulic pumps mounted between the central and two outer pontoons near the hinges. This hydraulic power can be used to drive a motor/generator (rated at 250–500 kW) or else to pressurise sea water for desalination using a reverse osmosis plant. A full size prototype was tested in Ireland in 1996. A new commercial device has been constructed (Hydam) (Fig. 15.9).

FIGURE 15.8 The Archimedes Wave Swing (Source: A.W.S.).

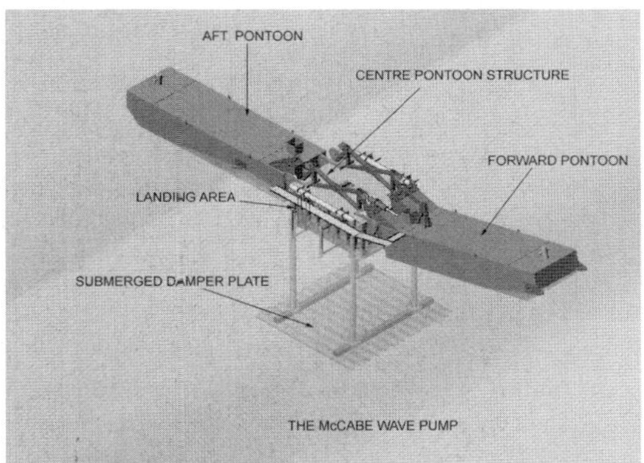

FIGURE 15.9 The McCabe Wave Pump (Source: IEA-Ocean Energy Systems Newsletter).

The PowerBuoy™

This is a development of a prototype system that was tested in the USA some time ago. It comprises a buoy that is submerged more than a metre below the water's surface. Inside, a piston-like structure moves as the buoy bobs with the rise and fall of the waves and this movement drives a generator on the ocean floor, producing electricity, which is sent to the shore by an underwater cable (Fig. 15.10).

AquaBuOY™

This device combines two technologies developed in Sweden, the IPS buoy and the Hosepump. The IPS buoy has a submerged vertical tube underneath, which is open to the sea at both ends and contains a piston attached to the hosepumps. In waves, the buoy and tube move relative to the piston, which alternatively stretches and relaxes the double action hose-pumps. This flexing alters the internal volume

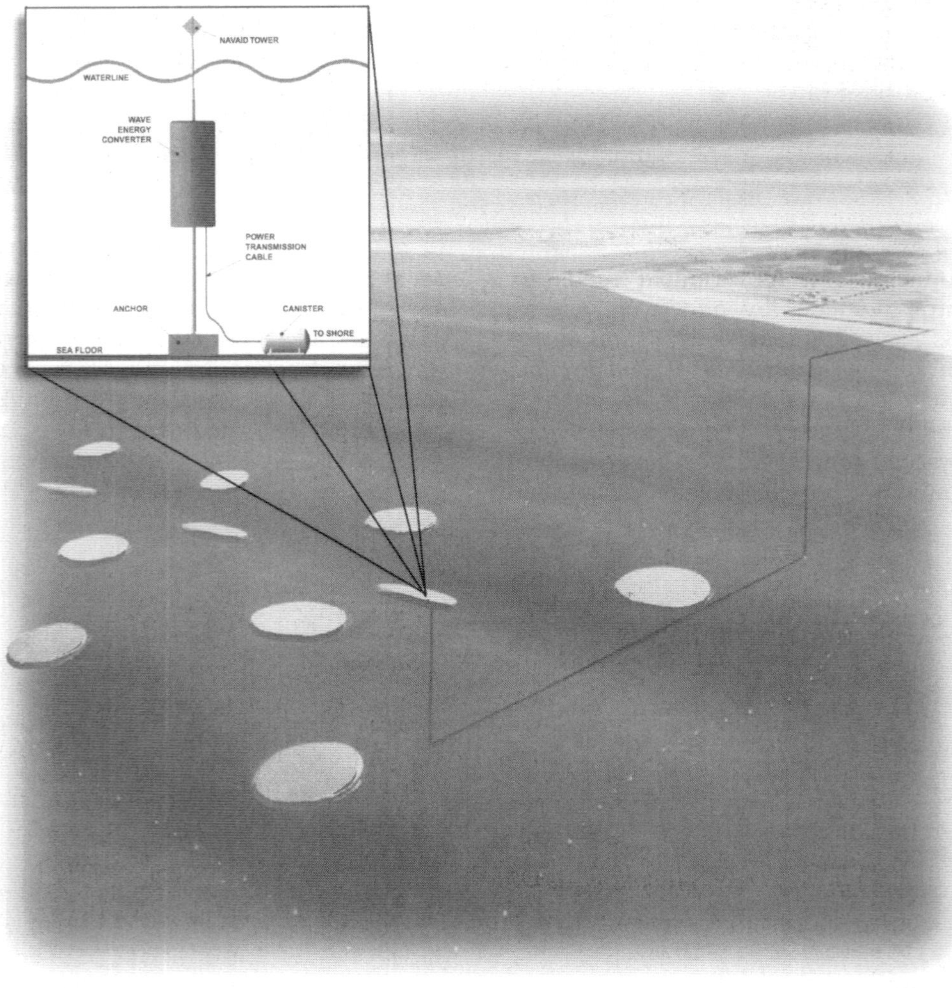

FIGURE 15.10 The PowerBuoy™ (Source: Ocean Power Technologies).

of the hosepump, sucking in sea water at the bottom and then pressurising it before feeding it to a Pelton turbine at the top of the buoy, which turns an electrical generator. This combination of two technologies appears not to have been tested at full size before and the company has plans for deploying four buoys supplying a total of 1 MW in the USA (Aqua) (Fig. 15.11).

Next Steps in Wave Energy Technology

The above devices represent those that are known to be deployed or are about to be deployed. There are other technologies that might also become operational in the near future and some of the more advanced designs have been included in the country notes that follow. Clearly, this diversity of technologies indicates that wave energy is far from mature. Nevertheless, some of these devices represent a significant improvement over earlier designs leading towards improved economics. This has led to an increase in government funds for development of this technology.

However, there are significant non-technical challenges to be faced by wave energy:

- Typically, wave energy is being developed by small companies with the total investment in each company ranging from US$ 3–10 million (one or two companies have exceeded this). This is a comparatively miniscule amount on which to research, develop and deploy a completely new technology, thereby increasing the chances of failures in early prototypes, which could lead to a loss in confidence;
- In many countries, there is a high cost associated with obtaining licences, gaining permits and carrying out environmental impact assessments, even for prototype schemes that small companies find difficult to meet. Some countries (see the UK country note) are taking steps to reduce or eliminate these high costs;
- Once deployed in free energy markets, wave energy has to compete with established renewable energy technologies that have benefited from billions of dollars of cumulative investment. Some countries (see the country note on Portugal) have addressed this issue.

The Cost of Wave Energy

There has been considerable debate about the actual costs of generating electricity from waves and the likely future cost. It is undisputed that the generating costs of prototypes are high, because all the high fixed costs associated with a wave energy scheme (permits, surveys, grid connection) are defrayed against the output of a single device. In addition, prototype devices are, by definition, immature and hence they will perform less well than follow-on schemes.

An assessment has been undertaken of the likely generating costs of arrays of mature devices located in promising wave energy sites (Wavenet, 2003b). Fig. 15.12 shows a steady reduction in predicted generating costs over the past decade, so that there are now a few devices with the potential to generate electricity at US 5–10 cents/kWh (at 8% discount rate including grid connection and all annual costs such as O&M and insurance). The costs of generation from prototypes are likely to be 2–3 times this.

Environmental Aspects

A recent study (Wavenet, 2003a) has concluded that the environmental impact of wave energy schemes is '...likely to be low, provided developers show sensitivity with appropriate site selection and planning authorities control deployment in sensitive locations'. Operational experience of the limited number of devices employed to date confirms this; the largest environmental impact having been noise from the Wells turbines employed in some OWCs, which was reduced by adding acoustic baffles.

FIGURE 15.11 The AquaBuOY™ (Source: AquaEnergy/Bengt-Olov Sjostrom).

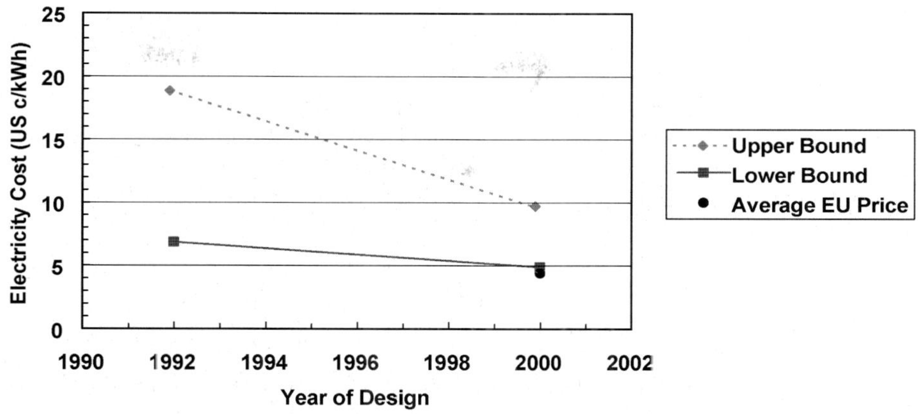

FIGURE 15.12 Wave generating costs.

Conclusions

This is a most interesting time for wave energy. There is still a plethora of ideas and designs for wave energy devices, many of which will not work or will be hopelessly uneconomic. Nevertheless, there are a few technologies that are ready to be deployed and which show considerable promise. The country notes present an overview of the various activities on wave energy around the world.

Tom Thorpe
Energetech Australia Pty,
Australia

References

Aqua www.aquaenergygroup.com/home.htm
Brooke, J., et al., 2003. Wave Energy Conversion, Elsevier Publications.
Clément, A., et al., 2002. Wave Energy in Europe: Current Status and Perspectives, Renewable and Sustainable Energy Reviews, V6. Elsevier Science, pp. 405–431.
Energetech www.energetech.com.au
Hydam www.wave-power.com
IEA www.iea-oceans.org/oes/about/index.htm
JAMSTEC www.jamstec.go.jp/jamstec/MTD/Whale/index.html
OPD www.oceanpd.com
OPT www.oceanpowertechnologies.com
Panicker, N.N., 1976. Power Resource Potential of Ocean Surface Waves, Proceeding of Wave and Salinity Gradient Workshop, Newark, Delaware, USA, pp. J1–J48.
Thorpe, T.W., 1999. A Brief Review of Wave Energy, ETSU Report R-122 for the UK Department of Trade and Industry.
Thorpe, T.W., 1992. A Review of Wave Energy. ETSU Report R-71 for the UK Department of Energy.
Wave Dragon www.wavedragon.net
Wavegen www.wavegen.com
Wavenet, 2003a. Final report of the European Thematic Network on Wave Energy available on www.wave-energy.net
Wavenet, 2003b. Wave Energy in Europe: Current Status and Perspectives, summary of the work carried out by the European Thematic Network on Wave Energy, produced by the Centre for Renewable Energy Studies, Greece, available at www.wave-energy.net

COUNTRY NOTES

The following Country Notes on Wave Energy have been compiled by Tom Thorpe and the editors. Every effort has been made to be comprehensive but omissions may have occurred. It is not an exhaustive list since information is difficult (if not impossible) to obtain on some countries. Given the large number of technologies to be included in a limited space, references have been given where the reader can find more detailed information. The Notes concentrate on those devices already deployed or intended for deployment in the near future. There are, however, many innovative ideas that could prove promising but which are at a less developed stage; a list of such developers can be found at www.poemsinc.org/links.html

Australia

Energetech Australia Pty Ltd is currently constructing a 500 kW nearshore OWC device at Port Kembla, New South Wales. This incorporates a parabolic wave collector (to compensate for the lower wave-power levels near shore) and a novel variable-pitch turbine that is predicted to have higher efficiencies that the turbines normally used in OWCs. (see www.energetech.com.au)

Canada

Canada has not traditionally been thought of as having an interest in wave energy. However, there have been several interesting developments in recent years, including Canada becoming a member of the IEA's Implementing Agreement on Ocean Energy Systems.

BC Hydro

In June 2001, BC Hydro announced the Vancouver Island Green Energy Demonstration project. Requests for proposals were sent out, which were rigorously assessed. As a result, two wave energy developers were selected: Energetech (Australia) and Ocean Power Delivery (UK). Subsequently, a change in the regulatory environment meant that BC Hydro could not proceed as planned with these projects.

Engineering Committee on Oceanic Research (ECOR)

In 1995 the international secretariat of ECOR was moved to the Ocean Engineering Research Centre at the Memorial University of Newfoundland. This organisation has overseen the production of a book on wave energy conversion, edited by J. Brooke (see References).

Wavemill

This new device has been developed and tested in lakes. It claims to be able to extract energy from the 'rise and fall of waves and wave surge'. (see www.wavemill.com)

China

Since the beginning of the 1980s, China's wave energy research has concentrated mainly on fixed and floating oscillating water column devices and also the pendulum device. By 1995, the Guangzhou Institute of Energy Conversion (GIEC) of the Chinese Academy of Sciences had successfully developed a symmetrical turbine wave-power generation device for navigation buoys (60 W). Over 650 units have been deployed in the past 13 years, mainly along the Chinese coast, with a few exported to Japan.

There are two main projects currently supported by the State Science and Technology Committee aiming to develop onshore wave-power stations:

- a shoreline OWC at Shanwei in Guangdong province consisting of a two-chambered device with a total width of 20 m, rated at 100 kW, began operating in September 1999;

- developing a 5 kW Backward Bent Duct Buoy (a floating OWC with the opening to the OWC chamber pointing towards the land) in association with Japan.

Denmark

Denmark has been a strong supporter of wave energy and is a member of the IEA's Implementing Agreement. In 1998 the Danish Energy Agency launched the Danish Wave Energy Programme 1998–2004. The Programme has a maximum of 80 million Danish Krone at its disposal for broadly supporting development projects initiated by inventors, private companies, universities etc., covering a wide range of possible converter principles. This provided developers with the facilities to have the basic research carried out on their devices.

Wave Dragon

The main development to come out of this Danish National Programme is the Wave Dragon, a wave concentrating and overtopping device. A 20 kW small-scale device is currently operating at Nissum Bredning, the Danish Wave Test Site. (see www.wavedragon.net)

Wave Plane

Another overtopping device has been developed using private funding by WavePlane International A/S and its parent company Dansk Bølgeenergi Udvikling. Model tests have been carried out in various locations throughout the world but without any electrical generation. The company also proposes to use this device to oxygenate sea water and has representatives in seven countries throughout the world. (see www.waveplane.com)

Europe

The European Commission has long funded work on wave energy, including building a pilot plant in the Azores and supporting the development of other schemes. It is a member of the IEA's Implementing Agreement. Its most recent major activity was to have representative bodies from many European countries included in the Thematic Network on Wave Energy. This carried out work in the following areas:

- *co-operation with power industry*: to induce a long-term co-operation with the power industry (e.g. electricity utilities, wind power industry) in order to involve the utilities and to learn from the experience of the wind power industry;
- *social, planning and environmental impact*: to identify the planning, legal and commercial barriers and the social benefit, energy and environmental impact arising from the expected development of wave energy schemes. To create recommendations for their development;
- *financing and economic issues*: to evaluate the financing, economics and monetary issues for developing wave energy schemes;
- *R&D on wave energy devices*: to identify the current status of wave and tidal energy device development. To determine the technical barriers to the commercial development of these devices at different time-scales. To develop a standard for assessment of existing and new devices. To develop a Strategy for Development and an Action Plan;
- *generic technologies*: to co-ordinate activities on generic technology issues concerning the utilisation of wave and tidal/current energies, so as to facilitate the exchange of experience and the transfer of knowledge. To promote knowledge and technology transfer from the offshore industry and coastal engineering. To promote studies on these issues;
- *promotion of wave energy*: to promote wave energy as a renewable source of energy, capable of a significant contribution to electricity production in Europe in the near future. This promotion will use several media in order to reach different areas of industry and society.

The results include a web site (www.wave-energy.net) that contains copies of the main reports produced by the project and useful links to other web sites. A similar follow-on activity is currently being planned.

Greece

R&D on wave energy has been carried out at the Centre for Renewable Energy Sources and various universities. During the 1990s Greece played a role in developing the European Wave Energy Atlas and has subsequently been involved with the EU DGXII MAST 3 Project: Eurowaves, a computerised tool for the evaluation of wave conditions at any European coastal location.

Daedalus

DAEDALUS Informatics, in coordination with the University of Patras, developed a new device (SEKE), which uses an array of water columns (usually built into a breakwater) to provide compressed air for power generation. Several experimental test scale models of the SEKE device have been developed. During the last 7 years their efforts have been focussed on developing a combinatorial system solution, able to harness simultaneously both wave and wind energy using compressed air (see http://195.170.12.01/daei/products/ret/general/retww1.html).

India

The Indian wave energy programme started in 1983 at the Institute of Technology (IIT) under the sponsorship of the Department of Ocean Development, Government of India. Initial research identified the OWC as most suitable for Indian conditions: a 150 kW pilot OWC was built onto the breakwater of the Vizhinjam Fisheries Harbour, near Trivandrum (Kerala), with commissioning in October 1991. The scheme operated successfully, producing data that were used for the design of a superior generator and turbine. An improved power module was installed at Vizhinjam in April 1996 that in turn led to the production of new designs for a breakwater comprised of 10 caissons with a total capacity of 1.1 MW_e.

The National Institute of Ocean Technology succeeded IIT and continues to research wave energy including the Backward Bent Duct Buoy (a variant of the OWC design).

Indonesia

In 1998, following experience gained from Norwave's demonstration plant near Bergen and a feasibility study, a Norwegian team coordinated by Indonor AS and including Norwave AS, Groener AS and Oceanor ASA won a contract to deliver a Tapchan wave-power plant. However, this appears not to have progressed and there are no plans to deploy this technology elsewhere. Research on wave energy is carried out at Indonesian universities.

Ireland

Wave energy research has been undertaken in Ireland since 1980, much of the work being conducted at University College Cork although other universities (such as Limerick) are now playing an increasing role. In addition to testing various devices, the College has also coordinated the European Wave Energy Research Programme and has collaborated in the development of the European Wave Energy Atlas and mapping the wave energy resource for Ireland. In 2003, the Marine Institute and Sustainable Energy Ireland completed a consultation study on a strategy for exploiting wave energy in Ireland (see www.irish-energy.ie/uploads/documents/upload/publications/wave.pdf). Ireland is a member of the IEA's Implementing Agreement.

McCabe Wave Pump

In 1996, Hydam Technology deployed a 40 m long prototype McCabe Wave Pump

(MWP) off the Irish coast. A commercial demonstration scheme was launched at Kilbaha, County Clare in 2003, which will undergo testing prior to being installed on location. (see www.wave-power.com)

Wavebob

A floating, self-reacting hydraulic device has been developed by Wavebob Ltd, in conjunction with researchers in other countries. Little information is publicly available on this system but it is thought to be ready for deployment in 2005.

Japan

Extensive research has been undertaken in Japan, which is a member of the IEA Implementing Agreement. Particular emphasis has been placed on the development of air turbines and on the construction and deployment of prototype devices (primarily OWCs), with numerous schemes having been built:

- a 40 kW OWC was deployed in 1983 on the shoreline structure at Sanze for research purposes. It has since been decommissioned;
- a five-chambered 60 kW OWC was built as part of the harbour wall at Sakata Port in 1989;
- ten OWCs installed in front of an existing breakwater at Kujukuri beach, Chiba Prefecture were operational between 1988 and 1997. The air emitted from each OWC was manifolded into a pressurised reservoir and used to drive a 30 kW turbine;
- a 130 kW OWC was mounted in a breakwater in Fukushima Prefecture in 1996. This uses rectifying valves to control the flow of air to and from the turbine;
- a floating OWC known as the Backward Bent Duct Buoy was deployed in Japan in 1987. This continues to be developed in co-operation with institutes in China and Ireland;
- the Pendulor wave energy device has been under investigation for over 15 years by the Muroran Institute of Technology. Wave action causes pendulum oscillations of a plate ('pendulor') at the entrance to a box, this movement being used in conjunction with a hydraulic power take-off to generate electricity;
- a 50 m long, 30 m wide, 12 m deep prototype floating OWC (the 'Mighty Whale') was deployed in 1998 by the Japan Marine Science and Technology Centre (JAMSTEC). It is the world's largest floating OWC (110 kW) and continues to operate outside the mouth of Gokasho Bay, Mie Prefecture. (see www.jamstec.go.jp/jamstec/MTD/Whale/index.html).

Maldives

The Government of the Maldives has announced that it intends to introduce wave energy to the islands. Sea Power International of Sweden has signed a letter of intent with the Government to supply a floating wave-power vessel.

Netherlands

The Archimedes Wave Swing has been constructed but difficulties were encountered in launching so that it currently sits in the harbour of Porto, Portugal, waiting for more funding. (see www.waveswing.com)

Norway

Research into wave energy has been centred on the Norwegian University of Science and Technology (NTNU), Trondheim for the past 25 years. Two commercial schemes (a 350 kW_e Tapchan and a 500 kW_e OWC) operated successfully for a prolonged period during the 1980s. Both schemes have ceased to function and subsequently NTNU has conducted extensive theoretical research into optimum control and phase control of wave-energy converters.

Since 1994 NTNU has collaborated with Brødrene Langset AS to develop the Controlled Wave-Energy Converter. In 1998 ConWEC AS was formed to undertake further technical development, demonstration and global marketing.

Oceanor-Oceanographic Company of Norway ASA has played a leading role in the development of Eurowaves, a computerised tool for the evaluation of wave conditions at any European coastal location.

Portugal

Since 1978 Portugal has played a significant role in wave energy R&D. This work has been undertaken at the Instituto Superior Técnico (IST) of the Technical University of Lisbon and the National Institute of Engineering and Industrial Technology (INETI) of the Portuguese Ministry of Economy. Most of the research on wave energy conversion has been devoted to OWCs and associated turbines. Early work concentrated on theoretical and experimental studies of the device hydrodynamics and the behaviour of Wells turbines (including monoplane and biplane rotors, as well as contra-rotating and variable-pitch designs). This included the building of a pilot 400 kW OWC plant on the island of Pico in the Azores, which was completed in 2000 with funding from the European Commission.

In addition to resource assessment studies on a national level, INETI co-ordinated two projects for the European Union:

- development of a common methodology for resource evaluation and characterisation, which led to:
- production of the European Wave Energy Atlas for the deep water resource.

Portugal is well placed to take a lead in wave energy:

- it is a member of the IEA Implementing Agreement;
- the Portuguese Government is attracting inward investment with enhanced prices paid for electricity from wave energy devices (approximately €0.22/kWh);
- in 2003 the Wave Energy Centre was set up with the objective of providing dissemination, promotion and support to the implementation of wave energy technology and commercialisation of devices. The Centre has a number of ongoing projects (see www.wave-energy-centre.org).

The first device attracted to Portugal was the Archimedes Wave Swing.

Spain

Little indigenous work on wave energy has been undertaken in Spain. Energetech (Australia) are in negotiations to install their OWC scheme in the port of Bilbao.

Sri Lanka

A 150 kW demonstration OWC has been installed in Sri Lanka, funded by the Ministry of Science and Technology.

Sweden

Interproject Service AB developed the concept of combining the IPS buoy and the Hose-Pump converter. This system is now being exploited by the Aqua Energy Group in the USA.

Sea Power International

This company has developed a floating overtopping device for mooring in deep water (see www.seapower.se/indexeng.html). It had won an opportunity to install a device in Shetland as part of the Scottish Renewables Order but little progress appears to have been made.

United Kingdom

At one time the UK had one of the largest government-sponsored R&D programmes on wave energy, covering a wide range of devices. This was greatly reduced in the early 1980s but research continued at several universities, in particular at Edinburgh and Queen's University, Belfast. The profile of wave energy has recently seen a resurgence, with the UK Government and other bodies supporting several initiatives.

Some indication of the change in fortunes for wave energy in the UK can be gathered from the following activities:

- a marine energy test centre has been established in the Orkney Islands, providing subsea cables, a monitoring station and other facilities for wave devices that operate in 50 m water depth. Its aim is to stimulate and accelerate the development of marine power devices (see www.emec.org.uk);
- the Carbon Trust has issued a Marine Energy Challenge, whereby device teams are assisted by working with engineering companies who can help them through a cost engineering exercise that will produce information relating to the technical viability and economics of their device. (see www.thecarbontrust.co.uk);
- the Supergen Initiative, launched in 2001 and formally inaugurated in November 2003, has set aside £2.3 million of funding in wave and tidal power research and development in universities (see www.see.ed.ac.uk/research/IES/supergen);
- Regen SW (the renewable energy agency for the South West of England) is proposing a strategic environmental assessment of offshore renewable energy developments, to facilitate the deployment of large-scale schemes in that region. They have also commissioned an initial review of the region's wave resources which will map wave and tidal resources and identify areas with good renewable energy potential. (see www.regensw.co.uk).

Many different devices at various stages of maturity continue to be developed in the UK under the above initiatives. This commentary only allows room for the leading developers to be discussed.

Ocean Power Delivery

OPD announced in February 2004 that it had completed the first full-scale Pelamis Converter. It is planned to install the device at the marine energy test centre in Orkney later in 2004. OPD have also linked up with a wind energy developer with a view to installing a wave energy scheme in the South West of England. (see www.oceanpd.com);

ORECon

The ORECon concept is an floating OWC employing a multiple oscillating water column configuration rather than the single chamber used by other OWC devices. By combining multiple columns within the collector component, the device can be tuned to resonate at multiple rather than single frequencies to capture energy over a much broader waveband. Efficiency has also been extended by use of a self-rectifying impulse turbine in place of the more common Wells turbine. A device has been tested in the sea but further R&D is being undertaken;

Wavegen

Wavegen's pioneering shoreline OWC continues to function on the island of Islay. Its output is fed into the local grid but the plant also serves as a test bed for new technology. Wavegen is developing its OWC concept for deeper waters. It has also entered into an agreement to investigate installing its OWC technology in a tunnel built into a cliff in the Faeroe Islands. (see www.wavegen.com).

United States of America

Many devices have been invented in the USA but most remain at the R&D stage. Therefore any overview of the activities in this country has to be selective and only those companies with clear plans for deployment in the near future have been included.

Aqua Energy Group

AEG are developing plans for the installation of a 1 MW scheme comprising four AquaBuOYs in Makah Bay, Clallam County, Washington State.
(see www.aquaenergygroup.com);

Energetech

Energetech is developing a scheme together with the Connecticut Clean Energy Fund to deploy an OWC near Point Judith, Rhode Island;

Ocean Power Technologies

It was announced in March 2004 that OPT had signed a Memorandum of Understanding to deploy a pilot project off the north coast of Spain. The project will begin with a 1.25 MW power plant comprising a number of PowerBuoys. If the project proves to be successful the power plant could be expanded to 100 MW.

Additionally, OPT has been actively working with the US Navy on the installation of a 1 MW power station in Hawaii. The initial phases of the project have commenced and in March 2004 it was expected that the first system would be installed shortly thereafter.
(see www.oceanpowertechnologies.com).

Chapter 16

Ocean Thermal Energy Conversion

World Energy Council
CONSEIL MONDIAL DE L'ENERGIE

COMMENTARY

Ocean Thermal Energy Conversion (OTEC) is a means of converting into useful energy the temperature difference between surface water of the oceans in tropical and sub-tropical areas, and water at a depth of approximately 1 000 m which comes from the polar regions. Fig. 16.1 shows the temperature differences in various parts of the ocean, and for OTEC a temperature difference of 20 °C is adequate, which embraces very large ocean areas, and favours islands (see Gauthier & Lennard, 2001) and many developing countries.

The continuing increase in demand from this sector of the world (as indicated by World Energy Council figures) provides a major potential market. Specifically the percentage of 'new' energies will grow—from a near-zero figure at the end of the 20th century to 6% by the year 2020, which translates into 'new' energies of some 12 000 MW a year averaged over the period from 2000 to 2020. The capital cost of OTEC plants, because of their low efficiency, is of the order of US$ 5 000–10 000/kW, some 10 times the capital cost of conventional power systems. The funding of 'new' energies therefore might add up to a total sum each year in the region of US$ 60–120 billion: by any standards this is very substantial business, and for the construction, operational and financing sectors, an activity of very considerable interest. But this business will only develop if it is economically attractive to the utilities that will invest in and operate it—and this situation is now rapidly approaching.

Whilst the thermal resource is relevant, particularly to many developing countries, there is a multitude of other factors to be considered before it can be said that a particular country or location is suitable for an OTEC installation. These include: distance from shore to the thermal resource; depth of the ocean bed; depth of the resource; size of the thermal resource within the exclusive economic zone (EEZ); replenishment capability for both warm and cold water; currents; waves; hurricanes; seabed conditions for anchoring; seabed conditions for power cables of floating plants; present installed power, and source; installed power per head; annual consumption; annual consumption per head; present cost per unit—including any subsidy; local oil or coal production; scope for other renewables; aquaculture potential; potable water potential; environmental impact—to name but a few.

2004 Survey of Energy Resources

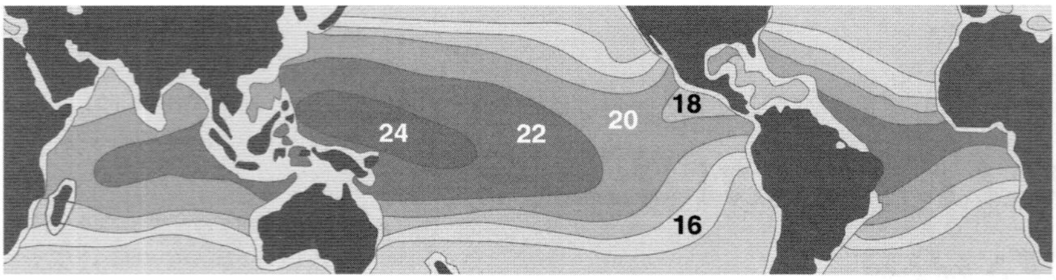

FIGURE 16.1 The area available for OTEC and the temperature difference. (Source: Xenesys Inc).

Types of OTEC Plant

Depending on the location of their cold and warm water supplies, OTEC plants can be land based, floating, or—as a longer term development—grazing. Floating plants have the advantage that the cold water pipe is shorter, reaching directly down to the cold resource, but the power generated has to be brought ashore, and moorings are likely to be in water depths of, typically, 2 000 m. The development of high voltage DC transmission offers substantial advantage to floating OTEC, and the increasing depths for offshore oil and gas production over the last decade mean that mooring is no longer the problem which it once was—but still a significant cost item for floating OTEC. Land-based plants have the advantage of no power transmission cable to shore, and no mooring costs. However, the cold water pipe has to cross the surf zone and then follow the seabed until the depth reaches approximately 1 000 m—resulting in a much longer pipe which has therefore greater friction losses, and greater warming of the cold water before it reaches the heat exchanger, both resulting in lower efficiency.

The working cycle of an OTEC plant may be closed or open, the choice depending on circumstances—Fig. 16.2 illustrates the closed version; Fig. 16.3, the open version. All these variants clearly develop their power in the tropical and sub-tropical zones shown in Fig. 16.1 for the benefit of countries in those parts of the world, but a longer term development—a grazing plant—allows OTEC energy use in highly developed economies which lie in the world's temperate zones. In this case the OTEC plant is free to drift in ocean areas with a high temperature difference, the power being used to split seawater into liquid hydrogen and liquid oxygen. The hydrogen, and in some cases where it is economic the oxygen too, is offloaded to shuttle tankers which take the product to energy-hungry countries. Also, the hydrogen may be an intermediate product, being used in turn to produce ammonia. At present, use of ammonia fertilisers is determined in part by production capacity from natural gas; the use of such fertilisers in the developing world—much of it in the tropical and sub-tropical zones where OTEC processes are available—could make a major contribution to world food production.

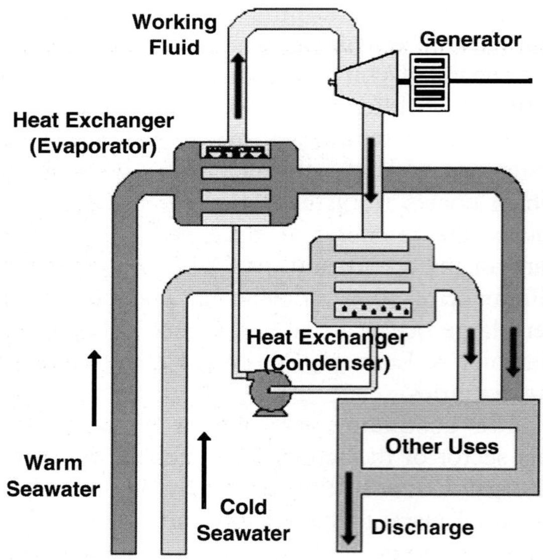

FIGURE 16.2 Closed cycle OTEC. (Source: Australian Sustainable Energy Centre, Murdoch University [image adapted from National Energy Laboratory Hawaii]).

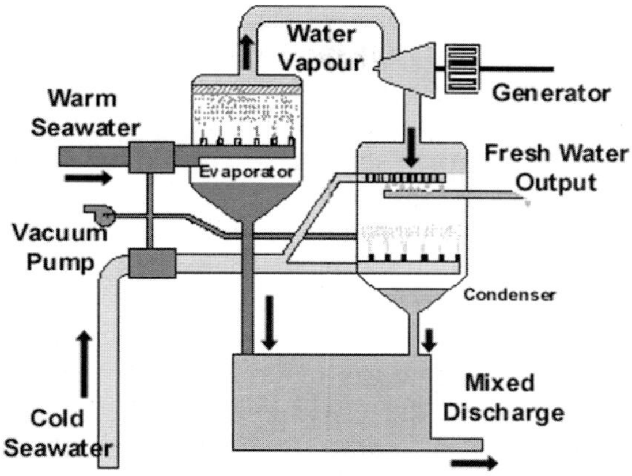

FIGURE 16.3 Open cycle OTEC. (Source: Australian Sustainable Energy Centre, Murdoch University [image adapted from National Energy Laboratory Hawaii]).

An especial benefit of OTEC is that, unlike most renewable energies, it is base-load—the thermal resource of the ocean ensures that the power source is available day and night, and with only modest variation from summer to winter. It is environmentally benign, and some floating OTEC plants would actually result in net CO_2 absorption. And a further unique feature of OTEC is the additional products which can readily be derived—food (aquaculture and agriculture); pharmaceuticals; potable water; air conditioning; etc. (see Fig. 16.4). Many of these arise from the pathogen-free, nutrient-rich, deep cold water. OTEC is therefore the basis for a whole family of deep ocean water applications (DOWA), which can additionally benefit the cost of generated electricity. Potable water production alone can reduce electrical generating costs by up to one-third, and is itself in very considerable demand in most areas where OTEC can operate.

In order to incorporate all these variables into an economic model it is necessary to assess:

- the objectives of each application;
- the state of the art;
- other fields of application for the technology;
- opportunities for further development.

In the 1990s the European Commission assembled a group experienced in OTEC and DOWA research which addressed these topics.

Economics and Finance

Although these additional products offer significant potential improvements to the economy of OTEC, a contributory reason for the lack of commercialisation of OTEC/DOWA to date is that the economic benefits of these products have generally still not been integrated

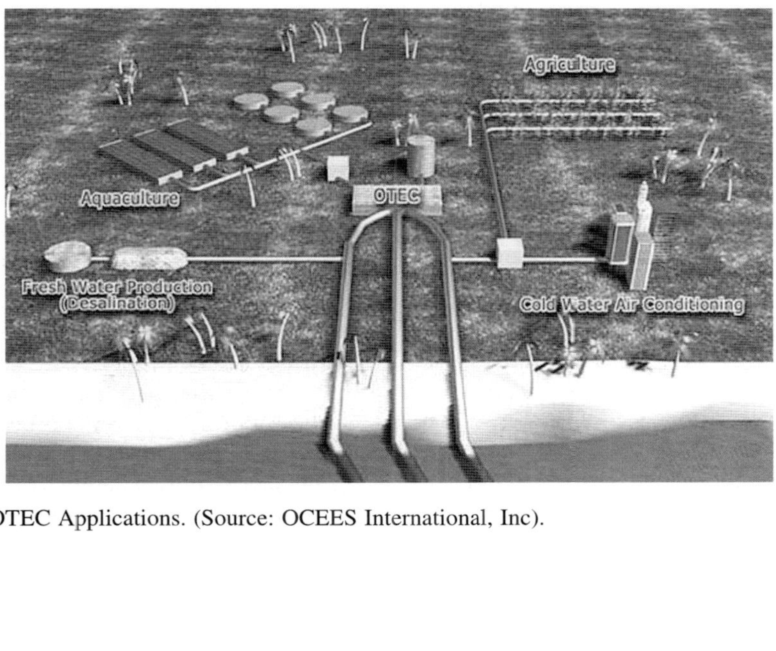

FIGURE 16.4 OTEC Applications. (Source: OCEES International, Inc).

into the scenarios of development. It is difficult at present to measure these benefits accurately, and only the potable water production benefit has been quantified. The relevance of environmental impact was given a considerable boost by the Rio and Kyoto summits, and follow-up actions have included a much greater emphasis on this aspect by a number of energy companies, including the impact of a possible Carbon Tax for fossil fuels, but to date this has not been applied. If the tax is brought into use—as may well be the case in the first decade of the 21st century—then all renewables, including OTEC, will benefit further in terms of competitiveness with hydrocarbons. However, calculations for generating costs already take increasing account of other 'downstream environmental factors'—for example the costs associated with CO_2 emissions. With such criteria included, OTEC/DOWA is becoming an increasingly attractive option.

Quite apart from this aspect, technological improvements—such as the much smaller heat exchangers now required—have contributed to significantly reduced capital expenditure. On top of these two factors, the worldwide trend to whole-life costing benefits all renewables when compared with those energy systems which rely on conventional fuels (and their associated costs), since the fuel for OTEC is totally free. Even when the higher initial maintenance costs of early OTEC/DOWA plants are taken into account, net benefits remain. As a result, when compared with traditional fuels the economic position of OTEC/DOWA is now rapidly approaching equality, and work in Hawaii at the Pacific International Center for High Technology Research (PICHTR) has contributed to realistic comparisons (see Vega, 1994), as well as component development.

Nations which previously might not have contemplated OTEC/DOWA activities have been given legal title over waters throughout the 200 nautical mile EEZ associated with the UN Convention on the Law of the Sea (UNCLOS). Prior to that, no investor—private or public—would seriously contemplate funding a new form of capital plant in such seas and oceans, but since UNCLOS a number of nations have worked steadily to prepare overall ocean policies and recent years have seen a number of these introduced—for example in Australia.

Despite the existence of EEZs, the low first costs of many 'traditional' energy resources in the recent past had not encouraged venture capital investment in OTEC/DOWA, but the currently higher costs of oil, plus the growing recognition of the environmental effects (and their costs) of some traditional fuels, are changing the economics of these in relation to OTEC/DOWA and other renewables.

It is *all* these factors which now place OTEC/DOWA within realistic reach of full economic commercialisation early in the 21st century. But, whilst a number of the components for an OTEC/DOWA plant are either available, or nearly so, the inherent simplicity of a number of key elements of these plants will still require refinement into an effective system, and this will need further RD&D investment. Before OTEC/DOWA can be realised, this RD&D must be completed to show clearly to potential investors, via a demonstration-scale plant, that the integrated system operates effectively, efficiently, economically, and safely.

A Typical OTEC Design

To put this into perspective, consider a specific design for an OTEC plant. The example described is a 10 MW closed-cycle floating OTEC plant, for application in a specific Caribbean or South Pacific island site. It was initially designed in the mid-1980s and has been progressively updated. Landed costs for fuel oil in these islands can be 75% higher than in continental locations—US$ 35 a barrel rather than the US$ 20 a barrel which has typified continental prices over the last decade.

Power generation is provided by two of the three 5 MW power 'pods'. The concept recognises that, as a 'demonstrator plant', reliability will be lower than for a production plant, and the third power pod is included both for development work and as a standby for times when a production pod is out of use for either regular service or unscheduled outage. The two sites chosen have the cold deep resource close to shore—in the Caribbean the 1 000 m depth being no more than 2.5 km from shore, and the minimum measured temperature difference between the surface and that depth being 21 °C, increasing to 23 °C at the warmest time of year. The 21 °C difference is used as the basis for calculation, which results in an overall efficiency of 2.7%. This compares with an efficiency value for diesel fuel power plants of 25–35%, and values at the upper end of that range for a modern fossil fuel power station.

Specific costs of individual components were calculated, and used as the basis for total capital costs, and then derived generating costs, the latter incorporating all operating, maintenance and insurance costs in addition. Contingencies were assessed, with the cold water pipe and moorings having the lowest confidence levels.

Total estimated cost for the plant, in 2 000 dollars, incorporating the target costs for components as a basis, is US$ 97 million which, depending on the contingencies, could increase by as much as 25% or decrease by up to 13%.

A discount rate of 5% was used, on the basis that this demonstrator plant was akin to a public sector project. Although the design life of the plant was 25 years, payback was taken as 10 years—a stringent assumption, with interest charged at 11%, which with present lower levels of interest charges worldwide, may also be unduly harsh. Annual inflation rates were assumed at 5% and again this is pessimistic in the present industrial climate.

Availability of the plant was assessed at 90%. This would be high for a normal demonstrator, but here the third pod is available as standby. The resulting calculated generating cost was 18 cents/kWh, with no allowance for potable water production since, as a floating plant, desalination can only be provided as a by-product of electrical generation. However, if the price of water is high enough, a financial credit will be obtained. Using the PICHTR calculations (see Vega, 1994) as a basis, the costs for this 10 MW sized plant would fall from 18 cents/kWh by approximately 4 and 7 cents/kWh, respectively, to 14 and 11 cents/kWh, corresponding to potable water credits of 40 and 80 cents/m^3. Since potable water in Pacific islands can cost from 40 cents/m^3 up to US$ 1.60/m^3, the generating cost of 11 cents/kWh—corresponding to a water credit of 80 cents/m^3—is considered realistic.

Other potential by-products, described earlier, are ignored because the quantities needed are small when compared with those available from the OTEC/DOWA plant, and initially will have only a small influence on the overall economy, although the human benefits of these by-products to a population may well be considerable. In the present calculations no benefit is claimed for these by-products in terms of reduced generating costs for electricity from the OTEC plant.

The remaining economic item to consider is 'environmental benefit'—or put the other way, the proposed 'Carbon Tax'. Such a tax would clearly benefit renewable energy systems, including OTEC. The proposed levels of such a tax have varied considerably, from as little as US$ 3/barrel to as much as US$ 13/barrel, which would result in a likely 'effective benefit' further to decrease OTEC/DOWA generating costs by between US 0.5 and 3 cents/kWh.

All these calculations have been for a demonstration plant. On the assumption that, without any benefits of major re-design, operating experience will refine detail design, manufacture and operation, the overall improvement in the system for the eighth production 10 MW floating plant is calculated to result in a significant 30% reduction in electrical generating cost; that is to US 12.6 cents/kWh for the basic OTEC plant, and to 9.8 and 7.7 cents/kWh, respectively, with water revenues at the levels of 40 and 80 cents/m^3.

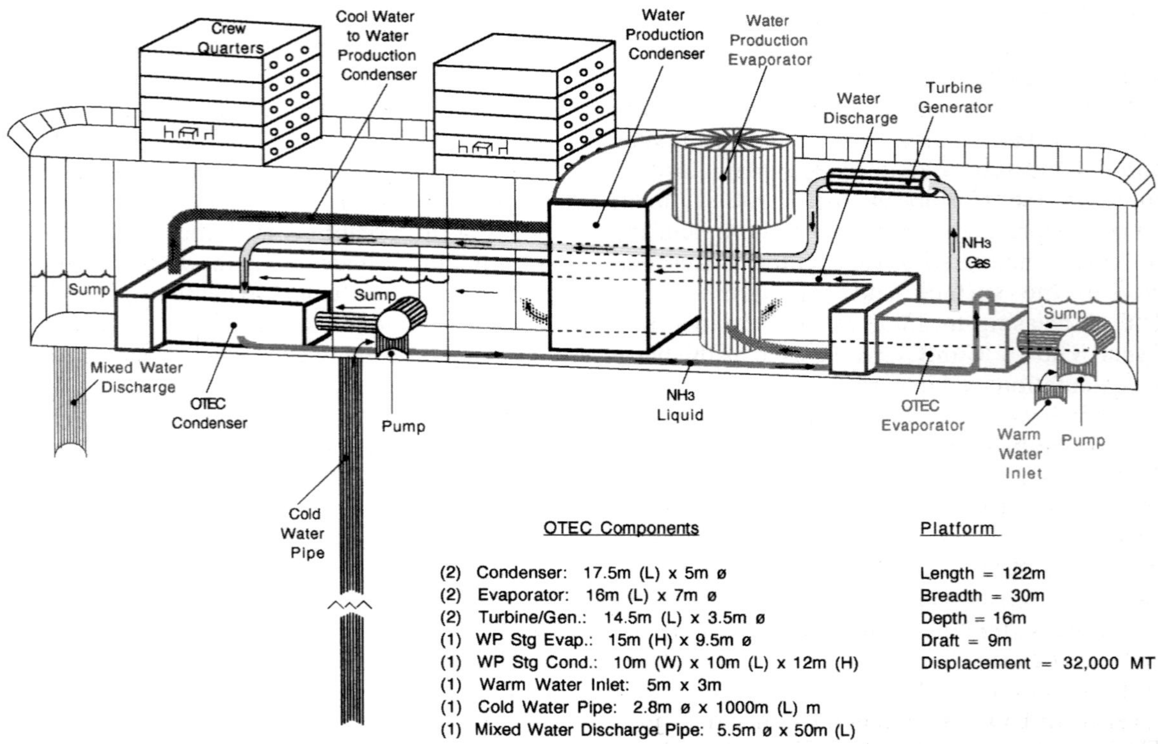

FIGURE 16.5 Schematic of 5 MW OTEC pre-commercial plant. (Source: Luis A. Vega, Project Director, PICHTR, Hawaii).

Whilst these generating costs are becoming interesting in comparison with conventional sources for electrical power generation, the OTEC plant must also be attractive to the utility that is to operate it. For the 10 MW plant described here, the rates of return are 20.4% (nominal) and 14.7% (real), which are reasonably attractive in terms of accepted commercial practice.

For this demonstration plant then, the prospects for both plant operator and the consumer of electricity are looking more competitive and more interesting than 10 years ago. On a simple costing basis OTEC is becoming attractive, with its DOWA and environmental benefits as a bonus, over and above the base economic case.

Fig. 16.5 provides a schematic illustration of a 5 MW pre-commercial OTEC plant, similar in general layout to the concept described above.

The Need for a Demonstrator Plant, Current Practice and the Market

Until a representative-scale demonstrator plant is built and successfully operated, conventional capital funds are unlikely to be available. Whilst, therefore, the establishment of renewable energy subsidiaries of energy companies is important, there is no doubt that the principal hurdle remaining for OTEC/DOWA is not economic or technical, but the convincing of funding agencies—such as the World Bank or the European Development Bank—that these techno-economic values are sufficiently soundly based for the funding of a demonstrator. A further potential source of funding is possible through the Lomé and Cotonou Agreements between the European Union and the 79 African-Caribbean-Pacific (ACP) States, many of which are prime candidates to use OTEC power—including Palau, referred to below.

Specific national activities are referred to in the Country Notes which follow, but special mention should be made here of Taiwan, China which, in addition to extensive studies at a number of candidate sites for OTEC and aquaculture on its east coast, still hosts The International OTEC/DOWA Association (IOA) (see IOA Newsletter), which has a worldwide membership of organisations and individuals dealing with all aspects of OTEC/DOWA.

Current practical OTEC activity is taking place in India and Palau, both in cooperation with Japanese technology. A further general indication of interest in DOWA, rather than OTEC alone, is provided in Japan where the industrial OTEC Association was succeeded by the Japan Association of Deep Ocean Water Applications.

In Europe both the European Commission and the industrially-based Maritime Industries Forum examined OTEC opportunities with relevance to DOWA in general rather than just OTEC, whilst in 1997 the UK published its Fore*sight* document for the marine sector (see Fore*sight*: Progress Through Partnership, 1997), looking 5–20 years ahead, with both OTEC and DOWA included in the energy section of the paper (see Fore*sight*: Report of the Working Group on Offshore Energies, 1998). It is significant that the emphasis in the recommendations from all three European groupings has, again, been on the funding and construction of a demonstrator in the 5–10 MW range.

Island opportunities have already been mentioned, and in addition to Japan and Taiwan, the European work has stressed these as the best prospects, and it is noteworthy that both Japanese and British evaluations continue to identify Fijian prime sites, one each on the two largest islands of that archipelago.

The worldwide market for renewables has been estimated (see Energy: The Next Fifty Years, 1999) for the timescales from 1990 to 2020 and 2050, with three scenarios, and all show significant growth. Within those total renewable figures, opportunities exist for the construction of a significant amount of OTEC capacity, even though OTEC may account for only a small percentage of global electricity generating capacity for some years. Estimates have been made by French, Japanese, British and American workers in the field, suggesting worldwide installed power for up to 1 000 OTEC plants by the year 2010, of which 50% would be no larger than 10 MW, and less than 10% would be of 100 MW size. On longer timescales the demand for OTEC in the Pacific/Asia region has been estimated at 20 GW in 2020 and 100 GW in 2050 (see Energy: The Next Fifty Years, 1999). It has to be said that some of these numbers seem optimistic, with realisation depending on the successful operation of the demonstrator at an early date.

Therefore, the key breakthrough now required for OTEC/DOWA is no longer technological or economic, but the establishment of confidence levels in funding agencies to enable building of a representative-scale demonstration plant. Given that demonstrator, the early production plants will be installed predominantly in island locations where conventional fuel is expensive, or unavailable in sufficient quantity, and where environmental impact is a high priority. Both simple OTEC and combined OTEC/DOWA plants will feature, depending on the particular requirements of each nation state.

It can now be realistically claimed that the economic commercialisation of OTEC/DOWA is potentially close—what is needed above all else is funding for the demonstrator plant to be built in the early years of this 21st century.

Don Lennard
Ocean Thermal Energy Conversion Systems Ltd, United Kingdom

References

Economics of Ocean Thermal Energy Conversion, Vega, IOA Conference, Brighton, UK, 1994.
Energy: The Next Fifty Years. Organisation for Economic Co-operation and Development, 1999 (includes results drawn from different scenarios studied by the International Energy Agency, the International Institute for Applied Systems Analysis, and the World Energy Council).
Fore*sight*: Progress Through Partnership 16; Marine. Office of Science and Technology, UK Government, 1997.
Fore*sight*: Report of the Working Group on Offshore Energies, Marine Panel, Office of Science and Technology, UK Government, 1998.
IOA Newsletter, International OTEC/DOWA Association, ITRI, Taiwan 31015, China—quarterly.
Ocean Engineering and Energy Systems (OCEES) International, Inc.: www.ocees.com
Ocean Thermal Energy Conversion: An Opportunity for the Maritime Industry with Early Application to Islands, Gauthier and Lennard, Crete, 2001.
XENESYS Inc.: http://xenesys.com

COUNTRY NOTES

The Country Notes on OTEC compiled for the WEC *Survey of Energy Resources 2001* have been revised, updated and augmented by the editors, using national sources and other information. Valuable inputs were provided by Don Lennard of Ocean Thermal Energy Conversion Systems Ltd.

Côte D'Ivoire

A French project to build two open-cycle onshore OTEC plants of 3.5 MW each in Abidjan (Côte d'Ivoire, at that time a French possession) was proposed in 1939. The experimentation was eventually undertaken after World War II, with the main research effort occurring during 1953–1955.

The process of producing desalinated water via OTEC proved to be uneconomic and the project was abandoned in 1958.

Cuba

This was the site of the first recorded installation of an OTEC plant and the island remains a very desirable location in terms of working temperature difference (in excess of 22 °C). Georges Claude, a French engineer, built an experimental open-cycle OTEC system (22 kW gross) at Matanzas in 1929–1930. Although the plant never produced net electrical power (i.e. output minus own use) it demonstrated that the installation of an OTEC plant at sea was feasible. It did not survive for very long before being demolished by a storm.

Fiji

This group of islands has been the subject of OTEC studies in the UK and in Japan. In 1982 the UK Department of Industry and relevant companies began work on the development of a floating 10 MW closed-cycle demonstration plant to be installed in the Caribbean or Pacific. The preferred site was Vanua Levu in Fiji.

At end-1990 a Japanese group undertook an OTEC site survey on the Fijian island of Vitu Levu. Design work on an integrated (OTEC/DOWA) land-based plant was subsequently undertaken.

The studies have not given rise to any firm construction project. However, when the tourist industry grows further, the Vanua Levu site will again be ideal, with cold deep water less than 1 km from shore. The development of the tourist industry will require substantial electrical power, potable water and refrigeration.

French Polynesia

Feasibility studies in France concluded that a 5 MW land-based pilot plant should be built with Tahiti as the test site. An industrial grouping, Ergocean and Ifremer (the French institute for research and exploitation of the sea) undertook extensive further evaluation (of both closed and open cycle) and operation of the prototype plant was initially expected at the end of the 1980s, but the falling price of oil caused development to be halted. Ifremer continues to keep the situation under review and has been active in the European Union.

Specifically, Ifremer with various partners has examined DOWA desalination, since a much smaller (1 m diameter) cold water pipe would be needed. Techno-economic studies have been completed but further development is on hold.

Guadeloupe

Experimental studies on two open-cycle plants were undertaken by France between the mid-1940s and the mid-1950s in Abidjan, Côte d'Ivoire—at that time a French possession. The results of these studies formed the basis of a project to build an OTEC plant in Guadeloupe (an Overseas Department of France) in 1958. This onshore 3.5 MW OTEC plant was intended to produce desalinated water but the process proved to be uneconomic and the project was abandoned in 1959.

India

Having an extremely long coastline, a very large EEZ area and suitable oceanic conditions, India's potential for OTEC is extensive.

Conceptual studies on OTEC plants for Kavaratti (Lakshadweep Islands), in the Andaman-Nicobar Islands and off the Tamil Nadu coast at Kulasekharapatnam were initiated in 1980. In 1984 a preliminary design for a 1 MW (gross) closed Rankine Cycle floating plant was prepared by the Indian Institute of Technology in Madras at the request of the Ministry of Non-Conventional Energy Resources. The National Institute of Ocean Technology (NIOT) was formed by the governmental Department of Ocean Development in 1993 and in 1997 the Government proposed the establishment of the 1 MW plant of earlier studies. NIOT signed a Memorandum of Understanding with Saga University in Japan for the joint development of the plant near the port of Tuticorin (Tamil Nadu).

During 2001 the Department of Ocean Development undertook an exercise to determine the actions required to maximise the country's potential from its surrounding ocean. The result was a Vision Document and a Perspective Plan 2015 (forming part of the 10th 5-year plan) in which all aspects of the Indian Ocean will be assessed, from the forecasting of monsoons through the modelling of sustainable uses of the coastal zone to the mapping of ocean resources, etc.

It has been postulated that most of India's future fully-commercial OTEC plants will be closed-cycle floating plants in the range 10–50 MW (although 200–400 MW plants are not ruled out). Working with Saga University, NIOT had planned to deploy the 1 MW demonstration plant in March/April 2003. However, mechanical problems prevented total deployment and the launch was delayed until the next available weather window. Once testing has taken place, it is planned to relocate the plant to the Lakshadweep Islands for power generation prior to full commercial operation from scaled-up plants.

Indonesia

A study was carried out in the Netherlands for a 100 kW (net power) land-based OTEC plant for the island of Bali, but no firm project has resulted.

Jamaica

In 1981 it was reported that the Swedish and Norwegian Governments, along with a consortium of Scandinavian companies, had agreed to provide the finance required for feasibility studies towards an OTEC pilot plant to be located in Jamaica.

In a reference to OTEC, the National Energy Plan (c. 1981) stated that 'a 10 MW plant was envisioned in the late 1980s'. Although this project never came to fruition, a plan remains in place for an offshore 10 MW plant producing energy and fresh water. For implementation to take place, purchasing agreements from the power and water utility companies need to be in place.

Japan

Research and development on OTEC and DOWA has been carried out since 1974 by various organisations (Ocean Thermal Energy Conversion Association of Japan; Ocean Energy Application Research Committee, supported by the National Institute of Science and Technology Policy; Japan Marine Science and Technology Center, Deep Seawater Laboratory of Kochi; Research Institute for Ocean Economics and

Toyama prefectural government; Saga University; Electrotechnical Laboratory and Shonan Institute of Technology).

Saga University conducted the first OTEC power generation experiments in late-1979 and in early-1980 the first Japanese experimental OTEC power plant was completed in Imari City.

During the summer months of 1989 and 1990 an artificial up-welling experiment was conducted on a barge anchored on the seabed at 300 m offshore in Toyama Bay.

With the establishment in 1988 of the OTEC Association of Japan, now the Japan Association of Deep Ocean Water Applications (JADOWA), the country has placed greater emphasis on products that use deep ocean water in the manufacturing process. Such products (food and drink, cosmetics and salt) have all proved commercially successful.

In March 1996, a Memorandum of Understanding was signed between Saga University and the National Institute of Ocean Technology of India. The two bodies have been collaborating on the design and construction of a 1 MW plant to be located off the coast of Tamil Nadu in India, expected to be commissioned at end-2003.

In mid-2003 Saga University's Institute of Ocean Energy (IOES) inaugurated a new research centre for the study of OTEC.

During 2003 it was reported that Saudi Arabia had shown great interest in working with Saga University to develop the Kingdom's OTEC potential.

If the OTEC projects the university is helping to implement are proved to be viable, the enormous potential of Japan's own EEZ could be exploited in the future.

Kiribati

During late-1990, an OTEC industrial grouping in Japan undertook detailed research (including the water qualities of the ocean, seashore, lagoon and lakes) on Christmas Island. Following on from this research, the basic concepts were improved but no developments have ensued.

Marshall Islands

In the early-1990s the Republic of the Marshall Islands invited proposals from US companies to undertake a detailed feasibility study for the design, construction, installation and operation of a 5–10 MW (net) OTEC power plant to be located at Majuro.

The contracted study was carried out by Marine Development Associates of California between April 1993 and April 1994 but no project resulted.

At a forum convened prior to the World Water Forum, (Kyoto, March 2003) by Japan's Saga University and the Government of Palau (a group of Pacific Islands to the east of the Marshall Islands), interest was renewed in the possibility for OTEC installations. The success of the planned project in Palau could well prove to be the impetus required for development in the Marshall Islands and other Pacific Islands.

Nauru

In 1981, the Tokyo Power Company built a 100 kW shore-based, closed-cycle pilot plant on the island of Nauru. The plant achieved a net output of 31.5 kW during continuous operating tests. This plant very effectively proved the principle of OTEC in practical terms over an extended period, before being decommissioned.

Netherlands Antilles

A feasibility study carried out by Marine Structure Consultants of the Netherlands and funded by the Dutch Government for the Netherlands Antilles Government examined the competitiveness of a 10 MW floating OTEC plant. No development ensued.

New Caledonia

Ifremer (the French institute for research and exploitation of the sea) has re-examined a previous proposal to establish a test site for OTEC/DOWA in New Caledonia.

Northern Marianas

It was reported in early 2003 that a Memorandum of Understanding had been signed by the Islands' Governor and Sea Solar Power International, a Maryland-based company, for the construction of an OTEC plant. Having identified a suitable site in the group of islands, Sea Solar Power would build and operate the plant designed to produce approximately 10 MW of net electricity and 3 million gallons/-day of fresh water.

Palau

In a plan to obviate a future need for diesel-generated electricity, Palau will switch to powering its electricity supply from its ocean thermal resource.

In Spring 2001 the Government of Palau, Japan's Saga University and Xenesys Inc. (a Japanese private company) entered into an agreement that resulted in research and feasibility studies being undertaken for the identification of suitable sites for OTEC installations. Seven such sites were located on the biggest island in Palau (Babeldaob). A pilot project is expected to have a capacity of 3 000 kW by 2005 and by 2015 OTEC capacity could reach 30 000 kW, an increase in excess of 50% from the current diesel-generated supply.

It is intended that, in addition to the production of power, the by-products of salt and fresh water could be used for organic farming.

Under the ACP-EU Partnership Agreement, the European Commission and the Government of Palau have drawn up a Country Strategy Paper and an Indicative Programme for the period 2002–2007. The EU will provide financial assistance to Palau in order to expand the utilisation of renewable energy sources.

Puerto Rico

A resource assessment conducted in 1977 studied the potential for a nearshore OTEC plant. In 1997 a new evaluation concluded that a closed-cycle, land-based OTEC plant of up to 10 MW was feasible, especially with the inclusion of DOWA. The headland of Punta Tuna on the south-east coast of the island satisfied the criteria for such a plant.

Saudi Arabia

It was reported in early 2003 that there had been high level governmental discussions between Japan and Saudi Arabia with a view to the Kingdom's OTEC potential being utilised for water desalination and electricity production. A delegation from Saudi Arabia attended the World Water Forum, Kyoto, March 2003 with the express purpose of examining the OTEC technology developed by Saga University. A joint venture has been established between Xenesys (a Japanese company with sole right of patents owned by Saga University), 40% and a group of Saudi investors (including Jamjoon, operator of desalination plants), 60%.

It is envisaged that the desalination plants would be placed at industrial sites along the Persian Gulf and the Red Sea and at end-May 2003 it was stated that construction of a 3 000 kW pilot plant (with an output of 2 000 tonnes of water per day) would be started 'within the year'.

Sri Lanka

Interest in OTEC and DOWA has been revived by the National Aquatic Resources

Agency in Colombo, in the context of making use of Sri Lanka's EEZ, which is some 27 times its land area.

Three submarine canyons (Panadura, Dondra and Trincomalee) have been identified as highly suitable sites for OTEC plants and the production of electricity. However, despite successful experiments conducted during 1994, a lack of funding has meant that any proposals have stagnated.

St. Lucia

In 1983, as a part of a commitment to develop alternative energy systems, the Government of St. Lucia welcomed the opportunity to be part of an OTEC initiative that included the design and construction of a 10 MW closed cycle floating OTEC demonstration plant off Soufriere. Hydrographic surveys in 1985 confirmed that the 1 000 m contour was less than 3 km from shore, with cold water in the volcanic canyon adjacent to Petit Piton and Gros Piton. This landfall was also close to the electrical grids. The surface temperature of the sea on that part of the west coast never falls below 25 °C, reaching 27/28 °C in summer.

The UK-designed plant was provided with a fully costed proposal by a merchant bank, which showed that with construction commencing in 1985, and operation from 1989, the OTEC plant would show a cost benefit over oil-fired plant from 1994, the higher capital cost of OTEC being balanced by the 'free fuel', whereas there were ongoing fuel costs for the diesel plant. However, the final decision was to go for a diesel plant, with the whole of the capital cost being funded by another country.

Taiwan, China

The seas off eastern Taiwan are considered to be highly favourable for OTEC development. Following preliminary studies during the 1980s, three nearshore sites were selected and the steeply shelving east coast was thought to be able to accommodate an onshore OTEC plant. However, only one site (Chang-Yuan) was deemed suitable for further investigation by the Institute of Oceanography.

In 1989, the Pacific International Center for High Technology Research in Hawaii prepared a development plan for the Taiwanese *Multiple Product Ocean Thermal Energy Conversion Project* (MPOP). The intention of the MPOP was to construct a 5 MW closed-cycle pilot plant for generating power and also the development of mariculture, desalinated water, air conditioning, refrigeration and agriculture. It was thought that the operating data obtained from the pilot plant could be used in the building of a 50–100 MW commercial plant. In 1993 it was assumed that 6 years would be required for site preparation and 5 years for construction (to be in operation by end-2003), with the plant having a 25-year life cycle.

During the 1990s the concept of MPOP changed to a *Master OTEC Plan for R.O.C.* (MOPR), with the objective of ultimately establishing eight 400 MW floating OTEC power plants.

With its positive interest, Taiwan was the initiator, in 1989, of the International OTEC/DOWA Association (IOA). A permanent Taiwanese secretariat has worked to ensure a higher international profile for OTEC/DOWA, but within the country, plans for OTEC have, at present, somewhat stagnated.

United States of America

Hawaii remains the focus of US activity in OTEC/DOWA, primarily through work carried out at the Natural Energy Laboratory of Hawaii (NELHA) facility at Keahole Point.

In 1979 'Mini-OTEC', a 50 kW closed-cycle demonstration plant, was set up at NELHA. It was the world's first net power producing OTEC plant, installed on a converted US Navy barge moored 2 km offshore: it produced 10–17 kW of net electric power.

In 1980 the Department of Energy constructed a test facility (OTEC-1) for closed-cycle OTEC heat exchangers on a converted US Navy tanker. It was not designed to generate electricity.

In the early 1980s a 40 MW OTEC pilot plant was designed. It was to be sited on an artificial island off the Hawaiian coast. However, funding was not forthcoming and the plant was not constructed.

An experimental 210 kW (gross electrical) open-cycle OTEC plant was designed and operated by the Pacific International Center for High Technology Research (PICHTR) at Keahole Point. It produced a record level of 50 kW of net power in May 1993, thus exceeding the 40 kW net produced by a Japanese OTEC plant in 1982. The plant operated from 1993 until 1998 and its primary purpose was to gather the necessary data to facilitate the development of a commercial-scale design. Following the experiments, the plant was demolished in January 1999.

A further PICHTR experiment at NELHA employed a closed-cycle plant to test specially developed aluminium heat exchangers. It used the (refurbished) turbine from 'Mini-OTEC' to produce 50 kW gross power. During initial operation in May 1996, corrosion leaks developed in the heat exchanger modules; the plant had to be shut down and the units re-manufactured. From October 1998, when the new units were received until end-1999—the end of the project—data were collected on the heat exchange and flow efficiencies of the heat exchangers and thus on the economic viability of competing types of heat exchangers.

In addition to research into ocean thermal energy, NELHA has established an ocean science and technology park at Keahole Point. Cold deep seawater is pumped to the surface and utilised for the production of energy, air-conditioning, desalination, fish farming, agriculture, etc. According to NELHA's FY 2002 Annual Report, deployment of the offshore pipelines and onshore pump station canisters for a new seawater system to serve the park was completed by June 2002. The onshore distribution system was to be completed during FY 2003.

Virgin Islands

The island of St. Croix has been found to be a suitable site for the development of OTEC-produced electricity and desalinated water.

In the early 1990s an agreement was drawn up between the US company GenOtec and the Virgin Islands Water and Power Authority (WAPA). The plan was to obtain 5 MW of OTEC-produced electricity and 1.5 million gallons/day of desalinated water from a land-based, closed-cycle OTEC plant. Additionally, various mariculture industries were planned. The project did not come to fruition.

Abbreviations and Acronyms

World Energy Council
CONSEIL MONDIAL DE L'ENERGIE

10^3	kilo (k)
10^6	mega (M)
10^9	giga (G)
10^{12}	tera (T)
10^{15}	peta (P)
10^{18}	exa (E)
ABWR	advanced boiling water reactor
AC	alternating current
ACP	African Caribbean Pacific
API	American Petroleum Institute
b/d	barrels per day
bbl	barrel
bcf	billion cubic feet
bcm	billion cubic metres
billion	10^9
BOO	build, own, operate
BOT	build, operate, transfer
bpsd	barrels per stream-day
bscf	billion standard cubic feet
Btu	British thermal unit
BWR	boiling light-water-cooled and moderated reactor
C	Celsius
CHP	combined heat and power
CIS	Commonwealth of Independent States
cm	centimetre
CNG	compressed natural gas
cP	centipoise
CSP	concentrating solar power
cusec	cubic foot per second
d	day
DC	direct current
DHW	domestic hot water
DOWA	deep ocean water applications
EAR	estimated additional resources
ECE	Economic Commission for Europe
EEZ	exclusive economic zone
EHO	extra-heavy oil
EIA	US Energy Information Administration
ETBE	ethyl tertiary butyl ether
F	Fahrenheit
FAO	UN Food and Agriculture Organization
FBR	fast breeder reactor
FSU	former Soviet Union
ft	feet
FY	fiscal year
g	gram
GEF	Global Environment Facility
GHG	greenhouse gas
GW_e	gigawatt electricity
GWh	gigawatt hour
h	hour
ha	hectare
HDR	hot dry rock
HWR	heavy water reactor
Hz	hertz
IAEA	International Atomic Energy Agency
IBRD	International Bank for Reconstruction and Development
IEA	International Energy Agency
IIASA	International Institute for Applied Systems Analysis
IMF	International Monetary Fund
IMO	International Maritime Organization
IPP	independent power producer
ISL	in-situ leaching
J	joule
kcal	kilocalorie
kg	kilogram
km	kilometre

2004 Survey of Energy Resources

km^2	square kilometre
kW_e	kilowatt electricity
kWh	kilowatt hour
kW_p	kilowatt peak
kW_t	kilowatt thermal
lb	pound (weight)
LNG	liquefied natural gas
LPG	liquefied petroleum gas
l/s	litres per second
l/t	litres per tonne
LWGR	light-water-cooled, graphite-moderated reactor
LWR	light water reactor
m	metre
m/s	metres per second
m^2	square metre
m^3	cubic metre
mb	millibar
MJ	megajoule
Ml	megalitre
mm	millimetre
MPa	megapascal
mPa s	millipascal second
MSW	municipal solid waste
mt	million tonnes
mtpa	million tonnes per annum
mtoe	million tonnes of oil equivalent
MW	megawatt
MW_e	megawatt electricity
MWh	megawatt hour
MW_p	megawatt peak
MW_t	megawatt thermal
N	negligible
n.a.	not available
NEA	Nuclear Energy Agency
NGLs	natural gas liquids
NGO	non governmental organisation
Nm^3	normal cubic metre
NPP	nuclear power plant
OAPEC	Organisation of Arab Petroleum Exporting Countries
OECD	Organisation for Economic Co-operation and Development
OPEC	Organisation of the Petroleum Exporting Countries
OTEC	ocean thermal energy conversion
OWC	oscillating water column
p.a.	per annum
PBMR	pebble bed modular reactor
PHWR	pressurised heavy-water-moderated and cooled reactor
ppm	parts per million
psia	pounds per square inch, absolute
PV	photovoltaic
PWR	pressurised light-water-moderated and cooled reactor
RAR	reasonably assured resources
R&D	research and development
RD&D	research, development and demonstration
RE	renewable energy
R/P	reserves/production
rpm	revolutions per minute
SCO	synthetic crude oil
SER	Survey of Energy Resources
SHS	solar home system
SWH	solar water heating
SR	speculative resources
t	tonne (metric ton)
tb/d	thousand barrels per day
tC	tonnes of carbon
tce	tonne of coal equivalent
tcf	trillion cubic feet
toe	tonne of oil equivalent
tpa	tonnes per annum
tpsd	tonnes per stream day
trillion	10^{12}
ttoe	thousand tonnes of oil equivalent
tU	tonnes of uranium
TWh	terawatt hour
U	uranium
U_3O_8	uranium oxide
UN	United Nations
UNDP	United Nations Development Programme
vol	volume
W	watt
WEC	World Energy Council

W_p	watts peak
wt	weight
WTO	World Trade Organization
WWER	water-cooled water-moderated power reactor
yr	year
–	unknown or zero
~	approximately
<	less than
>	greater than

Conversion Factors and Energy Equivalents

World Energy Council
CONSEIL MONDIAL DE L'ENERGIE

Basic Energy Units

1 joule (J) = 0.2388 cal

1 calorie (cal) = 4.1868 J

(1 British thermal unit [Btu] = 1.055 kJ = 0.252 kcal)

WEC Standard Energy Units

1 tonne of oil equivalent (toe) = 42 GJ (net calorific value) = 10 034 Mcal

1 tonne of coal equivalent (tce) = 29.3 GJ (net calorific value) = 7 000 Mcal

Note: the tonne of oil equivalent currently employed by the International Energy Agency and the United Nations Statistics Division is defined as 10^7 kilocalories, net calorific value (equivalent to 41.868 GJ).

Volumetric Equivalents

1 barrel = 42 US gallons = approx. 159 litres

1 cubic metre = 35.315 cubic feet = 6.2898 barrels

Electricity

1 kWh of electricity output = 3.6 MJ = approx. 860 kcal

Representative Average Conversion Factors

1 tonne of crude oil = approx. 7.3 barrels

1 tonne of natural gas liquids = 45 GJ (net calorific value)

1 000 standard cubic metres of natural gas = 36 GJ (net calorific value)

1 tonne of uranium (light-water reactors, open cycle) = 10 000–16 000 toe

1 tonne of peat = 0.2275 toe

1 tonne of fuel wood = 0.3215 toe

1 kWh (primary energy equivalent) = 9.36 MJ = approx. 2 236 Mcal

Note: actual values vary by country and over time.

Because of rounding, some totals may not agree exactly with the sum of their component parts.

2004 Survey of Energy Resources

Reserve/Resource Terminology—Recent Developments

World Energy Council
CONSEIL MONDIAL DE L'ENERGIE

INTRODUCTION

The systematic classification of reserves and resources dates back to the beginning of industrialisation, when organised mining began. At that time mining was mainly captive, and thus there was no need for uniform terminology across country borders. As a result, numerous classifications with different terminology and classification principles behind them were established worldwide.

Only later, when international trade and business began to play a more significant role, were attempts made to compare the various classifications in order to identify and understand similarities and differences between them. Some 25 years ago, a first effort to create a uniform supranational resource classification system was made by the United Nations. However, this effort did not succeed, as by then the national systems had already grown historical roots which could not easily be pulled out and replaced.

During the 1990s, as globalisation began to spread around the world, a new drive for harmonisation began to emerge nearly concurrently within the main sectors of the extractive industries: petroleum (oil and gas), uranium, coal and mineral commodities. Each of these sectors managed to reconcile differences within their respective terminologies.

The final move toward full harmonisation of classification of the three remaining resources has been accepted as the mandate of the United Nations Economic Commission for Europe (UNECE). In 2002, the UNECE Committee on Sustainable Energy set up a task force—an Ad Hoc Group of Experts representative of the relevant global industries, organisations and governments.

The current status of the classification process is outlined below, together with the desired outcome of the harmonisation of the remaining differences and their possible reconciliation with the classification used by the World Energy Council.

PETROLEUM (OIL AND GAS)

Since 1994 there has been an active and constructive evolution of concepts on reserve/resource classification and definitions in the petroleum sector. The Society of Petroleum Engineers (SPE), the World Petroleum Congresses (WPC) and the American Association of Petroleum Geologists (AAPG) have led this joint effort. Two structural elements have been

2004 Survey of Energy Resources

introduced to accommodate the need for an improved treatment of uncertainties associated with estimating the reserve/resource and the classification of reserves/resources by the status of the projects producing them. The United Nations Framework Classification (UNFC) has been extended to petroleum with the goal of providing a classification framework that would improve the basis for reporting by the extractive industries worldwide. These efforts have resulted in the SPE/WPC oil and gas reserves definitions issued in 1997, and the SPE/WPC/AAPG resource classification issued in 2000.

URANIUM

Since the mid-1960s the International Atomic Energy Agency (IAEA), in cooperation with the Nuclear Energy Agency (NEA)/OECD, has been trying to develop a comprehensive inventory of recoverable uranium resources to fuel the world nuclear electric power programme. An important activity in this cooperation was the development of a classification system that could be used for the preparation of the biennial inventory of uranium resources published in the world report on uranium 'Uranium Resources, Production and Demand', otherwise known as the 'Red Book'.

Projection of the future availability of uranium to meet present and future nuclear power requirements depends on the reliability of uranium resources estimates. Discrepancies in the definitions of the different classes of uranium resources and reserves between countries make the compilation and analysis of such information difficult. The problem was accentuated in the early 1990s with the entry of uranium producing countries from the former Soviet Union, Eastern Europe and China into the world uranium supply market. The need for an internationally acceptable resources/reserves classification system and terminology using market based criteria became obvious.

For these reasons, the IAEA has organised several meetings (since the early 1990s) to harmonise the terms and definitions used in uranium resources and reserves classification. At these meetings, similar activities in the resource classification undertaken by the UNECE were taken into consideration. It was found that the classification system used in the 'Red Book' was generally consistent with the UN International Classification System, which in addition emphasised the importance of the economic dimension.

A correlation between the UNFC and the IAEA/NEA classification systems was established, and reported in the Red Book (2003) in parallel with other country classifications.

COAL AND MINERAL COMMODITIES

Two initiatives to harmonise reserve/resource terminology of this sector took place almost concurrently.

In 1992, the UNECE launched its Framework Classification for Reserves/Resources (UNFC), which accommodates and incorporates all national systems to make them comparable and compatible. The UNFC was endorsed in 1997 by the United Nations Economic and Social Council (ECOSOC) and recommended for worldwide application. Since then it has been accepted by a growing number of nations and institutions.

The UNFC was created jointly with the participation of more than 50 nations with the objective of becoming an instrument that permits reserves/resources of solid fuels and mineral commodities to be classified in an internationally uniform system based on market economy criteria. The system directly reflects the procedures used in practice to investigate and evaluate reserves/resources which are carried out at consecutive stages of the geological, technical and economic evaluation of a given deposit. These assessment stages are believed to be basically similar for all energy and mineral commodities and might provide a path to permit direct comparison of the different reserve/resource classes, regardless of the terminology. This comparison is further simplified by a numerical codification introduced by the UNFC to provide an unambiguous identification of reserve/resource categories.

At the same time the Council of Mining and Metallurgy (CMMI) unified the differences in terminology between their member countries, resulting in a uniform set of CMMI definitions for reserves and resources.

Finally, in 1998 UNECE and CMMI experts agreed to integrate their respective definitions into a single, universally applicable set of definitions for coal and mineral commodities. The joint UN/CMMI definitions of mineral reserves and resources were finalised in 1999.

HARMONISATION OF PETROLEUM, URANIUM, COAL AND MINERAL TERMINOLOGY

In 2001, UNECE decided to consolidate the achievements made in the abovementioned three individual reserve/resource classifications and to extend the UNFC to petroleum and uranium to become a Framework Classification for Energy Reserves/Resources with the aim of providing inter alia an input to the World Energy Council's triennial Survey of Energy Resources (SER). The UNECE Ad Hoc Group of Experts on Harmonisation of Energy Reserves/Resources Terminology convened in June 2002 and a Task Force for each of the commodity groups: petroleum, uranium and coal, was established. The European Federation of Geologists, International Committee for Coal and Organic Petrology, IAEA/NEA, OPEC, SPE/WPC/AAPG, WEC, industry representatives and selected experts joined this intergovernmental group.

The possible avenue to achieving a final harmonisation of these three classifications is seen in the numerical codification introduced by UNFC in 1992, which has since proved useful in making national classifications comparable and compatible. The codification has the advantage of being simple and easily understandable. It can best be described by using, as an example, the most important reserve category which in the case of coal is termed 'proved reserve' and which is codified as 111.

The first digit represents the economy of a deposit, with the number 1 classifying its content as economically extractable. The second digit refers to the mining engineering, cost, legal assessment or feasibility assessment referred to as modifying factors by IMM. The number 1 stands for fully assessed, bankable.

The third digit represents the geological assessment. The number 1 covers deposits with all relevant characteristics established with a high degree of accuracy. Additional numbers are used to codify inferior economic aspects and lower assessment criteria.

The UNECE Committee on Sustainable Energy reviewed the results and recommendations of the Ad Hoc Group at its meeting in November 2003. It was agreed that the group should attempt to produce an updated description of the UNFC for adoption by the UN through its Economic and Social Council, at its annual session in July 2004. The International Accounting Standards Board's project on the extractive industries has joined this effort.

RECONCILIATION WITH WORLD ENERGY COUNCIL TERMINOLOGY

The World Energy Council's triennial Survey of Energy Resources (SER) provides an up-to-date review of the world's energy reserves and resources, on national, regional and global levels. The UNECE Ad Hoc Group of Experts can assist WEC by providing clear, concise and unambiguous definitions of the categories of energy reserves/resources data to be collected from WEC members. The prospect of summarising the individual categories used in the petroleum, uranium and coal sectors to reconcile them with those used by WEC appears to be good.

Dietmar Kelter
On behalf of
the UNECE Ad Hoc Group of Experts

Index

Index